T0396800

Advances in Geographical and Environmental Sciences

Series Editor

R. B. Singh, University of Delhi, Delhi, India

Advances in Geographical and Environmental Sciences synthesizes series diagnostigation and prognostication of earth environment, incorporating challenging interactive areas within ecological envelope of geosphere, biosphere, hydrosphere, atmosphere and cryosphere. It deals with land use land cover change (LUCC), urbanization, energy flux, land-ocean fluxes, climate, food security, ecohydrology, biodiversity, natural hazards and disasters, human health and their mutual interaction and feedback mechanism in order to contribute towards sustainable future. The geosciences methods range from traditional field techniques and conventional data collection, use of remote sensing and geographical information system, computer aided technique to advance geostatistical and dynamic modeling.

The series integrate past, present and future of geospheric attributes incorporating biophysical and human dimensions in spatio-temporal perspectives. The geosciences, encompassing land-ocean-atmosphere interaction is considered as a vital component in the context of environmental issues, especially in observation and prediction of air and water pollution, global warming and urban heat islands. It is important to communicate the advances in geosciences to increase resilience of society through capacity building for mitigating the impact of natural hazards and disasters. Sustainability of human society depends strongly on the earth environment, and thus the development of geosciences is critical for a better understanding of our living environment, and its sustainable development.

Geoscience also has the responsibility to not confine itself to addressing current problems but it is also developing a framework to address future issues. In order to build a 'Future Earth Model' for understanding and predicting the functioning of the whole climatic system, collaboration of experts in the traditional earth disciplines as well as in ecology, information technology, instrumentation and complex system is essential, through initiatives from human geoscientists. Thus human geosceince is emerging as key policy science for contributing towards sustainability/survivality science together with future earth initiative.

Advances in Geographical and Environmental Sciences series publishes books that contain novel approaches in tackling issues of human geoscience in its broadest sense—books in the series should focus on true progress in a particular area or region. The series includes monographs and edited volumes without any limitations in the page numbers.

More information about this series at https://link.springer.com/bookseries/13113

Narayan Chandra Jana · R. B. Singh
Editors

Climate, Environment and Disaster in Developing Countries

 Springer

Editors
Narayan Chandra Jana
Department of Geography
The University of Burdwan
Bardhaman, West Bengal, India

R. B. Singh ⓘ
Department of Geography
University of Delhi
New Delhi, Delhi, India

ISSN 2198-3542   ISSN 2198-3550   (electronic)
Advances in Geographical and Environmental Sciences
ISBN 978-981-16-6965-1   ISBN 978-981-16-6966-8   (eBook)
https://doi.org/10.1007/978-981-16-6966-8

This Springer imprint is published by the registered company Springer Nature Singapore Pte Ltd.
The registered company address is: 152 Beach Road, #21-01/04 Gateway East, Singapore 189721,
Singapore

Foreword

The book entitled *Climate, Environment and Disaster in Developing Countries* edited by Narayan Chandra Jana and R. B. Singh is an extremely important document focusing on the issues of contemporary relevance. In fact, climate change, environmental degradation and disaster occurrences are interlinked requiring integrated solutions.

No country in this world is free from climate change, environmental hazards and disasters but the scale and magnitude may vary from country to country. It is now almost impossible for any country to formulate a development strategy without considering climate variability, environmental problems and disasters. It may, however, be mentioned in this context that the people of the countries in tropical regions are the worst sufferers for all these problems although other regions are not untouched from the adverse consequences of climate change.

The present book is a collection and compilation of 27 research papers of diverse interests related to climate, environment and disaster in the South Asian countries and beyond. The issues focused in this volume are mainly the case studies from India,

Bangladesh, Sri Lanka, Thailand and Saudi Arabia. The range of issues on national, regional and local dimensions discussed here are not merely confined to the field of geography but also to the interest areas of allied disciplines as well.

I would like to appreciate and congratulate the editors for their hard work in bringing out this precious volume of contemporary relevance. This book will be an addition to the existing literatures and may attract the attention of geo-scientists and researchers of allied fields like climatology, environmental science, hydrology, disaster management, land use studies and agriculture. I am sure that this book will be of immense help to the researchers, scientists, planners and decision-makers engaged in dealing with the problems of climate, disasters and environmental aspects in developing countries.

July 2021   Prof. P. C. Joshi
                                                                Vice-Chancellor
                                                            University of Delhi
                                                                    Delhi, India

Introduction

No place in the world is free from disaster, though the frequency and magnitude may vary across regions. It may be noted that the world's worst disasters tend to occur in the countries between the Tropic of Cancer and the Tropic of Capricorn, which are densely populated and the poorer nations of the world. *Between 2000 and 2019, at least 45,000 people lost their lives around the world as a direct result of at least 11,000 extreme weather events, causing a loss of US$ 2.5 Trillion on PPP.* In South Asian countries, long coastal regions are prone to tropical cyclones, arid and semi-arid regions to persistent droughts, Himalayan mountain terrain and parts of the continental crust to earthquakes and landslides as well as perennial rivers to periodic floods. Recurrent natural disasters hinder developmental activities and aggravate poverty conditions. Every year crores of rupees are spent by the government of the respective countries in mitigating the effects of these disasters, which may have been utilized for development activities.

Keeping the above facts in the backdrop, the decade 1990–2000 was declared *International Decade for Natural Disaster Reduction* (IDNDR) by UN General Assembly with the focused objective to reduce natural disasters. *It is followed by the Hyogo Framework and the Sendai Framework of the Disaster Risk Reduction in March 2015.* Whatever may be the status of the nations, it underlines the necessity for a coordinated international action in order to strengthen all aspects of disaster *risk reduction and* management.

The present book covers 27 research papers of diverse interests related to climate, environment and disaster in South Asian countries and beyond. The issues covered at different levels are disaster risk reduction, climate change and extremes, environmental impacts of change in river morphology, effects of coastal erosion on coastal environment, vulnerability and exposures to landslides, water scarcity due to arsenic problem, biodiversity degradation, challenges of climate-resilient livelihoods, causes and effects of waterlogging, impacts of agrochemicals and pesticides, hydro-meteorological analysis of flood, assessment of droughts, mapping and reclamation of Wastelands in drought-prone areas, hydrological variability and landscape development, urban flooding and human response, estimation of soil erosion, impact of soil and water salinization, impact of water reservoir and irrigation canals, urban heat island, sustainable water resource development, global model rainfall forecasts, spatial pattern of arsenic in Holocene aquifers, fluctuation in seasonal groundwater level, watershed management, land-use land-cover dynamicity, LULC change on runoff and sediment yield. It is important to note that geography is a subject of inter-disciplinary nature. The range of issues on national, regional and local dimensions discussed here are not merely the sole areas of geography but also the interest areas of other disciplines as well. The diverse issues dealt with in detail in this proposed volume may attract the attention of geo-scientists and researchers of allied fields like climatology, disaster management, environmental science, hydrology, agriculture, land-use studies and so on. This book may be of immense help to the researchers, scientists, planners and decision-makers engaged in dealing with problems of climate, disasters and environmental aspects in developing countries.

Part I includes five articles related to climate.

In Chap. 1 entitled *Challenges of Climate Resilient Livelihoods and an Inquiry of Mitigation Strategies in India* by Babita Chatterjee and Amrita Dwivedi, the focused objectives of the study are to understand the role of climate on the way of life and to give a way out. The major findings are rainfall dispersion, heat-stressed highly influenced GDP and human development index (HDI). These three cumu-latively put stress on food security. Livelihood resilience can become a constructive 'boundary object' across disciplinary and policy boundaries, situating action around a common objective, i.e. anti-poverty and development policy. In doing so, it demands greater attention on the societal root causes underlying differences in vulnerability and resilience. Mostly we can say that to sustain in these specific circumstances, a few advanced extension services such as cropping patterns and production systems should be followed to manage the extremity.

Chapter 2 entitled *Analysing LULC Change on Runoff and Sediment Yield in Urbanizing Agricultural Watershed of Monsoonal Climate River Basin in West Bengal, India* by Arnab Ghosh, Malabika Biswas Roy and Pankaj Kumar Roy repre-sents an integrated approach based on remote sensing (RS), geographical informa-tion system (GIS) and hydrological modelling to assess the impact of land-cover change through runoff modelling and sediment yield. A rainfall-runoff model by SCS-CN and sediment yield model by MUSLE evolved to estimate the effect of the land-cover change on runoff volume with sediment yield through HEC-HMS v 4.3. Several different land-use scenarios were then simulated with the model, calibrating the impacts of land-use change on the hydrology of the watershed. The baseline test results of R2 and Nash–Sutcliffe model efficiency (NSE) values ranged between 0.61 and 0.88 across the calibration and validation periods, indicating that HEC-HMS accurately replicated the alluvial streamflow. The hypothetical scenario simulations revealed that sediment yield increased with runoff volume, and most of the concen-tration happens in the non-monsoonal period with declining runoff. The results of the real scenario simulations revealed that urbanization is the most substantial contrib-utor to changes in runoff volume and sediment yield. This study is also essential to understand how land use and industrialization change at the local level offer future planners to propose the rehabilitation program in watershed planning.

In Chap. 3 entitled *Remote Sensing-Based Analysis of Relationship Between Urban Heat Island and Land Use/Cover Type in Bhubaneswar Metropolitan Area, India* by Asad Ali Sarkar, an attempt has been made to understand the relationship between changing land use/cover (LULC) pattern and mode of urban heat island (UHI) in Bhubaneswar Metropolitan Area. It faced rapid urban growth in the last few decades and a remarkable impact on urban heat islands. Remote sensing data, especially TM, ETM thermal band and Landsat 8 OLI, provide a lot of information about the study of UHI. To detect changing land-use pattern and land surface temperature (LST), three Landsat satellite images were used at different times. The satellite images were classified into seven categories in Erdas imagine 2014 by unsupervised classification method. To understand the proper relation between land use/cover change (LULC) and intensity of urban UHI, different mathematical models were built using four indices such as Normalized Difference Vegetation Index (NDVI), Normalized Difference Built built-up Index (NDBI), Normalized Difference Water Index (NDWI) and Normalized Difference Bareness Index (NDBaI). It showed that dry land, bare land, impervious land and river sand contained high temperatures than other land-use types. Besides the higher temperature in the built-up area, a scattered pattern distribution of temperature was found in 1997. Later it has changed to a concentric pattern to built-up land in 2017. A negative correlation was found between LST and NDVI, NDWI, NDBaI, but a positive correlation was shown between LST and NDBI.

Chapter 4 entitled *Statistical Downscaling Method for Improving Global Model Rainfall Forecasts of Seasonal Rainfall over West Bengal (WB), India,* by Aminuddin Ali and Tirthankar Ghosh focuses on the forecasts of seasonal rainfall behaviour. The present study uses rainfall forecasts from six advanced experimental NMME (North-American Multimodel Ensemble) models from 1996 to 2010. The seasonal rainfall forecasts from the statistical downscaling and BLUE (best linear unbiased estimator) schemes are calibrated and verified over districts of West Bengal. The forecast performances and predicted seasonal rainfall distribution from the statistical downscaling and BLUE methods are the main objectives of this work. The seasonal rainfall prediction skills are compared among models before and after downscaling methods. The forecasts of downscaled global models are more accurate compared to real observations. The improvements of seasonal rainfall forecasts skills are also discussed. The CFSv2 model is more capable of forecasts at the district level than others. More improvement is expected with dense real observation.

Chapter 5 entitled *Identification of Climate Change Vulnerable Zones in Bangladesh Through Multivariate Geospatial Analysis* by Md. Golam Azam and Md. Mujibor Rahman focuses on the entire Bangladesh in their chosen field of investigation. To demonstrate spatial vulnerability to climate change, the present study retained unbiased weights of indicators for indexing in the geospatial environment. A total of 42 indicators, 12 from the biophysical category and 30 from the socio-economic category, have been incorporated with the IPCC framework through a Geographic Information System (GIS) raster database. For unbiased weights, principal component analysis (PCA) of normalized raster has been accomplished in the raster calculator of ArcMap 10.5. The overall vulnerability of the country from exposure, sensitivity, and adaptive capacity has been indexed in this study. The coastal region, part of the hilly region, riverine areas, and the haor basin are found highly vulnerable since these regions are more exposed as well as highly sensitive to climate change effects. The current study nevertheless is a new contribution from Bangladesh in climate vulnerability indexing since it incorporates multivariate geospatial analysis to quantify and visualize climate change vulnerability countrywide. Findings from this work, mainly maps, will be an important foundation in taking appropriate measures associated with the mitigation and adaptation of climate change negative impacts from local level measures to policy-making stages.

Part II covers 12 papers on different aspects of environments.

In Chap. 6, the paper entitled *Water Scarcity in Coastal Bangladesh: Search for Arsenic-safe Aquifer with Geostatistics*, by M. Manzurul Hassan, Anamika Shaha and Raihan Ahamed, seeks to explore arsenic-safe drinking water sources at different depths of shallow and deep aquifer using spatial geostatistical techniques with inverse distance weighting (IDW) method along with the geographical information systems (GIS). The relevant water quality information—mainly the arsenic concentrations in tube well water, tube well attributes, and spatial data—were collected from the field survey. Field testing kits (FTK) for water quality information, geographical positioning systems (GPS) for spatial data and questionnaire survey for tube well attributes were employed in this research. The relevant information was collected from the Magura *Union* of Satkhira district in the south-western part of coastal Bangladesh. The study site covers about 27.58 km^2 of the area with a total population of 20375 and a total of 2650 tube wells.

The IDW prediction method examined that about three-quarters of the study site are contaminated following the level of Bangladesh Drinking Water Standard (BDWS) of 50 µg/L. The study identified a few scattered areas for arsenic-safe zones at different depths, but it is mainly on the north-eastern side of the study site. The suitable safe-water 'pockets' were demarcated at some areas at different depths. A proper policy including safe water management can therefore be formulated for ensuring safe drinking water. The availability of safe water pockets in groundwater until the promotion of new technology for the community could be an option to achieve 'Clean Water and Sanitation' (Sustainable Development Goal 6) to some extent with exploring the safe aquifer.

The paper in Chap. 7 entitled *Biodiversity Degradation of Southwestern Region in Saudi Arabia* by Adel Moatamed aimed to monitor the degradation of biodiversity in one of the south-western regions of Saudi Arabia. The study area has a rich diversity of flora and fauna, particularly in the juniper trees ecosystem. The total area covered with natural vegetation was 4385.6 km^2 in 1980; it decreased to 3645.5 km^2 according to the satellite image of 2019. The population growth and urban sprawl were the main factors causing the degradation of natural vegetation cover in this region.

In the paper of Chap. 8 entitled *Causes and Effects of Water Logging in Dhaka City* by Mallik Akram Hossain, Sanjia Mahiuddin, Arif Uddin Ahmad and A. H. M. Monzurul Mamun, the main objective of the study is to determine the causes and consequences of waterlogging in Dhaka City. Twenty-seven vulnerable areas in the city have been identified through field visits and reports from the national daily newspaper. To understand the reasons and consequences of waterlogging, an intensive questionnaire survey among the respondents is conducted in 2019. Three hundred respondents have been interviewed from seven areas of the city. The findings of the research indicate that waterlogging is generated by both natural and man-made causes. Waterlogging also creates adverse social, economic, environmental and health impacts on the city dwellers. To mitigate the waterlogging problem, widening the drainage system, regular cleaning of the drains and awareness of the inhabitants are urgently required to achieve the goal of a livable city.

Keeping in view the growing water demand for domestic and other economic activities, the paper in Chap. 9 entitled *Water Resource Development and Sustainable Initiatives of India: Present and Future* by Jayant Kumar Routray attempts to make a review of water resource situation of India, its management and associated challenges, existing policies and focuses on the sustainable initiatives practised now and a perspective to the future.

In the paper of Chap. 10 entitled *Mapping and Reclamation of Wastelands in Drought-prone Purulia District of West Bengal, India using Remote Sensing and GIS* by Manoj Kumar Mahato and Narayan Chandra Jana, the main objectives of the present research are to study the spatial distribution of different types of wastelands and to suggest the appropriate measures for reclamation of various categories of wastelands. The wastelands of the Purulia have been identified and categorized through the SOI toposheets of 1:50,000 scale, SRTM DEM LISS-III, Landsat-8 OLI/TIRS C-1 Level-1 and Google Earth images by GIS software with rigorous field survey. Based on the analysis of secondary and primary data and information, the authors in the present context have given appropriate suggestions towards the reclamation of wastelands in the Purulia district.

In the paper (Chap. 11) entitled *Soil, Water Salinization and its Impact on Household Food Insecurity in the Indian Sundarbans* by Nabanita Mukherjee and Giyasuddin Siddique, the study is an attempt to comprehend the effects of soil and water salinity on household food security. Excess salinization of soil and water exacerbates significant long-term environmental risks in the coastal and deltaic environment, especially in the densely populated tropical deltas. This study focuses on the coastal regions of the Indian Sundarbans directly confronting the Bay of Bengal based on data collected from the household survey 2017–2020; soil and water survey during the pre- and post-monsoon period of 2019 to investigate the effects of soil and water salinity on human life and ecological sustainability of the deltas. The result reveals significant effects of soil and water salinization on a climate-sensitive subsistence economy. The growth of crops and plants gets affected by increased soil salinity, which further leads to soil sterility and poor seed germination. Water salinization leads to the toxicity of specific ions which directly affects those households dependent on freshwater ponds for irrigation and fishing activities. The findings clearly suggest policy intervention to deal with enhanced soil and water salinity in the context of changing climate.

The objective of the present paper in Chap. 12 entitled *Impact of Water Reservoir and Irrigation Canals on Land Use and Land Cover Changes in Upper Kumari River Basin, West Bengal, India* by Piya Bhattacharjee and Debasish Das is to assess the impact in the upper catchment area of Kumari River basin, Purulia, West Bengal taking a case study of two reservoirs, i.e. Kumari and Hanumata and related canals. Kumari irrigation scheme, Balarampur covers 3255 hectares during the Kharif season and 320 hectares for the Ravi season. However, Hanumata irrigation scheme, Balarampur covers 1265 hectares during the Kharif season and 460 hectares during the Ravi season. Many settlements have been evacuated due to the construction of reservoirs. Survey of India (SOI) toposheet, integrated remote sensing and geographical information (RS and GIS) technique along with necessary field studies have been adopted for proper accomplishment of the work. Many irrigated agricultural plots have been identified on both sides of the canals. The moisture content of the agricultural field here is 2.3–2.5% (dry season) and up to 8% (monsoon) and pH is 7.2–8.0. The agricultural fields lacking in recent irrigation show dry soil conditions.

In the paper of Chap. 13 entitled *Spatial Pattern of Arsenic Contamination in Floodplain Aquifers, Western Bank of Bhagirathi River, Lower Ganges Delta, West Bengal, India* by Sunam Chatterjee, the research based on a secondary database, collected from the public health engineering department (PHED) found that about 48.40% of tube well samples ($N = 1535$) have arsenic concentration >0.01 mg/L. The maximum concentration of dissolved arsenic in tube well water extends up to 0.618 mg/L. The focus of the present work includes spatial and depth-wise distribution of tube well 'As' contamination, mapping of 'As' concentration and seasonal variation. The association of contaminated tube wells with floodplain morphology and lithologs is also studied. Geospatial technology is adopted for mapping the pattern and realising the spatial association. Vertical distribution of arsenic contamination shows that 52.77% of samples ($N = 1535$) are restricted within shallow to intermediate depth zones, i.e. 40–80 m. At deeper depths (>100 m) the counts of tube wells exceeding >0.01 mg/L are few, though the amount of 'As' concentration at each depth class is extremely erratic in nature. Seasonal variations based on 20 tube wells find that about 70% of samples have exceeding 'As' concentration in the pre-monsoon season. The contaminated tube wells (>0.01 mg/L) are mostly occurring within the meander belt of Bhagirathi river and have been associated with different floodplain features like paleo-channels, meanders scar, back swamps, earlier levees, floodplain mashes etc. The lithological study reveals the Holocene aquifers consist of grey-coloured younger alluvial deposits of *Katwa* formation and the *present-day* deposits are found contaminated, while Pleistocene aquifers are almost safe. The research contributes to understanding convoluted distributional patterns of arsenic concentration in the western bank of the Bhagirathi floodplain.

Based on secondary data, the present paper in Chap. 14 entitled *Spatial Pattern of Groundwater Depletion, Its Access and Adaptive Agricultural Strategies in Barddhaman District, West Bengal, India* by Biswajit Ghosh and Namita Chakma intends to study the spatial pattern of groundwater level and natural groundwater recharge by using the soil moisture balance approach in five selected Community Development Blocks of Barddhaman district, West Bengal with different geohydrological settings.

It also documents groundwater depletion, resultant changes in irrigation technology and adaptation approaches taken by farmers to minimize groundwater exhaustion based on primary data collected from two 'semi-critical' blocks and 22 villages. It has been found that rainfall in the monsoon period is the key factor for the recharge of groundwater and the amount of annual recharge varies from 10% to 20% except in Jamuria. The deepening of bore wells due to groundwater depletion started in the 1980s and still continuing. The temporal pattern of change to submersible pump from centrifugal one represents an S-shaped curve. Water-saving methods are limited to a very small part of the total operational holding. Promotion of water-saving technologies and practices may reduce groundwater depletion and improve access to the smallholders.

In the paper of Chap. 15 entitled *Detection of Land Use/Land Cover Changes of Irga Watershed in the North-Eastern Fringe of Choto Nagpur Plateau, Jharkhand, India* by Ratan Pal and Narayan Chandra Jana, the study attempts a quantitative evaluation of the land-use/land-cover changes of the Irga Watershed for the last three decades using geo-spatial technology with ground truth verification. The whole study area has been divided into six major land use categories, viz., built-up area, vegetation cover, agricultural land, waterbody, river and barren land. The investigation revealed that the agricultural land, vegetation cover and built-up area are the main LULC categories, occupying around 90% of the total geographical land of the study area. The findings suggest that agricultural land and built-up areas are constantly expanding while vegetation cover and barren land decreasing continuously. The area under the waterbody and river remains more or less constant during the study period. The anthropogenic activities are found as the main culprit for LULC changes. The generated maps have reasonably good accuracy rates with Kappa coefficient 0.831, 0.858, 0.902 and 0.918 for the years 1990, 2000, 2010 and 2020, respectively.

Swetasree Nag, Malabika Biswas Roy and Pankaj Kumar Roy in their paper (Chap. 16) entitled *An Analytical Study on Interplay between Physiographic Condition and Land Use Land Cover Dynamicity* have tried to draw attention to such kind of behaviour among the topographical elements of the Kuya river basin which makes the topography more erosive and complex in nature. This paper also enumerates how the land-use land-cover practices of the study area can be influenced by the topographical complexity. To explore such a topographic complex zone, a weighted overlay analysis has been carried out using six major morphometric parameters. A multiple correspondence analysis (MCA) has also been executed which proves that human beings cannot overcome the adverseness of nature as they were well accustomed to the physiographical complexity by utilizing the land in a sustainable way.

In Chap. 17, the paper entitled *Assessment of Land Use and Land Cover Change Dynamics using Remote Sensing and GIS Techniques in Most Affected Parts of Rajpur-Sonarpur Municipality* by Bijay Halder, Papiya Banik and Jatisankar Bandyopadhyay focuses on the dynamicity of land-use and land-cover in a municipality adjacent to Kolkata metropolitan area. Some parts of Rajpur-Sonarpur municipality have a huge amount of population pressure, which are facing huge land degradation. Transport accessibility has significantly contributed to the bridge between urbanization and the population. Remote sensing and GIS technologies have created their platform for various types of investigation using satellite imageries. In the present context, multitemporal Landsat OLI data has been used to calculate the total land-use and land-cover change in Rajpur-Sonarpur municipality. The supervised classification technique has been used along with the maximum likelihood method for detecting the areal change in the years 2014 and 2019. The result shows that the built-up areas were mostly increased in the last five years around 193.595 ha area due to population pressure and the loss of total vegetation area to 166.244 h Many open spaces and agricultural land were converted into built-up area.

Part III: Disaster includes 10 papers in this volume
In Chap. 18, Tapati Banerjee in her paper entitled *Disaster Risk Reduction in the Changing Scenario* gives an overview of disaster risk reduction at the international level. The present world is facing a lot of challenges in its day-to-day activities. The report of the United Nations of Disaster Risk Reduction (UNDRR, formerly UNISDR) of 2015 reveals that since 1980 nearly 1.6 billion people lost their lives in disasters and also warned that the global average of annual loss will be increased up to US$ 415 billion by 2030 as the climate-related natural hazards are increasing in number and intensity giving rise to new vulnerabilities with differential spatial and socio-economic impacts on communities.

In the paper of Chap. 19 entitled *Exploring the Impacts of River Morphology Change Associated Natural Disasters on Teesta Riparian Environment of Bangladesh* by Mst. Rebeka Sultana and Shitangsu Kumar Paul, the objective of the study is to examine the effects of natural disasters such as flood, riverbank erosion, sedimentation and river channel shifting. The study also seeks to explore both the positive and negative impacts of river morphology change allied natural disasters on the environment. The study has been conducted in the downstream locations of the Teesta River within Bangladesh. Purposively selected seven union's adjacent to Teesta River has been chosen as the study area. A rigorous field survey has been conducted in a flooded environment to collect data and information regarding river morphology change-related natural disaster impacts. Using a simple random sampling technique, data has been collected through a household questionnaire survey. Data analysis and interpretation have been completed with SPSS and Excel software. The study is based on multiple response analysis. The study result unveils 33.1% responses on the negative impacts of the flood as it erodes river banks. Moreover, the positive impact of sedimentation on the environment shows 34.3% replies on silt increases soil fertility. Hence, the study explored significant positive and negative impacts on the environment due to river morphology change-related natural disasters. The present research advocates consciousness of community inhabiting in the Teesta riverine environment to reduce the negative impacts of natural disasters owing to river morphology change. Besides, the government should take initiatives towards minimization of river morphology change accompanying natural disaster impacts on the riparian environment using modern technology.

In Chap. 20, the paper entitled *An Assessment on Effects of Coastal Erosion on Coastal Environment: A Case Study in Coastal Belt between Kalu River mouth and*

Bologoda River mouth, Sri Lanka by P. Kirishanthan, the study mainly focused on an assessment on the effects of coastal erosion on the coastal environment in the coastal belt between Kalu River mouth and Bologoda River mouth in the south-western coastal zone, Sri Lanka. The data were collected through both primary and secondary sources, especially the thermal infrared sensor (TIRS) images (2003–2018) were collected through the United States Geological Survey (USGS). Further, a questionnaire survey and key informant interviews were also conducted. The questionnaire survey was statistically analysed, while data gathered from interviews were analysed thematically. The results of the study revealed that destroying coastal landforms, saltwater intrusion, loss of green belt and wildlife and coastal flooding were recognized as the key environmental issues due to coastal erosion. Among them, the loss of coastal landforms is identified as a crucial environmental threat in the study area. As a result of accelerating coastal erosion, coastal landforms such as sandbars, sandy beaches and estuaries are more prone to erosion in the study area, especially at Kalido beach, Kalutara. The current research clearly highlighted that accelerating coastal erosion severely affects the balance and well-being of the coastal environment. Thus, coastal conservation and protection bodies, plans and programs should immediately take necessary action to mitigate the environmental issues in the coastal belt.

In Chap. 21, the paper entitled *Vulnerability and Exposures to Landslides in the Chittagong Hill Tracts, Bangladesh: A Case Study of Rangamati Town for Building Resilience* by Md. Iqbal Sarwar and Muhammad Muhibbaullah attempts to identify the causes and consequences of the most recent landslides in the lives and livelihoods of local people, i.e. inhabitants of Rangamati hill tracts, after assessing the household coping mechanism. The underlying objective is to put forward some recommendations to build resilience for sustainable land management in the study area. Conducting a questionnaire survey with 250 respondents and secondary information, the study was carried out in five landslide-prone sites located in Rangamati Sadar. These respondents were selected randomly based on the landslides susceptible locations. Several focus group discussions (FGDs) have also been conducted in the study area for identifying the vulnerability, causes and impacts of the recent landslide. The findings of the research can be used to prioritize risk mitigation investments, measures to strengthen the emergency preparedness and response mechanisms for sustainable mountain development, and reducing the losses and damages due to future landslides events in the CHTs area.

Rarh Bengal, the part of the Lower Gangetic Plain (LGP), is one of the most affected flood-prone regions in the world as well as in India with immense social impact, including loss of human life and also major damage to infrastructure. The entire region is constituted of five major tributary systems of the Bhagirathi–Hooghly river (a distributary of the Ganges system) in the western part, viz., the Kopai-Mayurakshi-Dwarka system, the Ajay, the Damodar, the Dwarakeswar-Silabati-Rupanarayan system and the Kangsabati-Keleghai-Haldi system. The majority of rivers in this region is short and has small catchments, but these all often cause significantly high flooding in terms of unit peak discharges, even considerably higher than world large rivers (WLRs), Indian peninsular rivers (IPRs) and North Bihar rivers

(NBRs). If we unfold almost the last 60 years of flood history since 1960, it can be observed that the Rarh Bengal has experienced only 11 flood-free years. The 2015 Rarh Bengal flood was exceptional not only in terms of hydrometeorological aspects but also in terms of duration, damage and death toll. During the 2015 monsoon, the extreme rainfall in just 40 days (from the end of June to the beginning of August), more specifically the last spell (26 August to 2 September 2015) caused devastating riverine flooding in the history of Rarh Bengal. In this paper (Chap. 22) entitled *Hydrometeorological Analysis of the 2015 Rarh Bengal Flood in the Lower Gangetic Plain of India: Exceptional, Fast and Furious* by Soumen Chatterjee and Narayan Chandra Jana, an attempt has been made to analyse the hydrological variability of Rarh Bengal's rivers during the 2015 flood with detail analysis of synoptic conditions. This study has also explained the extension (inundated area) of the 2015 flood by using remote sensing data (MODIS and Landsat 8 OLI) with its devastating consequences in Rarh Bengal.

In this paper (Chap. 23) entitled *Application of Remotely Sensed Data for Estimation of Indices to Assess Spatiotemporal Aspects of Droughts in Bankura District of West Bengal, India* by Asraful Alam, Rajat Kumar Paul and Lakshminarayan Satpati, the study primarily aims to deliberate upon the use of RS and GIS for analysis of meteorological data to find out a drought risk assessment in Bankura district of West Bengal. The study incorporates a multispectral band ratio to estimate vegetation density and vegetation health for the evaluation of spatio-temporal aspects of drought conditions. Land surface temperature (LST) and Normalized Difference Vegetation Index (NDVI) have been worked out to measure three other indices, namely Vegetation Condition Index (VCI), Temperature Condition Index (TCI) and Vegetation Health Index (VHI). Moreover, an attempt has been made to assess the spatio-temporal issues of drought risks associated with agriculture through analysis of temporal images for NDVI and Standardized Precipitation Index (SPI) of the area. Three drought years, i.e. 2000, 2010 and 2018, have been selected for the assessment of drought in the district. Correlations have been done among NDVI, SPI and rainfall anomalies. In the year 1990, the VHI value exhibited that the drought situation of the district was ranging between moderate and mild, but in 2010 it appeared to be a case of severe drought. Overall, the results showed that the district has experienced moderate to severe drought situations in the recent past.

In the paper in Chap. 24 entitled *Temporal variability of discharge and suspended sediment transport in the Subarnarekha River Basin, eastern India: A geomorphic perspective* by Sunanda Banerjee, Arup Kumar Roy, Asraful Alam, the aim of this experimental work is to access the temporal variation of sediment yield and to identify the factors controlling the hydro-sedimentary responses. In the world, over the last 50 years, many rivers experienced a decreasing trend in runoff and sediment load. In the Subarnarekha river, after analysing the data of discharge, runoff and sediment load of hydrological stations, it is clear that Subarnarekha also has a declining trend. In the last few decades, the precipitation rate is also declining day by day. In 2010–2011 precipitation rate is very low in the Subarnarekha basin. Runoff, water discharge and sediment load are proportionately related to precipitation rate. Anthropogenic factors are also responsible for a declining rate in the Subarnarekha river. Before

1970 the changes in water discharge is closely related to the normal precipitation and monsoonal rainfall when it flew gigantically but after 2001 it gradually decreases due to human intervention like a barrage, dam construction and other activities. Time-series plot of total annual discharge (cumecs) variation in the Subarnarekha river during 1973 to 2013, the total annual discharge is quite variable and unpredictable during natural flow regime, i.e. pre-dam period. In Ghatshila station, after dam construction, peak suspended concentration and peak of flood hydrograph coincide with each other. Landsat satellite image shows the annual discharge as well as the sedimentation variability effects on the downstream of Subarnarekha river.

River Jiadhal is one of the major right bank tributaries of the river Brahmaputra. It originates from the Arunachal Himalaya, traverses a few kilometres through Arunachal Pradesh and enters the Brahmaputra valley (Assam part) near Jiadhalmukh of Dhemaji districts, Assam. Like all the right bank tributaries of the river Brahmaputra, Jiadhal inundates vast areas of its basin during the monsoon season and put a significant impact on the people and their livelihood. This river carries a huge amount of sediment load and debris triggered by continuous and heavy rainfall within the basin mainly in the upper catchment region causing severe loss of fertile agricultural land and infrastructure of the basin. The paper in Chap. 25 entitled *Assessment of Jiadhal River Basin using Sedimentary Petrology and Geospatial Approach* by Akangsha Borgohain, Kusumbor Bordoloi, Dhrubajyoti Sahariah, Anup Saikia, Ashok Kumar Bora aims to analyse the morphological characteristics of the River Jiadhal using sedimentary petrology, remote sensing techniques and geographic information system (GIS). The study gives a better prospect towards understanding the morphometric characteristics and sediment dynamics of the basin. It will help the concerned authorities for better planning and mitigation of the issues related to this river.

The present paper (Chap. 26) entitled *An Assessment of RUSLE Model and Erosion Vulnerability in the Slopes of Dwarka—Brahmani Lateritic Interfluve, Eastern India* by Sandipan Ghosh focuses on the hillslope erosion by water (i.e. gully erosion, sheet erosion, rill and inter-rill erosion) which is observed in the lateritic badlands of Dwarka-Brahmani Interfluve (eastern part of Rajmahal Basalt Traps). The experimental design and geomorphic methodology choose the basin or catchment of a gully as a fundamental unit of erosion study. Using Revised Universal Soil Loss (RUSLE) model and field measured data (developing dams and sedimentation pits), the annual erosion rates (i.e. 8.12–24.01 kg m^{-2} y^{-1}) of 18 sample slopes are estimated and validated using different quantitative and statistical techniques. The linear regression of experimental results shows a positive correlation (r = 0.72) and high increment between observed (X) and predicted (Y$_C$) erosion rate (Y$_C$ = 5.90 + 0.659 X; R^2 = 0.521). Based on 18 dam sites and sedimentation data (2016–2017), the average erosion rate (A$_P$) is 16.63 kg m^{-2} y^{-1} which is beyond the soil tolerance value of laterite (i.e. 1.0 kg m^{-2} y^{-1}), showing a high level of erosion vulnerability (i.e. loss of land, soil bareness, ferruginous crusting, deterioration of crop and other biomass). Alongside the dynamics and susceptibility of water, erosion is assessed here to emphasize the role of gullies in the badland evolution of laterite terrain. The study reveals that the head and sidewall collapse of the gully is a composite and cyclical process resulting from downslope creep, tension crack development, crack saturation by overland flow, head or wall collapse followed by debris erosion which facilities the next failure. In this region, 52.51% of gullies are developed by overland flow erosion while 27.96% belongs to landslide erosion or mass movement.

The rapid growth of Guwahati, the largest urban centre in India's northeast, in terms of area, population and functionality, is making the urban environment more complex. Although due to the influence of the monsoon the amount of rainfall in the city almost remains the same, it has undergone distributional change with a decrease in rainy days and an increase in high-intensity rain events. Variety of factors typical to its dynamic environment characterized by surrounding hills, the Brahmaputra flowing through it, and the ever-expanding concrete surface and high-rise buildings in the midst of somewhat rugged terrain across Guwahati and induces severe urban flooding problem. The gravity of the situation can be marked by the fact that moderate to heavy rainfall for about 2–3 h often results in high-intensity urban flooding during the monsoon season. Sometimes the situation becomes devastative in certain localities within the city and the life and living off the city-dwellers become deplorable. Therefore, the study in this paper in Chap. 27 entitled *Urban Flooding Scenario and Human Response in Guwahati, India* by Sutapa Bhattacharjee and Bimal Kumar Kar attempts to analyse the flooding pattern in Guwahati with respect to its intensity, identify the major causes and consequences associated with it, and understand the human response to deal with the resulting situation, primarily on the basis of field observation and investigations.

The editors are grateful to the authors of 27 chapters of this present book hail from India, Bangladesh, Sri Lanka, Saudi Arabia and Thailand for their valuable contributions on the different aspects of climate, environment and disaster in developing countries. We are thankful to Dr. Sujay Bandyopadhyay and Mr. Sasanka Ghosh of Kazi Nazrul University, Dr. Somasis Sengupta and Tapan Pramanick of Burdwan University, and Sri Kalikinkar Das of Gour Banga University, West Bengal, India for active cooperation in the preparation of this manuscript. We are also grateful to Springer Nature, Singapore for accepting this volume for publication.

Bardhaman, India   Narayan Chandra Jana
New Delhi, India   R. B. Singh

Contents

Editors and Contributors

About the Editors

Dr. Narayan Chandra Jana is an Applied Geographer with Post-Graduate and Doctoral Degrees in Geography, Post-Graduate Degree in Disaster Mitigation, PG Diploma in Sustainable Rural Development and Diploma in Tourism Studies. He has contributed more than 100 research papers published in various national and international journals and edited volumes. He has *authored three books* entitled (i) The Land: Multifaceted Appraisal and Management (with Prof. N. K. De), (ii) Transformation of Land: Physical Properties and Development Initiatives, (iii) Tsunami in India: Impact Assessment and Mitigation Strategies; *jointly edited five books* entitled (i) Disaster Management and Sustainable Development: Emerging Issues and Concerns (with Prof. Rajesh Anand and Dr. Sudhir Singh), (ii) Human Resources (with Prof. Sudesh Nangia and Prof. R. B. Bhagat), (iii) West Bengal: Geo-Spatial Issues, (iv) Resources and Development: Issues and Concerns (with L. Sivaramakrishnan and others) and (v) Population Dynamics in Contemporary South Asia: Health, Education and Migration (with Prof. Anuradha Banerjee and Dr. Vinod Kumar Mishra). Dr. Jana was also actively engaged in post-doctoral research and teaching in the Centre for the Study of Regional Development, Jawaharlal Nehru University, New Delhi for five years. He was the Coordinator (Eastern India) of the *International Geographical Union: Commission on Geography of Commercial Activities* (1992–96). He is the Life Member of 24 academic societies of repute. He was the Vice-President of *National Association of*

Geographers, India (NAGI), Delhi (2011–12, 2012–13, 2013–14 and 2019–20) and was the Convener of 33rd Indian Geography Congress, 2011; 35th Indian Geographers' Meet, 2013 and XIV IGU-INDIA International Conference, 2020. Dr. Jana was the Deputy Coordinator of the UGC-SAP-DRS Programme (2012–17) and was the Coordinator of the DST-FIST Programme (2012–13). He is a Member of the Editorial Board of the *Indian Journal of Landscape Systems and Ecological Studies*, Kolkata and a Member of the Advisory Board of the journal *Earth Surface Review*, Gorakhpur. He was the Secretary of the *Institute of Indian Geographers* (2016–19) and the Founder-Secretary of *Association of Bengal Geographers*. He is also the editor of a newly launched journal *Contemporary Geographer*. He is a member of the PG Board of Studies in Geography in Kazi Nazrul University, Asansol; Bankura University, and Coochbehar Panchanan Barma University, as well as BRS Member in Geography in the University of Gour Banga, Malda and Bankura University, West Bengal. Dr. Jana visited *Nepal (1994), Sri Lanka (2012), Bangladesh (2013), Thailand (2013), Russia (2014), China (2016), Japan (2016)* and *Thailand (2017)* for various academic purposes. He has delivered about 100 lectures in Academic Staff College, UGC sponsored national seminars and various academic departments of different universities. He has successfully guided 11 M.Phil. and 11 doctoral dissertations. He has completed one Major Research Project entitled Tribal Livelihood and Sustainable Development in Mayurbhanj, Orissa sponsored by ICSSR and one small research project on Wasteland sponsored by NRDMS of DST. He has also conducted one research methodology course sponsored by ICSSR. His areas of research interest cover applied geomorphology, hazards & disasters, environmental issues, land use and rural development. Prof. Jana has been recently nominated as Steering Committee Member in *IGU Commission on Research Methods in Geography*. Currently, Dr. Jana is a Professor (Former Head) in the Department of Geography & Coordinator, M.Sc. in Geospatial Science, The University of Burdwan, West Bengal, India.

Dr. R. B. Singh (Date of Birth: 3 February 1955) is an Ex. Professor since 1996, Coordinator UGC-SAP-DRS III (2014–2019) and Ex. Head in the Department of Geography, Delhi School of Economics, University of Delhi, Delhi-7 (2013–2016 and 2019–20). Earlier served as UGC Research Scientist-B/Reader (1988–1996), Lecturer (1985–1988), and CSIR Pool Officer (1983–1985).

Recently, Prof. Singh got elected as first Indian and second Asian Secretary-General and Treasurer of the International Geographical Union for 2018–22. Dr. Singh is formerly Chair, Research Council, CSIR-Central Food Technological Research Institute, Mysore; Ex-Member-Research Council-CSIR-Central Institute of Medicinal and Aromatic Plants, Lucknow; Member of International Science Council (earlier ICSU) Prestigious Scientific Committee-Health and Wellbeing in Changing Urban Environment-System Analysis Approach since 2016. Prof. Singh was Vice-President, International Geographical Union (IGU) since 2012 and is elected again for the second consecutive term (2016–18) of the highest world geographical body. He is invited by the IAP-Global Network of Science Academies to join the Working Group for a statement on science and technology for disaster risk reduction. He was unanimously elected President of the Earth System Science Section of the Indian Science Congress Association for 2019–20. NITI Aayog, Government of India invited him as a member of the prestigious committee for preparing Vision India-2035. Chair, Task Force, Landslides Awareness, National Disaster Management Authority, Government of India, 2018–2019. He is also International Science Council GeoUnions Standing Committee on Risk and Disaster Management. He is also associated with the prestigious programme such as Co-Chair ISC-CODATA-PASTED, Member Earth System Governance and IAP-Global Network of Science Academies representative on disaster risk reduction. He is one of the three founders of the Centre for Himalayan Studies, Univ. of Delhi and now an Advisory Committee Member.

He has to his credit 15 books, 39 edited research volumes and more than 247 research papers, including 123 published in national and international journals (i.e. Remote Sensing, Climate Dynamics, Current Science,

Singapore Jl. of Tropical Geography, Energies, Sustainability, Theoretical and Applied Climatology, Environmental Science and Policy, Physical Geography, Advances in Meteorology, Physics and Chemistry of the Earth, Agriculture, Ecosystem and Environment, Hydrological Processes, Mountain Research and Development, Journal of Mountain Science, Climate, Frontiers in Environmental Science, Advances in Earth Science, Advances in Limnology, European Jl. of Geography, Asian Geographer, Environmental Economics, Cities and Health, Tourism Recreation Research). He was Special Series Editor of prestigious journals like Sustainability, Advances in Meteorology, Physics and Chemistry of the Earth, NAM Today. He is an Editorial Committee Member of Jl. of Mountain Science. In 1988, the UNESCO/ISSC (Paris) awarded him Research and Study Grants Award in Social and Human Sciences. He was also associated with prestigious international collaborative research programs such as ICSSR-IDPAD, CIDA-SICI, DFID, Finland Academy of Sciences, UGC and Ministry of Agriculture. He has supervised 38 Ph.D. and 81 M.Phil students. He is Springer Series Editor-Advances in Geographical and Environmental Sciences; Sustainable Development Goals.

He was awarded the prestigious Japan Society for Promotion of Science (JSPS) Research Fellowship at Hiroshima in 2013 and Several Travel Fellowships/Support from UNEP, UNITAR, UNISDR, IAP, UNU, UNCRD, WCRP, IAHS, IGU, NASDA, INSA, UGC, SICI, MAIRS and University of Delhi etc. for participating and presenting papers, Chairing session and discussing research projects in about 45 countries. He was also associated with Nordic Inst. of Asian Studies, Copenhagen (Denmark) in 1998 and Visiting Professor for delivering invited Lectures at the University of Turku (Finland). He was one of the contributors to the famous The World Atlas-Earth Concise, Millennium House Ltd., Australia. He was invited by UGC for Preparing National Level CBCS Syllabus for Undergraduate Geography in 2015. He is also Chair of the UGC prestigious committee for preparing the learning outcome-based curriculum framework since July 2018. Recently, UGC-Consortium for Educational Communication invited him as Academic Advisory Council for

CEC MOOCS on SWAYAM. He has been Chairman-Governing Body of the two prestigious Delhi University Colleges, i.e. Kamla Nehru College, Shaheed Bhagat Singh College. He has been expert in the prestigious Committees of the Government of India, Ministry of Environment and Forests, Department of Science and Technology, National Disaster Management Authority (NDMA), ICSSR, CSIR, UGC, UGC-CEC, NCERT, UPSC and NIOS etc.

Contributors

Raihan Ahamed BIGD, BRAC University, Mohakhali, Dhaka, Bangladesh

Arif Uddin Ahmad Department of Geomatics, Patuakhali Science and Technology University (PSTU), Patuakhali, Bangladesh

Asraful Alam Department of Geography, Serampore Girls' College, University of Calcutta, Serampore, Hooghly, West Bengal, India;
Department of Geography, Rampurhat College, Rampurhat, Birbhum, West Bengal, India

Aminuddin Ali Department of Statistics, Ramkrishna Mission Vidyamandira, Belur, India

Jatisankar Bandyopadhyay Department of Remote Sensing and GIS, Vidyasagar University, Midnapore, India;
Centre for Environmental Studies, Vidyasagar University, Midnapore, India

Sunanda Banerjee Department of Geography, Kazi Nazrul University, Asansol, West Bengal, India

Tapati Banerjee Department of Science and Technology, Government of India, National Atlas and Thematic Mapping Organisation, Kolkata, West Bengal, India

Papiya Banik Department of Geography, University of Calcutta, Kolkata, India

Piya Bhattacharjee Department of Environmental Science, University of Kalyani, Kalyani, West Bengal, India

Sutapa Bhattacharjee Department of Civil Engineering, Indian Institute of Technology Guwahati, Guwahati, India

Malabika Biswas Roy Department of Geography, Women's College, Kolkata, West Bengal, India

Ashok Kumar Bora Department of Geography, Gauhati University, Guwahati, Assam, India

Kusumbor Bordoloi Department of Geography, Gauhati University, Guwahati, Assam, India

Akangsha Borgohain Department of Geography, Gauhati University, Guwahati, Assam, India

Namita Chakma Department of Geography, The University of Burdwan, Barddhaman, West Bengal, India

Babita Chatterjee Department of Humanistic Studies, IIT (BHU), Varanasi, India

Soumen Chatterjee Department of Geography, The University of Burdwan, Golapbag, West Bengal, India

Sunam Chatterjee Independent Researcher, Burdwan city, West Bengal, India

Debasish Das Department of Environmental Science, University of Kalyani, Kalyani, West Bengal, India

Amrita Dwivedi Department of Humanistic Studies, IIT (BHU), Varanasi, India

Arnab Ghosh School of Water Resources Engineering, Jadavpur University, Kolkata, India

Biswajit Ghosh Khorod Amina High School, Satgachia, Barddhaman, West Bengal, India

Sandipan Ghosh Department of Geography, Chandrapur college, West Bengal, Purba Bardhaman, India

Tirthankar Ghosh Department of Statistics, Visva Bharati, Santiniketan, India

Md. Golam Azam Environmental Science Discipline, Khulna University, Khulna, Bangladesh

Santonu Goswami Earth and Climate Science Area, National Remote Sensing Centre, Indian Space Research Organization, Hyderabad, India

Bijay Halder Department of Remote Sensing and GIS, Vidyasagar University, Midnapore, India

M. Manzurul Hassan Department of Geography and Environment, Jahangirnagar University, Savar, Dhaka, Bangladesh

Mallik Akram Hossain Department of Geography and Environment, Jagannath University, Dhaka, Bangladesh

Narayan Chandra Jana Department of Geography, The University of Burdwan, Golapbag, Bardhaman, West Bengal, India

Bimal Kumar Kar Department of Geography, Gauhati University, Guwahati, India

Pankaj Kumar Roy School of Water Resources Engineering, Jadavpur University, Kolkata, West Bengal, India

Manoj Kumar Mahato Department of Geography, The University of Burdwan, Bardhaman, West Bengal, India

Sanjia Mahiuddin Department of Geography and Environment, Jagannath University, Dhaka, Bangladesh

Adel Moatamed Department of Geography, College of Humanities, King Khalid University, Abha, Saudi Arabia;
Faculty of Arts, Department of Geography, Assiut University, Assiut, Egypt;
Prince Sultan Bin Abdul-Aziz Center for Environment and Tourism Research and Studies, King Khalid University, Abha, Saudi Arabia

A. H. M. Monzurul Mamun Geography and Environment, Pabna University of Science and Technology, Pabna, Bangladesh

Muhammad Muhibbullah Department of Geography and Environmental Studies, University of Chittagong, Chittagong-4331, Bangladesh

Md. Mujibor Rahman Environmental Science Discipline, Khulna University, Khulna, Bangladesh

Nabanita Mukherjee Department of Geography, The University of Burdwan, Bardhaman, India

Swetasree Nag School of Water Resources Engineering, Jadavpur University, Kolkata, West Bengal, India

Ratan Pal Department of Geography, The University of Burdwan, Bardhaman, West Bengal, India

Rajat Kumar Paul Department of Geography, University of Calcutta, Kolkata, India

Shitangsu Kumar Paul University of Rajshahi, Rajshahi-6205, Bangladesh

Kirishanthan Punniyarajah Department of Geography, University of Colombo, Colombo, Sri Lanka

Jayant Kumar Routray Asian Institute of Technology, Bangkok, Thailand

Arup Kumar Roy Department of Geography, Kazi Nazrul University, Asansol, West Bengal, India

Malabika Biswas Roy Department of Geography, Women's College, Calcutta, India

Pankaj Kumar Roy School of Water Resources Engineering, Jadavpur University, Kolkata, India

Dhrubajyoti Sahariah Department of Geography, Gauhati University, Guwahati, Assam, India

Anup Saikia Department of Geography, Gauhati University, Guwahati, Assam, India

Asad Ali Sarkar Department of Geography, Chandrakona Vidyasagar Mahavidyalaya, ChandrakonaPin-721201, West Bengal, India

Md. Iqbal Sarwar Department of Geography and Environmental Studies, University of Chittagong, Chittagong-4331, Bangladesh

Lakshminarayan Satpati Department of Geography and Director, UGC-HRDC, University of Calcutta, Kolkata, India

Anamika Shaha SKS School and College, Gaibandha Sadar, Gaibandha, Bangladesh

Giyasuddin Siddique Department of Geography, The University of Burdwan, Bardhaman, India

Rebeka Sultana University of Rajshahi, Rajshahi-6205, Bangladesh

Part I
Climate

Chapter 1
Challenges of Climate Resilient Livelihoods and an Inquiry of Mitigation Strategies in India

Babita Chatterjeeⓘ **and Amrita Dwivedi**ⓘ

Abstract Climate change and livelihood are different but interrelated concepts. The impact of dangerous climatic change falls disproportionately on the livelihood systems. Agriculture is one leading contributors to global warming. The gradually increasing environmental degradation reduces the comforts of local communities. It shrinks income opportunities that ultimately enhance the scope of greater food insecurity. The majesty of mitigation and prevention is a significant parameter to promote more productive, resilient livelihoods. It requires policy support, capacity building, and variation in agriculture. Land zoning and land-use management need to consider the spatial parameters of physical vulnerability. Essential strategies aim to strengthen the capacities and resilience of communities to protect lives. The focused objectives of the study are to understand the role of climate on the way of life and give way out. Significant findings are rainfall dispersion, heat-stress highly influenced GDP, and human development index (HDI). Doing so demands greater attention on the societal root causes underlying differences in vulnerability and resilience. Mostly we can say to sustain these specific circumstances, a few advanced extension services such as cropping patterns and production systems would be followed to manage the extremity.

Keywords Climate change · Environmental degradation · Resource management · Resilient livelihood · Mitigation

Abbreviations

WFS	World Food Summit
MDG	Millennium Development Goal
SD	Sustainable Development
CC	Climate Change

B. Chatterjee (✉) · A. Dwivedi
Department of Humanistic Studies, IIT (BHU), Varanasi, India
e-mail: babitachatterjee.rs.hss19@itbhu.ac.in

© The Author(s), under exclusive license to Springer Nature Singapore Pte Ltd. 2022
N. C. Jana and R. B. Singh (eds.), *Climate, Environment and Disaster in Developing Countries*, Advances in Geographical and Environmental Sciences, https://doi.org/10.1007/978-981-16-6966-8_1

SLM	Sustainable Land Management
CR	Climate Risk
WPP	Water pricing Pricing
CSS	Climate Sensitive Sectors
TN	Tamil Nadu
J & K	Jammu and Kashmir
DRR	Disaster Risk Reduction
CDM	Clean Development Mechanism
PPP	Privet Public Partnership
FNS	Food and Nutrition Service
SD	Sustainable Development
GW	Ground Water
SW	Surface Water
HA	Hectare
Chh	Chhattisgarh
MP	Madhya Pradesh
AP	Andhra Pradesh
NE	North East
HP	Himachal Pradesh
TG	Telangana
AI	Andaman and Nicobar Islands

1.1 Introduction

Livelihood consists of the ability of activities to get the required materials for present and future sustenance. Adequate livelihood option empowers the less-powered people. The perspective of proper livelihood options can be built beautifully for social bonding and social life. But, it becomes tougher for rapidly growing people to produce food for means. The most fundamental resources are fertile land and freshwater for food production. However, both of these are affected by volatile technological development. We are very much accustomed to the term climate change for the last two decades. All the creations of the universe are duly controlled by Mother Nature. Perhaps, the food production system is also formed by cultural processes. It helps to sustain all living planet. But, gradually the food production system has changed. Human beings are exploited by the past heritage inter-relationship of nature and human civilization. More than 700 million rural inhabitants rely on natural resource-based CSS. But, they are not little cared to them.

The sources of income of more or less three billion global inhabitants predominantly live on the smallholding. Rural natives highly depend on natural resources, around 70% of the population in developing nations live in rural areas (IPCC 2007). The per capita GDP of these huge, almost 68% of the country's people comes from farming. This is not a matter of happiness for any country. India is positioned at

130 and 129 among 189 countries in the 2018 and 2019 Human Development Index (HDI), respectively. It clearly expressed the economic and social health of her citizens. The economic crisis is laterally associated with public health. Overall, the whole of Indian agriculture is highly influenced by the vagaries of the climatic phenomenon, such as long duration heavy downpour, sudden cloud burst, and deep water scarcity in dry months with heat stress. The data on basic nutritional demand of human beings, 83% of the 697 kg of food consumed per person per year, 93% of the 2,884 kcal per day, and 80% of the 81 g of protein eaten per day coming from terrestrial production in 2013 (FAOSTAT 2018).

There are some food safety missions to enhance food security and finish worldwide hunger. Hence, the formation and targeted goal of WFS and some MDGs is to support the vulnerable class. Instead of all these, a huge number of the world population is still undernourished. Sometimes it crossed 1.27 billion global residents. There are three valuable certain goals to practice sustainable agriculture for integrating the capacity of regional institutions, namely—i. to keep into consideration environmental health, ii. profit maximizing, and iii. fairness in economic and social spheres. Therefore, the main focus should be mainstreaming marginal farming households and ensuring every assistance related to eco-friendly farming practices. Climatic diversity affects human health in other ways that interact with food utilization. Agriculture relies still on manual labor in numerous parts of the world. It is projected that heat stress will shorten the hours of work and advance their risk (Dunne et al. 2013). Some general meetings and committees are set up to combat the hazardous impact of adverse climate. The main focuses of all these were to give suitable earth and sustainable environment to all living creatures (Elbehri et al., FAO UN 2015). The term resilience is intensively interrelated with ecological, social else socio-ecological components of the human ecosystem. The functions of its components lead to ameliorate, check, and mitigate disastrous events or scenarios in a timely and eco-friendly manner.

1.2 Materials and Methods

1.2.1 Physical Dividends of India

The nation is located between 8°4′ N and 37°6′ N latitude and 68°7′E and 97°25′E longitude. It has a wide diversified altitude, like Himalayan mountain ranges, a storehouse of world-highest peaks, while lowest topographical locations along two broad coastlines. The Tropic of Cancer passes the middle portion of the country. Its existence separated the territory into two climatic zones. The temperate and semi-tropical climate in the northern half and tropical type in the southern half. Based on Köppen's classification, Indian climate is categorized into six major subtypes. We can get a clear-cut idea if we study its location. All the states do not experience temperatures less than 10 °C all over the year except Himalayan foothills states. The range of diurnal even seasonal is quite high in some north-western states. Eastern states moisture by Monsoon and whirlpools, whereas northern states get rain by retreating monsoon. It has an arable land area of 159.7 million hectares and is the second-largest in the world. The amount of gross irrigated crop area of 82.6 million hectares is the largest in the world. A total of 65% of cropped land is irrigated from the groundwater. Unlike the climate, it has immense agro-biodiversity.

1.2.2 Impacts of Climate Change Associated with India

Gradual degradation of the environment is somehow responsible for the intensive roughness of nature. We are familiar with the term 'Environmental Refugee', as so many hazardous climatic events occurred in the recent past. The loss and damage to those calamities are huge. It becomes a trend, huge people are ruined due to this every year. For instance, if we take India as an instance, the number of occurrence of floods and drought in many of the provinces are many. Here, we can mention some of the devastating floods in the twenty-first century. Uttarakhand flood in 2013, 2015, Gujarat flood, Assam flood in 2016, and flooding in Kaziranga National Park, yet again Gujarat was flooded in July 2017 with more than 200 deaths. Most devastated of this century are 2018 Kerala flood with over 445 deaths, Odisha flooded due to Cyclonic storm, Phani, resulting in huge property and human loss and the most recently, hazard in West Bengal due to Super Cyclone 'Amphun'. Due to these sudden calamities lakhs of thousands of people became displaced from their homeland. They lost every possessed land and became 'Environmental refugees'. It deteriorated the livelihoods of natives resulting in unwashed-upcropping field.

1.2.3 Food Security, Livelihood Risk, and Status of Indian Farmer

Climate change (CC) induced to increase the climate variability. It affects the food supply system in four dimensions—availability, accessibility, utilization, and stability. Side by side, CC affected both rural and urban life. It leads to a water crisis and hampers physical infrastructure. CC has intensive negativities on agrarian economy-based states. The major source of GDP is farming; food safety of common people is severely hampered and 2 to 4 °C of temperature increase highly influenced agricultural losses. Temperature asymmetry is a catalyst for biophysical staving. It has a ripple effect on all parameters of agro-ecology, the most demolishing recomposed in the hydrological process. The disequilibrium in temperatures impacted moisture by exhalation and forming a deficit in soil moisture (Parry and Swaminathan 1992). Therefore, this life-supporting sector seriously affects growth as well as cuts the absorption of labor resulting in the slowing down the poverty reduction. Furthermore, the marginalized communities are disproportionately affected such as the Below Poverty Line (BPL). Additionally, imbalanced rainfall is caused in dry-drought periods. It stressed new plants and fluctuates productivity, especially in the key periods of growth of the plant's life cycle (Gadgil 1995). Overall, CC highly impact 'public health'.

1.2.4 Crisis Management for Resilient Livelihood Option

The Organisation for Economic Co-operation and Development (OECD 2009) has developed a Policy. Additionally, four general steps for how to develop adaptation measures in different settings have been agreed upon in this Policy Guidance Those are:

i.   Assess the vulnerability,
ii. Integrating adaptation into development planning,
iii. Agricultural pricing,
iv. Risk management (contingency plans, insurance, seed banks, etc.).

1.2.5 Existing Plan and Policies

The Intergovernmental Panel on Climate Change (IPCC) plays a significant role in understanding the risk of anthropogenic issues for climate change. Strategies should be region-oriented and time-bound. International treaties like the UNFCC, Kyoto Protocol, CDM, and Marakesh are a few instances. The integration of the women population in this mission must be included under its goal. Ministry of Environment focused on eight missions combined—primary sectors, green energy, suitable habitat, and scientific knowledge to combat hazardous impacts. Moreover, there must be an assessment committee for measuring the susceptibility of policies and to train executives properly. The 2030 Agenda for SD emphasized transformational vision. Our changing world is bringing new obstacles. We must overcome those to live happily in a world without hunger.

Mission on Strategic Knowledge seeks to understand climate science and its challenges. It envisioned a new Research Fund for Climate Science to figure out climate modeling by international collaboration in PPP partnerships. The National Mission for a 'Green India' focuses on the plantation of six million hectares of degraded forest lands and the extension of forest cover from currently 23% to 33% of India's territory. It also included:

i.   Rural development and planning (guaranteed employment and planning for rural infrastructure),
ii. Rules and regulations (strategic environmental assessments).

The National Mission for Sustainable Agriculture supports acclimatization in cultivation by the advancement of climate-resilient crops with enhancing weather insurance mechanisms and innovative agricultural practices.

1.3   Objectives

The main objectives of our study are.

i.   to understand the role of climate to control way of life and to study the factors behind the ever-increasing food crisis,
ii. to enquire existing policies and to frame suitable strategies to mitigate the global issue.

1.4   Methodology

The present study on 'Challenges of Climate Resilient Livelihoods … in India' is used as an example of developing nations. The overall problem is more on less the same in all struggling countries. We framed and followed a transparent methodology base to fulfill the objectives of the research. By considering the fatal global issue of 'Climate Change' we tried to search out its negative externalities on human civilization. So, we thoroughly studied the related theories from specific journal reports from the website of Department of Environment, Public Welfare of Government of India, and some English daily. We used only secondary sources of data because one can easily get an overall outlook of the impacts of natural events over a wide region at a time, without paying much effort. We collected the related information from FAO, UNDP, IMD, WMD, etc. departmental websites and some renowned journals. And used both quantitative and qualitative techniques. We have used ArcGIS 10.4 version map-making software to give a generalization. Here, we used the India map before 2019. We prepared the chart by using Microsoft office excel 2007 version, and IBM SPSS 23 packages to show correlation. Also for regression analysis, there is some limitation in our work. There is some limitation in the work, as we used secondary data sources.

1.5   Results

1.5.1   Data Analysis

India had 24.56% forested or 807,276 sq. km. area in 2019 according to its total land area. The new policy of forestry targeted 33% of the country's geographical

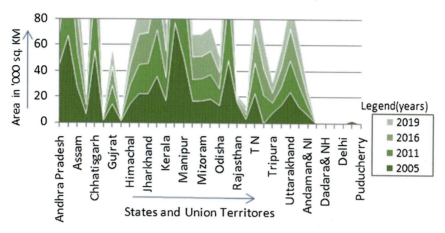

Fig. 1.1 Showing a total forest-covered area of Indian states and UTs in four different years: 2005, 2011, 2016, and 2019. *Data Source* India State of Forest Report (ISFR), 2016, Forest Survey of India (FSI)

area under the green cover. This ratio is not constant. Rather every year it changes. We studied the forested area in five years and three years gap for the recent year. In Fig. 1.1, the green area has been shown in vertical axes. For more clarification, we scaled it to 20,000 sq. km. States and UTs have shown on a horizontal scale.

According to the 2019 Forest Survey, the state of Madhya Pradesh has the largest forest cover but Mizoram has 85% forest coverage of the total state area. It includes all kinds of forestry that exist within the boundary of the state. The state of Andhra Pradesh and Punjab lost a significant amount, more than 60% from 2005 to 2019. Chhattisgarh and Jharkhand remained almost the same. The 'seven sister states' stayed unexploited. Rajasthan has lost more than 65% of vegetation since 2005. This scenario preceded the aridity and dryness of the local climate.

1.5.2 Rainfall Anomalies

Climate variations are also called anomalies. These are differences in the state of the climate system from normal conditions. Ranged from +1 to −1. The value closed to +1 signified less dispersion. Far from this symbolized a huge gap. It is calculated on average over many years, usually 30 years of the year.

Figure 1.2 shows the monsoonal rainfall anomalies both pos from 1st June to 30th September 2018. For a transparent outlook, we categorized the whole data set into five sects. The highest deviation, i.e., >−40% in two northeastern states, Manipur and Meghalaya, located in windward slope and islanded state Lakshadweep, shifting of pressure belt. The second category fell most of Eastern, northeastern states, and

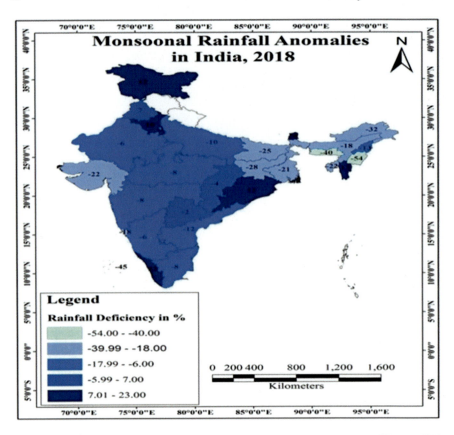

Fig. 1.2 Showing Monsoonal rainfall dispersion of India, 2018. *Data Source* Annual Report, 2018. Indian Meteorological Department

Gujrat. The third zone lied on nine western, central, and southern states; fourth zone hailed over six different states, not directly contacted. At least seven states got heavy moisture cloud burst, heavy downpour due to cyclonic storm and monsoonal winds. The level of the groundwater table is a good symbol for soil moisture wholly soil health. Figure 1.3 shows the peak proportion of wells water table fell in the dry season. We took into consideration 11 states all over the area, located in a diverse topographical zone.

Across the season water table fluctuation is most in three south Indian states Kerala, Karnataka, and Andhra Pradesh; Chhattisgarh also tolerated the same condition. Gujrat and Himachal belonged to the same level. Jharkhand faced a water crisis due to Plateau and terrain location. Irrigation is the most vital component in present-day cultivation. This ever-increasing tendency supports farmers for their survival in the lean season. There is a list of sources of irrigation: open-source and underground water. This system varies from state to state, based on the physiographical

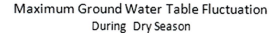

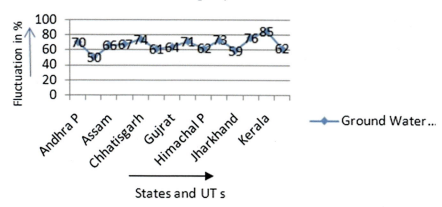

Fig. 1.3 Showing the maximum percentage of groundwater table fluctuation during dry months. *Data Source* Ground Water Year Book of India 2016–2017. Central Ground Water Board, Ministry of Water Resources

sites and scope of technology using capacity. The cost and effort of accessing open-source are much lower than using groundwater irrigation. Some types of irrigation depend on the nature of soil like drip, sprinkler, lateral move, center pivot, etc. practiced by Indian farmers. Sometimes it denominates the status of farmers. Figure 1.4 shows the irrigation potentiality of Indian farmers from minor sources like surface and underground sources, the area in 1,000 ha. The vertical axis scale interval is 6,000 ha of potential irrigation capability. It is displayed for 19 states together in Composite Bar Graph. We can explain that Madhya Pradesh and Andhra Pradesh

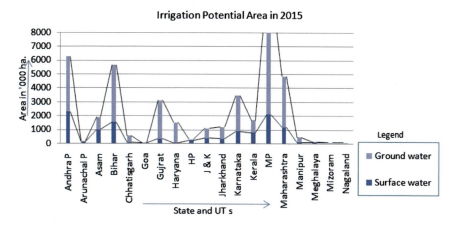

Fig. 1.4 Showing Irrigation potential area of some Indian states in 2015. *Data Source* Water and Related Statistics, CWC, 2015

have the highest proportion of micro-irrigation capacities in SW, i.e., >2,100,000 ha; Jharkhand, Karnataka, Kerala, and Gujarat have a moderate level of SW potentialities, i.e., >800,000 ha; J&K, Chh, HP, Haryana, and NE states are the least potentials for SW irrigation. This zone lied >400,000 ha of area. Bihar, Andhra Pradesh, and Madhya Pradesh have >3,900,000 ha of area for GW. MP is the highest potential for GW irrigation. Gujarat and Maharashtra have >2,800,000 ha GW sources, where HP, Kerala, Jharkhand, Chh, etc. states have moderate sources of GW. Again the NE states, Goa, and HP are positioned at the lowest in the GW storage. All these scenarios of micro-irrigation facilities are carried out with the level of development, growth and standard of particular states, and the status of small and marginal farmers.

1.5.3 Human Development Index

The Human Development Index (HDI) is a statistical composite index of life expectancy, education, and per capita income indicators. HDI by 2018 revised the UN method used. It helps to categorize all countries into four tiers of HDI. A country scores a higher HDI when the lifespan is higher, education level, and gross national income (GNI) per capita is higher. In Fig. 1.5 we plotted different values of HDI for the years 2000, 2010, and 2018. All the states and UTs are taken into consideration. India scored 0.498 in 2000, 0.583 in 2010, and 0.647 in 2018.The map marked the trend of the HDI of India in the years 2000, 2010, and 2018. The different data have been plotted in Comparative Bar Graph on a two-dimensional map of India to get an overall synoptic view across the states.

We can generalize the socio-economic variation across the Indian Territory over 29 states and seven UTs. The value of HDI nearness to one has been considered as more progressed than the others. We classified the HDI data into three separate blocks:

i.   Under 0.5,
ii. 0.51 to 0.70, and
iii. 0.71 and above values of the three years.

A total of seven states namely Odisha, Bihar, MP, UP, Andhra Pradesh, Rajasthan, and Meghalaya were in below zone 1 (<0.5). In 2000, India scored 0.498. These states reached a better condition, scored >0.50 in 2010. Though, until 2018 Bihar, Jharkhand, and Odisha belonged to the same standard.

In 2000 some states and all UTs were in class 2. Among them, Kerala, Lakshadweep, Puducherry, Goa, New Delhi, and AI reached class 3 in 2010. And HP, Haryana, Gujarat, Chandigarh, Karnataka, Arunachal, J&K, Tamil Nadu, and NE states were under class 2 in 2010. The BIMARU states and AP scored >0.60 in 2010, among them UP, Bihar, Jharkhand, and Odisha were in the same condition till 2018. The states and UTs which lied in class 3 in 2018 were Chandigarh, New Delhi, Haryana, Himachal Pradesh, Punjab, Goa, Gujarat, Goa, Kerala, Tamil Nadu, Sikkim, Mizoram, and

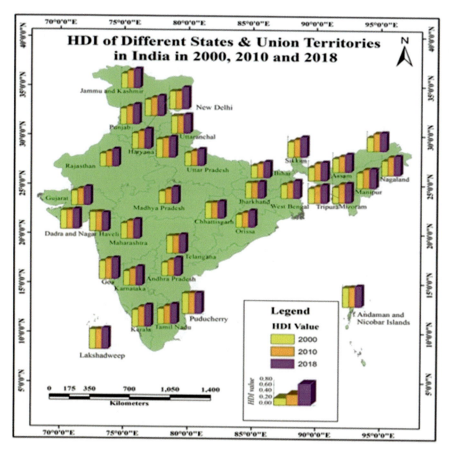

Fig. 1.5 Human Development Index of Indian states/UTs (by revised UN Method) for 2000, 2010, and 2018. *Data source* Human Development Report, 2018, Government of India

Lakshadweep. Kerala scored 0.79 and ranked highest in 2019 for two consecutive times. Chandigarh scored 0.775 and ranked second in 2019. While UP and Bihar score 0.596 and 0.576, respectively, and ranked the lowest to state-wise HDI of India (Tables 1.1 and 1.2).

From Table 1.3 it can be claimed that the R value of the correlation coefficient is 0.0555 and R^2 value after converting in percentage, i.e., 30.8%. Means per capita GDP shows (100–30.8)% = 69.2% of variation with per capita food grains availability in the selected years.

We tried to show a relationship between GDP per capita (PPP-based) in column two; GDP in PPP is a crude domestic product that transformed in international dollars using buying power level rates and divided by total population and annual food grains availability in kgs in colum three. And column one contained different years since 1971 to 2019 as in Table 1.1. Due to the unavailability of all year's data, here we used data of some years gap but we used specifically important years of major economic

Table 1.1 Showing the relationship between PPP, GDP, and Annual Food Grain Availability

Year	Per Capita GDP (in US$ PPP)	Per Capita Annual Availability of Food Grains in kgs.
1971	393	156.03
1981	632	166.02
1991	1,193	186.2
2001	2,130	151.9
2011	4,750	151.9
2012	5,037	170.9
2013	5,383	169.5
2014	5,814	178.6
2015	6,294	169.8
2016	6,761	177.9
2017	7,287	178.4
2018	7,824	180.3
2019	8,378	179.6

Data Source Directorate of Economics and Statistics, DAC and FW

Table 1.2 Correlation is significant at the 0.05 level (two-tailed)

		GDP Per Capita (inUS$ PPP)	Per capita net food grains availability kgs/year
GDP per capita (US$)	Pearson Correlation	01	0.555[*]
	Sig. (2-tailed)		0.049
	N	13	13
Per capita net food availability kgs/year	Pearson Correlation	0.555[*]	1
	Sig. (2-tailed)	0.049	
	N	13	13

[*] shows that the relationship between per capita GDP with per capita net food grain availability in Kgs/year

Table 1.3 Model Summary

Model	R	R^2	Adjusted R^2	Std. error of the estimate
1	0.555[a]	0.308	0.245	8.71619

[a]Predictor (Constant), GDP PPP (US$)

changes. From the result in Table 1.2, we can argue that there is a significant positive relationship between the variables. Here, Null hypothesis is rejected because p > 0.05 and that is 0.555. Alternative hypothesis is accepted.

Table 1.4 Coefficients[a]

		Unstandardized coefficients		Standardized coefficients	t	Sig.
		B	Std. error	Beta		
1	(Constant)	163.117	4.944		32.994	.000
	GDP PPP (USD)	0.002	0.001	0.555	2.210	0.049

[a]Dependent Variable: Per capita net food availability kgs/year

Table 1.4 shows data analysis based on Table 1.1. The constant means intercepts or A in regression coefficient, if per capita GDP PPP be zero then the variation in per capita food availability would be 163.11. And for GDP PPP, the slope is 0.002. We find very slow changes because there is a huge gap between the two variables. We mean to say that if US$ 1 of GDP PPP increase leads to 0.02 units of food grains, 2 US$ increase means 0.04 increase in food grains availability. The significance is equal to ANOVA significance. P value is 0.049 and almost the same like A value, i.e., 0.05. So, Null hypothesis is accepted. And we can emphasize the logic that GPD PPP effects on percapita capita food grain availability.

From the scatter diagram (Fig. 1.6) the value of $R^2 = 0.308$, depends on the forumula yc = a + bx. It predicted the significant inverse relation of GDP PPP with food grains availability. The increase in the GDP PPP leads to a shortage of food grains.

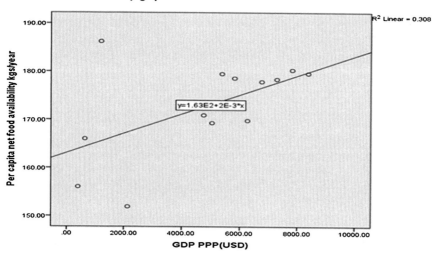

Relationship between Per capita GDP (US$) and Annual Availability of Food Grains (Kgs.)

Fig. 1.6 Prepared by authors using IBM SPSS 23 version package

1.6 Discussion

The term resilience is intensively interrelated with ecological and social else socio-ecological components of the human ecosystem. The functions of its components lead to ameliorate, check, mitigate the disastrous events or scenarios in a timely and eco-friendly manner. Water scarcity has its most immediate and future effects on the soil as well as public health. Unreliable monsoons and falling groundwater level, increasing the results of saline soils. Soil erosion and annexing rates of evapotranspiration rising temperature that effect on decreasing the marginal returns. Due to this, a section of farmers suffers from food security for subsistence a precarious situation. From Fig. 1.3, we argued that the groundwater table has fluctuated in some particular states of India. Of course, some of them are located in the semi-tropical, semi-arid climatic zone, or hilly terrain. Where human intervention on nature is more, it is clear that Andhra Pradesh, Chhattisgarh, and Kerala used the maximum of Surface Water and Ground Water (Fig. 1.4). North and NW part of Gujarat is situated in a semi-arid zone, closed to the Thar Desert and a Part of Relict mountain. The steady growth of HDI symbolizes the level of interrupting the water cycle (from Figs. 1.2 and 1.5). The ever-increasing amount of irrigation positively income and negative effect on GW table, SW storage. Therefore, the seasonal water level fluctuates. Monsoonal rainfall dispersion in 2018 is displayed in Fig. 1.2. It showed an unbalanced even inverse amount of getting rainfall during monsoon. Therefore, Kerala, AP, and Odisha got excessive rainfall, hampered by massive floods. While the eastern states and Gujarat faced extreme water scarcity. There is a strong relation between HDI and water used for irrigation purposes. Because Percapita Income is one of the three parameters of HDI. In Fig. 1.4 we can see that AP and Maharashtra had >3,000,000 ha of GW and the capacity of SW varies in different states. HDI score <0.65 in this year, it is a good indicator of the economy and also for society. The UTs are administered by the central government. That is a good sign for holistic development and national integration. Except for New Delhi, all other UTs are uninterrupted. MP has the highest green coverage area in 2019. The HDI score of the state was constantly < 0.6, but it scored 0.6 in 2018. The nature of the green-covered area had tremendously changed in AP, Punjab, TN, and Kerala. Here, the degradation of forests leads to an extension of cropping fields and industrialization. Because it is a trend of bringing more investment for economic growth. In Punjab and Haryana the proportion of vegetation to land area is <15%. Their SGDP and HDI are satisfactory. Because the Green Revolution was successful here. It was adopted in India in 1965s to cope with the food crisis and to import food grains. It was initiated in different agro-climatic regions. It also started in Rajasthan. The Indira Gandhi Canel is projected at length 600 km and 45 km width. But rice and wheat were major crops for the Green Revolution, and were not suitable in a semi-arid and arid climates. The project was not successful because of water scarcity and soil properties. Meanwhile, the whole planning of the revolution was a failure in Rajasthan. So, first of all, the government should go for different testing, justification before taking an initiative. Especially in climate-sensitive areas. From Tables 1.2 and 1.3, we determine that the two variables (rainfall variation and groundwater fluctuation) are interlinked and inversely related. Per capita GDP PPP reflects lower potentiality to gain food grains. The first variable increase leads to an increase in the scarcity of food grain shortages.

1.7 Major Findings

A significant proportion of the rural population, the children of the soil, face the result of climatic harshness directly. Both the weather and climatic phenomena are occurring in higher frequency. It is clear from Fig. 1.7 that how the uncontrolled and unplanned anthropogenic activities caused hazards. Many times it results in disasters, severe droughts, and massive floods. It highly affects the production system. And they cumulatively contribute to employment loss, i.e., rural poverty without a scope of recovery. This primary sector is already impacted by due stresses like scarcity of water, degradation of topsoil, biodiversity loss, and pollution. However, CC could give better production of some specific crops in mid-high latitudes. But, again it is not a solution for the negative externalities of CC. There is enough evidence of warming of the globe over five consecutive decades, caused largely by a human being, such as high exploitation of fossil fuels and rough land-use patterns. A large proportion of people are becoming marginal for the fragility of the environment. It is most welcome to maintain a proper food system by adopting indefinite adjustment strategies through practicing agriculture in an optimum way. The states Punjab, Haryana, UP, Maharashtra, Tamil Nadu, and Karnataka started to exploit natural resources after the Liberalization and Privatization strategies (1991s) were taken. The state WB, Gujarat, and Odisha followed the way but not in an attractive manner. Furthermore, we can genuinely say that these states are quite less hampered from the ecological loss aspect. But natural hindrances like physiographic location also cause a natural disaster. Those are closely related to food safety and loss of livelihood.

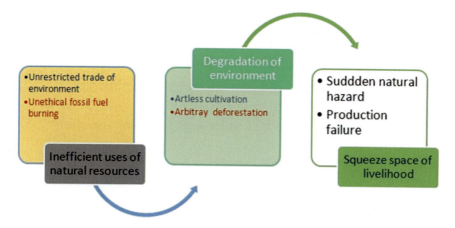

Fig. 1.7 Shows limitless anthropogenic activities to create an unstable economy

1.7.1 Feasible Way Out from Climatic Challenges

Synergies is badly needed between management and adaptation to conduct the strategies in an undisrupted way. Moreover, upraise the way of living level by giving a chance to make a better quality of life. Hence, all the subjects make useful for the welfare society by making a self-confident individual. FAO emphasized multi-hazard risk analyses from current reports to facilitate monitoring of concentrated risks. It contributed to a more robust and simplified monitoring capacity to inform decision-makers about climatic threats. The integration of primary activities is needed to balance food production and security.

Climate sensitive pathway is badly needed to adapt and redress holistic development by combining flexible innovative ideas. For severe CC it required some transformation without damages. Sustainable Land Management (SLM) of farming systems is important to address CC. While dealing with frightening food security required access to nutritious food to maintain healthy and active lives in Africa (AGRA 2017). A balanced strategy on Indian agriculture founded on SLM and varieties of land-use procedures for producing cash crops, biodiversity preservation, ecosystem services allotment, and ILK.

1.7.2 Major Focus Should Be Given to Fulfill the Targets of Sustainable Development Goals (SDGs) in This Very Context of CC to Get Resilient Earth

GOAL 15: Life on Land,   GOAL 13: Climate Action,
GOAL 2: Zero Hunger,   GOAL12: Responsible Consumption and Production,
GOAL 10: Reduced Inequality,   GOAL 11: Sustainable Cities and Communities,
GOAL 7:Affordable and Clean Energy, GOAL 9: Industry, Innovation, and Infrastructure

We emphasized the corresponding links of anthropogenic activities to build a resilient climate in Fig. 1.8. It portrayed how human interventions regulate environmental crises. Here, a proper elaboration of fundamental routes.

1.8 Model of Land Zoning for Sustainable Livelihoods

1.8.1 Multi-Dimensional Way to Reduce Extreme Variability

Strategies to address drivers of vulnerability are:

a. diversification of livelihoods and conservation of heritage indigenous culture;
b. reforestation and efficient natural resource uses;

Fig. 1.8 Shows Interrelated links of anthropogenic activities to build resilient climate

c. monitoring water quality or disaster risk management.
d. human sensitivity, evaluated by analyzing reliability economic sources for holistic development.

1.8.2 The Four Thematic Pillars

I.   Enabling Nature: Aims to implement and execute the exact policy framework to secure durable initiatives.
II. Patrolling in protection: Major aims to early forecasting and to harmonize FNS information for quality monitoring. So, precautions can be taken before natural threats.
III. Refraining Measures: It aims to reduce the level of crisis and to support by providing basic facilities, like advanced equipment and diversified areas of work.
IV. Readiness to reply: It is the preparation to tackle sudden unwanted disasters efficiently. Always keep stronger at all. The main aims are to support people in extreme natural calamities.

1.8.3 Strategic Partnerships

Partnerships including semi-permanent and permanent are needed in local and regional strata of society. For instance, neighboring countries groups and nations falling under the same climatic condition by creating a group for a common interest, such as IGAD, the Food Security Working Group in East Africa, and Regional Vulnerability Assessment Committees, as well as early warning, food security information systems. Organizations of the world level to be set up for coordination to achieve a common goal. Among those, the World Meteorological Organization is based on global climatic indicators. The International Food Safety Authorities Network (INFOSAN) promotes speed in exchanges of intelligence for food safety events also in global level emergencies.

1.8.3.1 Green Growth and Poverty

There would be welfare gains for decentralization 'on average' for a less fair country as a whole. Instead of heterogeneous population, people own different endowments or use factors of production.

Firstly, it is related to policies aiming at the flexibility of the prices of environmental capital to modify the externalities as well as market failures.

Secondly, a set of notable policies remarked to focus directly on investment purposes like low carbon or less environmentally hazardous.

Thirdly, Both public and private and joint climate-resilient investments are open to give motion to the mission.

1.8.3.2 Environmental Pricing and Regulation

Low Carbon and Other Environment Friendly Public Investments. Adaptation and Other Resilience Enhancing Investments.

1.8.3.3 Water Pricing

Water pricing and adequacy in water-related services are the main constituents for reducing pollution, waste, and more investment in infrastructural development. Like value for watershed management services. In many states free electricity supplied for irrigation in agriculture and in most of the Indian states, no such payment for hourly water fees or else. These facilities result in over-exploitation of natural resources. And have long-term implications—salinization, falling water levels, and even the presence of heavy metals.

1.9 Conclusion

We can claim that anthropogenic functions give commands to the Climate Changing process. As it impacts the natural root of ecosystems of India like developing nations. The contemporary natural disasters conflict leads to the way of diversity increase exposure to food-borne diseases, the longevity of food crises. Insufficient institutional capacity to react to the crises indicated high food insecurity. The negative effects of

CC are not always visible to unaware, less educated, or profit-making people. They should accept and genuinely follow existing policies, newly implemented programs, initiatives by the government. So that they can employ without harnessing the natural resource base. In this very situation, farmers should practice indigenous knowledge of cultivation with advance and viable extension facilities. They must be well-facilitated by adequate equipment. It is necessary to use mandatorily diversified scientifically approved quality seeds, manures, pesticides all over the country. Accurate soil testing availability in nearly location, land zoning according to soil properties badly needed. The agriculture development department at both central and local levels should look after this matter. A strong and strict policy execution committee has to be framed to look after medium and long-term policies. Farmer should be well-facilitated for loans and other financial services. Otherwise, the situation would be out of our control. Nature would be ruder. More strict punishment should be given to dishonest persons, responsible for maximum pollution. Furthermore, proper training programs should be organized at regular intervals for farmers directly. The government should fix the price of crops.

References

Alston M (2014) Gender mainstreaming and climate change. Womens Study Institute Forum. Issue 47:287–294. https://doi.org/10.1016/J.WSIF.2013.01.016
Alteri M et al (2017) Technological approaches to sustainable agriculture at a crossroads: an agroecological perspective. Sustainability (9), 349. https://doi.org/10.3390/su9030349
Asadullah MN et al (2018) Poverty reduction during 1990–2013: did millennium development goals adoption and state capacity matter? In: World development, vol 105. Elsevier, pp 70–82. https://doi.org/10.1016/j.worlddev.2017.12.010
Baez JE et al (2013) Rural households in a changing climate. The World Bank Research Observer,28 (2), pp. 267–289. Oxford University Press. UK.
Barghouti RN (2012) Ecological agriculture and sustainable adaptation to climate change: a practical and holistic strategy for Indian smallholders. J Sustain Develop 9(1):132–159. Columbia University, USA
Béné C, Tanner T (2013) Promoting resilient livelihoods through adaptive social protection: lessons from 124 programmes in South Asia. Development policy review
Dercon S (2014) Is green growth good for the poor? The World Bank Research Observer 29 (2), 163–185. Oxford University Press. https://www.jstor.org/stable/24582414
Dunee JP, Stouffer RJ, John JG (2013) Reduction in labor capacity from heat stress under climate warming. Nature clim chang 3(6):563–566
Elbehri A (eds) (2015) Climate change and food systems: global assessments and implications for food security and trade. Food and Agriculture Organization of the United Nations. Rome
FAO (2013) Resilient livelihoods – disaster risk reduction for food and nutrition security framework programme. Food and Agriculture Organization of the United Nations
FAOSTAT (2018) Food and agriculture organization corporate statistical database Rome. www.fao.org/faostat/en/#home
Gadgil S (1995) Climate change and agriculture – an indian perspective. Curr Sci 69(8):649–659
Ghosh S, Mahato K et al (2017) The resilience of agriculture reducing vulnerability to climate change in West Bengal. Current Adv Agricult Sci 9(2):170–177

Govt of India (2018) Ministry of statistics and programme implementation. Central Statistics Office (Social Statistics Division). www.mospi.gov.in http://dx.doi.org/ 10.1016/ j.crm.2015.11.005, http://www.ecologyandsociety.org/vol16/iss3/art3/

Govt of India (2019) Human development reports. hdr.undp.org. Retrieved 16 February 2019

IPCC (2007) Summary for policymakers. Climate change 2007. Impacts, adaptation, and vulnerability. contribution of working group ii to the fourth assessment report of the intergovernmental panel on climate change [Parry ML, Canziani OF, Palutikof JP, van der Linden PJ, Hanson CE (eds)]. Cambridge University Press, U, pp 7–22

Jayawardhan S (2017) Vulnerability and climate change induced human displacement. J Sustain Develop 17(I):103–142. Columbia University, USA.

Koushik G, Sharma KC (2015) Climate change and rural livelihoods- adaptation and vulnerability in Rajasthan. Global NEST J 17(1):41–49

List of Indian states and union territories by Human Development Index as of 2018. Google Wikipedia. Checked on 06/06/2020

Liu S et al (2016) Evaluating economic costs and benefits of climate-resilient livelihood strategies. Clim Risk Manag 12:115–129. Elsevier

Morna CL et al (eds) 2015.SADC Gender Protocol 2015 Barometer. Chapter 10 Gender, climate change, and sustainable development. http://www.jstor.com/stable/j.ctvgc60t9.17

Morton JF (2007) The impact of climate change on smallholder and subsistence agriculture. edited by William Easterling. Proc Natl Acad Sci 104:19680–19685. Pennsylvania State University, USA

Narain S, Ghosh P et al (2009) Climate change perspectives from India. United Nations Development Programme, India. Lasting Solutions for Development Challenges, UNDP, India

OECD (2009) Organization for economic co-operation and development (2009) Integrating climate change adaptation into development co-operation: Policy guidance. OECD

Parry M, Swaminathan MS (1992) Science and sustainable food security: selected papers of M S Swaminathan. IISc Cenetary lecture series; vol 3. Singapore

PocketBook of Agricultural Statistics (2017) Department of agriculture, cooperations and farmers, Govt of India

Power A (2010) Ecosystem services and agriculture: tradeoffs and synergies. Philos Trans R Soc B-Biol Sci 365:2959–2971

Sanghi A, Mendelsohn R (2008) The impacts of global warming on farmers in Brazil and India. Glob Environ Chang 18:655–665

Scheffran J et al (2012) Migration as a contribution to resilience and innovation in climate adaptation: Social networks and co-development in Northwest Africa. Appl Geogr 33:119–127

Selvaraju R et al (2011) Climate science in support of sustainable agriculture and food security. Clim Res 47(½,):95–110, CR special 25. Climate services for sustainable development. Inter-Research Science Center

Sivakumar MVK (2006) Climate prediction and agriculture: current status and future challenges. Clim Res 33(1):3–17. Inter-Research Science Center Stable

Sumba D (eds) (2017) Africa agriculture status report. The business of smallholder agriculture in Sub- Saharan Africa. Alliance for a Green Revolution in Africa (AGRA) (5), 180

Torre A et al (2014) Identifying and measuring land-use and proximity conflicts: methods and identification. SpringerPlus 3(1):85. The University of Paris. Paris. https://www.jstor.org/stable/ https://doi.org/10.2307/24869317

Chapter 2
Analysing LULC Change on Runoff and Sediment Yield in Urbanizing Agricultural Watershed of Monsoonal Climate River Basin in West Bengal, India

Arnab Ghosh⊙**, Malabika Biswas Roy**⊙**, and Pankaj Kumar Roy**⊙

Abstract Changes in rainfall also affect the runoff of a particular area with sediment yield. The idea of runoff and sediment yield is very essential to conserve water and soil as part of a specific watershed management plan. This paper represents an integrated approach based on remote sensing (RS), geographical information system (GIS), and hydrological modelling to assess the impact of land cover change through runoff modelling and sediment yield. A rainfall-runoff model (SCS-CN) and sediment yield model (MUSLE) evolved to estimate the effect of the land cover change on runoff volume with sediment yield through HEC-HMS v 4.3. Several different land-use scenarios were then simulated with the model, calibrating the impacts of land-use change on the hydrology of the watershed. The baseline test results of R^2 and NSE values ranged between 0.61 and 0.88 across the calibration and validation periods, indicating that HEC-HMS accurately replicated the alluvial streamflow. The hypothetical scenario simulations revealed that sediment yield increased with runoff volume, and most of the concentration happens in the non-monsoonal period with declining runoff and rapid urbanization.

Keywords HEC-HMS · Sediment yield · Runoff · SCS-CN · MUSLE

2.1 Introduction

Changes in land use are often non-linear and provide feedback that would put humanity in vulnerable conditions. Honestly said, the change in land use highlights not only the current context but also the acceptability of people to live in suitable

A. Ghosh · P. K. Roy
School of Water Resources Engineering, Jadavpur University, Kolkata 700 032, India
e-mail: pankaj.kroy@jadavpuruniversity.in

M. B. Roy (✉)
Department of Geography, Women's College, Kolkata, West Bengal, India

© The Author(s), under exclusive license to Springer Nature Singapore Pte Ltd. 2022   23
N. C. Jana and R. B. Singh (eds.), *Climate, Environment and Disaster in Developing Countries*, Advances in Geographical and Environmental Sciences, https://doi.org/10.1007/978-981-16-6966-8_2

and sustainable conditions in the future. The ever-growing population and their ever-increasing demand for food and shelter help to change the land-use pattern of one place to another. From the very beginning of civilization, man has been utterly dependent on the river for his livelihood. Considering the river and its productivity, it can be said that many times, the river has forced people to change their needs. River course change had a direct effect on the move in human necessities on land use. The amount of flood in a flood-prone area, changes in the course of the river, and changes in productivity make people eager to change their land-use pattern. Although the floodplain area increases agrarian productivity with excellent communication and transport system, there is a continuous increase in population, which in turn gradually changes the land use of the area. The basic needs of human beings are fulfilled through the character of the floodplain land use. People change land use to meet their own sustainable developmental needs, which indicates an increase in the amount of flooding in the floodplain zone (Ali and Boer 2007; Azari et al. 2015). The changing pattern of land use and rapid urbanization simultaneously increases the amount of flood which is often known as an urban flood. Changes in land use rationally increase the amount of pollution in the air, which in turn controls the number of clouds and precipitation and albeit to a lesser extent. The rapid urbanization method accelerates the amount of runoff with rainfall. Also, contaminated water effuses into the river through sewers. The pliers of these two methods erode the land, and the erosive material accumulates in the stream as sediment. Although we know that floodplain flow naturally increases the amount of deposit in the river, the change in land use rapidly accelerates the amount of sediment in the river terrain. Climate change over time alters the capacitance of different atmospheric elements. Land-use changes with human activity through climate, also increase the possibility of detachment and transport of surface soil towards the river. Many previous researchers have tried to establish a relationship between sediment yield and land use based on soil nature, quality, and sediment transport. In the regions of the varying land use condition with the agrarian economy, a proportional relationship of land use with sediment yield has been observed (Delgado et al. 2015; Erskine et al. 2002; Jain and Das 2010; Sathya and Thampi 2020; Younis and Ammar 2017; Roy et al. 2015).

The HEC-HMS (Hydrologic Engineering Centre-Hydrologic Modelling System) model can easily show changes in land use as well as rainfall-runoff and sediment yield. Previous work has shown that the sediment yield has changed in different river basins with the change in land use and has been beautifully represented by the HEC-HMS model (Asselman et al. 2003; Chu and Steinman 2009; Halwatura and Najim 2013; Pak and Scharffenberg 2008). The HEC-HMS model (HEC-GeoHMS) used to create the Watershed model can be splendidly represented through GIS. With the help of RS (Remote Sensing) and GIS (Geographical Information System), a watershed model can create in HEC-HMS that illuminates the change or hydrological response in future land (Nakil and Khire 2015; Santillan et al. 2010). The surface soil is not very erosive as the slope is shallow in the floodplain area with the dense agrarian nation. Unscientific farming and bank erosion were observed in the floodplain region, which erodes a small amount of surface soil. But in the rainy season, the amount of this erosion increases rapidly. The MUSLE (Modified Universal Soil Loss Equation) model is used in floodplain areas to determine the amount of sediment yield, as most of the soil in the floodplain area is of penetrating nature (Scharffenberg et al. 2018; Usda 1972; Smith et al. 1984).

Bhagirathi-Hooghly River is a distributary channel of River Ganges, flowing from his bifurcation stage at Mithipur, Murshidabad district, West Bengal, towards the Bay of Bengal encompassing Kolkata (Bera et al. 2019). This river system is the lifeline in the southern part of West Bengal for supplying water in irrigation, industrialization, and human habitation to an enormous scope (Laha 2015). From the twentieth century, several studies are created based on the meandering of altering the course and effect of the dominant variable of fluvial-dynamics prediction, estimation, and migration on the Bhagirathi-Hooghly river system (Bandyopadhyay et al. 2015, 2014; Basu 2005). Bhagirathi-Hooghly River is entirely dependent on Farakka Barrage for its water. Many researchers believe that the amount of water flow is decreasing due to the immutability of water agreement with Bangladesh. Due to the diminishing rate of discharge, the nature of this distributary metamorphosed from braiding to meandering in character with modifications through spatial geometry. Previous researchers also have put more emphasis on spatial changes on the meandering loop through release in channel instability and meandering (Bandyopadhyay et al. 2014; Rudra 2010, 2014; Guchhait et al. 2016). A flood happens because the carrying capacity of this river has severely reduced due to sediment accumulation. Although most of the previous researchers have discussed the cause of flooding in this river as shifting motion, reduced navigability, or heavy rainfall, they have not considered changing nature of land-use pattern with varying rainfall-runoff amount and sediment yield from the last few decades. The present article deals with an integrated approach to identify and analyse the past and current land-use change of sub-catchment basin through RS, GIS, and hydrologic modelling which calibrated the changing nature on surface runoff and sediment yield through validation. This paper may be beneficial to local stakeholders by issuing information, which will be helpful for regional planning and management strategies in areas of expanding demographic expansion.

2.2 Materials and Methods

2.2.1 Study Area

The selected sub-catchment basin of the middle course of Bhagirathi-Hooghly river is up to 161 km long from Nabadwip (upstream) to Kalyani (downstream) comprising of a portion of Burdwan, Nadia, and Hooghly district of West Bengal, India (23°24'45" N and 88°22'45" E to 22°58'01" N and 88°24'27" E). This area is geologically located in the Rarh region, the lower portion of the moribund deltaic part of the Bengal Basin, and is composed of recent deposits of the Pleistocene period. Initially, the area was covered with sandy clay and sand along the course of the river and fine silt, sandy

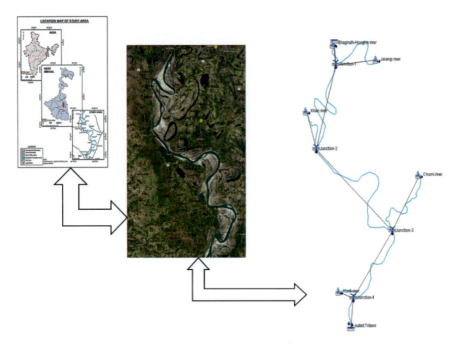

Fig. 2.1 Location map of study area with basin model in HEC-HMS

loam, and loamy soil have been found in the flat portion of the plain (Fig. 2.1). Most of the soils in this sub-catchment basin belong to Hydrologic Soil Group C (sandy clayey loam) with moderate runoff capacity. Hydrologic Soil Group D (clay loam, silty clay, and sandy clay) with high runoff capacity concentrates in riverine areas. The entire study area consists of the wet and dry seasons on a seasonal basis and most of the rainfall occurs here from June to October.

2.2.2 Data Used

This study was prepared through two different ortho-rectified satellite imageries (LANDSAT V TM and VIII-OLI) to observe changing land-use pattern between 2000 and 2018 (There is 18 years gap in between data collection of imageries, to show the extensive variation in land-use change of this region. All these images projected through Universal Transverse Mercator projection with zone number 45 (UTM 45 N) and World Geodetic System 1984 (WGS 1984) datum in Arc GIS v 10.4.1 software. As the images were retrieved from different sources and not geo-rectified for accuracy, a geometric correction was introduced with a root mean square error (RMSE) of ±0.5 pixels to correct the image-to-image registration technique. This sub-catchment basin was analysed through the Shuttle Radar Topographic Mission (SRTM) Digital

Table 2.1 Sensor detail of satellite imageries

Year	Sensor	Path/Row	Acquisition date	Spatial resolution (m)	Source of image
2000	LANDSAT 7 ETM + (Enhanced Thematic mapper)	138/44	17 November 2000	30	USGS Earth Explorer (https://earthexplorer.usgs.gov/)
2018	LANDSAT VIII (OLI) (Operational Land Imager)	138/44	28 February 2018	30	USGS Earth Explorer (https://earthexplorer.usgs.gov/)
2014	SRTM DEM	138/44	23 September 2014	30	USGS Earth Explorer (https://earthexplorer.usgs.gov/)

Elevation Model (DEM). The soil group of the study area was extracted from the soil map provided by NBSS-LUP (National Bureau of Soil Survey and Land Use Planning). Except that, the entire climatological data and river's daily discharge data were provided by IMD (India Meteorological Department), Pune, and Kolkata Port Trust, West Bengal, India, respectively.

Table 2.1 There is 18 years gap in between data collection of imageries, to show the extensive variation in land-use change of this region. All these images projected through Universal Transverse Mercator projection with zone number 45 (UTM 45 N) and World Geodetic System 1984 (WGS 1984) datum in Arc GIS v 10.4.1 software. As the images were retrieved from different sources and not geo-rectified for accuracy, a geometric correction was introduced with a root mean square error (RMSE) of ± 0.5 pixels to correct the image-to-image registration technique. This sub-catchment basin was analysed through the Shuttle Radar Topographic Mission (SRTM) Digital Elevation Model (DEM). The soil group of the study area was extracted from the soil map provided by NBSS-LUP (National Bureau of Soil Survey and Land Use Planning). Except that, the entire climatological data and river's daily discharge data were provided by IMD (India Meteorological Department), Pune, and Kolkata Port Trust, West Bengal, India, respectively.

2.2.3 Methodology

2.2.3.1 Image Analysis and Change Detection

Ten land cover classes have been identified from the images through maximum likelihood classifier (MLC) and visual interpretation. These classes include cultivation land, waterbody, urban and rural area, industrial sector, plantation, brick kiln,

Table 2.2 Most important causes of urban flooding in Guwahati

Land Cover type	Area in year 2000 (ha)	Percentage of total area in 2000	Area in year 2018 (ha)	Percentage of total area in 2018	Percentage increase/decrease
Cultivation land	89,706	75.48	89,776	75.53	0.1
Industrial land	52	0.04	194	0.16	300.0
Transportation	946	0.4	915	0.385	−3.8
Urban Settlement	3326	2.80	9,274	7.8	178.6
Rural settlement	9181	7.72	5,189	5.18	−32.9
Sand deposition	2210	1.86	409	0.34	−81.7
Brick field	249	0.21	647	0.54	157.1
Old river course	2037	1.71	2,643	2.22	29.8
Waterbody	11,148	9.38	7,523	6.33	−32.5
Plantation	0	0	13.2	1.11	N/A

and many more. Representative samples of each type have been collected from the images through supervised image classification. The sub-catchment basin was also processed through DEM by the HEC-Geo HMS extension of ArcGIS. Soil map was analysed through hydrological soil group (HSG) of the study area and precipitation data was collected and performed as an input element of the model in HEC-HMS (Fig. 2.2).

2.2.3.2 Rainfall-Runoff Modelling Through HEC-HMS

HEC-HMS is a rainfall-runoff lumped model. By this experimental method, the amount of precipitation was converted to the volume of the runoff. The hydrographs generated through this method provide insights into flood forecasting, future urbanization, and floodplain development. Drainage paths and drainage watersheds have been created with the help of HEC-Geo HMS, an extension of ArcGIS, by developing a digital terrain model (DTM). The runoff was calibrated through this DTM by transferring the data from HEC-Geo HMS to HEC-HMS. Rainfall-runoff modelling of this sub-catchment was performed through the Soil Conservation Service-Curve Number (SCS-CN) model. SCS-CN is an established and simple hydrologic engineering method based primarily on soil nature, LULC (Land Use-Land Cover), surface condition, and antecedent moisture. It is expressed in the following formula (Santillan et al. 2010; Scharffenberg et al. 2018; Usda 1972)

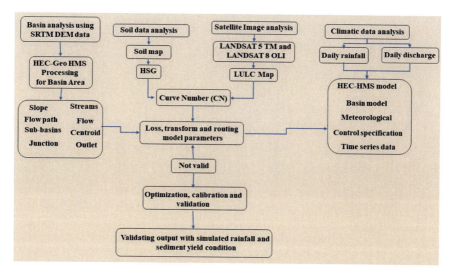

Fig. 2.2 Schematic diagram of methodology of present study area (Santillan et al. 2010)

$$Q = \frac{(P - I_a)^2}{P - I_a + S} \text{ for, } I_a \leq S, \text{ otherwise } Q = 0 \qquad (2.1)$$

$$I_a = \lambda S \text{ where } S = \frac{25400}{CN} - 254 \qquad (2.2)$$

where P is the total rainfall, I_a is the initial abstraction, Q is the direct runoff, S is the potential maximum retention (ranging between 0 and ∞), and λ (unit less parameter) is the initial abstraction coefficient. The short-term loss such as evaporation and infiltration, included through λ, and its ratio is described through S. The value of λ can be estimated and calibrated through the field, but SCS maintained the amount in 0.2. S depends on land cover and soil infiltration through Curve Number (CN), which in turn depends on the antecedent moisture condition of the sub-catchment basin. CN value is obtained from the curve number generation procedure from LULC and soil map by ArcGIS and a higher value indicates higher runoff potential (Fig. 2.3). The rainfall-runoff modelling was calibrated through HEC-HMS by discharge and rainfall data covering from 15 March to 15 November 2000 and 2018 to compare in the outlet section. As the accumulating rainfall value in the middle of the year is greater than 50 mm, the moisture condition of the area indicates AMC III value.

2.2.3.3 Sediment Yield Modelling by HEC-HMS

Many land cover and runoff changes have the most significant impact on sediment yield in the sub-catchment basin. MUSLE (Modified universal soil loss equation) equation is used to determine the sediment yield in the current study area

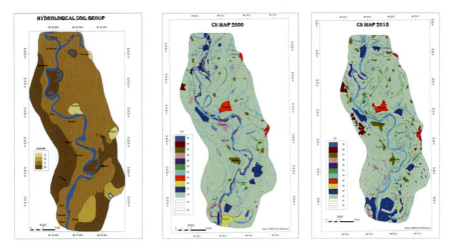

Fig. 2.3 Hydrological soil group map of study area with changing CN value in 2000 and 2018

(Scharffenberg et al. 2018; Usda 1972)

$$S_y = R'.K.L.S.C.P \text{ where } R' = a\left(Q.q_p\right)^b \qquad (2.3)$$

where S_y is the sediment yield in metric tonnes in specific rainfall event, R' is runoff erosivity factor, Q is the volume of runoff (m^3), q_p is the peak flow rate (m^3s^{-1}), and a and b are the location-dependent coefficients that can be estimated through calibration with measured sediment yield. The terms $K, L, S, C,$ and P are the standard MUSLE factors that represent the soil erodibility, slope length, slope steepness, cover management, and support practices, respectively. Among these factors, cover management is essential as it determines the percentage of soil cover in an area, which, in turn, affects sediment yield. As in a floodplain formation, the slope factor rarely varies from one place to another. The amount of slope is very minimal in this area and reflects the gentle demarcation of slope in DEM. HEC-Geo HMS's algorithm calculates the mean MUSLE factors using SRTM DEM, soil map, and LULC map. The SCS-CN-based rainfall-runoff model helps to determine the sediment yield with the help of the MUSLE model, where Q and q_p play an essential role. The daily calibration of the sediment yield model has been done in the outlet of the sub-catchment area to validate the model between 15 March and 15 November 2018.

2.3   Results and Discussions

2.3.1   Land Use-Land Cover Change Analysis

LANDSAT images 2000 and 2018 have supervised classifications based on the maximum likelihood classifier (Fig. 2.4).

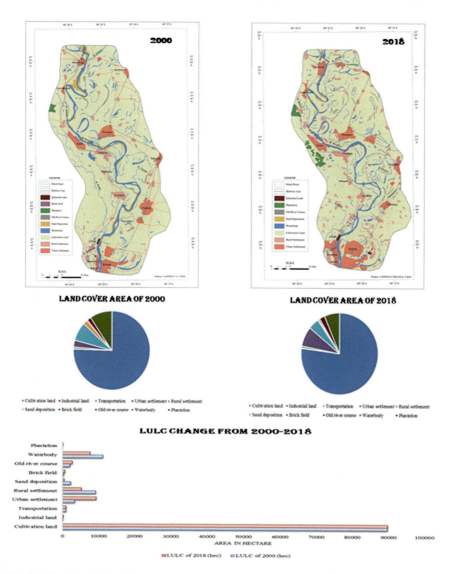

Fig. 2.4 Results of LULC change analysis of study area in between 2000 and 2018

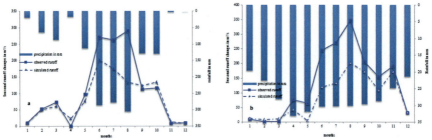

Fig. 2.5 Rainfall-runoff model calibration of **a** 2000 and **b** 2018

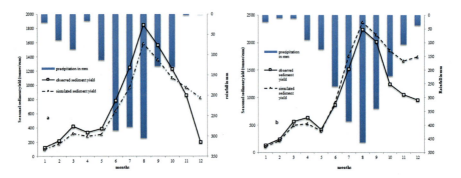

Fig. 2.6 Sediment yield model calibration of **a** 2000 and **b** 2018

There are mainly ten classes, namely: waterbody, urban settlement, rural settlement, plantation, industrial land, brick kiln, sand deposit, old river course, cultivation land, and old river course. First, all the land cover areas of 2000 were calculated, and the different regions in the 2018 land cover were compared with 2000. The overall accuracy of all calculations was highlighted with the help of error matrix generation. Error matrix has been created with the help of various points from Google imagery. Then these points have been overlaid on the images of 2000 and 2018 for validation. The overall accuracy of these two LULC maps is 81.4% and 86.3%, respectively. Kappa coefficients have been used by ArcGIS to fix statistical classification accuracy. The Kappa coefficients of 2000 and 2018 maps are 0.845 and 0.887, respectively (Tables 2.1 and 2.2).

2.3.2 Calibration and Validation of Rainfall-Runoff Model

Calibration of the HEC-HMS model on runoff was performed using observed precipitation data in the year 2000 and 2018. Figure 2.5 shows the graphical distribution of observed and simulated runoff of the years 2000 and 2018. The character of

Table 2.3 Areal changes in LULC classification

Land Cover type	Area in year 2000 (ha)	Percentage of total area in 2000	Area in year 2018 (ha)	Percentage of total area in 2018	Percentage increase/decrease
Cultivation land	89,706	75.48	89,776	75.53	0.1
Industrial land	52	0.04	194	0.16	300.0
Transportation	946	0.4	915	0.385	−3.8
Urban Settlement	3326	2.80	9,274	7.8	178.6
Rural settlement	9181	7.72	5,189	5.18	−32.9
Sand deposition	2210	1.86	409	0.34	−81.7
Brick field	249	0.21	647	0.54	157.1
Old river course	2037	1.71	2,643	2.22	29.8
Waterbody	11,148	9.38	7,523	6.33	−32.5
Plantation	0	0	13.2	1.11	N/A

rainfall also shifted the observing pattern from August to October–November in the session of 2018. There was a good match between validation and calibration in both observed and simulated values developed by Eqs. 2.1 and 2.2. The calibration has been performed on both automatic and manual basis. Peak weighted RMSE and univariate gradient algorithm choose to observe and optimize the method in the calibration process. The performance of the calibration is validated through (Table 2.3) percentage error in simulated volume (PEV), the percentage error in simulated peak (PEP), Nash–Sutcliffe efficiency (NSE), coefficient of determination (R^2), and percentage bias (PBIAS). According to these, the performance of the model is sound and simulated runoff volume is estimated through the peak rainfall distribution in both 2000 and 2018.

Table 2.4 Comparison of observed and simulated runoff during validation

Performance measure	2000		2018	
	Calibration	Validation	Calibration	Validation
NSE	0.76	0.66	0.85	0.78
R^2	0.85	0.74	0.88	0.81
PEV (%)	−1.25	−6.14	−1.71	−5.23
PEP (%)	16.53	21.56	17.85	26.85
PBIAS	1.52	−11.25	1.69	−9.62

2.3.3 Calibration and Validation of Sediment Yield Model

The sediment yield data of 2000 and 2018 was performed and calibrated through MUSLE by Eq. 2.2. The HEC-HMS changed the observed sediment concentration from g/l to tonnes/cu m. The daily simulated sediment load was computed for the calibration period from 15 March to 15 November in both 2000 and 2018 (Fig. 2.6). The calibrated result was validated through R^2, NSE, and PBIAS values for graphical measurement of observed and simulated sediment yield data as the slope factor has rarely varied in the sub-catchment basin (Table 2.4). However, there was a good match between the observed and simulated sediment yield. The performance of the sediment yield model is satisfactory. The amount of sediment yield is increasing in the present condition with a rapid land cover variation—the simulated sediment yield variation was estimated through peak rainfall distribution in both 2000 and 2018.

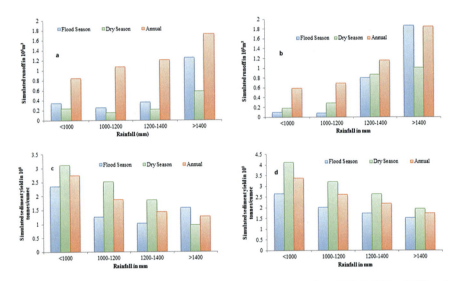

Fig. 2.7 Average annual runoff changes and sediment yield change in **a** 2000, **b** 2018, **c** 2000, and **d** 2018 with land-use change

Table 2.5 Comparison of observed and simulated sediment yield during validation

Performance measure	2000		2018	
	Calibration	Validation	Calibration	Validation
NSE	0.63	0.61	0.84	0.72
R^2	0.82	0.77	0.86	0.80
PBIAS	19.63	32.51	21.54	33.6

2.3.4 Changes in Runoff and Sediment Yield with Land Use

To identify the impact of land-use change through runoff and sediment yield, the annual precipitation was divided into four parts, namely, < 1000, 100–1200, 1200–1400 mm, and above 1400 mm in this sub-catchment basin of the year 2000 and 2018 (Fig. 2.7). The runoff diminished in the flood season, but due to rapid urbanization, it increased with a high volume of rainfall over the catchment area. The urban structure prevents the infiltration of the rainfall water and flows through the basin as runoff in the present condition.

The conversion of the agricultural area into the urban structure is mainly responsible for increasing runoff volume here. The sediment yield also increases with precipitation, but there is a variation of sediment concentration than runoff yield. Sediment concentrates with low velocity in the dry season instead of monsoon time. The rapid dilution with increasing flow velocity effused the sediment heave from the river bed and transported it towards the outlet. After the monsoon, the flow velocity diminished and the sediment yield also enriched in the river bed. The amount of sediment yield varied with rapid urbanization in the present time, due to the effused water from the industrial treatment plant and sewage system. The amount of sand bar in the river also increased in the current time, and this happens only when the sediment yield becomes higher in the dry season.

2.4 Conclusion

Based on the land use map and calibration value of runoff and sediment yield in two different periods, this study develops an integrated RS GIS-based hydrologic modelling approach to identify the relationship between land-use change with runoff and sediment yield. The analysis expanded with detecting changes in LULC with the parameterization of rainfall-runoff and sediment yield. This study revealed some impressive results with the application of HEC-HMS.

- Land use has experienced a drastic change from 2000 to the present. Rapid urbanization and industrialization enhance to enrich the sediment in the river by converting the agricultural ground into building material. An increasing amount of sand bar proves the highest sediment yield in the river in the dry season.
- Rapid urbanization also changes the character of the spatial distribution of CN value with the AMC III scenario in two different periods. The average amount of CN increased in the downstream part with the accelerating growth of industry and urban land. These factors affected the hydrological process by increasing the proximity of floods.
- The SCS-CN method exhibited the condition of the changing nature of runoff from 2000 to 2018. Runoff shows an increase of 33% from 2000 to 2018 with high flow in monsoon time. The extension of observed and simulated data ranges differentiated from 2000. The rainfall pattern showed changes due to anthropogenic and

climatic factors. So, the runoff volume accelerated in September and October of 2018. The model validated with NSE (0.66–0.78), R^2 (0.74–0.81), PEV (-6.14 to -5.23), PEP (21.56–26.85), and PBIAS (-11.25 to -9.62).

- The sediment yield model through MUSLE shows the accelerated growth of sediment load from 2000 to 2018. The average annual increase of sediment yields 1,452 tonnes/cu m in 2000 to 2,134 tonnes/cu m in 2018. The observed and simulated value expressed the increasing amount of load in September and October period in 2018. The problem of runoff variation affects the growth of sediment load in the total sub-catchment basin. This model was validated through NSE (0.61–0.72), R^2 (0.77–0.80), and PBIAS (32.51–33.6).
- Human-induced land-use changes also have a quantitative effect on runoff and sediment yield. Sediment yield and runoff changed by altering the character of land use from 20 to 30%. With the help of varying precipitation, the runoff changes were inversely proportionate with sediment yield in different land-use patterns seasonally. The effect of climate change exhibited the sediment heaving on the river bed by decreasing runoff volume in the dry period.

The government should take serious action plans towards protecting the lifeline of many stakeholders. Industrial policymakers may decide to develop an effluent treatment plant in the industrial area and adequately maintain effluent conditions. Farmers should control the use of fertilizer and pesticides in agricultural fields. A governmental action plan should be essential here to eradicate the future river sedimentation problem. The eroded material from the river bank may increase the sediment yield in bottom sediment, which originated from the agricultural field. After the immersion procedure, the government should take proper action plans against the painted idol's immersion and remove it from the riverbed. The dredging process should be regulated appropriately as the depth and flow of the river sustained adequately. Adequate future research is required to analyse the future projection of sediment yield with runoff volume in this sub-catchment basin with varying land cover parameterization. The amount and character of the flood may highlight the future sediment yield scenario of this area.

References

Ali KF, De Boer DH (2007) Spatial patterns and variation of suspended sediment yield in the upper Indus River basin, northern Pakistan. J Hydrol 334(3):368–387. https://doi.org/10.1016/j.jhydrol.2006.10.013

Asselman NEM, Middlekoop H, van Dijk PM (2003) The impact of changes in climate and land use on soil erosion, transport and deposition of suspended sediment in the River Rhine. Hydrol Process 17(16):3225–3244. https://doi.org/10.1002/hyp.1384

Azari M, Moradi HR, Saghafian B, Faramarzi M (2015) Climate change impacts on stream flow and sediment yield in the North of Iran. Hydrol Sci J 61(1):123–133. https://doi.org/10.1080/02626667.2014.967695

Bandyopadhyay S, Kar NS, Das S, Sen J (2014) River systems and water resources of West Bengal: a review. In: Special publication of the geological society of India, vol 3, pp 63–84. https://doi.org/10.17491/cgsi/0/v0i0/62893

Bandyopadhyay S, Das S, Kar NS (2015) Discussion: "changing river courses in the western part of the Ganga-Brahmaputra delta" by K. Rudra (2014), Geomorphology 227:87–100. Geomorphology 250:442–453. https://doi.org/10.1016/j.geomorph.2015.02.037

Basu SR (2005) Recent findings on the river dynamics of Bengal: Post-Farakka condition of the off-take and bank erosion of the River Bhagirathi-Hugli. Geograph Rev India 67:315–346

Bera B, Bhattacharjee S, Roy C (2019) Estimating stream piracy in lower Ganga plain of a quaternary geological site in West Bengal, India applying sedimentological bank facies, log and geospatial technique. Curr Sci 117(4):662–671

Chu X, Steinman A (2009) Event and Continuous Hydrologic Modelling with HEC-HMS. J Irrig Drain Eng 135(1):119–124. https://doi.org/10.1061/(ASCE)0733-9437(2009)135:1(119)

Delgado MI, Gaspari FJ, Kruse EE (2015) Land use changes and sediment yield on a hilly watershed in central-east Argentina. Soil Water Res 10(3):189–197. https://doi.org/10.17221/49/2014-SWR

Erskine WD, Mahmoudzadeh A, Myers C (2002) Land use effects on sediment yields and soil loss rates in small basins of Triassic sandstone near Sydney. CATENA 49:271–287. https://doi.org/10.1016/S0341-8162(02)00065-6

Guchhait SK, Islam A, Ghosh S, Das BC, Maji NK (2016) Role of hydrological regime and flood-plain sediments in channel instability of the Bhagirathi River, Ganga-Brahmaputra Delta, India. Physical geography http://doi.org/https://doi.org/10.1080/02723646.2016.1230986

Halwatura D, Najim MMM (2013) Application of the HEC-HMS model for runoff simulation in a tropical catchment. Environ Model Softw 46:155–162. https://doi.org/10.1016/j.envsoft.2013.03.006

Jain MK, Das D (2010) Estimation of sediment yield and areas of soil erosion and deposition for watershed prioritization using GIS and remote sensing. Water Resour Manag 24:2019–2112. https://doi.org/10.1007/s11269-009-9540-0

Laha C (2015) Oscillation of meandering Bhagirathi on the alluvial flood plain of Bengal Basin, India; as controlled by the Palaeo-geomorphic architecture. Int J Geomat Geosci 5(4):564–572

Nakil M, Khire M (2015) Effect of slope steepness parameter computations on soil loss estimations: review of methods using GIS. Geocarto Int 31:1078–1093. https://doi.org/10.1080/10106049.2015.1120349

Pak JH, Scharffenberg WA (2008) Soil erosion and sediment yield modeling with the hydrologic modeling system (HEC-HMS). In: Proceedings of the world environmental and water resources congress, pp 1–10. https://doi.org/10.1061/40976(316)362

Ponce VM, Hawkins RH (1996) Runoff curve number: has it reached maturity? J Hydrol Eng 1:11–19. https://doi.org/10.1061/(ASCE)1084-0699(1996)1:1(11)

Roy PK, Samal NR, Roy MB, Mazumdar A (2015) Integrated assessment of impact of water resources of important river basins in Eastern India under projected climate conditions. Global Nest J 17(X):XX–XX

Rudra K (2010) Dynamics of the Ganga in West Bengal, India (1764–2007) – Implications for science-policy interaction. Quaternary international, vol 227, pp 161–169. https://doi.org/10.1016/j.quaint.2009.10.043

Rudra K (2014) Changing river courses in the western part of the Ganga-Brahmaputra delta. Geomorphology 227:87–100. https://doi.org/10.1016/j.geomorph.2014.05.013

Santillan JR, Makinano MM, Paringit EC (2010) Integrating remote sensing, GIS and hydrologic models for predicting land cover change impacts on surface runoff and sediment yield in a critical watershed in Mindanao, Philippines. Int Arch Photogr, Remote Sens Spatial Inf Sci 38(8):436–441

Sathya A, Thampi SG (2020) Impact of projected climate change on streamflow and sediment yield – a case study of the Chaliyar River Basin, Kerala. Roorkee Water Conclave 2020:1–13

Scharffenberg B, Bartles M, Brauer T, Fleming M, Karlovits G (2018) Hydrologic modelling system HEC-HMS user's manual. U.S. Army Corps of Engineer, Hydrologic Engineering Centre, Davis, CA 53–247:295–603

Smith SJ, Williams JR, Menzel RG, Coleman GA (1984) Prediction of sediment yield from Southern plains grasslands with the Modified Universal Soil Loss Equation. J Range Manag 37:295–297

SCS USDA (1972) Section 4: Hydrology. National Engineering Handbook, Washington, D.C., Soil Conservation Service.

Williams JR (1975) Sediment routing for agricultural watershed. Water Resour Bull 11:965–974. https://doi.org/10.1111/j.1752-1688.1975.tb01817.x

Younis SMZ, Ammar A (2017) Quantification of impact of changes in land use-land cover on hydrology in the upper Indus Basin, Pakistan. The Egyptian Journal of Remote Sensing and Space Sciences https://doi.org/10.1016/j.ejrs.2017.11.001

Chapter 3
Remote-Sensing-Based Analysis of Relationship Between Urban Heat Island and Land Use/Cover Type in Bhubaneswar Metropolitan Area, India

Asad Ali Sarkar

Abstract Urbanization is a significant phenomenon in third world country like India. It has a massive impact on urban climate such as surface temperature increased by the rapid rate of urbanization mainly due to the increase of impervious surface. In this paper, I report the relationship between changing land use/cover (LULC) pattern and mode of urban heat island (UHI) in Bhubaneswar Metropolitan Area, India. To detect changing LULC and land surface temperature (LST) three Landsat satellite images were used for different times. To understand the proper relation between LULC and intensity of UHI different mathematical models were built using four indices such as Normalized Difference Vegetation Index (NDVI), Normalized Difference Built built-up Index (NDBI), Normalized Difference Water Index (NDWI), and Normalized Difference Bareness Index (NDBaI). It showed that dry land, bare land, impervious land, and river sand has contained high temperature than other LULC types. Besides, higher temperature in the built-up area, a scattered pattern distribution of temperature was found in 1997, later it has changed to concentric pattern to built-up land in 2017. The negative correlation was found between LST and NDVI, NDWI, and NDBaI, but the positive correlation was shown between LST and NDBI.

Keyword Remote sensing · Urban heat island · Land surface temperature · Land use-land cover · Urbanization

3.1 Introduction

A metropolitan city is comparatively warmer than its rural surroundings, this is known as urban heat island (UHI). The factors by which this phenomenon occurred in a metropolitan area are the presence of buildings and street surface materials, mainly cement and asphalt, which have a high heat-absorbing capacity. So these

A. A. Sarkar (✉)
Department of Geography, Chandrakona Vidyasagar Mahavidyalaya, Paschim Medinipur, ChandrakonaPin-721201, West Bengal, India

© The Author(s), under exclusive license to Springer Nature Singapore Pte Ltd. 2022   39
N. C. Jana and R. B. Singh (eds.), *Climate, Environment and Disaster in Developing Countries*, Advances in Geographical and Environmental Sciences, https://doi.org/10.1007/978-981-16-6966-8_3

urban features captured much more heat in the daytime and released heat energy at a very slow rate at night (Khan and Chatterjee 2016). Besides these differential heat absorbing and releasing rates of the materials, the extensive urbanized impervious surface has changed the energy and water balance capacity of the urbanized area and directly strike on the dynamics of air movement (Oke 1987). The complexity of urban structure has an immense impact on the land surface energy system, such as land surface albedo, emissivity, and containing heat capacity (Rizwan et al. 2008). The direct result of this thermal energy structure in an urban area creates a significant difference in the local climate between urban and suburban area, i.e., the UHI effect. Nowadays UHI has become common evidence in urban climate and asserts the direct impact of cultural landscape on air temperature (Arnfield 2003). Urban climatology has a long history of its evolution. It began with Luke Howard's pioneering work related to urban climate. In 1815 Howard conducted the first-ever systematic urban climate study measuring UHI effect based on thermometers in the city of London and nearby (Howard 2012). Nevertheless, the first systematic study of UHI was conducted in the urban area of Singapore in 1964 (Nieuwolt 1966).

In comparison with the traditional method to study UHI phenomena, remote sensing has become one of the most effective methods, for its quick acquisition of up-to-date information over a large geographical area (Lee 2011). The progress of thermal remote sensing offered a new approach for research of urban climate to Urban Climatologist (Pongrácz et al. 2010). It is very effectively, efficiently, and quantitatively observed the distributional pattern of UHI as well as it offered thermal information of UHI across different LULC of an urban area (Grimmond 2007). At the first-hand stage, the UHI study was done through the use of moderate resolution imaging spectrometer (MODIS) and a very high-resolution radiometer (AVHRR). These remote sensing data provide different information on the general distribution of surface UHI (Imhoff et al. 2010). Chen et al. (2006) reported the combination with Landsat and Thematic Mapper (TM) technology to assess the LULC change and spatio-temporal distributional nature of UHI in Pearl River Delta under the impact of rapid urbanization (Chen et al. 2006).

Thematic Mapper (TM) data are one of the most useful materials for environmental studies. TM is composed of seven bands, among them, six are visible and near-infrared, and the rest one band is thermal infrared. TM band 1 is visible blue and has 0.45–0.52 μm wavelength. This band is mainly used for the study of coastal water. The wavelength of TM band 2 is 0.52–0.60 μm and is used to study the health of vegetation. Normalized Difference Vegetation Index is calculated with the help of TM band 3 and TM band 4 having wavelengths of 0.63–0.69 μm and 0.76–0.90 μm, respectively. TM band 5 and TM band 7 having a wavelength of 1.55–1.75 μm are extensively used to study snow, ice, cloud, and discrimination of geological formation. Lastly, TM band 6 with wavelength 10.40–12.50 μm is used for the retrieval of LST (Sobrino et al. 2004; Ramachandran Justice Abrams 2010).

Nowadays, Landsat 8 OLI is more useful to analyze the relationship between UHI and LULC. It has been reported that Landsat 8 was successfully launched on 11th February 2013. In addition, Landsat 8 contains two extensive scientific instruments such as the operational land imager (OLI) system composed with nine spectral bands and thermal infrared sensor (TIRS) with two spectral bands in LWIR (Kamran et al. 2015).

Monjit et al. reported the changes in the relationship between LULC and UHI of Guwahati Metropolitan Area using Landsat TM and Landsat ETM + (Borthakur and Nath 2012). Lee et al. reported the relationship between UHI and LULC by Landsat TM and ETM + satellite images of Xuzhon city of Japan (Lee 2011). All these researches revealed that the growth of urban area reduced urban green space and transferred it into impervious lands surface which is directly related to UHI.

In these papers, I have used different indices like Normalized Difference Vegetation Index (NDVI), Normalized Difference Water Index (NDWI), Normalized Difference Built-up Index (NDBI), and Normalized Difference Bareness Index (NDBaI). NDVI provides an approximate estimation of vegetation health and a method of monitoring vegetation change over time, while NDWI delineates open water features and detects vegetation water content. NDBI and NDBaI are developed for the identification of built-up land and bare land, respectively. These indices can well represent LULC and the integrated use of them can help to explore and build up a model of the relationship between UHI and LUCC.

The main objectives of this paper are: (a) classification of land use/cover types; (b) calculation of brightness temperature from thermal band of respective images; (c) calculation of NDVI, NDWI, and NDBI and NDBaI from different bands of Landsat TM and Landsat 8 OLI satellite images; and (d) finally investigation of the relationship between LULC and UHI.

3.2 Study Area and Data

Bhubaneswar, the modern capital of Odisha, a state of India, is popularly known as the "Temple city of India". Bhubaneswar is located between $21°12'N–20°25'N$ latitude and $85°44'E–85°55'E$ longitude. Coming under the Khurda district, Bhubaneswar is bounded by Bhargabi river in the east, Khandagiri and Udayagiri in the west, Kuakhai River in the north, and Dhaulagiri and Daya river in the south (Fig. 3.1).

The municipality is bounded on the north by the villages Patia, Rokat, and Mancheswar; on the east by the villages Koradakanta, Keshura, Bankual, Basuaghai, Mahabhoi Sasan, and Raghunathpur; on the south by the villages Kakudaghai, Orakala, Ebaranga, and Bahadalpur; and on the west by the villages Jadupur, Begunia, Dumduma, Jokalandi, Bharatpur, Andharua, and Jagannathprasad. Designed by Otto Königsberger in 1946, Bhubaneswar, along with two other cities Jamshedpur and Chandigarh, is known as the first planned cities in India. Due to the presence of reserve forests in the north-western part and flood plains in the eastern part, the city has grown more toward the south-west. The population of it was 113,095 on the 1st April of 1971 (Chatterjee et al. 2016). The Bhubaneswar Municipal Corporation was established in 1994. It contained 67 administrative wards and 46 Revenue

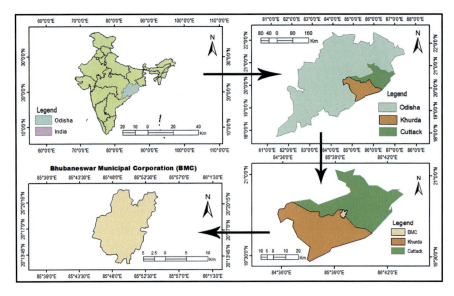

Fig. 3.1 Location map of Bhubaneswar Municipal Corporation (BMC)

villages. The area under the jurisdiction of Bhubaneswar Municipal corporation (BMC) covers 135 km². Bhubaneswar has a population of 837,737 (2011 census of India) and decadal growth rate is 29.4%.

To extract the changing LULC pattern and LST of this study area, three Landsat satellite images were selected at the time of onset monsoon. Among three images, two are Landsat 5 (1997-04-21, 2008-04-03, and 2011-04-12) and one Landsat 8 OLI image (2017-04-12). The spatial resolution is 30 m for the image Landsat 5 of the band 1–5 and 120 m for the thermal band (band 6). In Landsat 8 images there are two thermal bands, i.e., Band 10 and Band 11. These TIRS bands are acquired at the 100-m resolution but are resampled to 30 m in delivered data product.

3.3 Methodology

3.3.1 Method for Land Use/Cover Classification

The main aim of this paper is to establish the relationship between changing land use pattern with UHI. So, various variables were needed to derive LULC change and LST. Three satellite images were used. The first and prior step is to create LULC map. To prepare the LULC map from satellite images, a suitable classification method is to be needed. The number of LULC classes would be determined based on the requirement of a particular application (Arora and Mathur 2001). In this study, seven LULC classes were prepared, these are vegetation zone, waterbody, fallow land,

Table 3.1 Description of LULC type in BMC

Land cover types	Description
Vegetation zone	Forest, trees, road-side trees, gardens, parks, grassland
Water body	River, ponds, cannels, marshy land
Fallow land	All types of open space without any construction and vegetation
Built-up land	All types of settlement, road, airport runway, commercial and industrial area, man-made structure
Agricultural land	Crop field
Dry land	Open land having minimum moisture
River sand	Sand located in river

built-up area, agricultural land, dry land, and river sand. The description of these classes is given in Table 3.1. Unsupervised classification method was performed with K-Means algorithm in ERDAS imagine 2014.

Accuracy assessment is very necessary for a map generated from satellite images, and an error matrix is a way to prove the authentication of the classified image (Reis 2008). Accuracy assessment was done with the help of visual interpretation of the satellite-based image as well as Google Earth. Different parameters such as overall accuracy, user's accuracy, and producer's accuracy and Kappa statistics were extracted from error matrix, as given in Table 3.4.

3.3.2 Method of Brightness Temperature Retrieval

The radiance at the top of the atmosphere (TOA) measured by satellite TIR band and brightness temperature can be extracted by using Plank's low (Dash et al. 2002). Water vapor is an important element that has a vital role to control the temperature of the atmosphere. It is assumed that it remains uniformly over a small area. Hence, the impact of water vapor on the radiance temperature of the atmosphere could be ignored (Kato and Yamaguchi 2005). The main focus of the study is to know the spatial distributional pattern of surface temperature with relation to LULC, so there is no need to extract the real LST. Hence, in this study, I used brightness temperature (BT) as LST. The method of extracting BT is different from Landsat 5 and Landsat 8 OLI images. These are stated below.

3.3.2.1 Brightness Temperature Retrieval of Landsat 5 Images

At least three methods to extract LST from thermal infrared data, mainly Landsat 5 images, are currently available. The first is called Radioactive Transfer Equation which needs in situ radio-sounding data when the satellite passes. The other two are mono-window algorithm (Qin et al. 2001) and single-channel method, both of which use limitedly accessible land surface emissivity (Jiménez-Muñoz and Sobrino 2003).

Chen et al. developed an algorithm to retrieve brightness temperature from the Landsat TM images. Although retrieval methods are different due to the differences in TM and ETM + sensors, they could be expressed by the same two steps. Firstly, convention from DN value to radiation luminance via the following Eq. (3.1) (Yun-hao et al. 2002).

$$L\lambda = 0.0056332 \times DN + 0.1238 \tag{3.1}$$

Secondly, convention from radiation luminance to BT in Kelvin (K) via Eq. (3.2)

$$K = \frac{K1}{Ln\left(\frac{K2}{L\gamma} + 1\right)} \tag{3.2}$$

where K1 = 1,260.56 K and K2 = 60.766 mW*cm^{-2}*sr^{-1} μm^{-1} for Landsat TM images.

And then Kelvin BT converted to degree Celsius (°C) via Eq. (3.3).

$$C = K - 273.15 \tag{3.3}$$

3.3.2.2 Brightness Temperature Retrieval of Landsat 8 OLI Image

In this paper, I retrieve brightness Temperature from Landsat 8 OLI image with the help of the following formula. Retrieving LST from Landsat 8 OLI have some major steps these are stated below.

Top of Atmospheric Spectral Radiance

The satellite-based digital number is converted to at-sensor spectral radiance (Lλ) using the following Eq. (3.4) (Walawender et al. 2014).

$$L\lambda = M_L Q_{cal} + A_L \tag{3.4}$$

where Lλ-Top of Atmospheric Radiance in watts/(m^2*srad*μm), M_L-Band specific multiplicative rescaling factor (radiance_mult_band_10/11), Q_{cal}-Band 10/11 image, A$_L$-Band specific additive rescaling factor (radiance_add_band_10/11).

Thermal constant	Band 10	Band 11
K1	774.89	480.89
K2	1,321.08	1,201.04

Table 3.2 Value of (K1 and K2)

Conversion to At-Satellite Brightness Temperature

Inverted Planck's Law is used to transform the at-sensor spectral radiance to at-sensor brightness temperature as follows in Eq. (3.5).

$$TB = \frac{K_2}{Ln\left[\left(\frac{K_1}{L_\gamma +1}\right)\right]} - 273.15 \tag{3.5}$$

where TB-At-satellite brightness temperature (°C), Lλ-Top of Atmospheric Radiance in watts/(m^2*srad*μm), K1 and K2-thermal conversion constant, and it varies for both TIR bands given in Table 3.2.

3.3.3   Retrieval of Difference Indices

The NDVI, NDWI (Gao 1996), NDBI (Zha et al. 2003), and NDBaI indices are very useful to characterize LULC pattern of a study region and to extract the proper relation between UHI and LULC pattern quantitatively. To know the vegetation density, the most frequently used index is NDVI (Purevdorj et al. 1998), extract with Eq. (3.6). The different band combination of Landsat 5 TM and Landsat 8 OLI images are given in Table 3.3.

$$NDVI = \frac{NIR - Red}{NIR + Red} \tag{3.6}$$

Presence of water in a region is an important factor for LST of a region. NDWI indicates the water content of vegetation, extract with Eq. (3.7), it has also known as leaf area water-absent index (Gao 1996; Jackson et al. 2004; Zarco-Tejada et al. 2003; Maki et al. 2004) and water state of vegetation.

Band	Landsat 5	Landsat 8
NIR	Band 4	Band 5
Red	Band 3	Band 4
SWIR1	Band 5	Band 6
TIRS	Band 6	Band 10

Table 3.3 Band combination of Landsat 5 and Landsat 8 OLI images

Table 3.4 Accuracy assessment of LULC classification of BMC

Year	Type	Vegetation	Water	Fallow	Built	Agriculture	Dry zone	River sand	Producer accuracy (%)	User accuracy (%)	Kappa
1997	Vegetation	100	0.00	1.64	0.00	4.00	0.00	0.00	100	94.34	96.25
	Water	0.00	100	0.00	0.00	0.00	0.00	0.00	100	100	
	Fallow	0.00	0.00	93.44	2.00	0.00	0.00	0.00	93.44	96.61	
	Built	0.00	0.00	4.92	95.00	0.00	0.00	0.00	95	96.94	
	Agriculture	0.00	0.00	0.00	0.00	96.00	4.00	0.00	96	96.00	
	Dry zone	0.00	0.00	0.00	3.00	0.00	96.00	0.00	96	94.12	
	Sand	0.00	0.00	0.00	0.00	0.00	0.00	100	100	100	
2011	Vegetation	96.49	0.00	0.00	0.00	0.00	0.00	0.00	96.49	100	96.04
	Water	3.51	97.96	0.00	2.00	0.00	1.96	0.00	97.96	90.57	
	Fallow	0.00	0.00	96.00	0.00	0.00	0.00	0.00	96.00	100	
	Built	0.00	0.00	0.00	96.00	0.00	5.88	0.00	96.00	96.97	
	Agri	0.00	0.00	0.00	0.00	100	0.00	0.00	100	100	
	Dry zone	0.00	2.04	0.00	2.00	0.00	92.16	1.96	92.16	92.16	
	Sand	0.00	0.00	4.00	0.00	0.00	0.00	98.04	98.04	96.15	
2017	Veg	98.00	0.00	0.00	0.00	2.00	1.96	0.00	98	96.08	95.88
	Water	0.00	100	0.00	0.00	0.00	0.00	0.00	100	100	
	Fallow	0.00	0.00	100	8.00	0.00	0.00	0.00	100	86.21	
	Built	0.00	0.00	0.00	90.00	0.00	1.96	0.00	90	98.90	
	agriculture	2.00	0.00	0.00	0.00	98.00	0.00	0.00	98	98.00	
	Dry zone	0.00	0.00	0.00	2.00	0.00	96.08	0.00	98	96.08	
	Sand	0.00	0.00	0.00	0.00	0.00	0.00	100	100	100	

$$NDWI = \frac{NIR - SWIR1}{NIR + SWIR1} \qquad (3.7)$$

Another important index used in this study is NDBI (Zha et al. 2003), in Eq. (3.8), which is sensitive to the built-up area. Normalized Difference Built-up Index has been very helpful for mapping urban built-up area (Kato and Yamaguchi 2005).

$$NDBI = \frac{SWIR - NIR}{SWIR + NIR} \qquad (3.8)$$

To know the bareness of a region NDBal (Zhao and Chen 2005), Eq. (3.9) is very helpful and used in this study.

$$NDBal = \frac{SWIR1 - TIRS}{SWIR1 + TIRS} \qquad (3.9)$$

3.4 Result and Discussion

3.4.1 Changes of Land Use/Cover (LULC) Pattern of BMC

Landsat images are very reliable sources to prepare a LULC map because DNs are extracted based on the threshold approach in combination with Boolean operators system (Kato and Yamaguchi 2005). In this study, Landsat images were classified to prepare LULC map and have achieved very high accuracy. Table 3.4 represents the error matrix and accuracy assessment of the classified images. The sample points were taken in very special care across all over the study area and maintained uniformity of sample points from every LULC type. Each image has achieved above 95% overall accuracy.

In this study, three satellite images were selected with the range of 20 years time span and the selected images were classified into seven LULC categories, as given in Table 3.1. After the preparation of LULC maps, it is obvious that the urban built-up area has spread out from Bhubaneswar airport toward all directions of BMC.

A reserved forest is located in the north-western part of the city. With increasing built-up area, fallow land, dry land, and agricultural land were gradually decreased. To show the comparison of the LULC change from 1997 to 2017, the LULC maps were prepared and given in Figs. 3.2a, 3.3a, and 3.5a.

Table 3.1 shows the different types of LULC of BMC. Vegetation zone is the area covered by forest land, trees along roadsides, and natural vegetation. The waterbody is included with rivers, ponds, and channels, and marshy land. Fallow land is the area under BMC which is an open space without any construction and vegetation. Actually, dry land is one type of fallow land but it is different from later with the attaining of water content and due to lack of water content it reflects maximum energy

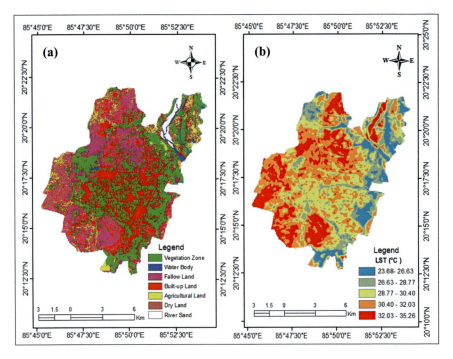

Fig. 3.2 Land use/Cover Pattern and Temperature in BMC retrieve from Landsat 5 image dated on 21 April 1997, **a** is Land use land cover pattern and **b** is LST

and in the satellite image it appears as the white portion. Built-up land is characterized by all types of settlement, road, airport runway, commercial and industrial area, and all other forms of man-made construction. Actually, agricultural land is cropland but in April most of the agricultural land is without any crop. Last LULC type is river sand, and it is located in the river.

Over all accuracy are 96.84, 96.65, and 96.51% of images on 1997, 2011, and 2017, respectively.

3.4.1.1 LULC of BMC in 1997

From Fig. 3.2a it is clear that in 1997 the built-up area of BMC is minor LULC feature in respect to a total area of the city because it holds only 21.85% of the total area of BMC and most of the built-up area is located in the northern part of Bhubaneswar airport. Bapuji Nagar and Bhauma Nagar are the major built-up areas at that time, but these areas were highly concentrated with trees. Besides these areas, a minor portion of built-up land was located in the western part of Kuakhai River. The south-west part of the reserved forest was occupied by fallow land and dry land. The reserved forest of north-western part is a major feature of BMC, besides other vegetation areas were located in the eastern part of Kuakhai River and western part of Bhargabi river. From Fig. 3.4, it is clear that agricultural activities were not major economic activities of this city because agricultural land occupied only 5.34% of the total area of BMC. Agricultural land was located in the eastern part of Khakhai river to a great extent and it was located in a scattered pattern on the boundary side of BMC. Figure 3.4 shows the percentage of different LULC of BMC of three selected images. In 1997 in BMC two major LULC features were fallow land and dry land. Apart from the northern part of Bhubaneswar airport, these two features were located in all parts of this city.

3.4.1.2 LULC of BMC in 2011

Figure 3.3a has shown the LULC pattern of BMC in the year 2011. The built-up land area had expanded by 8.43% from the time of 1997 and held 30.49% out of a total area of BMC. Significantly the vegetation land was reduced by 7.87% from the previous study. Dry land and fallow land exchange their area at this time. Due to urbanization, all LULC types were reduced in its size at this time.

In this time the urbanization was spread in the western part of Kuakhai River, the southern part of Bhubaneswar airport, and south-west part of reserved forest.

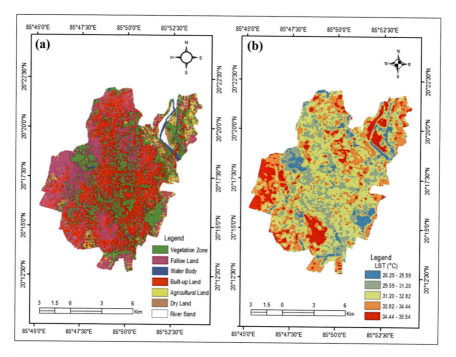

Fig. 3.3 Land use Land Cover Pattern and Temperature in BMC retrieve from, Landsat 5 image dated on 12 April 2011, **a** is Land use land cover pattern and **b** is the distribution of LST

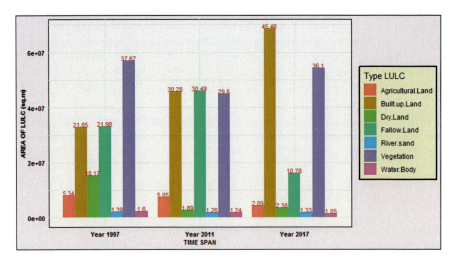

Fig. 3.4 Area utilized in different types of LULC from 1997 to 2017 in BMC. Value at top of the bar represents percentage of each LULC type

Some areas around Kuakhai River previously occupied by vegetation was converted in agricultural land in 2011. So, the LULC map of 2011 indicates the starting of the rapid urbanization in BMC.

3.4.1.3   LULC of BMC in 2017

From Fig. 3.5a, it is clear that the built-up land is spread out all over the BMC and occupied 45.46% of the whole study area in 2017. The impression of built-up land is shown in new areas such as southern edge part, the dry and fallow land of south-west part, surrounded by part of Kuakhai River and southern part Bhubaneswar airport. The agricultural land has been reached its half in size within 20 years and now holds 2.89% of the total area of BMC and is located on the boundary side of BMC. The fallow land and dry land are now converted into the built-up area and reduced its size in half within 20 years.

BMC is a planning city, hence the reflection of rules and regulations of construction are shown everywhere. For this reason, the area of the green zone of the city remains near about constant of its size from 1997 to 2017.

3.4.2   Temperature Variation Over Different LULC in BMC

To analyze the changing pattern of surface temperature over different LULC, the LST and information of LULC were retrieved from selected images. To study the

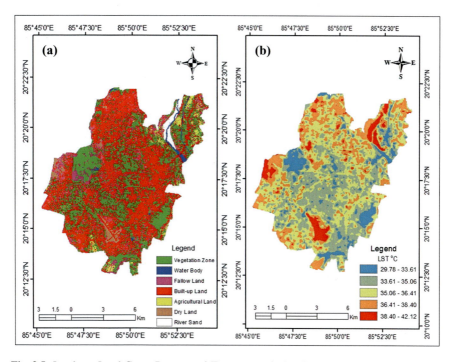

Fig. 3.5 Land use Land Cover Pattern and Temperature in BMC retrieve from, Landsat 8 OLI image dated on 12 April 2017, **a** is Land use land cover pattern and **b** is LST

proper relationship between them, the mean LST of different LULC were derived by averaging all corresponding pixel values. Figures 3.2b, 3.3b, and 3.5b showed the results of LST derived from three selected images. It is clear from Figs. 3.2b, 3.3b and 3.5b that the distribution of a relatively LST of the BMC was changed from scattered pattern to concentrated pattern within 20 years time span.

3.4.2.1   LST of BMC in 1997

Figure 3.2b showed the distribution of LST in 1997 in BMC. The highest and lowest temperatures of BMC at that time were 23.68 °C and 35.26 °C, respectively, so the temperature difference between the hottest and coldest part of BMC was 11.58 °C. Figure 3.2b has shown five thermal zones of BMC. In the first thermal zone, LST varies from 23.68 °C to 26.67 °C and this thermal zone was located in the bank side of Kuakhai River which was filled up with dense trees and in the waterbody. In the second thermal zone, LST varied from 26.67 to 28.77 °C and was located in the green zone of vegetation, mainly along the street side which was covered with trees and some parts were north-west side and eastern part of reserved forest. The third thermal zone held 28.77–30.42 °C LST and was situated in a major part of the reserved forest and central part of the built-up area. The central part of the built-up area came under the third zone because this area was mixed with urban buildings and dense trees. The LST in the fourth thermal zone was 30.42–32.05 °C and was located in the newly built-up area of the northern part and south-western part of dry and fallow land. The fifth thermal zone held 32.05–35.26 °C and stayed on Bhubaneswar airport, river sand, the middle-western part of built-up land, and south-west part of dry and fallow land. The fifth thermal zone holds 32.05–35.26 °C and stayed on Bhubaneswar airport, river sand, the middle-western part of built-up land, and south-west part of dry and fallow land.

3.4.2.2 LST of BMC in 2011

The distribution of LST in 2011 was depicted by Fig. 3.3b. The maximum and minimum LST were 26.25 °C and 39.55 °C, respectively. The difference between these two extreme temperature situations was 13.3 °C which is 3 °C more than in 1994. This temperature map also has five thermal zones. The first thermal zone was holding less than 29.59 °C and was located in the scattered pattern which was previously a concentric pattern along bankside of Kuakhai River and in the waterbody, but in 2011 it was situated in some part of reserved forest, along river eastern part of BMC. The second thermal zone of LST was classified with 29.59–31.20 °C and situated in the rest of the reserved forest and along the street of BMC which is rich in trees. The third thermal zone was characterized with 31.20–32.82 °C LST and located in the built-up area of the middle portion of the city. The fourth thermal zone was with 32.82–34.44 °C LST and associated with south-western dry and fallow land, middle north built-up land, and the northern part of newly built-up land. The fifth thermal zone characterized with 32.82–34.44 °C LST Bhubaneswar airport, a newly built-up area of the south-western part, and northern part and middle part of built-up area and river sand.

3.4.2.3 LST of BMC in 2017

The scattered patterned of UHI of 1997 in BMC are strongly associated with build-up area in 2017 that reveal with the Fig. 3.5b. The minimum and maximum temperatures of BMC are 29.78 °C and 42.12 °C, respectively, which are retrieved from Landsat 8 OLI image dated on 12 April 2017. The difference between extreme LST is 12.34 °C which is maximum within the study period. This LST map also has classified into five thermal zones like the two previous LST maps. The first thermal zone of LST is characterized by 29.78–33.61 °C and is located by entire reserved forest, waterbody, and vegetation land. With 33.61 °C of minimum and 35.06 °C LST second thermal zone is characterized by land located in the middle part of built-up land which has dense trees. The third thermal zone is with 35.06–36.41 °C LST and captured a major portion of the newly built-up area. The fourth thermal zone is bearing 36.41–38.40 °C LST. This thermal zone is closely related to the major part of the northern

Table 3.5 Temperature difference between different LULC types in BMC (°C)

Year	B-V	B-W	B-F	B-A	B-S
1997	0.55	1.75	−0.79	0.21	0.26
2011	0.28	0.25	−0.45	−0.39	−0.96
2017	0.27	0.33	0.16	0.52	0.54

side which is bearing densely settlement. The maximum thermal zone of the entire study is characterized by 38.40–42.12 °C and is located with Bhubaneswar airport, river sand, and the western and northern part of newly built-up area.

So, it is clear that the intensity of UHI is gradually increased in relation to the increase in impervious land. The mean LST in 1997 was 29.47 °C and it has reached 35.95 °C in 2017. So, the decadal growth rate of LST in BMC is 3.24 °C. To analyze the proper impact of LULC change on LST, I have prepared different mean LST on different LULC category and Table 3.5 has been showing the result. The table has been giving the message that the LST in a built-up area is maximum from other LST situated to specific LULC types in 2017 because of the positive deviation of LST of the built-up area from other LULC types. But it negatively deviates in 1997 and 2011 except for waterbody.

B-V represents the temperature difference between built-up land and vegetation zone, B-W represents the temperature difference between built-up land waterbody, B-F represents the temperature difference between built-up land and fallow land, B-A represents temperature difference between built-up land and agricultural land, B-S represents temperature difference between built-up land and river sand.

3.4.3   Relationship Between LST and Different Indices

Figure 3.6 shows the map of different indices such as NDVI, NDBI, NDWI, and NDBaI which were retrieved from selected Landsat 8 OLI image dated on 12 April 2017. Based on these indices map, I analyze the relationship between LST and LULC types, particularly image on 12 April 2017. Previous studies have extracted the proper relationship between LST and different types of LULC and their spatial distribution concerning LULC. LST is strongly controlled by various factors of a region, among them the presence of vegetation and availability of surface moisture are the most important.

Previous studies (Carlson et al. 1994; Carlson et al. 1995; Owen et al. 1998) about the relationship among different variables which controlled the LST have introduced a new methodology: "Triangle method". This method analyzes the relationship among vegetation, LST, and surface moisture. But in this study, I performed correlation and regression analyses instead of "Triangle Method". To extract the relationship between LST and different indices a scatter diagram was plotted in Fig. 3.7.

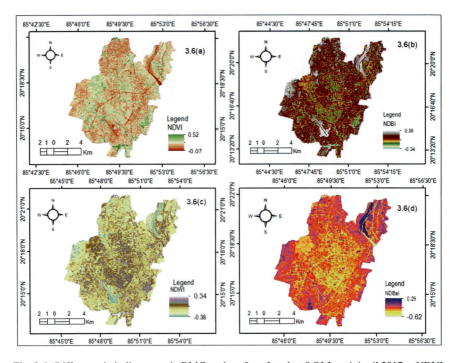

Fig. 3.6 Difference in indices map in BMC retrieve from Landsat 8 OLI on 4 April 2017. **a** NDVI, **b** NDBI, 3.6.c NDWI, and 3.6.d NDBaI

The previous study (Lee 2011) about the relationship between NDVI and LST showed that the relationship between them may be positive or negative, determined on the value of NDVI. When NDVI > 0 it is negatively correlated with LST. But when NDVI < 0, LST positively correlated with NDVI. Figure 3.7 shows the correlation between different indices and LST in the image of 2017. The NDVI was 0.68 to −0.347826 in the image dated in 1997. The correlation between LST and NDVI was 0.99915 when NDVI < 0 but when NDVI > 0 the r was −0.9997 and the R^2 between them are 0.9014 and 0.9995 when NDVI < 0 and NDVI > 0, respectively. Equations (3.10) and (3.11) are the regression equations between LST and NDVI when NDVI < 0 in 1997. Equations (3.12) and (3.13) have shown the regression between NDVI and LST when NDVI > 0 in 1997.

$$LST = 6.723 NDVI + 25.963 \tag{3.10}$$

$$LST = 23.49\,NDVI^2 + 15.47\,NDVI + 26.395 \tag{3.11}$$

$$LST = -13.27 NDVI + 32.503 \tag{3.12}$$

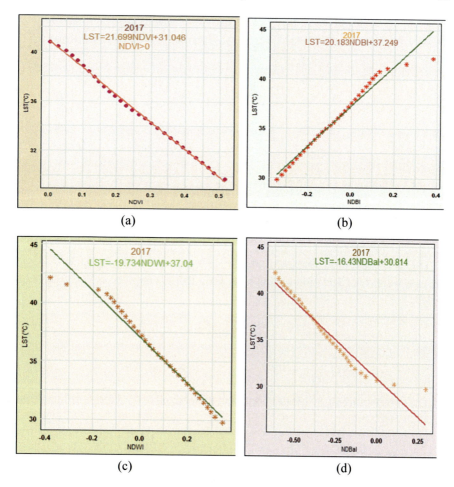

Fig. 3.7 Relationship between LST and different indices in BMC retrieve from Landsat 8 OLI, **a** NDVI and LST, **b** NDBI and LST, **c** NDWI and LST, and **d** NDBaI and LST

$$LST = 0.7754\, NDVI^2 - 13.803\, NDVI + 32.565 \tag{3.13}$$

The correlation between NDWI and LST was -1 and R^2 was 1 in 1997. The regression analysis between them is given in Eqs. (3.14) and (3.15) of 1997.

$$LST = -13.41 NDWI + 29.071 \tag{3.14}$$

$$LST = 2E - 08\, NDWI^2 - 13.411\, NDWI + 29.071 \tag{3.15}$$

The correlation between NDBI and LST in 1997 image was 1 and R^2 is 1. The regression analysis between them is given in Eqs. (3.16) and (3.17).

Table 3.6 Correlation (r) and R^2 between LST and different indices from 1997 to 2017

Year	Parameter	NDVI < 0	NDVI > 0	NDBI	NDWI	NDBal
1997	r	0.99915	−0.9997	1	−1	−1
	R^2	0.9014	0.9995	1	1	1
2011	r	0.99915	−0.99836	0.99631	−1	−0.99801
	R^2	0.9983	0.9967	0.9926	1	0.996
2017	r	0.99554	−0.99904	0.98327	−0.98224	−0.964320
	R^2	0.9911	0.9981	0.9668	0.9648	0.9299

$$LST = 13.411 NDBI + 29.869 \tag{3.16}$$

$$LST = -2E - 8\,NDBI^2 + 13.41\,NDBI + 29.071 \tag{3.17}$$

The correlation of NDBal and LST of the image in 1997 was −1 and R^2 between them is 1 and the regression equation between them is express in Eqs. (3.18) and (3.19).

$$LST = -11.574 NDBal + 26.157 \tag{3.18}$$

$$LST = 7E - 9\,NDBal^2 - 11.574\,NDBal + 26.157 \tag{3.19}$$

The correlation and R^2 between different indices and LST of the three images are given in Table 3.6.

From Table 3.6 it is clear that when NDVI < 0 then it positively correlates with LST and r was 0.99554 and R^2 was 0.9911 but when NDVI > 0 then it is negatively correlated with LST and r was −0.99904 and R^2 was 0.9981 in 2017. The correlation with NDBI and LST was 0.98327 and R^2 was 0.9668 in 2017. The correlation and R^2 between LST and NDWI were −0.98221 and 0.9648, respectively, in 2017. NDBal is negatively correlated with LST and r was −0.96432 and R^2 was 0.9299. Equations (3.20) and (3.21) show regression equation between LST and NDVI < 0 and Eqs. (3.22) and (3.23) are shown in the regression analysis between LST and NDVI < 0 in 2017.

$$LST = 17.089\,NDVI^2 + 27.87\,NDVI + 31.225 \tag{3.20}$$

$$LST = 17.089 NDVI + 31.022 \tag{3.21}$$

Equations (3.22) and (3.23) have shown the regression between NDBI and LST in 2017.

$$LST = 20.183 NDBI + 37.249 \tag{3.22}$$

$$LST = -14.47\,NDBI^2 + 19.315 NDBI + 37.696 \qquad (3.23)$$

Equations (3.24) and (3.25) show the regression analysis between NDWI and LST in 2017.

$$LST = -19.734NDWI + 37.04 \qquad (3.24)$$

$$LST = -14.55\,NDWI2 - 19.2 NDWI + 37.52 \qquad (3.25)$$

The regression analysis between LST and NDBal is expressed in Eqs. (3.26) and (3.27) in 2017.

$$LST = -16.43\,NDBal^2 - 9.588\,NDBal + 30.8722 \qquad (3.26)$$

$$LST = -16.43NDBal + 30.814 \qquad (3.27)$$

A previous study about the impact of fallow land on surface temperature showed the negative relationship between them (PRiCE 1990; Lee 2011). In this study, I have found a strong negative relation between NDBal on LST throughout 20 years of this study area. The impact of built-up area on surface radiance temperature in Toronto indicated a positive relationship between them. Similarly, this study shows the strong positive impact of NDBI on LST (Rinner and Hussain 2011). Ahmed et al. (2013) and Mukherjee and Singh (2020) find out the negative impact of surface moisture ability on land surface temperature. In this study NDWI represents the surface moisture, correlation and regression analysis show the strongly negative impact of NDWII on LST.

3.5 Conclusion

In this study, to extract the proper relation between changing LULC pattern and UHI, both qualitative and quantitative analyses were used. After the successful study, some conclusions were made. 1. The distribution of UHI has been changed from scattered to concentric pattern within the study period. 2. The effect of UHI intensity has been reduced by the presence of dense vegetation in this metropolitan city. 3. The UHI intensity mostly occurred in the newly built-up area. 4. The intensity of UHI is observed in the industrial area rather than the residential area. 5. Urbanization has taken place mainly in the fallow and dry land. 6. Besides having variation between different land uses, there was a maximum difference of LST existing in a LULC, as shown in qualitative analysis between LST and difference indices.

The climatic condition of the region depends on various factors, therefore only satellite image is not sufficient to study the UHI. I have analyzed the influential factor of changing LULC pattern on UHI at the local level. But the result of correlation and also regression analysis has reassured the reliability of satellite image to study the spatio-temporal variation of UHI at the local level.

It is too much needed for methodological development to get the fruitful result in the future study. Firstly, instead of BT real land surface temperature will be extracted. Secondly, the method of satellite image classification would be improved. Thirdly, the effect of UHI on the ecosystem would be studied. Fourthly, the method to mitigate the harmful effect of UHI on biosphere is needed to be researched. Fifthly, it should be important to look into the effect of human activities and other influential factors in decreasing the contributions of UHI to global warming and finally, the errors caused by different conditions in every LULC type need to be more accurately estimated or removed. In this paper, I have only established the relation between LST and different indices on statistical-based. My focus will be traced on the impact of vegetation on UHI and simulation of urban climate on statistical-based.

Acknowledgements I am grateful to the United States Geological Survey (USGS) for offering Landsat images without any charge. I also convey my regards to anonymous reviewers and editors for their helpful and constructive comments, which help me to improve the paper.

References

Ahmed B, Kamruzzaman M, Zhu X, Rahman SM, Choi K (2013) Simulating land cover changes and their impacts on land surface temperature in Dhaka, Bangladesh. Remote Sens 5:5969–5998; https://doi.org/10.3390/rs5115969

Arnfield AJ (2003) Two decades of urban climate research: a review of turbulence, exchanges of energy and water, and the urban heat island. Int J Climatol J Royal Meteorol Soc 23.1:1–26

Borthakur M, Nath BK (2012) A study of changing urban landscape and heat island phenomenon in Guwahati metropolitan area. Int J Sci Res Publ 2.11:1–6

Carlson TN, Gillies RR, Perry EM (1994) A method to make use of thermal infrared temperature and NDVI measurements to infer surface soil water content and fractional vegetation cover. Remote Sens Rev 9(1–2):161–173

Carlson TN, Gillies RR, Schmugge TJ (1995) An interpretation of methodologies for indirect measurement of soil water content. Agric for Meteorol 77(3–4):191–205

Chatterjee ND, Chatterjee S, Khan A (2016) Spatial modeling of urban sprawl around Greater Bhubaneswar city, India. Model Earth Syst Environ 2.1:14

Chen X-L et al (2006) Remote sensing image-based analysis of the relationship between urban heat island and land use/cover changes. Remote Sens Environ 104.2:133–146

Dash P et al (2002) Land surface temperature and emissivity estimation from passive sensor data: theory and practice-current trends. Int J Remote Sens 23.13:2563–2594

Gao B-C (1996) NDWI—A normalized difference water index for remote sensing of vegetation liquid water from space. Remote Sens Environ 58(3):257–266

Grimmond SUE (2007) Urbanization and global environmental change: local effects of urban warming. Geogr J 173.1:83–88

Howard L (2012) The climate of London: deduced from meteorological observations, vol 1. Cambridge University Press

Imhoff ML et al (2010) Remote sensing of the urban heat island effect across biomes in the continental USA. Remote Sens Environ 114.3:504–513

Jackson TJ et al (2004) Vegetation water content mapping using Landsat data derived normalized difference water index for corn and soybeans. Remote Sens Environ 92.4:475–482

Jiménez-Muñoz JC, Sobrino JA (2003) A generalized single-channel method for retrieving land surface temperature from remote sensing data. J Geophys Res Atmos 108.D22

Kamran KV, Pirnazar M, Bansouleh VF (2015) Land surface temperature retrieval from Landsat 8 TIRS: comparison between split window algorithm and SEBAL method. Third international conference on remote sensing and geoinformation of the environment (RSCy2015), vol. 9535. International Society for Optics and Photonics

Kato S, Yamaguchi Y (2005) Analysis of urban heat-island effect using ASTER and ETM + Data: separation of anthropogenic heat discharge and natural heat radiation from sensible heat flux. Rem Sens Environ 99:44–54

Khan A, Chatterjee S (2016) Numerical simulation of urban heat island intensity under urban–suburban surface and reference site in Kolkata, India. Model Earth Syst Environ 2(2):71

Kumar Arora M, Mathur S (2001) Multi-source classification using artificial neural network in a rugged terrain. Geocarto Int 16.3:37–44

Lee L et al (2011) Use of Landsat TM/ETM+ data to analyze urban heat island and its relationship with land use/cover change. In: 2011 International conference on remote sensing, environment and transportation engineering. IEEE

Maki M, Ishiahra M, Tamura M (2004) Estimation of leaf water status to monitor the risk of forest fires by using remotely sensed data. Remote Sens Environ 90(4):441–450

Mukherjee F, Singh D (2020) Assessing land use-land cover change and its impact on land surface temperature using LANDSAT data: a comparison of two urban areas in India. Earth Syst Environ. https://doi.org/10.1007/s41748-020-00155-9

Nieuwolt S (1966) The urban microclimate of Singapore. J Tropical Geogr 22:30–37

Oke TR (1987) Boundary layer climates, 2nd edn. Methuen, Routledge, London, pp 435

Owen TW, Carlson TN, Gillies RR (1998) An assessment of satellite remotely-sensed land cover parameters in quantitatively describing the climatic effect of urbanization. Int J Remote Sens 19(9):1663–1681

PRiCE JC (1990) Using spatial context in satellite data to infer regional scale evapotranspiration. IEEE Trans Geosci Remote Sens 28.5:940–948

Pongrácz R, Bartholy J, Dezső Z (2010) Application of remotely sensed thermal information to urban climatology of Central European cities. Phys Chem Earth Parts a/b/c 35(1–2):95–99

Purevdorj TS et al (1998) Relationships between percent vegetation cover and vegetation indices. Int J Remote Sens 19(18):3519–3535

Qin Z, Karnieli A, Berliner P (2001) A mono-window algorithm for retrieving land surface temperature from Landsat TM data and its application to the Israel-Egypt border region. Int J Remote Sens 22(18):3719–3746

Ramachandran B, Justice CO, Abrams MJ (eds) (2010) Land remote sensing and global environmental change: NASA's earth observing system and the science of ASTER and MODIS, vol. 11. Springer Science & Business Media

Reis S (2008) Analyzing land use/land cover changes using remote sensing and GIS in Rize, North-East Turkey. Sensors 8(10):6188–6202

Rinner C, Hussain M (2011) Toronto's urban heat island-Exploring the relationship between land use and surface temperature. Remote Sens 3(6):1251–1265

Rizwan AM, Dennis LYC, Chunho LIU (2008) A review on the generation, determination and mitigation of Urban Heat Island. J Environ Sci 20.1:120–128

Sobrino JA, Jiménez-Muñoz JC, Paolini L (2004) Land surface temperature retrieval from LANDSAT TM 5. Remote Sens Environ 90(4):434–440

Walawender JP et al (2014) Land surface temperature patterns in the urban agglomeration of Krakow (Poland) derived from Landsat-7/ETM+ data. Pure Appl Geophys 171.6:913–940

Yun-hao CHEN, Jie W, Xiao-bing LI (2002) A study on urban thermal field in summer based on satellite remote sensing. Remote Sens Land Resour 14.4: 55–59

Zarco-Tejada PJ, Rueda CA, Ustin SL (2003) Water content estimation in vegetation with MODIS reflectance data and model inversion methods. Remote Sens Environ 85.1:109–124

Zha Y, Gao J, Ni S (2003) Use of normalized difference built-up index in automatically mapping urban areas from TM imagery. Int J Remote Sens 24(3):583–594

Zhao H, Chen X (2005) Use of normalized difference bareness index in quickly mapping bare areas from TM/ETM+. Int Geosci Remote Sens Symp 3

Chapter 4
Statistical Downscaling Method for Improving Global Model Rainfall Forecasts of Seasonal Rainfall Over West Bengal (WB), India

Aminuddin Ali and Tirthankar Ghosh

Abstract This study uses rainfall forecasts from six advanced experimental NMME (North-American Multi-model Ensemble) models from 1996 to 2010. The seasonal rainfall forecasts from the statistical downscaling and BLUE (Best Linear Unbiased Estimator) schemes are calibrated and verified over districts of West Bengal. The forecast performances and predicted seasonal rainfall distribution from the statistical downscaling and BLUE methods are the main objectives of this work. The seasonal rainfall prediction skills are compared among models before and after downscaling methods. The forecasts of downscaled global models are more accurate compared to real observations. The improvements of seasonal rainfall forecasts skills are discussed. The CFSv2 model is more capable of forecasting at the district level than others. More improvement is expected with dense real observation.

Keywords Statistical Downscaling (SD) · Seasonal Rainfall · Skill

Acronyms and Abbreviations

BLUE	Best Linear Unbiased Estimator
CC	Correlation Coefficient
CFSv2	Climate Forecast System Version 2
CMC	Canadian Meteorological Centre
COLA	Center for Ocean–Land–Atmosphere Studies
CPC	Climate Prediction Center
CPC-URD	Climate Prediction Center -Unified Rain gauge Dataset
ECMWF	European Center for Medium Range Weather Forecasting
EM	Ensemble Mean

A. Ali (✉)
Department of Statistics, Ramkrishna Mission Vidyamandira, Belur 711202, India

T. Ghosh
Department of Statistics, Visva-Bharati, Santiniketan 731235, India

© The Author(s), under exclusive license to Springer Nature Singapore Pte Ltd. 2022
N. C. Jana and R. B. Singh (eds.), *Climate, Environment and Disaster in Developing Countries*, Advances in Geographical and Environmental Sciences, https://doi.org/10.1007/978-981-16-6966-8_4

GFDL   Geophysical Fluid Dynamics Laboratory
GMAO   Global Modeling and Assimilation Office
IRI   International Research Institute
ISMR   Indian Summer Monsoon Rainfall
JJAS   June-July–August-September
MAE   Mean Absolute Error
MOS   Model output statistics
NASA   National Aeronautics and Space Administration
NCAR   National Center for Atmospheric Research
NCEP   National Centers for Environmental Predictions
NOAA   National Oceanic and Atmospheric Administration
NWP   Numerical Weather Prediction
PP   Perfect prognosis
OLS   Ordinary Least Squares
RMSE   Root Mean Square Error
RSMAS   Rosenstiel School of Marine and Atmospheric Science
SS   Sum of square
WG   Weather generators

4.1 Introduction

The impact of seasonal climate variability has always importance for agricultural production, water management transportation, etc. in India. Naturally, monsoon (Indian Summer Monsoon Rainfall-ISMR) has a vital role in agricultural production here. A reliable monsoon forecast may result in great dividends for the farmers and the Indian economy. However, the prediction of ISMR is a very challenging task. Fortunately, after the computer revolution weather forecast has improved noticeably. Now the uses of computer models are to simulate weather forecasts from next few hours to the next few days. The method is called NWP or dynamical weather models. However, the rainfall forecast skills of dynamical methods are not always encouraging. This is because of variations in spatial scale and latent restrictions of these dynamical methods. In the dynamical method, it is not possible to include small-scale factors due to computational limitations. To overcome these limitations ensemble forecasting methods were introduced. Ensemble methods were introduced by many including the European Center for climate forecasts in early 1990 (Zhang and Krishnamurti 1997; Molteni et al. 1996; Toth and Kalnay 1997). Ensemble techniques are of two types viz. Single model and multi-model ensemble (MME). Multi-model ensembles, further, may be classified into two kinds. Those are simple (equally weighted) ensemble and unequally weighted ensemble. Du (2007) and Qi et al. (2014) have shown that the simple ensemble method improves the forecasts' skill. In unequally weighted ensemble method weights for different models are different.

Most analyses of model skills are generally useful for the study of the broad spatial and temporal resolutions, but it is limited in the case of analyzing the regional or seasonal rainfall forecasts (Chakrabarty and Krishnamurti 2006). Slater et al. (2016) mentioned that the skill of simple ensemble mean is better than any single individual model. However, there is scope for applying different model averaging techniques with an unequal weighting of individual models for generating improved forecasts.

The weighted ensemble methods are more useful to improve forecasts skills. Krishnamurti et al. (1999) has shown improvement in seasonal weather and climate forecasts from the multi-model ensemble. Various institutions like NCEP (Toth and Kalnay 1993) and the Canadian Meteorological Center (CMC) have developed forecasts using multi-model ensemble (MME) methods for their usefulness. Krishnamurti et al. (1999, 2000a, b, 2009) have used the principle of multiple linear regression to develop an MME, namely, super ensemble. It provides improved forecasts in seasonal weather, global weather, and hurricane track and intensity. Krishnamurti et al. (2009) have developed a multi-model ensemble to improve forecasts using bias and threat scores. Ensemble forecasts are advantageous for probabilistic forecasts as well (Zhu et al. 2013). Zhou and Du (2010) have also shown that the forecasts from ensemble methods are superior to any single model forecasts for both deterministically as well as probabilistically. The precipitation forecasts of ECMWF ensemble methods are always skillful. The advantages of ensemble prediction for forecasting weather and climate have emphasized over the years (Bauer et al. 2015). From the past studies, it may be said that ensemble prediction has more advantages for forecasting weather and climate (Bauer et al. 2015). It is important to find out more useful information from a large amount of forecast data. Therefore, many efforts have been reported in the literature to construct mass ensemble products that can improve forecasts (Tebaldi and Knutti 2007; Leutbecher and Lang 2014; Smith et al. 2014). Kumar and Ghosh (2018) used the concept of BLUE to construct a downscaled MME for forecasting seasonal rainfall for the Indian region.

4.1.1 Downscaling

The impacts of global warming play a vital role in climate change on small scale. So it becomes necessary to develop and apply the methodology to specific issues of a region (Cohen 1990). Decision-makers have to face significant challenges while framing policies due to a lack of impact assessment at desired scales (Nieto and Wilby 2005). Global models forecast at low resolutions which make it difficult to assess impacts at appropriate scales (Fowler et al. 2007; Wilby et al. 1998). The process of estimating local scale (finer resolutions) unknown variables from large scales atmospheric variables is termed as Downscaling (Najafi et al. 2011). Many downscaling studies have been executed so far for atmospheric variables rainfall and temperature, etc. (Leung et al. 2003). Dynamical Downscaling (DD) and Statistical Downscaling (SD) are there in literature (Lenart 2008). Dynamical Downscaling

(DD) methods are more complex and computational cost is high (Salathe 2007). The skills of forecasts from such models are currently inadequate.

4.1.2 Statistical Downscaling (SD)

The skills of forecasts from Statistical Downscaling (SD) techniques are sufficiently accurate. It establishes a statistical relationship between local scales (finer resolution information) and GCMs output (Fowler et al. 2007). The SD techniques have gained popularity among climate modelers and researchers. The establishment of stable statistical relationships at different spatial scales is the main assumption of Statistical Downscaling (SD) methods (Wilby et al. 1998). Statistical downscaling methods may be divided into three categories PP, MOS and WG methods (Maraun et al. 2010). MOS methods are widely applied to downscale GCM by Wood et al. (2004), Ndiaye et al. (2011), Landman and Beraki (2012), Yoon et al. (2012), Abatzoglou and Brown (2012), Tian et al. (2012).

4.2 The Study Area and Dataset Used

4.2.1 Study Area

According to the research objectives three districts of WB and the whole wb have been used as the study area of this work (Fig. 4.1).

4.2.2 Dataset Used

For this work, the following datasets have been used for selected regions.

The NMME forecasts data: The latest version of NMME forecasts system was considered in this work. It is an experimental seasonal forecasting system that includes coupled Global Circulation Models (GCMs) from The NOAA/GFDL, NOAA/NCEP, NCAR, the IRI for Climate and Society, NASA, and CMC. The rainfall parameter from six NMME models has been considered in this study. The details of the six models are presented in Table 4.1. The NMME system models are monthly forecasts at approximately 100 km ($1° \times 1°$ latitude longitude resolutions). The monthly forecasts of rainfall from all selected NMME models were converted to seasonal averages using four-month moving averages techniques. This work aims to downscale (at resolution $0.25° \times 0.25°$) and compare the six selected models for the JJAS season only at lead time one month over selected regions.

Fig. 4.1 Study area and
geographical map of WB

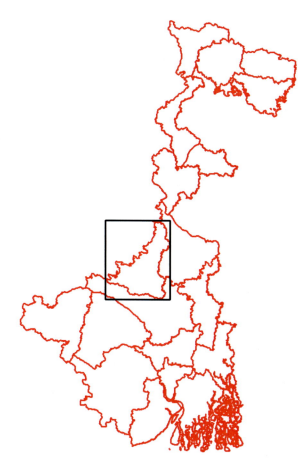

4.3   Methodology

The conventional models used in the multi-model are EM and BLUE.

4.3.1   BLUE Method

It is arduous to determine the weights for each spatial grid point for all models. Xie
and Akrin (1996) have given an error adjustment technique for this purpose. It may be
used to construct new estimators which provide seasonal forecasts with better skills
for the Indian region. For each spatial and temporal grid point, an error E_i (say) may
be defined from a bunch of N models. Then the forecast errors of the models may be
presented as follows.

Table 4.1 Detail descriptions of all global models participated in this study

Institute Name and References	Models	Analysis period	Ensemble size	Lead time (months)	Model explanation	Model resolution (atmosphere)	Model resolution (Ocean)
NCEP Saha et al. (2014)	CFSv2	1982–2010	24	0–9	Climate Forecast System version 2	T126L64	MOM4L40 0.25° EQ
Canadian Meteorological Centre (CMC) Merryfield et al. (2013)	CMC1-CanCM3	1981–2010	10	0–11	Canadian Meteorological Centre (CMC)	CanAM3T63L31	CanOM4L40 0.94° EQ
Canadian Meteorological Centre (CMC) Merryfield et al. (2013)	CMC2-CanCM4	1981–2010	10	0–11	Canadian Meteorological Centre (CMC)	CanAM3T63L35	CanOM4L40 0.94° EQ
GFDL Delwoth et al. (2006)	GFDL CM 2.1	1982–2010	10	0–11	Geophysical _Fluid Dynamics_ Laboratory Climate Models, Version 2	2 × 2.5° L24	MOM4L50 0.3° EQ
NASA Vernieres et al. (2012)	NASA-GEOSS2S Goddard _Earth_ Observing System Model, Version 2	1981–2010	10	0–11	National_ Aeronautics _Space_ Administration	1 × 1.25° L72	MOM4L40 0.25° EQ
NCAR Kirtman and Min (2009)	NCAR-CESM1 Community_Climate_System Model, Version 1	1982–2010	10	0–11	National _Center_ Atmospheric Research	T85L26	POPL42 0.3° EQ

$$E_i(x, y, z, t) = O(x, y, z, t) - T_i(x, y, z, t) \quad \text{for } i = 1 \, (1) \, N \qquad (4.1)$$

where $O(x, y, z, t)$ = Observed rainfall at time t and $T_i(x, y, z, t)$ = Predicted rainfall from the ith model. Then an average of squares of errors for a bunch of n historical observations is calculated. This is named as the estimate of error (V_i) for the ith model.

$$V_i(x, y, z, t) = \frac{1}{n} \sum_{i=1}^{n} E_i^2(x, y, z, t) \qquad (4.2)$$

$$w_i(x, y, z) = \frac{V_0}{V_i}$$

The weather parameter is estimated by (best estimate) by X and the member models predict it by T. Then the adjusted residual V at time $(t + 1)$ is written as follows

$$V_i(x, y, z) = \hat{X}(x, y, z) - T_i(x, y, z)$$

Then the residual equations may be written in matrix form as

$$V = \alpha \hat{X} - T$$

$$V = [V_1, V_2, V_3, \ldots\ldots, V_N]^T, \alpha = [1, 1, 1, \ldots\ldots, 1]^T, T = [T_1, T_2, T_3, \ldots\ldots, T_N]^T$$

Now the principle of least square technique has been employed to minimize the Residual SS ($V^T W V$) with weight W.

$$W = \begin{bmatrix} w_1 & 0 & \ldots & 0 \\ 0 & w_2 & \ldots & 0 \\ \vdots & \vdots & \ldots & \vdots \\ 0 & 0 & \ldots & w_N \end{bmatrix}$$

The solution may be written as follows

$$\hat{X} = (\alpha^T W \alpha)^{-1} \alpha^T W T$$

$$\hat{X} = \frac{\sum_{i=1}^{N} w_i(x, y, z) * T_i(x, y, z)}{\sum_{i=1}^{N} w_i(x, y, z)} \qquad (4.3)$$

$\hat{X}$ is nothing but advanced BLUE. Advanced BLUE is the multi-model ensemble technique used here. Kumar and Ghosh (2018) used BLUE as multi-model for forecasting weather parameters as if M_1, M_2,... ....., M_N are N estimators which are independent and unbiased for μ with $var(M_i) = \sigma_i^2$.

$$T = \sum_{i=1}^{N} \alpha_i M_i$$

Then T will be said to be BLUE where

$$\alpha_i = \frac{\frac{1}{\sigma_i^2}}{\sum_{i=1}^{N} \frac{1}{\sigma_i^2}} \quad \text{for all } i = 1(1) \text{ N}$$

σ_i^2 is the error variance of the primary model forecasts.

4.3.2 Simple Ensemble (Arithmetic Mean) Method

Now minimize the residual SS ($V^T V$) without weight W. Then the solution of $\hat{X}$ is reduced to simple EM.

$$\hat{X} = \frac{\sum_{i=1}^{N} T_i(x, y, z)}{N} = EM$$

This ensemble methodology is the simple average of all model forecasts (Becker et al. 2014). For each grid point, the predicted result from the ith model at the time t is denoted by T_{it}, $i = 1(1) N$, where N stands for a number of models.

4.3.3 BCCA Method

Bias Correction and Constructed Analogues (BCCA) method is a popular spatial downscaling method as BCSD under MOS. BCCA method was described by Hidalgo et al. (2008) and Maurer et al. (2010). BCCA is an analogue method similar to BCSD based on historical observations. This method estimates spatial weather information at local (unknown locations) scales establishing a linear relationship which relies on historical analogues weather information for spatial Statistical Downscaling of coarse global models outputs (Lorenz 1969). The best analogue is constructed using coarse resolution historical records and the pattern is used to downscale future fine resolution data (Timbal et al. 2003; van den dool et al. 2003; Diez et al. 2005). The best analogue is determined using the best linear combinations of historical records. GCMs produce biased weather information at $1° \times 1°$ resolutions. BCCA can produce promising downscaling results for the prediction of weather variables from GCM outputs and the assessment was made for effects of those hydrological factors on climate changes (Maurer et al. 2010; Hidalgo et al. 2008). Hidalgo et al. (2008) and Maurer et al. (2010) have described BCCA method in details.

4.3.4   The BCCA Procedure May be Presented in the Following Steps

Suppose Y is the coarse resolution forecasts from GCM of a variable of interest. X denotes the fine resolution data from observations. Let $Y_C^G = (Y_1, Y_2, Y_3, ..., Y_C)'$ is to be downscaled to fine resolution as $X_F^{SD} = (Y_1, Y_2, Y_3, ..., Y_F)'$. X_F^{obs} is coarse resolution observations and Y_C^{obs} is coarse resolution observations. F and C are number of grids for fine and coarse resolution observation.

Step One: Find the T number of years which has RMSE between observed and GCMs outputs through C number of coarse grids.

$$RMSE_t = \sqrt{\frac{\sum_{i=1}^{C} \left(y_{GCM,i} - y_{obs,i}^t \right) \cdot \left(y_{GCM,i} - y_{obs,i}^t \right)}{C}}$$

where $t = 1(1)T$, T is a number of years.

Step Two: Find the S ($< T$) number of years which has the smallest RMSE between observed and GCMs outputs through the C number of coarse grids. Then select the observations from those years as analogues years. Here $S = 7$ has been taken. Hidalgo has considered $S = 30$ in 2008.

Step Three: Estimate α, the least square solution coefficient using Multiple Linear Regression (MLR) as follows.

Multivariate Regression: MLR

The Multiple Linear Regression (MLR) considers independent variables as $x_i^{(1)}, x_i^{(2)}, ..., x_i^{(m)}$ where (m) stands for the number of predictors. The regression equation may be written as.

$$Y_i = \alpha_0 + \alpha_1 x_i^{(1)} + \alpha_2 x_i^{(2)} + ... + \alpha_m^{(m)} + \varepsilon_i$$

where it is assumed that ε_i follows normal distribution with mean zero and variance sigma square.

$$Y_C^G = (Y_1, Y_2, Y_3, \ldots\ldots\ldots\ldots, Y_C)',$$

Now from Least square theory the coefficient vector α may be computed as

$$\alpha = \left[\left(\left(X_C^{obs} \right)' * X_C^{obs} \right)^{-1} * \left(X_C^{obs} \right)' \right] * Y_C^G$$

where α is the $(S + 1) \times 1$ dimensional vector.

where Y_C^G is the $C \times 1$ dimensional vector of coarse resolution target GCM data and X_C^{obs} is the $C \times (S + 1)$ dimensional matrix selected from GCM outputs of the observed period from the second step.

$$X_C^{obs} = \begin{bmatrix} 1 & X_{obs,1}^{(1)} & \cdots & X_{obs,1}^{(S)} \\ 1 & X_{obs,2}^{(1)} & \cdots & X_{obs,2}^{(S)} \\ \vdots & \vdots & \cdots & \vdots \\ 1 & X_{obs,C}^{(1)} & \cdots & X_{obs,C}^{(S)} \end{bmatrix} \quad Y_C^G = \begin{bmatrix} Y_1 \\ Y_2 \\ \vdots \\ Y_C \end{bmatrix}$$

Step Four: Estimate the fine resolution spatially downscaled data from coefficient vector α as

$$X_F^G = X_F^{obs} * \alpha$$

where X_F^{obs} is the $F \times (S + 1)$ dimensional matrix of observed fine resolution data.

$$X_F^{obs} = \begin{bmatrix} 1 & X_{obs,1}^{(1)} & \cdots & X_{obs,1}^{(S)} \\ 1 & X_{obs,2}^{(1)} & \cdots & X_{obs,2}^{(S)} \\ \vdots & \vdots & \cdots & \vdots \\ 1 & X_{obs,F}^{(1)} & \cdots & X_{obs,F}^{(S)} \end{bmatrix}$$

Step Five: Repeat Steps one—four for all the seasons to estimate until all the fine resolution data.

4.4 Comparison of Results and Discussion

4.4.1 Rainfall Forecast Skills Over WB Region

In this work, BLUE and Statistical Downscaling (SD) method has been applied for the development of forecasts from the six primary models CMC1, CFSv2, GFDL, CMC2, and NASA. The improvements of JJAS seasonal rainfall forecasts over West

Table 4.2 The performance skills for the forecasts of JJAS seasonal rainfall over WB in Verification and Calibration. Mean of rainfall prediction skills for different GCMs at one month lead time

Skills	Methods	CFSv2	CMC1	CMC2	GFDL	NASA	NCAR	EM	BLUE
R	Direct	0.15	−0.64	−0.48	−0.47	0.41	0.35	−0.39	−0.27
	SD	0.73	0.70	0.71	0.70	0.65	0.65	0.70	0.70
NSE	Direct	−16.25	−7.76	−12.02	−19.08	−21.00	−30.86	−16.63	−15.64
	SD	−0.68	−0.83	−0.45	−1.29	−2.01	−1.54	−0.86	−0.79
RMSE	Direct	5.24	3.73	4.62	5.56	5.79	7.20	5.24	5.08
	SD	2.20	2.18	2.08	2.55	3.03	2.56	2.34	2.25
MAE	Direct	4.83	3.05	3.90	5.09	5.45	6.93	4.77	4.61
	SD	1.92	1.79	1.76	2.14	2.68	2.20	2.05	1.93

Bengal (WB), and, especially for three districts of WB have been presented. The details of selected regions are provided in Table 4.1. The JJAS seasonal rainfall has been predicted using six primary models and two conventional models for all grid points ($0.25° \times 0.25°$ Longitude-Latitude) over WB and three districts—Birbhum, Murshidabad, and Burdwan. The skill matrices have been calculated using observed and predicted rainfall to assess the ability of forecasting models. The Pearson's product-moment correlation coefficients (r), Nash–Sutcliffe Efficiency (NSE), Root Mean Square Error (RMSE), and Mean Absolute Error (MAE) have been computed for forecastability of models. Results of all model performances have been tabulated in Tables 4.2, 4.3, 4.4, 4.5, 4.6, 4.7, 4.8, 4.9, 4.10, 4.11, 4.12, 4.13 and 4.14. All computations and plots are done using software R-4.0.0. Correlation coefficient (r) and RMSE are two metrics to assess the performance of the model forecasts and the results by comparing with the observations (Wilks 2011). The r, NSE, MAE, and RMSE are computed for the WB and three districts of WB. The RMSE reflects the total bias of the simulations and the results compared to the observations. Tables 4.2, 4.3, 4.4, 4.5 and 4.6 indicates that the skills value is higher for CFSv2 model compared to any others. The skills values from individual models are also shown in graphical format in Figs. 4.2, 4.3, 4.4, 4.5, 4.6, 4.7, 4.8, 4.9, 4.10, 4.11 and 4.12. It, further, may be observed that the skills (RMSE, MAE, etc.) for the selected method is uniformly lower for all 15-year period as compared to primary models and other conventional models (before downscale). The MAE reflects the average errors between the forecasts from models and the observations.

4.4.2 Observed and the Forecasted Rainfall Fields

The significant features of this work are to compare the geographical distribution of rainfall forecasts from member models with the observed rainfall fields from Precipitation Estimation for Remotely Sensed Information using ANN (PERSIANN)

Table 4.3 Spatial Correlation Coefficients (r) between observations and forecasts (before and after SD) from member models and conventional models for JJAS seasonal rainfall over WB in Verification and Calibration

Year	CFSv2	CMC1	CMC2	GFDL	NASA	NCAR	EM	BLUE
1996	0.26	−0.78	−0.49	−0.62	0.63	0.67	−0.32	−0.75
1997	0.27	−0.66	−0.45	−0.81	0.31	−0.39	−0.39	−0.71
1998	0.56	−0.93	−0.28	0.63	0.64	0.89	0.07	−0.60
1999	0.03	0.08	0.08	−0.41	−0.33	0.05	−0.08	−0.78
2000	0.83	−0.23	0.52	−0.48	0.57	0.13	0.59	−0.32
2001	0.17	−0.95	−0.71	−0.35	0.58	0.22	−0.55	−0.40
2002	0.12	−0.81	−0.62	−0.74	0.71	0.68	−0.36	0.53
2003	0.14	−0.96	−0.71	−0.46	0.91	0.89	−0.47	−0.20
2004	0.58	−0.54	−0.09	−0.86	0.63	0.85	0.19	0.30
2005	0.13	−0.81	−0.66	−0.58	0.39	0.33	−0.66	−0.34
2006	−0.77	−0.23	−0.63	0.03	−0.82	0.02	−0.67	0.05
2007	−0.74	−0.19	−0.77	0.01	−0.01	0.26	−0.89	0.42
2008	−0.11	−0.81	−0.93	−0.74	0.81	0.26	−0.93	−0.22
2009	0.38	−0.86	−0.76	−0.92	0.38	−0.53	−0.62	−0.58
2010	0.40	−0.90	−0.75	−0.77	0.77	0.86	−0.81	−0.51
SD								
2006	0.29	0.18	0.29	0.23	0.11	0.01	0.22	0.21
2007	0.85	0.86	0.87	0.88	0.88	0.86	0.86	0.85
2008	0.82	0.75	0.80	0.75	0.65	0.70	0.80	0.76
2009	0.80	0.76	0.69	0.72	0.76	0.74	0.72	0.76
2010	0.90	0.93	0.92	0.94	0.86	0.94	0.91	0.92

developed at University of California for different seasons. The JJAS seasonal forecasts of 2006–2010 are illustrated as examples in Figs. 4.2, 4.3, 4.4, 4.5 and 4.6. The observed rainfall and forecasted rainfall from member models and conventional models have been included in each illustration at 0.25° × 0.25° resolutions (downscaled). Figure 4.2 illustrates the results for PERSIANN and forecasted JJAS seasonal rainfall (mm/day) from CFSv2, CMC2, CMC1, NASA, GFDL, EM, and BLUE models for JJAS seasonal forecasts after four-month moving average valid for 2006. The spatial correlation coefficient between observations and forecasts (SD) from different models has been presented in Tables 4.2, 4.3, 4.4, 4.5, 4.6, 4.7, 4.8, 4.9, 4.10, 4.11, 4.12, 4.13 and 4.14. The observed rainfall distribution shows the rainfall estimate for the forecasted year 2006 at WB and three districts with a peak of 7.7–15.4 mm/day (WB), 7.7–13.8 mm/day (Birbhum), 7.7–12 mm/day (Burdwan) and 7.7–12 mm/day (Murshidabad). The main interest of this work is to note that the member models and the conventional models forecasts rainfall with correlations 0.24, −0.77, −0.61, −0.38, 0.14, 0.02, −0.50, and -0.33 whereas SD method can

Table 4.4 Nash Efficiency (NSE) between observations and forecasts (before and after SD) from member models and conventional models at 0.25° × 0.25° resolutions for JJAS seasonal rainfall over WB in Verification and Calibration

Year	CFSv2	CMC1	CMC2	GFDL	NASA	NCAR	EM	BLUE
1996	−12.23	−6.19	−7.67	−12.20	−17.32	−25.13	−12.50	−11.49
1997	−56.55	−23.46	−46.48	−79.24	−83.33	−114.26	−62.52	−60.50
1998	−2.51	−1.24	−2.13	−2.65	−3.49	−5.15	−2.65	−2.55
1999	−47.32	−34.38	−37.41	−56.82	−60.98	−83.14	−51.80	−49.57
2000	−5.95	−2.35	−4.81	−10.93	−8.12	−13.87	−7.07	−7.17
2001	−5.87	−1.77	−5.11	−9.89	−7.20	−16.00	−6.56	−6.06
2002	−10.29	−2.51	−7.30	−12.30	−12.16	−18.07	−9.44	−8.84
2003	−2.71	−1.01	−2.17	−3.24	−3.47	−5.29	−2.62	−2.40
2004	−9.97	−5.00	−6.54	−15.42	−13.01	−21.76	−11.00	−10.51
2005	−4.30	−0.89	−1.38	−2.01	−4.25	−8.85	−2.69	−2.23
2006	−12.18	−3.90	−10.64	−9.90	−13.24	−24.33	−11.38	−9.81
2007	−59.48	−29.39	−36.65	−54.69	−69.68	−96.55	−55.11	−50.61
2008	−7.13	−2.13	−4.60	−7.88	−6.95	−14.16	−6.34	−5.62
2009	−6.58	−1.59	−6.83	−8.65	−11.11	−14.17	−7.11	−6.71
2010	−0.64	−0.63	−0.64	−0.45	−0.64	−2.11	−0.60	−0.49
SD								
2006	−1.60	−2.45	−1.60	−3.70	−2.66	−4.39	−1.92	−1.96
2007	0.11	−1.67	0.04	−1.32	−0.80	−0.84	−0.36	−1.48
2008	0.12	−0.16	0.30	−0.08	−2.42	−2.30	0.18	−0.30
2009	0.54	−0.42	−0.16	−0.38	−0.87	−0.02	0.35	−0.23
2010	−2.59	0.55	−0.82	−0.96	−3.29	−0.17	−2.57	0.03

able to enhance that correlations to 0.29, 0.18, 0.29, 0.23, 0.11, 0.01, 0.22, and 0.21 for that forecasts for 2006 over WB (Fig. 4.2 and Table 4.3). All model forecasts after downscaled are wet more than 6 mm/day over the northern part of WB which includes districts of Cooch Behar, Alipurduar and Jalpaiguri. However, the downscaled models forecasts are almost similar to observed rainfall for any other parts of WB. The second illustration shown in Fig. 4.3 is seasonal downscaled forecast valid for 2007. The models forecasts (SD) rainfall with a correlation (spatial) of 0.85, 0.86, 0.87, 0.88, 0.88, 0.86, 0.86, and 0.85 compared to those of models before downscaling are 0.31, −0.86, −0.46, −0.31, 0.41, 0.41, −0.41, and −0.16. It may be observed that the skills are consistently higher for the correlation from downscaled models as compared to the skills of models before downscaling. The third example presented in Fig. 4.4 is forecast valid for 2008. Here again, it may be observed the similar improvement for forecasting seasonal rainfall from models after downscaling. The spatial correlations are 0.82, 0.75, 0.80, 0.75, 0.65, 0.70, 0.80, and 0.76 from

Table 4.5 Root Mean Square Error (RMSE) between observations and forecasts (before and after SD) from member models and conventional models at 0.25° × 0.25° resolutions for JJAS seasonal rainfall over WB in Verification and Calibration

Year	CFSv2	CMC1	CMC2	GFDL	NASA	NCAR	EM	BLUE
1996	5.57	4.11	4.51	5.56	6.56	7.83	5.63	5.41
1997	5.34	3.48	4.85	6.31	6.47	7.56	5.61	5.52
1998	6.03	4.82	5.70	6.15	6.82	7.99	6.15	6.07
1999	7.16	6.12	6.38	7.83	8.11	9.44	7.48	7.32
2000	5.18	3.59	4.73	6.78	5.93	7.57	5.58	5.61
2001	3.82	2.43	3.60	4.81	4.17	6.01	4.01	3.87
2002	5.05	2.82	4.33	5.48	5.46	6.57	4.86	4.72
2003	4.58	3.37	4.24	4.90	5.03	5.97	4.53	4.39
2004	4.64	3.43	3.85	5.68	5.24	6.68	4.85	4.75
2005	4.10	2.45	2.75	3.09	4.08	5.59	3.42	3.20
2006	5.00	3.05	4.69	4.54	5.19	6.92	4.84	4.52
2007	7.22	5.12	5.70	6.93	7.81	9.17	6.96	6.67
2008	5.71	3.54	4.74	5.96	5.64	7.79	5.42	5.15
2009	3.98	2.33	4.05	4.49	5.04	5.63	4.12	4.02
2010	5.25	5.24	5.25	4.94	5.25	7.24	5.18	5.01
SD								
2006	2.27	2.61	2.27	3.05	2.69	3.26	2.40	2.42
2007	1.67	2.90	1.74	2.70	2.38	2.41	2.07	2.80
2008	1.39	1.60	1.24	1.55	2.75	2.70	1.35	1.70
2009	1.28	2.23	2.02	2.20	2.56	1.90	1.51	2.07
2010	4.38	1.54	3.13	3.24	4.79	2.51	4.37	2.27

Table 4.6 Time series of spatial Correlation Coefficients (r) between observations and forecasts (SD) from member models and conventional models at 0.25° × 0.25° resolutions for JJAS seasonal rainfall over the Birbhum district in Verification

Year	CFSv2	CMC1	CMC2	GFDL	NASA	NCAR	EM	BLUE
2006	0.51	0.50	0.51	0.48	0.35	0.50	0.48	0.44
2007	0.83	0.91	0.92	0.93	0.91	0.85	0.93	0.91
2008	0.82	0.57	0.78	0.56	0.60	0.68	0.79	0.62
2009	0.34	0.44	0.24	0.50	0.52	0.44	0.33	0.44
2010	0.79	0.88	0.92	0.95	0.95	0.86	0.92	0.89

Table 4.7 RMSE between observations and forecasts (SD) from member models and conventional models 0.25° × 0.25° resolutions at 0.25° × 0.25° resolutions for JJAS seasonal rainfall over the Birbhum district in Verification

Year	CFSv2	CMC1	CMC2	GFDL	NASA	NCAR	EM	BLUE
2006	0.85	1.09	0.85	1.75	0.89	1.22	1.36	0.92
2007	1.24	2.43	1.44	2.76	1.97	1.59	1.77	2.15
2008	0.57	0.69	0.41	0.68	1.47	1.95	0.49	1.26
2009	1.94	3.24	3.26	3.49	2.96	0.87	2.43	2.56
2010	4.08	1.01	2.63	2.03	4.36	2.96	3.60	2.30

Table 4.8 MAE between observations and forecasts (SD) from member models and conventional models at 0.25° × 0.25° resolutions for JJAS seasonal rainfall over Birbhum district in Verification

Year	CFSv2	CMC1	CMC2	GFDL	NASA	NCAR	EM	BLUE
2006	0.70	0.90	0.70	1.57	0.75	1.04	1.18	0.79
2007	1.16	2.41	1.39	2.74	1.94	1.54	1.74	2.13
2008	0.50	0.60	0.31	0.57	1.30	1.78	0.42	1.16
2009	1.84	3.20	3.19	3.45	2.90	0.69	2.39	2.46
2010	4.05	0.94	2.61	2.02	4.33	2.92	3.57	2.27

Table 4.9 Spatial Correlation Coefficients (r) between observations and forecasts (SD) from member models and conventional models at 0.25° × 0.25° resolutions for JJAS seasonal rainfall over Burdwan district in Verification

Year	CFSv2	CMC1	CMC2	GFDL	NASA	NCAR	EM	BLUE
2006	0.87	0.60	0.87	0.68	0.35	0.35	0.63	0.69
2007	0.47	0.16	0.65	0.56	0.12	0.12	0.61	−0.15
2008	0.73	0.64	0.70	0.52	0.16	0.16	0.71	0.47
2009	−0.04	−0.19	−0.22	−0.49	−0.47	−0.47	−0.35	0.12
2010	−0.50	0.10	−0.15	0.37	−0.11	−0.11	0.16	−0.21

Table 4.10 RMSE between observations and forecasts (SD) from member models and conventional models at 0.25° × 0.25° resolutions for JJAS seasonal rainfall over the Burdwan district in Verification

Year	CFSv2	CMC1	CMC2	GFDL	NASA	NCAR	EM	BLUE
2006	0.48	0.84	0.48	1.54	0.74	0.74	1.09	0.73
2007	1.79	3.30	2.06	3.30	2.76	2.76	2.55	3.16
2008	1.53	1.32	0.91	1.29	2.49	2.49	1.26	1.85
2009	0.70	1.99	1.76	2.11	1.45	1.45	1.31	1.28
2010	4.04	0.94	2.57	2.29	3.93	3.93	3.29	2.19

Table 4.11 MAE between observations and forecasts (SD) from member models and conventional models at 0.25° × 0.25° resolutions for JJAS seasonal rainfall over Burdwan district in Verification

Year	CFSv2	CMC1	CMC2	GFDL	NASA	NCAR	EM	BLUE
2006	0.39	0.68	0.39	1.45	0.65	0.65	0.94	0.63
2007	1.71	3.25	2.01	3.27	2.69	2.69	2.50	3.09
2008	1.44	1.26	0.82	1.20	2.43	2.43	1.18	1.80
2009	0.56	1.90	1.55	1.99	1.34	1.34	1.12	1.21
2010	4.01	0.86	2.55	2.26	3.89	3.89	3.26	2.15

Table 4.12 Spatial Correlation Coefficients (r) between observations and forecasts (SD) from member models and conventional models at 0.25° × 0.25° resolutions for JJAS seasonal rainfall over the Murshidabad district in Verification

Year	CFSv2	CMC1	CMC2	GFDL	NASA	NCAR	EM	BLUE
2006	0.87	0.93	0.87	0.91	0.92	0.92	0.96	0.84
2007	0.79	0.93	0.89	0.91	0.84	0.89	0.88	0.91
2008	0.91	0.89	0.93	0.92	0.89	0.86	0.93	0.81
2009	−0.10	−0.16	−0.31	0.03	−0.28	−0.29	−0.52	0.26
2010	0.74	0.85	0.85	0.94	0.85	0.91	0.86	0.90

Table 4.13 RMSE between observations and forecasts (SD) from member models and conventional models at 0.25° × 0.25° resolutions for JJAS seasonal rainfall over the Murshidabad district in Verification

Year	CFSv2	CMC1	CMC2	GFDL	NASA	NCAR	EM	BLUE
2006	1.47	1.77	1.47	2.60	1.15	1.91	2.11	1.20
2007	0.54	1.91	0.77	2.22	1.15	0.96	0.78	1.51
2008	0.64	0.34	0.51	0.28	1.13	1.25	0.48	0.49
2009	1.30	2.48	2.20	3.03	2.36	1.27	1.50	2.26
2010	5.43	1.70	3.63	2.33	5.39	3.74	4.78	3.01

Table 4.14 MAE between observations and forecasts (SD) from member models and conventional models at 0.25° × 0.25° resolutions for JJAS seasonal rainfall over the Murshidabad district in Verification

Year	CFSv2	CMC1	CMC2	GFDL	NASA	NCAR	EM	BLUE
2006	1.40	1.76	1.40	2.58	1.11	1.81	2.10	1.12
2007	0.48	1.89	0.67	2.18	1.05	0.86	0.67	1.46
2008	0.51	0.29	0.37	0.21	1.01	1.04	0.44	0.40
2009	1.03	2.31	1.88	2.94	2.10	1.15	1.26	2.10
2010	5.39	1.60	3.58	2.29	5.31	3.72	4.69	2.99

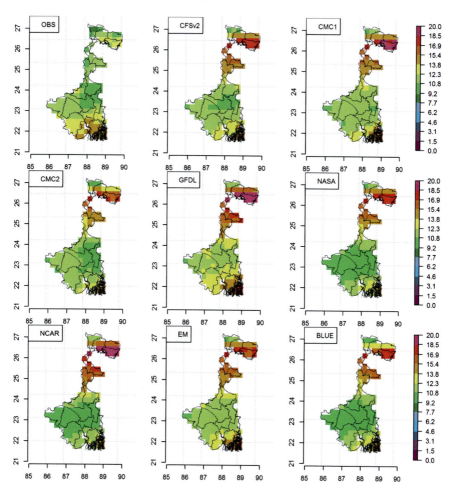

Fig. 4.2 The observed (PERSIANN) and forecasted (SD) rainfall (mm/day) from CFSv2, CMC2, CMC1, NASA, GFDL, EM and BLUE models for JJAS seasonal forecasts after four-month moving average valid for 2006

downscaled models while direct models provides forecasts rainfall with spatial correlation of 0.20, −0.54, −0.66, −0.20, 0.63, 0.26, −0.37, and −0.10 which is highest among all. The forecasts of the entire test period were examined in this manner. The improvement is very similar in skills for the models. It may also be noted that the skills from CFSv2, CMC1 are little higher than any other models.

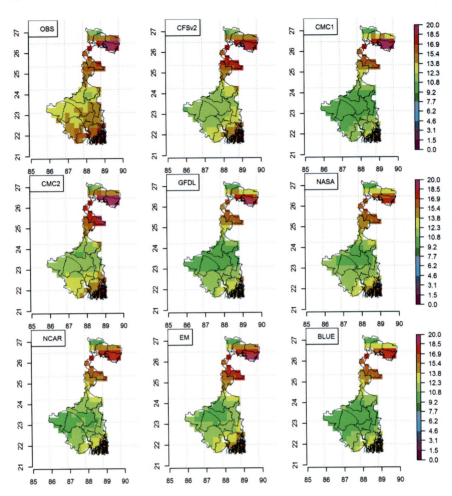

Fig. 4.3 The observed (PERSIANN) and forecasted (SD) rainfall (mm/day) from CFSv2, CMC2, CMC1, NASA, GFDL,EM, and BLUE models for JJAS seasonal forecasts after four month moving average valid for 2007

4.4.3 Analysis of the Forecasted Rainfall Fields

Based on the six primary models and two conventional methods described in the methodology, the results are calibrated and verified using the observed and simulated rainfall values from the 15-year period data sets (1996–2010). Figures 4.2, 4.3, 4.4, 4.5 and 4.6 shows the spatial distribution of seasonal (JJAS) rainfall forecasted from the 15-year period (four month moving averages) for both observed and the model predictions. Figures 4.2, 4.3, 4.4, 4.5 and 4.6 depicts that CFSv2 measured rainfall forecasts values are closer to the observed rainfall as compared to rest.

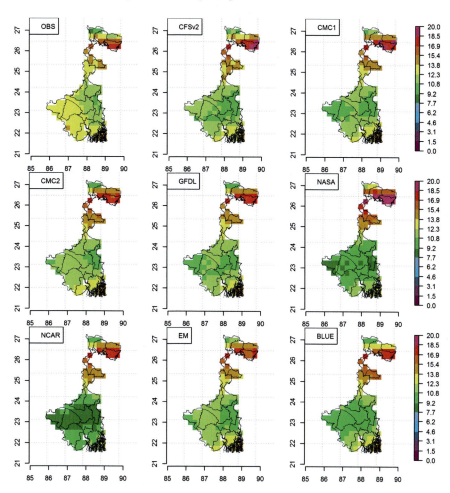

Fig. 4.4 The observed (PERSIANN) and forecasted (SD) rainfall (mm/day) from CFSv2, CMC2, CMC1, NASA, GFDL, EM, and BLUE models for JJAS seasonal forecasts after four-month moving average valid for 2008

4.5 Summary of the Results for WB

The highest skill in rainfall forecast is observed at the shortest lead time (one month) and declines rapidly thereafter. For seasonal rainfall forecasts over the districts of WB the skill of the CFSv2 model is better than any other models. The larger value of r and lower value of RMSE, MAE are observed in CFSv2 than any other method.

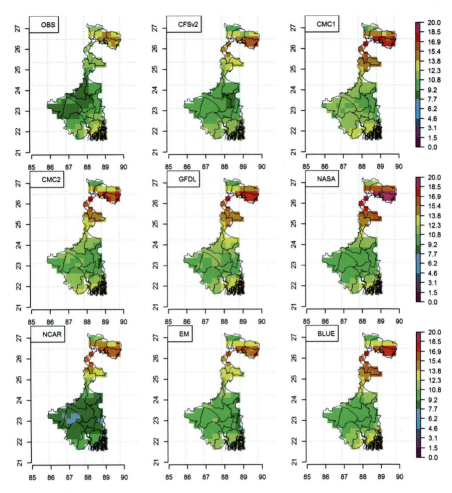

Fig. 4.5 The observed (PERSIANN) and forecasted (SD) rainfall (mm/day) from CFSv2, CMC2, CMC1, NASA, GFDL, EM, and BLUE models for JJAS seasonal forecasts after four-month moving average valid for 2009

4.6 Verification of Models for WB Rainfall Forecast

In the present study, 15 years of data are used to develop the seasonal rainfall forecasts of GCMs over WB region. However, verification has been undertaken for the years (2006–2010) for southwest monsoon season to examine the performance of the best performing model among primary models and conventional models in operational situations after statistical downscaling. Hence, the calibration is done first using the forecasted and observed data from the 10 years mentioned earlier. Then, they are applied to the forecasts result of test years 2006–10. It can be seen from the graph (Figs. 4.2, 4.3, 4.4, 4.5 and 4.6) that the predicted values of the rainfall from

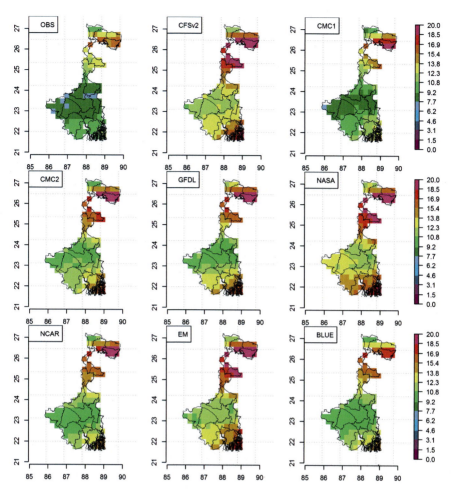

Fig. 4.6 The observed (PERSIANN) and forecasted (SD) rainfall (mm/day) from CFSv2, CMC2, CMC1, NASA, GFDL, EM, and BLUE models for JJAS seasonal forecasts after four-month moving average valid for 2010

the CFSv2 are very similar to the observed rainfall for most of the region over WB. But the CFSv2 has wet biases for some parts of the district of Malda, Cooch Behar, Uttar, and Dakshin Dinajpur. The r, RMSE, MAE are considered as statistical measures. Though the CFSv2 forecasts with skill r that ranges from 0.65 to 0.73 for verification period (downscaled). The improvement is observed as a skill over the years compared to member models. The model verification results for the years 2006–2010 demonstrated that the CFSv2 model may provide improved seasonal rainfall forecasts in the operational situation as well.

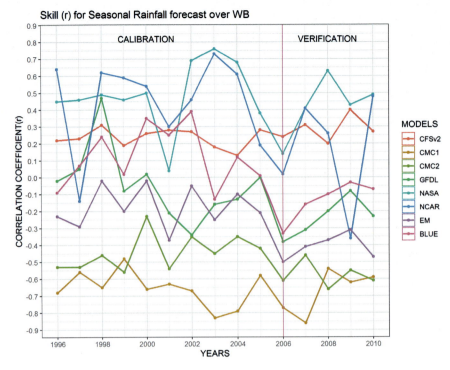

Fig. 4.7 Time series of the spatial Correlation Coefficients (*r*) between observations and forecasts for JJAS seasonal rainfall over WB from six primary models, EM and BLUE

To study in more detail, the performance of models on seasonal rainfall forecasts over WB region, the model was separately tested for districts of WB as shown in Tables 4.6, 4.7, 4.8, 4.9, 4.10, 4.11, 4.12, 4.13 and 4.14.

The BLUE model validations for the four selected monsoonal belt regions are performed for the years 2011–2014. It is noticed that the CFSv2 model outperformed for the Birbhum, Burdwan and Murshidabad districts of WB. The statistical values also represent similar indications in terms of r, RMSE, NSE, and MAE as given in Tables 4.2, 4.3, 4.4, 4.5, 4.6, 4.7, 4.8, 4.9, 4.10, 4.11, 4.12, 4.13 and 4.14. It is noticed that higher r and NSE values are obtained for the all districts (Tables 4.6, 4.7, 4.8, 4.9, 4.10, 4.11, 4.12, 4.13 and 4.41). This analysis shows that the CFSv2 model is performing well for WB including the districts level of WB. It may be inferred that CFSv2 model can be recommended for forecasting seasonal rainfall for the WB summer monsoon season.

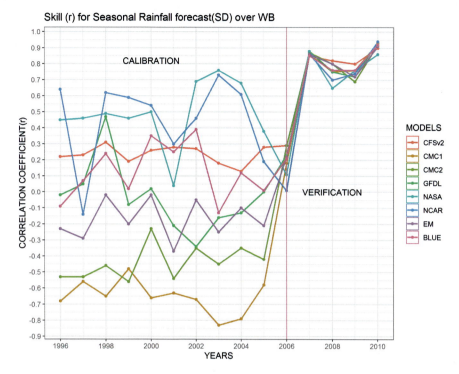

Fig. 4.8 Time series of the spatial Correlation Coefficients (*r*) between observations and forecasts (SD) for JJAS seasonal rainfall over WB from six primary models, EM and BLUE

4.7 Conclusions and Further Work Possibilities

The seasonal rainfall has been considered in this study from 15 years of data from selected models and observed values. On investigating the methods, it can be said that CFSv2 provides the most promising forecasts for WB region and districts of WB. It was also studied that the performance of multi-model weighted ensemble models to predict the ISMR forecasts. The spatial correlation coefficient (*r*), RMSE, MAE is used to measure forecast accuracy. The forecast skills of the said methods are investigated and discussed here. Out of eight undertaken models, CFSv2 provides the best forecasts in case of ISMR predictions. In future study, an effort will be to reduce the wet biases in CFSv2 over some parts of the North Bengal region.

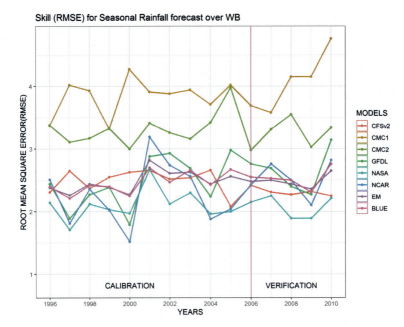

Fig. 4.9 Time series of RMSE between observations and forecasts for JJAS seasonal rainfall over WB from six primary models, EM and BLUE

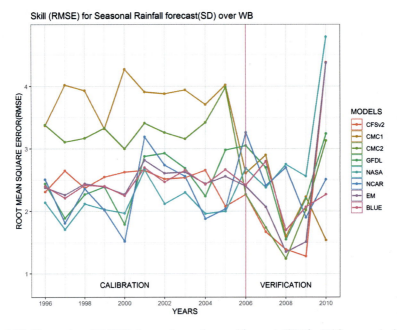

Fig. 4.10 Time series of RMSE between observations and forecasts (SD) for JJAS seasonal rainfall over WB from six primary models, EM and BLUE

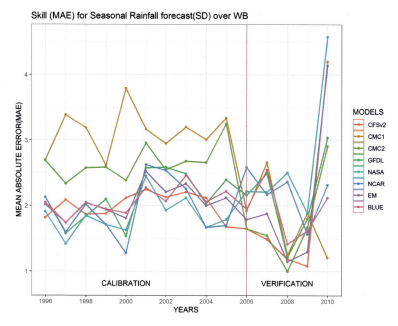

Fig. 4.11 Time series of MAE between observations and forecasts (SD) for JJAS seasonal rainfall over WB from six primary models, EM and BLUE

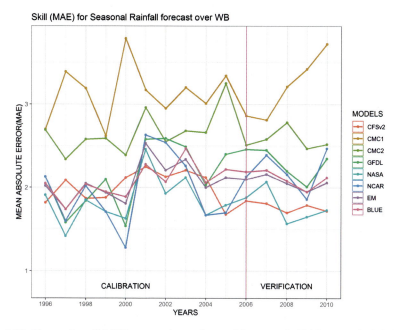

Fig. 4.12 Time series of MAE between observations and forecasts for JJAS seasonal rainfall over WB from six primary models, EM and BLUE

Conflicts There is no conflict.

References

Abatzoglou JT, Brown TJ (2012) A comparison of statistical downscaling methods suited for wildfire applications. Int J Clima 1–32:772–780

Bauer P et al (2015) The quiet revolution of numerical weather prediction. Nature 525:47–55

Becker E et al (2014) Predictability and forecast skill in NMME. J Clim 27(15):5891–5906

Chakrabarty A, Krishnamurti TN (2006) Improved seasonal climate forecasts of south Asian summer monsoon using a suite of 13 coupled ocean-atmospheric models. Monthly Weather Rev 134:1697–1721

Cohen SJ (1990) Bringing the global warming issue closer to home: the challenge of regional impact studies. Bull Am Metrorol Soc 71:520–526

Delworth TL et al (2006) Geophysical Fluid Dynamics Laboratory's global coupled climate models: formulation and simulation characteristics. J Climatol 19:643–674

Diez E et al (2005) Statistical and dynamical downscaling of precipitation over Spain from DEMETER seasonal forecasts. Tellus 57:409–423

Du J (2007) Uncertainty and ensemble forecasting. NOAA

Fowler HJ, Blenkinsop S, Tebaldi C (2007) Linking climate change modelling to impacts studies recent advances in downscaling techniques for hydro mode. Int J Climato 27:1547–1578

Hidalgo HG et al (2008) Downscaling with constructed analogues: daily precipitation and temperature fields over the United States. California Clim Chan Center Rep Ser Num 2007-027

Kirtman BP, Min D (2009) Multimodel ensemble enso prediction with CCSM and CFS. Monthly Weather Rev 137:2908–2930

Krishnamurti TN, Kishtawal CM, LaRow TE, Bchiochi DR, Zhang Z, Williford CE, Gadgil S, Surendran S (1999) Improved weather and seasonal climate forecasts from multimodel superensemble. Sci 285:1548–1550

Krishnamurti TN, Sagadevan AD, Chakraborty C, Mishra AK, Simon A (2009) Improving multi-model weather forecast of monsoon rain over China using FSU superensemble. Adv Atmospher Sci 26:813–839

Krishnamurti TN, Kishtawal CM, Shin DW, Williford CE (2000a) Improving tropical precipitation forecasts from a multianalysis superensemble. AMS: J Clim 13:4217–4227

Krishnamurti TN, Kishtawal CM, Shin DW, Williford CE (2000b) Multimodel ensemble forecasts for weather and seasonal climate. AMS: J Clim 13:4196–4216

Kumar V, Ghosh T (2018) Performance of multimodel schemes for seasonal precipitation over India region. Adv Meteorol Vol 2018, Article ID 5874270, pp 14. https://doi.org/10.1155/2018/5874270

Landman WA, Beraki A (2012) Multimodel forecast skill for mid summer rainfall over southern Africa. Int J Climato 32:303–314

Lenart M (2008) Downscaling techniques. The University of Arizona

Leung LR, Mearns LO, Giorgi F, Wilby RL (2003) Regional climate research. Bull Am Meteorol Soc 84:89–95

Leutbecher M, Lang STK (2014) On the reliability of ensemble variance in subspaces defined by singular vectors. Quart J Royal Meteorol Soc 140:1453–1466

Lorenz EN (1969) Atmospheric predictability as revealed by naturally occurring analogues. J Atmos Sci 26:636–646

Maurer EP et al (2010) The utility of daily large scale climate data in the assessment of climate change impacts on daily stremflow in California. Hydro and Earth Sys Sci 14(6):1125–1138

Maraun D, Wetterhall F, Ireson AM, Chandler RE, Kendon EJ, Widmann M (2010) Precipitation downscaling climate change: recent developments to bridge between dynamical models and end user. Rev Geophys 48

Merryfield WJ et al (2013) The Canadian seasonal to interannual prediction system models and initialization. Monthly Weather Rev 141:2910–2945

Molteni et al (1996) The ECMWF ensemble prediction system. Quart J Royal Meteorol Socie 122:73–119

Nieto JD, Wilbys RL (2005) A comparis of statisti downaling and climate change factor method impacts on lowflows in the River Thames, United Kingdoom. Clim Change 69:245–268

Najafi MR, Moradkhani H, Wherry SA (2011) Downscaling of precipitation using machine learning with optimal predictor selection. J Hydrol Eng 16

Ndiaye O et al (2011) Predicta of seasonal sahel rainfall using GCMs and lead time improve through the use of a coupled model. J Clima 24:1931–1949

Qi LB et al (2014) Selective ensemble-mean technique for tropical cyclone track forecast by using ensemble forecasting system. Quar J Royal Meteorol Socie 140:805–813

Saha S et al (2014) The NCEP climate forecast system version 2. J Climatol 27:2185–2208

Salathe EP (2007) Comparson of various precip downscalig methods for the simulation of streamflow in a rainshadow river basin. Int J Clim 23:887–901

Slater J et al (2016) Evaluation of the skill of North–American Multi-model ensemble (NMME) Global climate model in predicting average and extreme precipitation and temperature over the continental USA. Climate dynamics

Smith LA et al (2014) Probabilistic skill in ensemble seasonal forecasts. Quar J Roy Meteorol Socie 141:1085–1100

Tebaldi C, Knutti R (2007) The use of multi-model ensemble in probabilistic climate projection. Philosophic Trans Roya lSociet a: Math Physica Eng Sci 365:2053–2075

Timbal B et al (2003) An estimate of future climate change for western France using a statistical downscaling technique. Clim Dyn 20:807–823

Tian D, Martinez CJ (2012) Forecast reference evapotranpiration using retrospective forecast analogs in the southastern United States. J Hydrometeoro 13:1874–1892

Toth Z, Kalnay E (1993) Ensemble forecasting at NMC the generation of perturbations. Bull America Meteorol Socie 74:2317–2330

Toth Z, Kalnay E (1997) Ensemble forecasting at NCEP and breeding method. Monthly Weather Rev 125:3297–3319

Van den dool H et al (2003) Performance and analysis of the constructed analogue method applied to US soil moisture over 1981-2001. J Geophys Res Vol-108 (D16)

Vernieres G et al (2012) The GEOS-iODAS, description and evaluation. NASA technical report series

Wilks DS (2011) Statistical methods in atmospheric sciences. Proc Int Geophys Ser 100:676

Wilby RL et al (1998) Model low frequency rainfall events using airflow indices weather patterns and frontal frequencies. Water Resour Res 213(1-4):380–392

Wood AW, Leung LR, Sridhar V and Lettenmaier DP (2004) Hydrolo implicati of dynamic and statistic approaches to downscaling climate model outputs. Climat Chan 62:189–216

Xie PP, Arkin A (1996) Analyses of global monthly precipitation using gauge observations, satellite estimates and numerical model predictions. J Clim 9:840–856

Yoon et al (2012) Comparison of dynamically and statistically downscaled seasonal climate forecasts for the cold season over the United States. J Geophys Res 117 D21109

Zhang Z, Krishnamurti TN (1997) Ensemble forecasting of hurricane tracks. Bull Americ Meteorol Soc 78:2785–9275

Zhou BB, Du J (2010) Fog prediction from a multimodel mesoscale ensemble prediction system. Weather Forecast 25:303–322

Zhu JS et al (2013) A regional ensemble forecast system for stratiform precipitation events in the northern China region. Part II: Seasonal evaluation for summer 2010. Adv Atmos Sci 30:15–28

Chapter 5
Identification of Climate Change Vulnerable Zones in Bangladesh Through Multivariate Geospatial Analysis

Md. Golam Azam and **Md. Mujibor Rahman**

Abstract Being geographically susceptible, Bangladesh poses an exorbitant number of risks to the anticipated impacts of climatic fluctuations and extreme events. Therefore, a detailed study of climate change vulnerability (CV) covering the whole country is imperative to facilitate proper measures toward adaptation. To demonstrate spatial vulnerability to climate change in the Geospatial environment, a total of 42 indicators, 12 from the biophysical category and 30 from the socioeconomic category, have been incorporated with the IPCC framework through a Geographic Information System (GIS) raster database. For unbiased weights, Principal Component Analysis (PCA) of normalized raster has been accomplished in ArcGIS. The overall vulnerability of the country has been indexed from exposure, sensitivity, and adaptive capacity. The coastal region, part of the hilly region, riverine areas, and the haor basin are found highly vulnerable since these regions are more exposed as well as highly sensitive to climate change effects. The current study is a new contribution from Bangladesh in climate vulnerability indexing since it incorporates multivariate geospatial analysis to quantify and visualize CV countrywide. Findings from this work will be an important foundation in taking appropriate measures to mitigation and adaptation of climate change impacts, from local level measures to policymaking stages.

Keywords Climate change vulnerability · Bangladesh · Geographic information system · Multivariate spatial analysis · Vulnerability mapping

5.1 Introduction

In the present century, climate change is allegedly the greatest risk to our planet Earth affecting almost all parts of living systems including weather systems, hydrology, ecology, and the environment (Rahman and Lateh 2017). Almost every least developed country, where livelihoods are mostly natural resource-dependent, is readily at

Md. Golam Azam · Md. Mujibor Rahman (✉)
Environmental Science Discipline, Khulna University, Khulna 9208, Bangladesh

© The Author(s), under exclusive license to Springer Nature Singapore Pte Ltd. 2022 89
N. C. Jana and R. B. Singh (eds.), *Climate, Environment and Disaster in Developing Countries*, Advances in Geographical and Environmental Sciences, https://doi.org/10.1007/978-981-16-6966-8_5

risk to the negative impacts of forthcoming events credited to climate change (Heltberg and Bonch-Osmolovskiy 2011). Society and its interaction with the climate affect the climate change impact along with the biophysical characteristics of a certain area. According to the Second Assessment Report, Socioeconomic systems are more vulnerable in developing countries as the economic and institutional circumstances are not strong enough (IPCC, Climate Change 1995). The report also describes that vulnerability is highest where sensitivity is high and adaptive capacity is low. Further, in its Fourth Assessment Report (AR4), the IPCC explains vulnerability as the intensity of a system's susceptibility to, or inability to adjust, negative results of climatic change, while change includes variability in climatic parameters and recurring extreme events. After the AR4 of IPCC, researchers have focused mainly on how to adapt and mitigate climate change effects resulting from the introduction of human welfare-based vulnerability study of a certain area (Ibarraran et al. 2008). Bangladesh is in the cluster of extremely vulnerable countries to face dangerous effects from change and variability of the climate. Extreme events from climatic fluctuations have become typical in Bangladesh (Ahmed et al. 2013; DoE 2012). With 147,570 km^2 of the area (BBS 2011) and a quick-expanding population now she is at over 160 million (DoE 2012; Dewan et al. 2012) and anticipated to be added by another 20 million by 2025 (Shaw 2015). The density of the population of Bangladesh is 1015 per km^2 with a growth of 1.37% annually (Shaw 2015). By 2025, population density could be over 1200 per km^2 (Shaw 2015). Bangladesh, especially the coastal region next to the Bay of Bengal, is highly susceptible to extreme climatic phenomena, with changes of irregular disasters namely storm surges from cyclone, tidal floods, riverbank erosion, and increasing trend of sea-level rise, saline water intrusion, and many more natural calamities (Dasgupta et al. 2014).

Some scholarly works have been conducted in different places, using the vulnerability index (Cinner et al. 2013; Bjarnadottir et al. 2011), to measure the climate change vulnerability. Some studies concentrating on the socioeconomic aspect of the vulnerability index have also been performed before in Bangladesh as well as other countries (Ahsan and Warner 2014; Hagenlocher et al. 2013; Chen et al. 2013; Yoon 2012), mainly through household surveys. A study found the southern region especially the southeastern unions relatively more vulnerable (Ahsan and Warner 2014). Climate change vulnerability evolved from the quality of potable water and its management has been measured in rural zones of the coastal belt of Bangladesh (Delpla et al. 2014; Sarkar and Vogt 2015). The water vulnerability concerning global climate change provides the community an inclusive overview of the lack of water availability to identify the specific areas, groups, and sectors where attention is needed (Plummer et al. 2013) since the adequate safe drinking water supply is essential for economic, social, and sanitary reasons (Pagano et al. 2014). Contrarily, infrastructure-related vulnerability includes detail of a range of geographical, physical, functioning, and socioeconomic features (Grubesic and Matisziw 2013; Tang et al. 2013). This vulnerability includes the economic value of infrastructure, and residential and commercial building values (Thatcher et al. 2013).

Vulnerability is the factor for risk level assessment as well as building resilience (Salinger et al. 2005). Another way it is understood is how much a system is either susceptible or incapable to stand with the bad results of climate change, with climate variability as well as extreme events (IPCC, Summary for Policymakers 2014; IPCC 2001). The vulnerability can be demarcated as the capacity or incapacity of persons or communities to respond to, in terms of coping with, recover from, or adjust to, any external pressure imposed on their incomes and well-being (Kelly and Adger 2000). On the other hand, Barker et al. (2007) define vulnerability in a different concept, where the nature, intensity, and degree of climate fluctuation that determine a system's exposedness, responsiveness, and adaptability. Understanding of vulnerability has been described both hypothetically and practically through conceptual models, frameworks, and assessment techniques (Cutter et al. 2008). Adger (2006) conceptualizes vulnerability as a function of adaptive capacity, exposure, and sensitivity as per the definition of IPCC. The degree of climate stress upon a particular unit of analysis can be termed as exposure (Comer et al. 2012). Further, exposure can be defined as the experiences of disturbances in the internal and external system (Abson et al. 2012). The reaction of an environmental setting to climate hazards is called sensitivity (Preston and Stafford-Smith 2009). Sensitivity is variable as it depends on location, sectors, and population. According to Gallopin (Gallopín 2003), sensitivity is understood as the changing intensity of a system is or how the system is affected impacted by an external or internal disturbance or several disturbances. Both socioeconomic and ecological conditions outline the responsiveness of any system to climatic influences as well as determine the degree of a group's affectedness by environmental stress (SEI 2004). According to IPCC (IPCC 2001), the magnitude a system is usually affected either positively or negatively by climate stress is sensitivity. On the contrary, the term sensitivity is understood as the degree of experiencing positive or negative consequences by a system as a result of a direct or indirect climatic phenomenon (IPCC 2001). Adaptive capacity on the other hand is the capacity to overcome exposure and sensitivity to climatic influences (Preston and Stafford-Smith 2009). Capacity is usually determined by resource availability as well as by the institutional and governance networks that exist to deploy those resources, whereas socio-political barriers may inhibit successful adaptation to Climate change impacts (Deressa et al. 2009). A significant factor in characterizing vulnerability is adaptive capacity. According to Brooks (2003), the ability to modify social characteristics or behavior to better cope with current or predicted external stresses, and also shifts in external circumstances is adaptive capacity. IPCC (IPCC 2001) describes the adaptive capacity of a system, region, or community as its potential or ability to adapt to the effects or impacts of climate change (including climate extremes and variability). The capability of a system to manage extreme climate variability and to lessen the potential social and environmental costs is termed adaptive capacity (IPCC 2001; Adger 2006; Brooks 2003; Burton et al. 2002; Gallopín 2006; Yohe and Tol 2002; Gerlitz et al. 2014). Nonetheless, this phenomenon is context-specific and varies spatially and temporally.

Though assessment vulnerability is not a new initiative, still it appears in the science and policy application of climate very often (Füssel and Klein 2006) which is the starting point in decreasing the impact of the future extreme climate on the socio-ecological system (Adger 2006; Howden et al. 2007). Risk characterization can be done very easily by assessing vulnerability which makes the foundation for identifying measures to cope up with the negative impacts of climate change. This work will help stakeholders as well as policymakers to mark the most vulnerable locations in Bangladesh to climate change.

5.2 Materials and Methods

5.2.1 Study Area

The spatial extent of Bangladesh, the study area, is between 20° 34′ N to 26° 38′ N and 88° 01′ E to 92° 41′ E (Fig. 5.1) with an area of 144,000 km². The characteristic climate of Bangladesh is subtropical humid with extensive seasonal differences in precipitation, moderately warm temperature, and high humidity. Bangladesh has broadly three seasonal categories; dry winter/post-monsoon season from November to February, pre-monsoon hot summer season from March to May, and rainy monsoon season from June to October (Rashid 1991).

The historic average temperature of Bangladesh is 25.75 °C, monthly average ranging from 18.85 to 28.75 °C; the minimum temperature is 21.18 °C monthly average ranging from 12.5 to 25.7 °C; and maximum temperatures is 30.33 °C, monthly average ranging from 25.2 to 33.2 °C (BMD 2013). January is the coldest and April and May are the hottest months in Bangladesh. The historical average precipitation of the country is 2428 mm per year and the rainfall is very much seasonal in Bangladesh, which varies from 1400 to 4400 mm (BMD 2013). June, July, and August is the wettest season in Bangladesh. More than 75% of the entire precipitation in Bangladesh happens during the monsoon season, triggered by winds gusting from the Southern Hemisphere between mid-May and September, which accumulates moisture and deposits copious amounts of precipitation over the South Asian continent. With respect to global warming and climate change, this country is considered among the most vulnerable nations in the world due to its least capacity to address the devastating impacts (IPCC, Climate Change 2013). Recently, Bangladesh is experiencing higher temperatures, more variability in rainfall, more extreme weather events, and sea-level rise. Bangladesh is highly at risk, since it is low-lying, situated on the Bay of Bengal in the Ganges–Brahmaputra-Meghna delta, and also densely populated.

5.2.2 Conceptual Framework for Climate Change Vulnerability

The concept of vulnerability is used to depict the complex relationship of climate change impacts and the weakness of a system to its results. There exist manifold

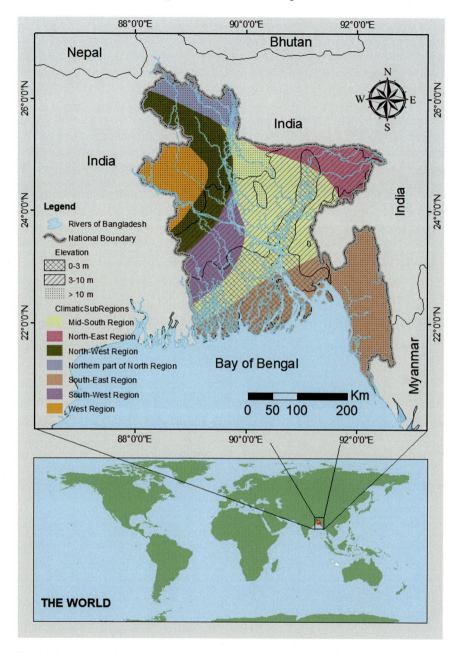

Fig. 5.1 Location of the study area, Bangladesh, in the context of the World along with climatic sub-regions and elevation of the country. Recreated from the climatic sub-region map of Rashid (1991) & USGS's elevation data

definitions and methods of operationalizing this concept. The Intergovernmental Panel on Climate Change (IPCC) sought to elaborate and advance an approach for understanding vulnerability in its AR4 as the intensity of a system's susceptibility and inability to adapt to the dangerous effects of climate change which comprises climate variability and climate-related extreme events (IPCC, Climate Change 2007). IPCC also portrayed vulnerability with character, magnitude, and rate of variation and change in the climatic factors that control a system's exposure, sensitivity, and adaptive capacity (IPCC, Climate Change 2007). There are three conceptual approaches for assessment of vulnerability Socioeconomic approach: focusing on socioeconomic and political variations within the society, but not environmental factors; Biophysical approach: focusing on physical damage done by environmental factors on the social and biological systems; Integrated assessment approach: combines both socioeconomic and biophysical approaches (Deressa et al. 2008). This study adopts the integrated assessment approach and uses the indicator method to assess the vulnerability of Bangladesh. With the increase in exposure, i.e., increase in the change in temperature and rainfall and also increase in the occurrence of natural hazards the people will be more vulnerable to climate change, especially farmers as their livelihood depends on it. Sensitivity increases the effect of exposure on the people and will have a more negative impact on them. Sensitivity will include factors like casualties and damage caused by natural hazards as well as human and environmental factors that make them more susceptible to natural hazards and climate variability. The combined effect of exposure and sensitivity will increase the vulnerability while adaptive capacity will decrease it.

The United Nations Framework Convention on Climate Change (UNFCCC 2008) mentions two types of vulnerability assessment framework: impacts (top-down) and adaptation (bottom-up). Impacts frameworks are also referred to as "first generation." They were mainly designed to understand the potential long-term impacts of climate change. The main elements of the Impacts framework are the baseline socioeconomic and environmental scenarios, climate change scenarios. Biophysical impacts (sensitivity) are assessed based on them, thus vulnerability can be estimated. After that climate adaptation policy can be examined (UNFCCC 2008). On the other hand, the adaptation frameworks also referred to as "second generation," which have developed in recent years, focus on involving stakeholders and addressing adaptation. The framework contains technical papers, engaging stakeholders, assessing vulnerability, assessing current and future climate risks, assessing changing socioeconomic conditions based on indicators, then assessing adaptive capacity. Finally formulating a climate adaptation strategy, and continuing the adaptation process. As different frameworks have different strengths, the adaptation framework emphasized stakeholders' involvement more than others did and therefore it played a significant role to select this framework.

5.2.3 Indicator Selection and Data Collection

Many studies have been conducted using social, economic, or biophysical indicators to assess vulnerability (Liu et al. 2008; Metzger and Schröter 2006; Stelzenmüller et al. 2010). However, for the present study, 30 socioeconomic indicators have been considered for this study, which have been obtained from the Bangladesh Bureau of Statistics (BBS 2011, 2015; BBS, District Statistics 2011). All socioeconomic data were incorporated into the GIS database in order to generate maps. A brief description of the selected indicators is given in Table 5.1. The 12 biophysical indicators for the present study have also been classified based on the gathered information from various sources and literature. A brief description of the biophysical indicators is provided in Table 5.1. The coefficient of temperature and precipitation variability has been extracted from the work of the Institute of Water and Flood Management (IWFM 2014) and mapped according to the climatic sub-regions produced by Rashid (Rashid 1991). A five-class map depicting the drought of the entire country has been reproduced from the Comprehensive Disaster Management Program (CDMP 2006). The cyclone risk map involved in this work, a four-class comparative map, has been accepted by the Center for Environment and Geographic Information Services (CEGIS 2006). The sea-level risk map has been reproduced from the digital elevation map acquired from the United States Geological Survey (USGS). Different types of flood risk maps have been reproduced from the maps of the Bangladesh Agricultural Research Council (BARC 2001) and the Bangladesh Water Development Board (BWDB, Annual Flood Report 2010). Erosion-prone areas with relative risks (BWDB, Annual Flood Report 2010) and salinity intrusion map of 1–5 ppt salinity line (SRDI 2010) also have been recreated in this study. Finally, a general hazard class map covering all over the country, with a 1–5 relative hazard proneness, has been adopted from Bangladesh Center for Advanced Studies (BCAS 2008).

5.2.4 Normalization of the Indicators

Some variables have a higher range of variance and some of them have a lower range of variance. Therefore, normalization is imperative for multivariate statistical operations, since this technique involves the transforming of a dataset containing all variables into a specific range (0–1). Normalization also creates a robust relationship among the dataset and removes the influence of one variable to another (Quackenbush 2002). Similar methods have been introduced in developing the human development index as well asthe life expectancy index earlier (Coulibaly et al. 2015; Piya et al. 2012; UNDP 2007). However, Eq. (5.1) has been followed to normalize the dataset using the raster calculator of ArcMap.

$$X_d = \frac{X_d - X_{min}}{X_{max} - X_{min}} \qquad (5.1)$$

Table 5.1 Selected indicators from different sectors of vulnerability

No.	Indicators	Units	Data sources	Vulnerability sectors	Vulnerability components
1	Literacy rate	Percent of people	District statistics BBS, District Statistics (2011)	Social	Adaptive capacity
2	Dependency ratio	Percent of people			
3	Irrigation	Percent of agricultural land covered		Infrastructural	
4	School	No. of Govt. primary school per 1000 people			
5	Shelter	No. of Cyclone and/or flood shelter per 1000 people			
6	Roads	Km of road per 1000 people			
7	Health institutes	No per 1000 people			
8	Electricity	Percent of HHs with connection	National report BBS (2011)		
9	Tube well	Percent of HHs with tube well			
10	Drinking water source	Percent of HHs with drinking water source within 200 m			
11	Away population	Per 1000 people		Economic	
12	Household	Total number of households			
13	Poverty	Percent of people below poverty line			
14	Radio	Per 1000 people		Information	
15	Television	Per 1000 people			
16	Agriculture dependency	Percent of HHs depending on farming	District statistics BBS, District Statistics (2011)	Ecological	Sensitivity
17	Fuelwood dependency	Percent of HHs using wood for cooking			
18	Disability	Percent of people		Human	

(continued)

Table 5.1 (continued)

No.	Indicators	Units	Data sources	Vulnerability sectors	Vulnerability components
19	Female HH head	Percent of HHs	National report BBS (2011)		
20	Population density	People per square km			
21	Injury in NH	No of people injured in 2009–2014	Bangladesh disaster-related statistics report BBS (2015)	Shocks to Natural Hazard	
22	Crop damage	Acre of cropland damaged in 2009–2014			
23	Household damage	No of HHs destroyed in 2009–2014			
24	Tornado affected HHs	No of HHs affected in 2009–2014			
25	Drought affected HHs	No of HHs affected in 2009–2014			
26	Storm affected HHs	No of HHs affected in 2009–2014			
27	Salinity affected HHs	No of HHs affected in 2009–2014			
28	Cyclone affected HHs	No of HHs affected in 2009–2014			
29	Flood affected HHs	No of HHs affected in 2009–2014			
30	Erosion affected HHs	No of HHs affected in 2009–2014			
31	Maximum temperature	Coefficient of change in 1960–2009	Report IWFM (2014)	Climatic variables	Exposure
32	Minimum temperature	Coefficient of change in 1960–2009			
33	Precipitation	Coefficient of change in 1960–2009			

(continued)

Table 5.1 (continued)

No.	Indicators	Units	Data sources	Vulnerability sectors	Vulnerability components
34	Drought	Relative risk map	Map (CDMP 2006)	Extreme events	
35	Hazard class	Relative risk map generalized hazards	Map (BCAS 2008)		
36	Tidal flood	Relative risk map	Map (BARC 2001)		
37	Sea-level rise	Coastal elevation (m)	DEM, USGS		
38	Cyclone	Relative risk map	Map (CEGIS 2006)		
39	Salinity intrusion	Relative risk of ppt of saline intrusion	Map (SRDI 2010)		
40	Flush flood	Relative risk map	Map (BARC 2001)		
41	River flood	Risk map based on inundation height	Map (BWDB, Annual Flood Report 2010)		
42	Erosion	Relative risk map	Map (BWDB, Annual Flood Report 2010)		

where,

X_d is raster value;
X_{max} is the highest value in a raster;
X_{min} is the lowest value in a raster.

For the normalization of raster datasets in ArcMap 10.5, a raster calculator was used which is a widely used tool under Map Algebra of the Arc toolbox. For each raster dataset, the following expression was used (Eq. 5.2);

$$\left("x" - "x" \cdot \text{minimum}\right)/\left("x" \cdot \text{maximum} - "x" \cdot \text{minimum}\right) \qquad (5.2)$$

where,

x = Raster name.

All of the created normalized raster data are then stored in a new database for further analysis.

5.2.5 Principal Component Analysis (PCA)

The principal component analysis is used by many scientific disciplines. PCA is a linear algebraic procedure that applies an orthogonal transformation in order to convert seemingly correlated variables into linearly uncorrelated altered variables. In a computer environment, this transformation is performed in such a manner that the first component has the highest possible variance, and all of the next components have the highest possible variance following the limit that it is orthogonal to the preceding components. Mathematically, the PCA depends on the Eigenvector-based multivariate analysis (Abdi and Williams 2010). This analysis can be performed by an Eigenvalue breakdown of a data covariance (or correlation) matrix or singular value breakdown of a data matrix. Outputs of a PCA are usually expressed in the form of component scores, also expressed as factor scores (the altered values of variable connected to a specific data point), and factor loadings (weights used to grow each normalized original variable to get the component score) (Wold et al. 1987).

5.2.6 Accumulation of Indicators to Assess Vulnerability

The study uses the integrated assessment approach using indicators, which is among the most common methods, to analyze vulnerability. According to IPCC Fourth Assessment.

Report vulnerability may be formulated as:

$$\text{Vulnerability} = \text{Exposure} + \text{Sensitivity} - \text{Adaptive Capacity} \qquad (5.3)$$

A higher adaptive capacity is associated with lower vulnerability while higher exposure and sensitivity are associated with higher vulnerability.

However, before indexing the overall vulnerability of the country, indicators from exposure, sensitivity, and adaptive capacity according to Table 5.1, have been accumulated after weighting. After the calculation of all three components through the weighted average, the vulnerability has been determined by following Eq. (5.3) in the raster calculator of ArcMap.

5.2.7 Classification of Raster

The cluster analysis encompasses a set of algorithms and methods to identify structures within the data as a homogeneous group of cases. Cluster analysis has been applied to a wide range of research problems (Anderberg 2014; Duran and Odell 2013; Tan et al. 2013). In this study, the Jenks natural break is performed for the classification of spatial vulnerability. The Jenks natural breaks Classification, also known as optimization, is a sort of data clustering technique intended to adjust the arrangement values into "natural" classes, the most optimal class range in a dataset. Values with similar features that form a "natural" cluster within a dataset are termed as a class or cluster range (Dent 1999). This clustering method tries to lessen the average distance from the cluster while increasing the distance from the means of the other clusters. In a word, this method decreases the variance within clusters increasing the variance between clusters (Jenks 1967). This phenomenon is termed as the goodness of variance fit (GVF), which equals the subtraction of SDCM (sum of squared deviations for class means) from SDAM (sum of squared deviations for array mean) (ESRI 2016; Slocum 1999).

5.3   Results and Discussion

Assessment of vulnerability, according to the IPCC framework, requires two multivariate statistical approaches before the aggregation of indicators. They are,

(i)   Normalization of indicators to reduce the heterogeneity of variance among indicators; and
(ii) Principal Component Analysis (PCA) to retain unbiased weights.

All of the 42 indicators in raster form have been analyzed in the raster calculator to retain multiband raster layers of each principal component (PCs). A text file has also been produced from which the PCA results have been visualized in charts.

On the other hand, the overall vulnerability assessment requires the accumulation of indicators based on the sectors occupied by the basic three components of vulnerability depicted in the AR4 of IPCC. Climatic variables and all extreme event risk maps have been accumulated through a weighted average to produce an exposure map of the country. Indicators of ecological, human, and natural hazard shocks are weighted and accumulated to sensitivity. All social, infrastructural, economic, and information indicators have been accumulated to produce the adaptive capacity map.

5.3.1   Retention of Unbiased Weights of Indicators

Principal component analysis has been conducted to reduce variable quantities into a smaller number of variables-called principal components (PCs) as well as to provide useful information of the original dataset as mentioned by Thompson (Thompson 2004) and Williams, et al. (Williams et al. 2012). A pairwise correlation matrix has been used in the PCA process visualizing the relations between respective variables to reduce the initial matrix to a subset of non-highly correlated metrics as described by Abson et al. (Abson et al. 2012). The Cattell scree test (Cattell 1966), in Fig. 5.3, and Kaiser's criterion with greater Eigenvalue (Kaiser 1960), in Fig. 5.4, have been performed to extract the number of principal components (PCs). In this study 6 principal components have been considered, tabulated in Table 5.2. Figures 5.3 and 5.4 are demonstrating the cumulative percentage of variance is 75% for 6 principal components (PCs) having greater Eigenvalues. Unbiased weights have been found for each indicator after carrying out PCA shown in Table 5.2 in which the heaviest Eigenvectors have been marked as shaded.

5.3.2 Climate Change Vulnerable Zones of Bangladesh

According to the IPCC's fourth assessment report (AR4), vulnerability has three components, exposure, sensitivity, and adaptive capacity. Here, all the three components have been indexed through raster analysis following Eq. (5.2) then Eq. (5.3). In the whole process of vulnerability assessment, the procedures demonstrated in Fig. 5.2 have been followed.

Exposure. Exposure is simply the climate variability or shift in climatic characters along with occurrences in climate-induced natural hazards. In Bangladesh, both of these sections of indicators are evident. The temperature and precipitation variability, mapped before according to their variability coefficient, is one of the major sectors of Exposure. On the other hand, biophysical maps namely cyclone risk map, salinity map, drought-prone map, flood-prone map, river, and coastal erosion map, etc. are accumulated to get the overall climate change exposure of the whole country.

The coastal region of Bangladesh is highly exposed to climate change since most of the extreme climatic events occur in this region including tropical cyclones, storm surges, sea-level rise, salinity intrusion, etc. (Fig. 5.5). Moreover, the variability in climatic indicators, especially maximum temperature, is also proved to be high in variability which also makes the coastal region of Bangladesh more exposed to climate change than any other region. Besides coastal regions, riverine areas all over the country are found highly exposed to climate change probably due to river erosion and flooding.

Most of the floodplain areas and depressed haor areas are found moderately exposed to climate change (Fig. 5.5). Since the lower elevation of the country always carries a significant level of flood proneness, the exposure level may be controlled by the elevation. However, the rest of the country especially the hilly areas, Barind tract, Madhupur tract, and plain areas are less exposed to climate change impacts and effects (Fig. 5.5).

Sensitivity. Sensitivity is the level of susceptibility of a region to climatic extreme events or variability-induced effects. Some socioeconomic indicators namely population density, disability percentage, literacy of the female population, etc. are considered to be controlling factors of climate change sensitivity. Other than these socioeconomic indicators, there are indicators familiar as hazard shock indicators, which can be a mingling of both biophysical and socioeconomic characters. Household damage, crop damage, injury in previous natural hazards are mainly three hazard shock indicators that are widely familiar. Moreover, in the present study, the number

Table 5.2 Selected PC layers with Eigenvectors. Shaded cells indicate the highest eigenvectors which are the weights for the input layers

Layer no	Input raster	PC layer 1	PC layer 2	PC layer 3	PC layer 4	PC layer 5	PC layer 6
1	Storm affected HHs	0.110	0.155	0.028	−0.135	0.079	0.049
2	Tornado affected HHs	0.107	−0.055	0.096	0.020	0.042	0.057
3	Disability	−0.005	0.040	−0.052	0.191	−0.056	0.121
4	Salinity affected HHs	0.051	0.980	0.006	−0.153	−0.046	0.032
5	Injury in NH	0.124	0.039	0.016	0.064	0.101	0.258
6	School	−0.104	0.082	−0.060	−0.129	0.248	−0.002
7	Shelter	0.072	0.167	−0.150	0.085	0.207	−0.030
8	Tube well	0.098	−0.098	0.208	0.069	0.135	0.196
9	Crop damage in NH	0.132	−0.017	0.032	−0.080	0.100	0.336
10	Cyclone affected HHs	0.107	0.152	0.042	−0.095	0.001	0.089
11	Roads	−0.101	0.123	−0.145	−0.139	0.146	0.037
12	Household damage in NH	0.071	0.121	0.085	−0.064	0.120	0.137
13	Irrigation	0.064	−0.157	−0.066	0.034	0.076	0.061
14	Flood affected HHs	0.057	−0.065	0.259	0.151	0.122	−0.011
15	Population density	−0.002	−0.024	−0.141	0.043	−0.022	−0.017
16	Fuelwood dependency	−0.113	−0.152	0.083	0.246	−0.018	−0.329
17	Away population	−0.005	−0.102	−0.134	0.062	−0.111	0.009
18	Dependency	0.042	0.050	0.143	0.184	0.170	−0.367
19	Household	0.035	−0.029	−0.183	0.071	0.037	0.005
20	Poverty	0.069	−0.037	−0.162	0.185	0.135	−0.034
21	Radio	−0.066	0.051	−0.054	−0.218	0.251	−0.085
22	Television	−0.018	−0.051	−0.256	0.009	0.182	−0.004
23	Drinking water source	0.087	−0.280	0.229	−0.167	0.046	0.172
24	Health institute	−0.023	−0.088	−0.143	−0.023	0.148	−0.014

(continued)

Table 5.2 (continued)

Layer no	Input raster	PC layer 1	PC layer 2	PC layer 3	PC layer 4	PC layer 5	PC layer 6
25	Erosion affected HHs	0.114	0.032	0.388	0.063	−0.087	0.101
26	Literacy rate	−0.203	0.081	−0.263	0.110	−0.167	0.061
27	Agriculture dependency	−0.063	−0.037	0.020	0.011	0.352	0.123
28	Electricity	−0.117	−0.070	−0.400	0.024	0.089	−0.033
29	Drought affected HHs	0.155	−0.067	0.103	0.072	0.104	0.008
30	Female HH head	0.090	0.012	−0.233	0.161	0.076	−0.259
31	Tidal flood	0.157	0.456	0.136	0.152	−0.405	0.090
32	Sea-level rise	0.263	0.342	−0.156	−0.182	0.054	−0.088
33	Drought	0.538	−0.073	−0.006	−0.098	−0.020	−0.076
34	Erosion	0.077	0.014	0.745	0.063	0.061	0.059
35	Flush flood	0.176	−0.009	0.244	0.101	−0.202	−0.047
36	Hazard class	0.196	0.279	0.013	0.151	0.179	0.079
37	Maximum temperature	0.400	−0.233	−0.046	0.353	−0.052	0.107
38	Minimum temperature	0.373	−0.210	0.129	−0.114	0.040	−0.161
39	Precipitation	0.388	−0.269	0.017	0.371	−0.004	0.199
40	River flood	0.164	−0.100	0.369	0.303	−0.356	−0.005
41	Cyclone	0.115	0.362	−0.075	−0.074	0.108	−0.005
42	Salinity intrusion	−0.161	0.228	−0.047	−0.172	0.019	−0.136

of households affected by previous natural hazards namely tropical cyclone, storm surge, flood, riverbank erosion, coastal erosion, tornado, drought, salinity intrusion, etc. are also been considered as sensitivity inducing indicators.

Most of the coastal region, part of the haor region, and the floodplain of the Brahmaputra are found highly sensitive to climate change (Fig. 5.6). Now, for the coastal region, there may be more records of previous damages to structures and loss of wealth and lives which make this region more sensitive to climate change. Floods and flash floods from the past have made the haor region highly sensitive. The Brahmaputra floodplain is characteristic of river flood, erosion, and meteorological drought. Moreover, tornado visits this region very frequently. Therefore, this region experienced much more damages and destruction from those above-mentioned hazards which make this region highly sensitive to climate change.

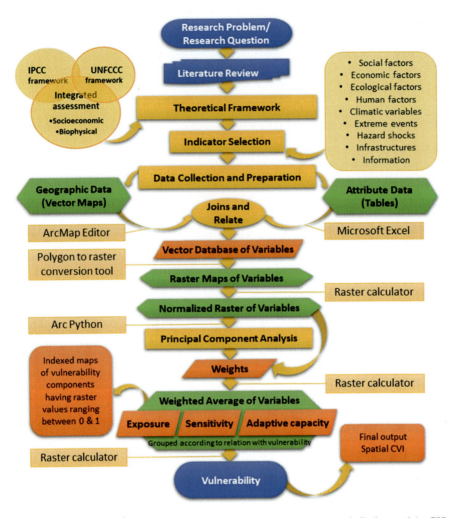

Fig. 5.2 Workflow diagram form a comprehensive vulnerability assessment in Python and ArcGIS

Mainly the hilly region and different small parts of the country are moderately sensitive while the northern and central parts of the country are found less sensitive to climate change impacts (Fig. 5.6).

Adaptive Capacity. Adaptive capacity to climate change depends on the infrastructural, information, human, economic, and ecological features of a country. Infrastructure includes roads, buildings for shelters, health institutes, irrigation facilities, drinking water facilities, etc. Information technology including radio and television access also contribute to enhancing the adaptive capacity. Economic and human indicators include away population, household number, poverty, female household head, etc. Nonetheless, the adaptive capacity is depicted to be the image of a country's

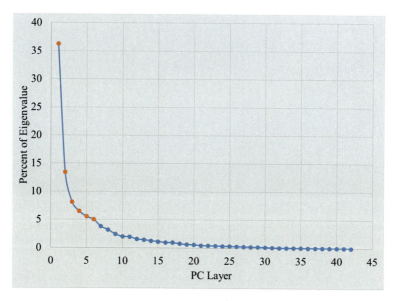

Fig. 5.3 Scree plot for the percent of the Eigenvalue of the dataset. The 7th component shows a different nature in the line which, according to Catell (1966), indicates that the first six PCs (orange dots) are responsible for most of the variability

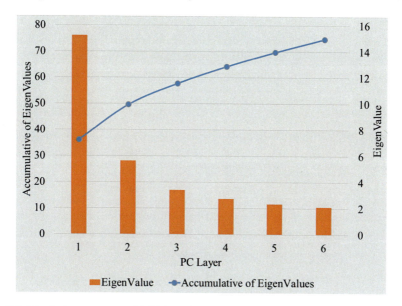

Fig. 5.4 Principal components with higher Eigenvalues, responsible for almost 75% of the total variability. The first six components are considered for weight and profile retention according to Kaiser's criterion (1960)

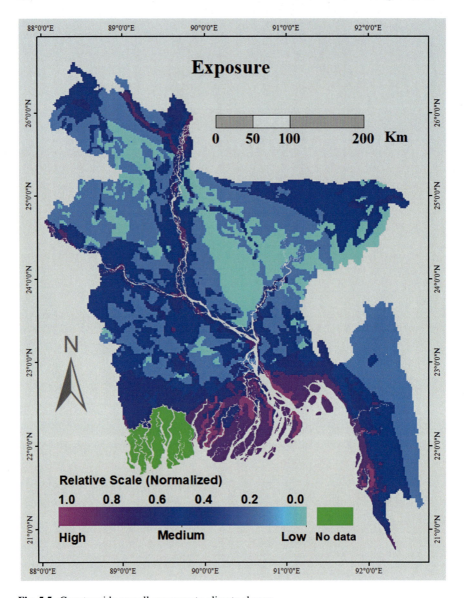

Fig. 5.5 Countrywide overall exposure to climate change

development scenario. As Bangladesh is rapidly developing the adaptive capacity of the country is also increasing. However, this study found most of the country is very much adaptive only haor region and the hilly region of Bandarbans is less adaptive (Fig. 5.7).

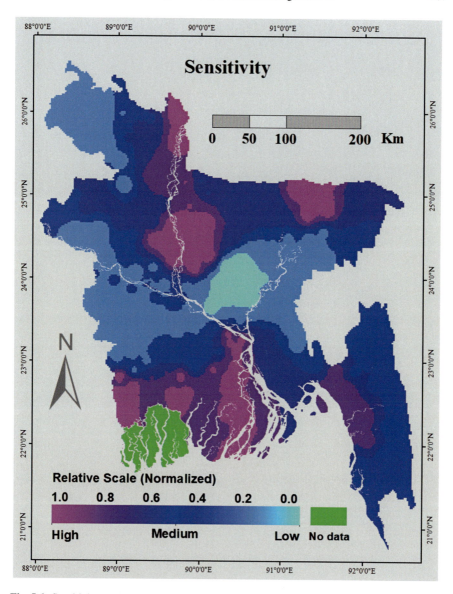

Fig. 5.6 Sensitivity to climate change of Bangladesh

Vulnerability. Vulnerability is the function of exposure, sensitivity, and adaptive capacity where adaptive capacity is subtracted from the aggregation of exposure and sensitivity.

All of the coastal regions, most of the hilly regions, riverine areas, and the haor basin are found highly vulnerable to climate change in this study (Fig. 5.8). The coastal region is characteristic of different extreme climatic events which make this

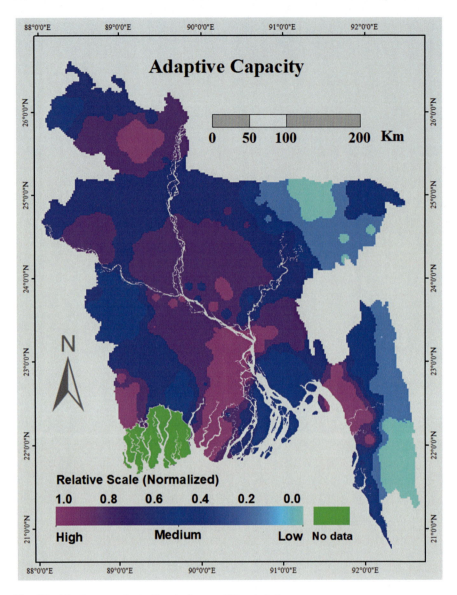

Fig. 5.7 Adaptive capacity to climate change of Bangladesh

region more exposed (Fig. 5.5) as well as highly sensitive (Fig. 5.6) to climate change effects. Though adaptive capacity is found quite good all over the country, the level of vulnerability of coastal regions is controlled by exposure and sensitivity.

The haor basin is moderate to highly exposed and sensitive to climate change (Figs. 5.5 and 5.6). However, the lower level of adaptive capacity makes this region highly vulnerable to climate change (Fig. 5.8). Riverine areas are generally found

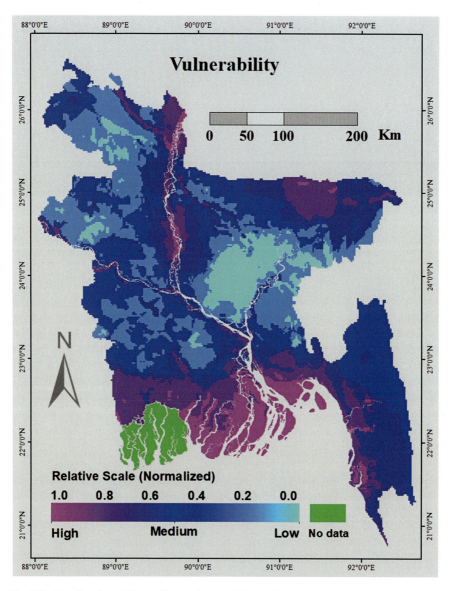

Fig. 5.8 Overall vulnerability to climate change of Bangladesh

highly vulnerable probably due to flood and riverbank erosion. Nevertheless, the Brahmaputra river basin particularly has more highly vulnerable areas than other riverine areas (Fig. 5.8).

Upper parts of the Chittagong hilly region, outer Brahmaputra floodplain, and most of the Ganges flood plain are moderate to low vulnerable (Fig. 5.8). Figures 5.5 and 5.6 are depicting that a lower level of exposure and sensitivity of these regions

Table 5.3 Classification of 64 districts according to vulnerability components

Index	Level	Jenk's natural break	No of districts
Exposure	Low	0.04–0.23	22
	Moderate	0.23–0.40	24
	High	0.40–0.74	18
Sensitivity	Low	0.06–0.22	26
	Moderate	0.22–0.31	20
	High	0.31–0.47	18
Adaptive capacity	Low	0.34–0.39	16
	Moderate	0.39–0.44	20
	High	0.44–0.54	28
Vulnerability	Low	0.0–0.02	26
	Moderate	0.02–0.25	26
	High	0.25–0.66	12

make them moderate to low vulnerable, as the adaptive capacity of these regions is moderate too (Fig. 5.7).

Concisely, a huge portion of the country is highly vulnerable to climate change due to its geography as well as socioeconomic features, as the whole country is frequently visited by various hazards and extreme events. The number of districts for a different level of vulnerability is also shown in Table 5.3 and Fig. 5.9. Regions, where the vulnerability level is low, are mainly due to the socioeconomic development of the country.

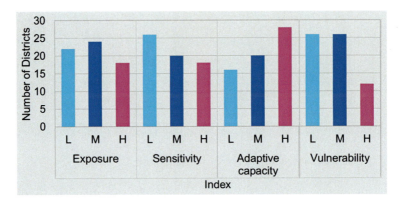

Fig. 5.9 Classification of 64 districts according to vulnerability components

5.4   Conclusion

The present study findings eloquently express the spatial climate change vulnerability for the whole country. The coastal region, part of the hilly region, riverine areas, and the haor basin are found highly vulnerable since these regions are more exposed as well as highly sensitive to climate change effects. Though adaptive capacity is found quite good all over the country, the level of vulnerability of coastal regions is controlled by exposure and sensitivity. The haor basin is moderate to highly exposed and sensitive to climate change, yet the lower level of adaptive capacity makes this region highly vulnerable. Riverine areas are found highly vulnerable probably due to flood and riverbank erosion, especially the Brahmaputra river basin found to be more vulnerable than other riverine areas. Upper parts of the Chittagong hilly region, outer Brahmaputra floodplain, and most of the Ganges flood plain are moderate to low vulnerable; lower levels of exposure and sensitivity of these regions make them moderate to low vulnerable having moderate adaptive capacity.

There are different factors of vulnerability that are dictating in different parts of the country, forthcoming climate change impact mitigation and adaptation measures are being proposed to be sector-specific. The present study has been completed with a comprehensive framework and a rigorous methodology, which presented the countrywide vulnerability of climate change. Exposure, sensitivity, adaptive capacity, and vulnerability hotspots have been identified based on a normalized relative scale. However, the present study findings, which are mainly maps, have scopes of being references for further studies. This study can also be an essential tool in measures related to the adaptation and mitigation of climate change impacts from the root level to policymaking level.

This study has considered socioeconomic data back from 2011 from the Bangladesh Population and Housing Census since countrywide socioeconomic data is only available from BBS. Though the data is eight years old, it is accepted for the present study, which is the main limitation of the study, because present data for the whole country is beyond reach.

References

Abdi H, Williams LJ (2010) Principal component analysis. Wiley Interdisc Rev: Comput Stat. https://doi.org/10.1002/wics.101

Abson DJ, Dougill AJ, Stringer LC (2012) Using principal component analysis for information-rich socio-ecological vulnerability mapping in Southern Africa. Appl Geogr. https://doi.org/10.1016/j.apgeog.2012.08.004

Adger WN (2006) Vulnerability. Glob Environ Chang. https://doi.org/10.1016/j.gloenvcha.2006.02.006

Ahmed N, Occhipinti-Ambrogi A, Muir JF (2013) The impact of climate change on prawn postlarvae fishing in coastal Bangladesh: socioeconomic and ecological perspectives. Mar Policy. https://doi.org/10.1016/j.marpol.2012.10.008

Ahsan MN, Warner J (2014) The socioeconomic vulnerability index: a pragmatic approach for assessing climate change led risks—a case study in the south-western coastal Bangladesh. Int J Disaster Risk Reduct. https://doi.org/10.1016/j.ijdrr.2013.12.009

Anderberg M (2014) Cluster analysis for applications: probability and mathematical statistics: a series of monographs and textbooks. Academic Press/Elsevier Science, Burlington

BARC (2001) Flood risk map of Bangladesh, Dhaka

Barker T et al (2007) Climate Change 2007: Mitigation, Geneva

BBS (2012) Bangladesh Population and Housing Census 2011: National Report, vol 4, Dhaka

BBS (2016) Bangladesh Disaster Related Statistics 2015: climate change and natural disaster perspective, Dhaka

BBS (2013) District Statistics 2011, Dhaka

BCAS (2008) Climate Change: responses, Dhaka

Bjarnadottir S, Li Y, Stewart MG (2011) Social vulnerability index for coastal communities at risk to hurricane hazard and a changing climate. Nat Hazards. https://doi.org/10.1007/s11069-011-9817-5

BMD (2013) "data," Dhaka

Brooks N (2003) Vulnerability, risk and adaptation: a conceptual framework. Tyndall Centre for Climate Change Research. Working Paper 38, Norwich

Burton I, Huq S, Lim B, Pilifosova O, Schipper EL (2002) From impacts assessment to adaptation priorities: the shaping of adaptation policy. Clim Policy. https://doi.org/10.3763/cpol.2002.0217

BWDB (2010) Annual Flood Report 2010, Dhaka

Cattell RB (1966) The scree test for the number of factors. Multivariate Behav Res. https://doi.org/10.1207/s15327906mbr0102_10

CDMP (2006) Component 4B of comprehensive disaster management programme, Dhaka

CEGIS (2006) Draft final report of impact of sea level rise on land use suitability and adaptation options in southwest region of Bangladesh, Dhaka

Chen W, Cutter SL, Emrich CT, Shi P (2013) Measuring social vulnerability to natural hazards in the Yangtze River Delta region, China. Int J Disaster Risk Sci. https://doi.org/10.1007/s13753-013-0018-6

Cinner JE et al (2013) Evaluating social and ecological vulnerability of coral reef fisheries to climate change. PLoS One. https://doi.org/10.1371/journal.pone.0074321

Comer PJ et al (2012) Climate Change vulnerabilitty and adaptaion strategies for natural communities: piloting methods in the Mojave and Sonoran desert, Arlington

Coulibaly JY, Mbow C, Sileshi GW, Beedy T, Kundhlande G, Musau J (2015) Mapping vulnerability to climate change in Malawi: spatial and social differentiation in the Shire River Basin. Am J Clim Chang. https://doi.org/10.4236/ajcc.2015.43023

Cutter SL et al (2008) A place-based model for understanding community resilience to natural disasters. Glob Environ Chang. https://doi.org/10.1016/j.gloenvcha.2008.07.013

Dasgupta S et al (2014) Cyclones in a changing climate: the case of Bangladesh. Clim Dev. https://doi.org/10.1080/17565529.2013.868335

Delpla I et al (2014) A decision support system for drinking water production integrating health risks assessment. Int J Environ Res Public Health. https://doi.org/10.3390/ijerph110707354

Dent B (1999) Cartography: thematic map design. McGraw-Hill, New York

Deressa T, Hassan R, Ringler C (2008) Measuring Ethiopian farmer's vulnerability to climate change across regional states, Washington, DC

Deressa T, Hassan R, Ringler C (2009) Assessing household vulnerability to climate change. The case of farmers in the Nile basin of Ethiopia, Washington, DC

Dewan A, Yamaguchi Y, Rahman Z (2012) Dynamics of land use/cover changes and the analysis of landscape fragmentation in Dhaka Metropolitan, Bangladesh. GeoJ 77(3):315–330. https://doi.org/10.1007/s10708-010-9399-x

DoE (2012) Environmental outlook, Dhaka

Duran B, Odell P (2013) Cluster analysis: a survey, vol 100. Springer Science and Buisness Media

ESRI (2016) What is the Jenks optimization method? Technical Support. https://support.esri.com/en/technical-article/000006743. Accessed 15 Mar 2019

Füssel HM, Klein RJT (2006) Climate change vulnerability assessments: An evolution of conceptual thinking. Clim Change. https://doi.org/10.1007/s10584-006-0329-3

Gallopín G (2003) A sistemic synthesis of the relations between vulnerability, hazard, exposure and impact at policy identification. Econ Comm Lat Am. Caribb (ECLAC). Handb Estim Socio-Economic Environ Eff Disasters. ECLAC

Gallopín GC (2006) Linkages between vulnerability, resilience, and adaptive capacity. Glob Environ Chang. https://doi.org/10.1016/j.gloenvcha.2006.02.004

Gerlitz J-Y, Banerjee S, Brooks N, Hunzai K, Macchi M (2014) An approach to measure vulnerability and adaptation to climate change in the Hindu Kush Himalayas. In: Handbook of climate change adaptation

Grubesic TH, Matisziw TC (2013) A typological framework for categorizing infrastructure vulnerability. GeoJ. https://doi.org/10.1007/s10708-011-9411-0

Hagenlocher M, Delmelle E, Casas I, Kienberger S (2013) Assessing socioeconomic vulnerability to dengue fever in Cali, Colombia: statistical vs expert-based modeling. Int J Health Geogr. https://doi.org/10.1186/1476-072X-12-36

Heltberg R, Bonch-Osmolovskiy M (2011) Mapping vulnerability to climate change, Washington DC

Howden SM, Soussana JF, Tubiello FN, Chhetri N, Dunlop M, Meinke H (2007) Adapting agriculture to climate change. Proc Natl Acad Sci USA. https://doi.org/10.1073/pnas.0701890104

Ibarraran M, Malone E, Brenkert A (2008) Climate chang e vulnerability and resilience: current status and trends for Mexico, Washington, DC

IPCC (2001) Climate change 2001: impacts, adaptation and vulnerability, Cambridge

IPCC (2007) Climate Change 2007: Mitigation. Contribution of Working Group III to the Fourth Assessment Report of the Intergovernmental Panel on Climate Change

IPCC, Climate Change 1995 (1996) Impacts, adaptation and mitigation of climate change: scientific technical analyses. Cambridge University Press, Cambridge

IPCC (2013) Climate Change 2013: The Physical Science Basis, Contribution of Working Group I. 2013 IPCC, Climate Change 2013: The Physical Science Basis, Contribution of Working Group I

IPCC (2014) Summary for Policymakers, In: Climate Change 2014, Mitigation of Climate Change. Contribution of Working Group III to the Fifth Assessment Report of the Intergovernmental Panel on Climate Change

IWFM (2014) Impact of climate change on rainfall intensity in Bangladesh, Dhaka

Jenks GF (1967) The data model concept in statistical mapping. Int Yearb Cartogr

Kaiser HF (1960) The application of electronic computers to factor analysis. Educ Psychol Meas. https://doi.org/10.1177/001316446002000116

Kelly PM, Adger WN (2000) Theory and practice in assessing vulnerability to climate change and facilitating adaptation. Clim Change. https://doi.org/10.1023/A:1005627828199

Liu J et al (2008) A spatially explicit assessment of current and future hotspots of hunger in Sub-Saharan Africa in the context of global change. Glob Planet Change. https://doi.org/10.1016/j.gloplacha.2008.09.007

Metzger MJ, Schröter D (2006) Towards a spatially explicit and quantitative vulnerability assessment of environmental change in Europe. Reg Environ Change. https://doi.org/10.1007/s10113-006-0020-2

Pagano A, Giordano R, Portoghese I, Fratino U, Vurro M (2014) A Bayesian vulnerability assessment tool for drinking water mains under extreme events. Nat Hazards. https://doi.org/10.1007/s11069-014-1302-5

Piya L, Maharjan K, Joshi N (2012) Vulnerability of rural households to climate change and extremes: analysis of Chepang households in the Mid-Hills of Nepal

Plummer R, de Grosbois D, Armitage D, de Loë RC (2013) An integrative assessment of water vulnerability in First Nation communities in Southern Ontario, Canada. Glob Environ Chang. https://doi.org/10.1016/j.gloenvcha.2013.03.005

Preston B, Stafford-Smith M (2009) Framing vulnerability and adaptive capacity assessment: discussion paper

Quackenbush J (2002) Microarray data normalization and transformation. Nat Genet. https://doi.org/10.1038/ng1032

Rahman MR, Lateh H (2017) Climate change in Bangladesh: a spatio-temporal analysis and simulation of recent temperature and rainfall data using GIS and time series analysis model. Theor Appl Climatol. https://doi.org/10.1007/s00704-015-1688-3

Rashid H (1991) Geography of Bangladesh. University Press Ltd., Dhaka

Salinger MJ, Sivakumar MVK, Motha R (2005) Reducing vulnerability of agriculture and forestry to climate variability and change: workshop summary and recommendations. https://doi.org/10.1007/s10584-005-5954-8

Sarkar R, Vogt J (2015) Drinking water vulnerability in rural coastal areas of Bangladesh during and after natural extreme events. Int J Disaster Risk Reduct. https://doi.org/10.1016/j.ijdrr.2015.09.007

SEI (2004) Choosing methods in assessments of vulnerable food systems, Stockholm

Shaw J (2015) Vulnerability to climate change adaptation in rural Bangladesh. Clim. Policy 15(3):410–412. https://doi.org/10.1080/14693062.2015.1007837

Slocum TA (1999) Thematic cartography and visualization

SRDI (2010) Salinity risk map of Bangladesh, Dhaka

Stelzenmüller V, Ellis JR, Rogers SI (2010) Towards a spatially explicit risk assessment for marine management: assessing the vulnerability of fish to aggregate extraction. Biol Conserv. https://doi.org/10.1016/j.biocon.2009.10.007

Tan P-N, Steinbach M, Kumar V (2013) Data mining cluster analysis: basic concepts and algorithms

Tang HS et al (2013) Vulnerability of population and transportation infrastructure at the east bank of Delaware Bay due to coastal flooding in sea-level rise conditions. Nat Hazards. https://doi.org/10.1007/s11069-013-0691-1

Thatcher CA, Brock JC, Pendleton EA (2013) Economic vulnerability to sea-level rise along the Northern U.S. Gulf Coast. J Coast Res. https://doi.org/10.2112/si63-017.1

Thompson B (2004) Exploratory comfirmatory factor analysis: understanding concepts and applications, New York

UNDP (2007) Human development report 2007/2008: fighting climate change: Human solidarity in a divided world, London

UNFCCC (2008) Compendium on methods and tools to evaluate impacts of, and vulnerability and adaptation to climate change. UNFCCC Secretariat. http://unfccc.int/files/adaptation/nairobi_workprogramme/compendium_on_methods_tools/application/pdf. Accessed 9 Mar 2019

Williams B, Onsman A, Brown T (2012) Exploratory factor analysis: a five-step guide for novices EDUCATION. Australas J Paramed

Wold S, Esbensen K, Geladi P (1987) Principal component analysis. Chemom Intell Lab Syst. https://doi.org/10.1016/0169-7439(87)80084-9

Yohe G, Tol RSJ (2002) Indicators for social and economic coping capacity—moving toward a working definition of adaptive capacity. Glob Environ Chang. https://doi.org/10.1016/S0959-3780(01)00026-7

Yoon DK (2012) Assessment of social vulnerability to natural disasters: a comparative study. Nat Hazards. https://doi.org/10.1007/s11069-012-0189-2

Part II
Environment

Chapter 6
Water Scarcity in Coastal Bangladesh: Search for Arsenic-Safe Aquifer with Geostatistics

M. Manzurul Hassan⊙, Anamika Shaha, and Raihan Ahamed

Abstract Water scarcity is a common phenomenon in coastal Bangladesh due to the elevated level of arsenic and salinity concentrations in groundwater. Since arsenic is a documented carcinogen, this paper mainly seeks to explore arsenic-safe drinking water source at different depths of shallow and deep aquifer. Inverse Distance Weighting (IDW) method along with the Geographical Information Systems (GIS) was used to fulfil the objectives of this research. The relevant arsenic concentrations in tubewell water were collected with field-testing kits (FTK), tubewell attributes with field observation and questionnaire survey, and spatial data with Geographical Positioning Systems (GPS) from Magura *union* of Satkhira district in the south-western part of coastal Bangladesh. The study site covers about 27.58 km^2 of area with a total population of 20,375 and a total of 2650 tubewells. About three-quarter of the study site are contaminated following the level of Bangladesh Drinking Water Standard (BDWS) of 50 µg/L. The study identified a few scattered areas for arsenic-safe water "pockets" at different depths in the northeastern part of the study site. The findings could be helpful in formulating a policy to achieve "Clean Water and Sanitation" (Sustainable Development Goal 6) at some extent with exploring the safe aquifer.

Keywords Water scarcity · Groundwater arsenic · Geostatistics · GIS · Indicator Kriging · Spatial discontinuity · Aquifer · Bangladesh

M. M. Hassan (✉)
Department of Geography and Environment, Jahangirnagar University, Savar, Dhaka, Bangladesh

A. Shaha
SKS School and College, College Road, Uttar Horin Singha, Gaibandha Sadar, Gaibandha, Bangladesh

R. Ahamed
BIGD, BRAC University, Mohakhali, Dhaka, Bangladesh

© The Author(s), under exclusive license to Springer Nature Singapore Pte Ltd. 2022
N. C. Jana and R. B. Singh (eds.), *Climate, Environment and Disaster in Developing Countries*, Advances in Geographical and Environmental Sciences, https://doi.org/10.1007/978-981-16-6966-8_6

6.1 Background

Water scarcity can be addressed as a shortage of sufficient safe water or lack of access to safe water supplies. Water scarcity, in this paper, is defined as the shortage of safe water for drinking and cooking purposes. Water scarcity is a common experience in coastal Bangladesh due to the elevated level of arsenic and salinity concentrations in groundwater as well as a lack of water availability to satisfy the individual and community demand. Coastal belts of Bangladesh are mainly characterized by saline water in both the surface and groundwater sources. It is apparent that shallow aquifer in coastal area is highly contaminated with arsenic and deep aquifer with salinity (Hassan 2018). In addition, surface water sources are mainly characterized by suspended sediments and pathogens which is unsafe for drinking and cooking purposes.

Groundwater contamination with inorganic arsenic is one of the most critical environmental problems in the world. The groundwater plays a foremost contributing role in the livelihood of people and the economy of countries as well. Groundwater is purportedly the main source of untreated pathogen-free safe drinking water for more than one-third of the total population on the globe (WHO 2015). However, Bangladesh has many water-related problems and groundwater systems in Bangladesh experience significant pressure on their quantity and quality due to an increased water abstraction for irrigation and domestic purposes and consequent lowering of water levels (Hassan 2018). It is ironic that so many tubewells installed for providing pathogen-free drinking water are found to be contaminated with toxic levels of arsenic that threaten the health of millions of people in Bangladesh (Hassan and Atkins 2007, 2011). About 85 million people in Bangladesh are exposed to inorganic arsenic more than 50 μg/L. In addition, most of the people in coastal Bangladesh are facing severe health problems with diarrhoea, hypertension, skin lesions, arsenic-related diseases, and so on (Hassan 2015).

There is a complicated pattern of spatial arsenic discontinuity in groundwater with differences between neighbouring wells at different scales and changes in aquifer depth (Hassan and Atkins 2006; Bhuiyan et al. 2016; Islam et al. 2017; Peters and Burkert 2008). In a study from Bangladesh, Yu et al. (2003) found an insignificant positive correlation between the concentrations level of arsenic and the density of shallow tubewell. Mukherjee and Fryar (Mukherjee and Fryar 2008) observed elevated dissolved arsenic (>10 μg/L) in a majority of the deep groundwater samples in West Bengal, India. Guo et al. (2012) show a large spatial variation of groundwater arsenic concentrations in the Hetao Basin of Inner Mongolia, and Pi et al. (2016) inspected the pattern of spatial variability of total arsenic concentrations in groundwater ranging from 5.47 to 2690 μg/L with an average of 697 μg/L in Datong basin, northern China. Analyzing with multivariate geostatistical modeling, Andrade and Stigter (2013) inspected the spatio-temporal arsenic occurrence in shallow alluvial groundwater under agricultural land in central Portugal.

Spatial heterogeneity of aquifers and their complex physical processes as well as chemical reactions may be the cause of contamination to change its physical state that finally change the degree of contamination in an aquifer (Civita 2010). "Aquifer vulnerability," therefore, refers to the sensitivity of the quality of groundwater to a contaminant based on the aquifer's characteristics. These aquifer characteristics can be changed largely with the excessive withdrawal of groundwater for irrigation, industrial, and domestic purposes. "Aquifer vulnerability mapping" can be used to monitor the geographic extent of groundwater contamination and environmental risk as well as to formulate sustainable spatial policies for preventing aquifer pollution from groundwater contaminants (Kourgialas and Karatzas 2015; Rebolledo et al. 2016; Sikdar and Chakraborty 2017).

Spatial mapping for arsenic concentrations can be performed with geostatistical simulations. Geostatistics has the advantage of considering the spatial continuity of data and capacity of presenting the uncertainty associated with the modeling that can be accomplished through semivariogram (Olea et al. 2017). In one of our previous papers (Hassan and Ahamed 2017), we applied Ordinary Kriging (OK) for searching arsenic-safe aquifer in an area of Bagerhat district. The paper is exceptional with different method and different area for exploring arsenic-safe drinking water source at different depths of aquifer. As we have already mentioned that groundwater in the study site is heavily concentrated with sodium chloride and inorganic arsenic. Therefore, shortage of safe drinking water is a burning problem in the study site. This paper presents IDW prediction method and GIS operations in analyzing the water quality data for investigating "arsenic-safe" aquifer along with the mapping of "spatial arsenic discontinuity."

6.2   GIS and Geostatistics in Safe Groundwater Investigation

Hassan (2018) defined GIS as "decision by position." GIS can be applied for monitoring the groundwater contaminants and their spatial discontinuity as well as for identification of demarcation of safe water zones in groundwater sources. GIS with geostatistical methods can be used for modeling the spatial distribution of groundwater quality at local and regional scales. Several authors have attempted GIS-based groundwater models to understand the pattern of contamination (Sikdar and Chakraborty 2017; Bonsor et al. 2017; Gorgij et al. 2017)). Hossain and Piantanakulchai (Hossain and Piantanakulchai 2013) attempted to predict the risk of groundwater arsenic contamination using a GIS. Hassan (2005) developed a PPGIS (Public Participation GIS) combining GIS and participatory rural appraisal (PRA) for assessing the mitigation needs of arsenic-affected communities.

Groundwater is an essential resource that provides drinking water to billions of people around the world, and assurance of drinking water safety is important to a community. With the combined effects of population growth, intense groundwater abstraction has caused widespread depletion and degradation of groundwater resources (Gorgij et al. 2017; Chen et al. 2017). Therefore, groundwater quality assessments have become increasingly important, and several criteria have been developed for assessing water quality, for instance, the WHO guidelines for a range of drinking water pollutants (Amiri et al. 2014).

There is an increasing interest in geostatistical techniques for exploring the spatio-temporal variations of chemical properties and contamination of groundwater (Belkhiri and Mouni 2014; Dehghanzadeh et al. 2015; Mizan et al. 2017; Narany et al. 2014; Yazdanpanah 2016). Geostatistical approaches can be utilized to achieve a sustainable use of groundwater resources as well as to evaluate surface water quality. Geostatistics can be used to design water quality monitoring at different aquifers with hydrochemical evaluation by spatial mapping (Machiwal et al. 2012; Ou et al. 2012). Geostatistics can also emphasize spatial variation of major factors that influence groundwater quality (Belkhiri and Narany 2015; Masoud 2014).

Geostatistics and GIS have been used as a decision-making and management tool in the spatial discontinuities of groundwater quality as well as groundwater arsenic concentration (Cinti et al. 2015; Delbari et al. 2016; Flanagan et al. 2016; Golia et al. 2015; Marko et al. 2014). Geostatistics relies on both statistical and mathematical methods to create surfaces for the concentration levels of groundwater arsenic (Liu et al. 2004). GIS is also considered an "automated decision-making system" (Achour et al. 2005; Berke 2004) with mapping capabilities for the geographically referenced information in preparing spatial mapping for investigating the historical and currently existing arsenic situations in groundwater.

Spatial discontinuity of arsenic concentration has been reported in Bangladesh (Hassan and Atkins 2011; Radloff et al. 2017), West Bengal in India (Biswas et al. 2014), China (Cai et al. 2015; Guo et al. 2014; Ma et al. 2016), Chianan Plain of Taiwan (Sengupta et al. 2014), Mekong Delta of Vietnam (Wilbers et al. 2014), the southern Pampa of Argentina (Díaz et al. 2016), the Duero River Basin of Spain (Pardo-Igúzquiza et al. 2015), Nova Scotia in Canada (Dummer et al. 2015), New England in the USA (Yang et al. 2012), the Águeda watershed area in Portuguese district of Guarda and the Spanish provinces of Salamanca and Caceres (Antunes et al. 2014), and so on.

Geostatistical algorithms can be used to assess the uncertainty of the predicted values instead of the traditional interpolation methods (Delbari et al. 2016; Al-Omran et al. 2017). Using the Indicator Kriging (IK) prediction method, Goovaerts et al. (2005) studied the spatial variability of arsenic in groundwater in southeast Michigan. Elumalai et al. (2017) deployed OK and IDW methods for spatial interpolation mapping for groundwater contamination in a coastal city of Richards Bay in South Africa. Beg et al. (2011) used IDW method for illustrating spatial prediction of fluoride concentration in unsampled locations in the groundwater of Chhattisgarh, India. Mehrjardi et al. (2008) compared the performance of IDW and kriging in analyzing groundwater quality with spatial distribution pattern. Javed et al. (2017) used OK for spatial assessment of groundwater quality in Jhelum city, Pakistan. Using IK prediction model, Hassan and Atkins (2011) analyzed the spatial discontinuity of groundwater arsenic from southwest Bangladesh. Using OK interpolation method, Al-Abadi et al. (2017) mapped the groundwater potential in the AltunKupri Basin in the northeast Kirkuk Governorate of Iraq. Using two deterministic methods (e.g.,

IDW and Radial Basis Function) and two stochastic interpolation methods (e.g., OK and Universal Kriging), Adhikary and Dash (Adhikary and Dash 2017) (WHO 2004) predicted the spatial variation of groundwater up to a depth of 20 m in Delhi, India.

6.3 Water Right

Drinking water quality is assessed as "microbial, chemical and physical water constituents may affect the appearance, odor or taste of the water, and the consumer will evaluate the quality and acceptability of water on the basis of these criteria" (Safe Drinking Water Foundation 2020). Drinking water should not have any substantial risk to health over a lifetime consumption (WHO 2015) and this quality is also required for all usual domestic purposes. It is mentioned that water was not explicitly mentioned as a human right in many international convention (e.g., the 1948 Universal Declaration of Human Rights, the 1966 International Covenant on Economic, Social and Cultural Rights, and the 1966 International Covenant on Civil and Political Rights). It was, however, implied through other human rights, such as the right to life, right to an adequate standard of living, and the right to health (CESCR 2002).

Nevertheless, in 2002, the United Nations officially adopted water as a human right stating that "the human right to water entitles everyone to sufficient, safe, acceptable, physically accessible and affordable water for personal and domestic uses (UNECE 2012)." This means that countries who have endorsed the International Treaty on Economic, Social, and Cultural Rights must ensure fair and equal access to safe drinking water (CESCR 2002). In 2002, the United Nations Committee on Economic, Social, and Cultural Rights (CESCR) recognized access to water as an independent human right as stated that, "the right to water clearly falls within the category of guarantees essential for securing an adequate standard of living, particularly since it is one of the most fundamental conditions for survival (UNECE 2012)."

In 2002, the CESCR on the Right to Water, has noted that a core content of the right to water is that "water required for personal or domestic use must be safe, therefore free from micro-organisms, chemical substances, and radiological hazards that constitute a threat to a person's health" (UNECE 2012). The CESCR has also mentioned that, "… water should be treated as a social and cultural good, and not primarily as an economic good. The manner of the realization of the right to water must also be sustainable, ensuring that the right can be realized for present and future generations (UNECE 2012)." In addition, the CESCR in 2002 adopted its general comment on the right to water stating that, "the human right to water entitles everyone to sufficient, safe, acceptable, physically accessible and affordable water for personal and domestic uses (UNECE 2012)."

The United Nations Economic Commission for Europe (UNECE) Protocol on Water and Health mentions central aspects of a human right to water, stating that "… equitable access to water, adequate in terms of both quantity and quality, should be provided for all members of the population, especially those who suffer a disadvantage or social exclusion." Access to water and sanitation services are reinforced in Article 6.1, which provides that "the Parties shall pursue the aims of: (a) access to drinking water for everyone; and (b) provision of sanitation for everyone" (UN 2010). Access to safe drinking water is now considered a basic human right in protecting human dignity and one of the SDG is linked with safe water—Goal 6 (clean water and sanitation), which mainly focuses on "achieving universal and equitable access to safe and affordable drinking water for all by 2030" (http://www.un.org).

At the international level, Bangladesh has recognized the human right to safe drinking water and sanitation on several occasions. The country accepted the UN General Assembly resolution 64/292 (The human right to water and sanitation) of 28th July 2010, which recognizes "the right to safe and clean drinking water and sanitation as a human right that is essential for the full enjoyment of life and all human rights" (UNHRC 2012). The Government of Bangladesh declared some 61 parameters for potable water within the "Environmental Conservation Rules 1997" which indicates the vision of the Government of Bangladesh in recognizing the right to safe drinking water. As a member of the Human Rights Council (HRC), Bangladesh adopted various resolutions affirming that the human right to safe drinking water and sanitation is derived from the right to an adequate standard of living (GOB 2013). Bangladesh has taken this right most seriously. In 2013, it adopted a Water Act (Act 14 of 2013) that declared water for drinking, sanitation, and hygiene as "the highest priority right" (BBS 2011).

6.4 Materials and Methods

6.4.1 Spatial Data

The tubewell locations for water samples were collected through extensive field visits with GPS (Model: Garmin GPSMAP 62STC). This GPS has high-sensitivity receiver with the facilities of preloaded base map with topographic features (Hassan 2015). Apart from geographical location identification of each sample tubewell, this device has the facilities for automatic routing with electronic compass and barometric altimeter (Hassan and Ahamed 2017). More spatial information was collected from small-scale mouza (the lowest level administrative unit in Bangladesh with Jurisdiction List number) sheets having the scale of 1:3960. The Google Earth and Open Street Map were also used for land-base and facility-base information. In addition, the position of each tubewell was plotted on the mouza sheets to check the accuracy of the GPS positional data. Figure 6.1 shows the flow diagram of the methodological procedure of this study.

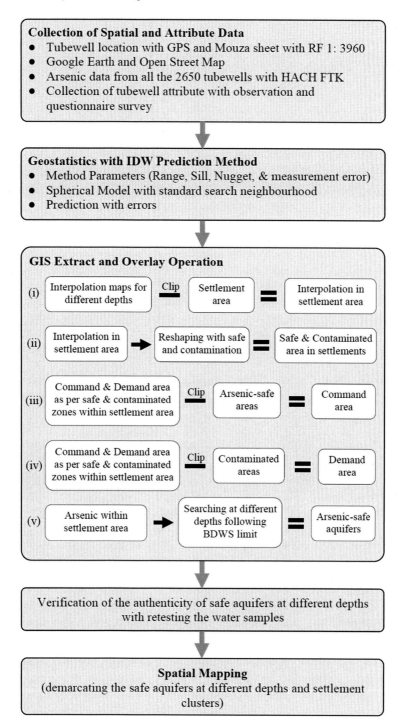

Fig. 6.1 Schematic diagram for the methodology for this research

6.4.2 Arsenic Data and Tubewell Attribute Data

Tubewell screening is the priority work for water quality data collection. In collection of arsenic data, all the 2650 tubewells from the study site were selected for this study. Since inorganic arsenic is a toxic element and is known as a documented carcinogen as well as arsenic concentration in groundwater is uneven and high paradoxical, all the tubewells were selected for screening arsenic concentration. The relevant arsenic data were mainly collected directly from the field with HACH field-testing kits (FTK) in 2014. Moreover, tubewell attributes in terms of tubewell depth, installation year, users, etc. were collected with observation and face-to-face questionnaire surveys. The relevant data were collected from Magura *union* (the fourth order local government administrative unit in Bangladesh after division, district, and sub-district) in Tala Upazila (sub-district) of Satkhira district in the coastal belt of southwest Bangladesh (Fig. 6.2).

The area comprises nine administrative wards covering an area of 27.58 km^2 with a total population of 20,375 (Dowling et al. 2002). The area is geologically a part of the quaternary deltaic sediments of the Ganges alluvial plains and tidal plains (Umitsu 1993) and the sub-surface geology of the area has complex inter-fingerings of coarse and fine-grained sediments from numerous regressions and transgressions throughout geological time (DPHE, JICA. 2010). The aquifer of this region is mostly unconfined to leaky confined and groundwater occurs within a few meters of the surface. Two distinct aquifers occur all over the area: the deep aquifer is at depths of $\geq$145 m and the shallow aquifer at depths of <145 m. The aquitard separating the shallow aquifer from the deep aquifer is a continuous one and increases in thickness toward the south and at places the thickness is about 100 m (Serón et al. 2001).

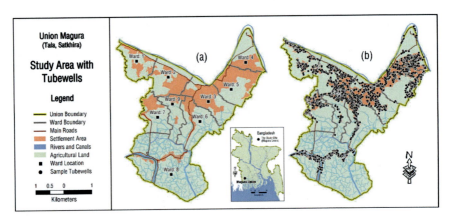

Fig. 6.2 The study area for arsenic-safe aquifer investigation: **a** physical features of the study area; and **b** location of all the 2650 selected tubewells in the study site

6.4.3   Data Analysis: Geostatistics with IDW

Map interpolation with GIS extract and overlay operations can be useful in demarcating arsenic-safe aquifer. GIS has strong spatial extract and overlay capabilities because of its mathematical and programming facilities which allow different map data to be combined in determining different "problem regions" of water scarcity in the study site. Therefore, overlaying different interpolation maps within settlement areas facilitate the generation of information on water "problem regions." Apart from GIS operation, we have applied geostatistics to estimate contaminant concentrations in unsampled areas.

IDW approach is the non-geostatistical method of interpolation which assumes that grades vary in a deposit according to the inverse of their separation (raised to some power). This method does not account for "nugget variance" or other aspects of the "variogram." Spatial interpolation with IDW method can be suitable to analysis spatial pattern of groundwater arsenic concentrations because of its exact interpolation capability (Hassan 2018). The IDW interpolator is a point estimation technique based on the weighting of a random function for a cell node of a grid (Ashraf et al. 1997). It assumes that each input point has a local influence that diminishes with distance. It weights the points closer to the processing cell greater than those further away, hence the name IDW interpolation or Inverse Squared Distance (ISD) interpolation (Longley et al. 2001). Therefore, the minimum and maximum values in the interpolated surface can only occur at sample points in IDW interpolation method (Johnston et al. 2001). The prediction value for location in IDW method can be calculated using the following equation (Tsanis and Gad 2001):

$$\hat{Z}(S_0) = \sum_{i=1}^{N} \lambda_i Z(S_i) \qquad (6.1)$$

where, $\hat{Z}(S_0)$ is the prediction value for location S_0, N is the number of measured sample points surrounding the prediction location, λ_i the weight assigned to each measured point, and $Z(S_i)$ observed value at location S_i. Therefore, in determining the weights, the following formula option is used (Tsanis and Gad 2001):

$$\lambda_i = d_{i0}^{-p} \bigg/ \sum_{i=1}^{N} d_{i0}^{-p} \sum_{i=1}^{N} \lambda_i = 1 \qquad (6.2)$$

Since the distance becomes larger, the weight is reduced by a factor of p. The quantity d_{i0} is the distance between prediction location S_0, and each of the measured locations S_i. In IDW method, the surface is driven by local variation and calculation depends on power parameter and neighborhood search strategy. The power parameter controls the influence of surrounding points upon the interpolated value. A higher power results in less influence from distant points, while the optimal power value is determined by minimizing the Root-Mean-Square Prediction Error (Isaaks and Srivastava 1989). If the power value is 0, there is no decrease with distance.

In addition, "search neighborhood" in the IDW is used in prediction of an unmeasured location. The shape of search neighborhood is based on an understanding of spatial locations and spatial autocorrelation of the dataset (Hassan and Atkins 2011). Generally, the search area of the IDW method is a circular weighting, but the area could be elliptical or even directional in order to remove the strong influence of local anomalous values due to clustered data surrounding the estimation point (Ashraf et al. 1997; Dokou et al. 2015). The spatial arsenic interpolation maps produced with IDW method are based on the weighting of a random function for the sample study points (Hassan 2018).

In the prediction maps of arsenic concentrations, cross-validation was used to compare the prediction performance of different interpolation algorithms. The cross-validation can be applied to improve the quality of geostatistical model and thus the results of the spatial analysis. Cross-validation can produce a low-bias estimator for the generalization abilities of a statistical model, and therefore provides a sensible criterion for model selection and performance comparison (Dokou et al. 2015). The differences between measured and predicted values provide a quality control for the model of computation for arsenic concentrations (Hassan 2018).

Cross-validation can be performed by eliminating sequentially each data point from the data set and then estimating it using the constructed model (Yeşilkanat et al. 2015; Lark 2000). The kriged values $\hat{Z}(x_i)$ are compared with the observed ones $Z(x_i)$, and various statistical measures are calculated in order to compare the goodness of fit of each model (Oliver and Webster 2015; Tiwari et al. 2017). Figure 6.3 shows the levels of prediction-error for IDW prediction for spatial arsenic discontinuity.

6.5 Results

6.5.1 Contamination Pattern

In analyzing a total of 2650 functional tubewells in Magura union, it was found that arsenic concentrations are very erratic and the pattern of concentrations ranging of 0–1500 μg/L with the mean concentration of 155.81 $\pm$ 173.75 μg/L. A slightly more than a quarter (27.7%) of the total tubewells were found to be safe following the BDWS limit of 50 μg/L; while slightly less than three-quarters (72.3%) of the tubewells were chalked out as arsenic contaminated from moderate to severe levels (Table 1). In the safe category, some 13.8% tubewells were analyzed for arsenic-safe water with WHO permissible limit of 10 μg/L.

Arsenic concentration in the study site is high paradoxical with spatial dimension. Elevated levels of arsenic were found in Wards 1, 2, 3, 7, 8, and 9; while low levels of arsenic were found in Wards 4, 5, and 6 (Fig. 6.4 and Table 6.1). The highest mean

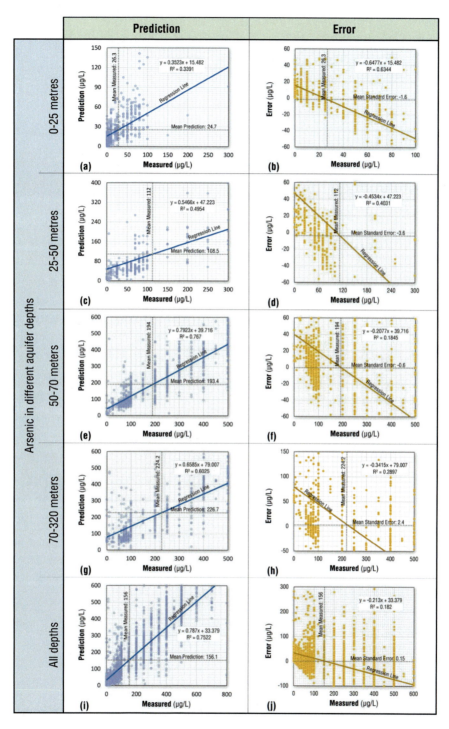

Fig. 6.3 Properties of cross-validation and prediction-error for IDW prediction for spatial arsenic discontinuity

Table 6.1 Descriptive statistics of arsenic concentrations in the study site

Wards	Concentration level (ng/L)					Number of Tubewells	Mean with Std. deviation
	WHO Limit ($\leq$10)	BDWS Limit ($\leq$50)	Moderate (51–100)	High (101–300)	Severe (>300)		
Ward 1	3	35	101	97	28	264 (9.96%)	164.45 ± 113.41
Ward 2	6	60	218	57	38	379 (14,30%)	142.70 ± 184.69
Ward 3	5	13	146	137	30	331 (12.49%)	171.62 ± 124.67
Ward 4	128	103	94	2	4	331 (12.49%)	39.90 ± 53.71
Ward 5	199	72	135	46	5	457 (17.25%)	53.20 ± 69.50
Ward 6	13	71	172	4	0	260 (9.81%)	66.23 ± 27.61
Ward 7	1	0	1	20	164	186 (7.02%)	453.23 ± 132.21
Ward 8	4	2	82	65	159	312 (11,77%)	323.69 ± 203.23
Ward 9	7	12	73	21	17	130 (4.91%)	142.73 ± 131.66
Total	366 (13.8%)	368 (13.9%)	1022 (38.6%)	449 (16.9%)	445 (16.8%)	2650 (100%)	155.81 ± 173.75

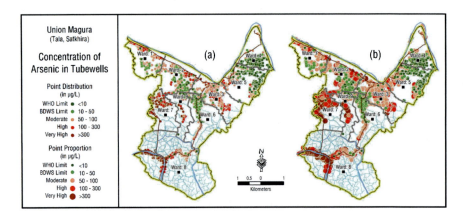

Fig. 6.4 The pattern of arsenic concentrations in tubewells under screening: **a** different symbology for different concentration levels; and **b** proportion of arsenic concentrations in each tubewell

arsenic concentrations were recorded in Ward 7 (453.23 µg/L) followed by Ward 8 (323.69 µg/L), Ward 3 (171.62 µg/L), Ward 1 (164.45 µg/L), and so on (Table 1). On the other hand, low level of mean arsenic concentrations were calculated for Ward 4 (39.90 µg/L) followed by Ward 5 (53.20 µg/L) and Ward 6 (66.23 µg/L) (Fig. 6.4).

Our analysis shows that there are 2085 (78.7%) shallow tubewells (STW) and 33 (1.2%) deep tubewells (DTW) in Magura. More than two-third (1409, 67.6%) of the STW were found to be elevated levels of arsenic; while about one-third (676, 32.4%) of the STW were found to be safe and of them half (343, 16.5%) of the STW retain the WHO permissible limit and about another half (333, 15.9%) retain the BDWS limit. The DTW were found to be safe following the BDWS limit except a few. In addition, in the intermediate aquifer (50–150 m depth), about 95% tubewells are concentrated with elevated levels of arsenic.

6.5.2 Interpolation Pattern

The IDW prediction method shows the spatial pattern of arsenic concentrations in the study site (Fig. 6.5). Almost three-quarter of this study area are found to be contaminated with elevated levels of inorganic arsenic. Safe zones are mainly concentrated

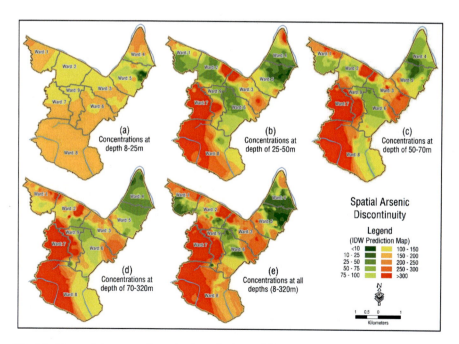

Fig. 6.5 The spatial pattern of arsenic discontinuity at different depths in the study site

in the northeastern part of the study area in a scattered manner (Fig. 6.5); while the contaminated zones are concentrated into the west, northwest and south-eastern parts of the study site. The contaminated zones are found everywhere in the study area but with a decrease in the degree of contamination from west to east. The west and northwest of the study area are heavily contaminated, while the southwest part of the study area is contaminated in a highly irregular pattern (Fig. 6.5).

The safe areas were identified in Wards 4 and 5 and these safe zones jointly cover about 13.4% (507 ha.) of the total study area. There are very few safe tubewells in rest of the administrative Wards. Arsenic-safe tubewells occur mainly in the northeastern parts of the study area. The high and severe contamination zones cover about 43.1% (1189 ha.) of the study area; while zones of moderate contamination cover about 38.5% (1062 ha.). It is noteworthy that the mean arsenic concentration in the study area is more than three times higher (155.8 µg/L) than the BDWS limit and 15 times higher than the WHO permissible limit.

Arsenic-safe zones identified using the IDW prediction method cover about 13.4% (87 ha.) within the settlement area in Magura in place of total area of the study site (Table 2) and the safe command areas are in the northeastern parts of the study area (Fig. 6.6). Only 18.9% of the tubewells (502 out of 2650) with different depths conform to this safe zone. The safe areas are especially located in Wards 4 and 5 in the study site, but very few safe tubewells are in irregular pattern in rest of the administrative Wards. Moreover, arsenic concentration in groundwater is high uneven

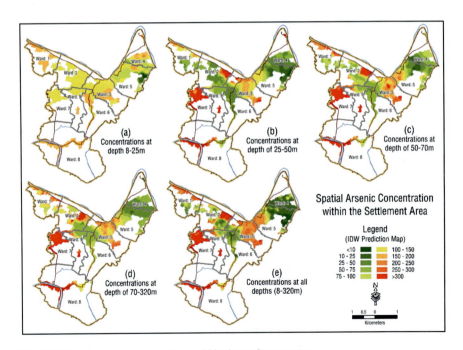

Fig. 6.6 Spatial arsenic concentrations within the settlement area

Table 6.2 Existing pattern of safe water facilities (by area): safe water command area and demand areas in the study site

Wards	Area (in hectares)	Settlement area		Safe water command area		Safe water demand area	
		Area (in hectares)	% of total area	Gross[a] (in ha.)	Net (in %)	Gross[a] (in ha.)	Net (in %)
Ward 1	218	62	9.0	3 (4.8)	1.5	59 (95.2)	12.0
Ward 2	271	97	14.1	19 (19.6)	9.8	78 (80.4)	15.9
Ward 3	283	128	18.7	0	0	128 (100)	26.0
Ward 4	209	76	11.1	69 (90.8)	35.6	7 (9.2)	1.4
Ward 5	298	132	19.2	87 (65.9)	44.8	45 (34.1)	9.1
Ward 6	311	52	7.6	16 (30.8)	8.2	36 (69.2)	7.3
Ward 7	219	56	8.2	0	0	56 (100)	11.4
Ward 8	784	49	7.1	0	0	49 (100)	10.0
Ward 9	165	34	5.0	0	0	34 (100)	6.9
Total	2758	686	24.9	194	28.3	492	71.7

[a] Figures in the parentheses indicate the percent of gross areas

over depth and space. The pattern of arsenic concentrations varies considerably and unpredictably over distances of a few meters. About 60% of tubewells are located within 35 m of each other, but within this distance there are remarkable variations. The overall pattern of arsenic concentrations in groundwater within the settlement area in Magura shows a moderate contamination running along the south bank of the Kobadakriver to the central part of the study site (Fig. 6.6).

6.5.3   Safe Water Command and Demand Areas

Which areas are treated as water scarcity and which areas will get priority for safe drinking water? The answers to the questions can detect the "command areas" and "demand areas" of safe water. Safe-water "command area" in this paper refers to the areas where arsenic-safe water is available from groundwater sources and people can have the access easily with this water. At the same time, safe water "demand area" refers to areas with elevated level of arsenic concentrations and no other alternative safe water option into the vicinity of the people within the contaminated areas. It is argued that people who are living within different contamination zones (492 hectares, 71.7%) within the settlement areas are needed for safe water options (Table 6.2). Both the STW and DTW are not suitable in the identified demand areas. It is noted that people living within the safe zones or safe water command areas do not need any safe water options for them. Figure 6.7 and Table 6.2 show the distribution pattern of arsenic-safe "command areas" and safe water "demand areas" in different administrative Wards.

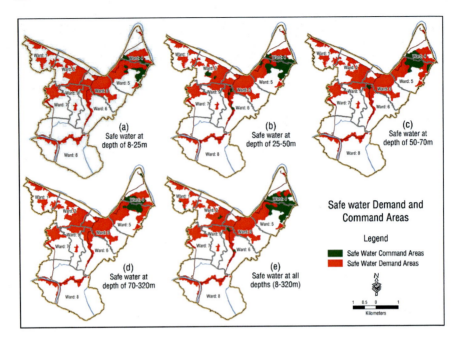

Fig. 6.7 The "command areas" and "demand areas" of safe water in the study site

Arsenic-safe command areas are mainly identified in Wards 1, 2, 4, 5, and 6 (Table 2). The maximum command areas are identified in Ward 5 (87, 44.8%) followed by Ward 4 (69, 35.6%), Ward 2 (19, 9.8%), Ward 6 (16, 8.2%), and Ward 1 (3, 1.5%) (Table 2). At the same time, there are no arsenic-safe areas in Wards 3, 7, 8, and 9 in the study site (Table 6.2). Moreover, arsenic-safe existing command areas in the study site provide safe water to a small number of people in the study site, while the maximum number of people in the study site are beyond of any arsenic-safe water (Fig. 6.7).

6.5.4 Suitable for Arsenic-Safe Aquifers

Suitable arsenic-safe area identification at different depths of aquifers is the main objective for this study. Suitability analysis is a process of systematically identifying or rating potential locations with respect to a particular use (Hassan and Ahamed 2017). We have already identified safe water "command areas" and safe water "demand areas" using GIS and geostatistical analysis. The IDW prediction method has identified the spatial determination for suitable areas for tubewell installation with aquifer depths and water-tables. It is noted that only the technological option for both the STW and DTW was considered for this Spatial Decision-Support System (SDSS). In considering the "command areas" and "demand areas" of safe water, it was our intention to identify the suitable areas for installing tubewells at different aquifer depths. Accordingly, we identified some suitable locations for installing tubewells for arsenic-safe water.

We considered both the shallow and deep aquifers for suitable area identification for safe water through tubewell technology. Figure 6.8 shows the suitable areas for arsenic-safe water at different depths of aquifers. At the depth of 8–25 m, there are areas located mainly in the north-eastern parts of Magura and parts of Wards 4 and 5 are within this category (Fig. 6.8). It is noted that we have identified suitable areas for installation of STW for arsenic-safe water. Several settlement clusters (*para* in Bangla, some 40–50 households on an average live in a *para*) have been identified in the study site and Wards 4 and 5 are dominating in this connection. Table 3 shows in detail about the suitable areas for arsenic-safe water sources at different depths of aquifers at different settlement clusters. Arsenic-safe water is frequently available from all depths of aquifer in Wards 4 and 5. Here, we have utilized the IDW prediction methods in different administrative wards and filter out the contaminated areas and existing safe water option in delineating the boundary line of suitable areas for arsenic-safe water.

At the depth of 8–25 m, arsenic-safe water can be available in Wards 4 and 5 and they are more specific in Chargram Village (Parr para and Sheikh para) and in Magura village (Hindu Para, Khan Para, Sardar Para, Sheikh Para, and Vogomani Para) in Ward 4. In Ward 5, arsenic-safe groundwater sources can be found in Baruipara

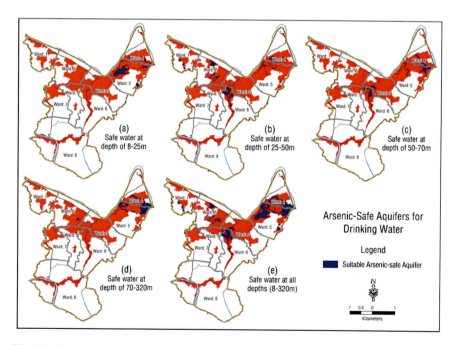

Fig. 6.8 Suitable arsenic-safe aquifer within the settlement area in Magura

Table 6.3 Suitable area selection for arsenic-safe tubewell in the study site

Depth (in meter)	Wards	Mouza	Village	Settlement cluster (Para)
8–25	Ward 4	Chorgram	Chorgram	Khan Para, Hindu Para
	Ward 5	Baruipara	Baruipara	Baruipara, Das Para, Maath Para
		Dholunda	Dholunda	Dakshin Para
25–50	Ward 1	Bolorampur	Bolorampur	Dakshin Para
	Ward 2	Magura	Chadkathi	Gazi Para, Morol Para
	Ward 3	Magura	Magura	Baag Para
	Ward 4	Chorgram	Chorgram	Hindu Para, Khan Para, Parr Para
	Ward 5	Dholunda	Dholunda	Das Para, Thakur Para
	Ward 6	Magura	Magura-Danga	Khan Para, Nath Para, Nomosuddar Para, Sadar Para
50–70	Ward 3	Magura	Magura	Baag Para
	Ward 4	Chorgram	Chorgram	Parr Para, Vogomani Para
	Ward 5	Baruipara	Baruipara	Baroi Para, Thakur Para
		Dholunda	Dholunda	Rishi Para
	Ward 6	Magura	Magura-Danga	Khan Para, Naath Para
	Ward 9	Roghunathpur	Roghunathpur	Sheikh Para
70–320	Ward 2	Magura	Chadkathi	Moddho Para
	Ward 4	Chorgram	Chorgram	Hindu Para, Khan Para, Parr Para
	Ward 5	Baruipara	Baruipara	Baroi Para, Thakur Para
		Dholunda	Dholunda	Rishi Para

village and in Dholunda village in Ward 5 (Fig. 6.8 and Table 6.3). At the depth of 25–50 m, arsenic-safe water can be obtained in several villages in a number of Wards (e.g. Bolorampur in Ward 1, Chadkathi in Ward 2, Magura in Ward 3, Chorgram in Ward 4, Dholunda in Ward 5, and Magura-Danga in Ward 6). At the depth of 50–70 m, there are safe water option in Wards 3, 4, 5, 6, and 9. Moreover, at the depth between 70 and 320 m, arsenic-safe water can be tapped from Wards 2, 4, and 5 (Table 6.3).

Deep aquifer is the most safe and reliable source for arsenic-safe and pathogen-safe water. The deep aquifer, more than 100 m depth, were found almost safe from arsenic contamination. There are sporadic arsenic-safe areas in Wards 2, 4, and 5 in the study site. A significant portion of areas in Wards 4 and 5 were identified as arsenic-safe areas and the rest of the Wards with small proportions were recognized as safe zones (Fig. 6.7 and Table 6.3). DTW can be installed in the areas for arsenic and pathogen-safe potable water. The study identified a very few scattered "pockets" of arsenic-safe zone at the shallow aquifer as well as in the deep aquifer as mentioned in Fig. 6.8. It is noted that we didn't consider sub-surface geology, but it would be better to consider sub-surface geology and other chemical elements for identifying safe water "command areas" and safe water "demand areas."

6.6 Discussion

In this section we would like to discuss the results of our research and its position with relation to other relevant study outcomes. In our study, we have identified some "arsenic-safe pockets" in groundwater sources with a combination of GIS and geostatistical methods. We have reviewed a plethora of literature, but we found few research papers for identifying or investigating the suitable areas for arsenic-safe groundwater sources. Tiwari et al. (Said and Yurtal 2019) developed hydrochemistry mapping to identify the spatial distribution of groundwater quality parameters in KBNIR area in Rajasthan of India. They observed some 36.02% of the populated area depend on exceptional groundwater quality and 23.56% on good groundwater quality, while poor and moderately poor groundwater quality are available for 6.26% and 34.16% respectively of the populated area. But they did not consider arsenic in the groundwater. In our present study we have identified that some 13.4% (87 hectare) of the settlement area has been demarcated for the safe drinking water option. Hassan and Ahamed 2017 investigated arsenic-safe aquifer in Bagerhat district in Bangladesh with OK prediction model and they found only an exceedingly small portion of area within the settlement zone as the possible source of safe drinking water.

Based on the groundwater data, Said and Yurtal (2019) assessed groundwater quality for drinking water purposes with OK and prepared interpolation maps for pH, electrical conductivity (EC) parameters, and total dissolved solids (TDS) at Puntland in Somalia. Using multiple statistical analysis and a geostatistical interpolation approach, Belkhiri and Narany (2015) explored the spatial variability of groundwater quality with determining factors and mechanisms controlling this variability on the Ain Azel Plain in Algeria. With the advection dispersion model, Radloff et al. (2017) explored the distribution of inorganic arsenic concentration with depth and time in a shallow Holocene aquifer at Araihazar in Bangladesh. Using multivariate analyses and geostatistics, Cai et al. (2015) investigated the pattern of spatial distribution and source of arsenic and heavy metals in southeast China and Han et al. (2013) examined the spatial and temporal patterns of groundwater arsenic in the shallow and deep groundwater of Yinchuan Plain, China. Andrade and Stigter (2013) used multivariate geostatistical modeling to assess the pattern of arsenic concentrations in shallow alluvial groundwater within the agricultural land in central Portugal. All the above mentioned literatures show the pattern of water quality distribution, but not the suitable area investigation for safe groundwater sources.

Mohana et al. (2020) analyzed risk-based water quality situation in rural areas in Bangladesh. They mainly focussed on water quality index and there was no spatial analysis in identifying risk-based areas. Using multivariate statistical methods (Principal Component Analysis-PCA and Hierarchical Clustering Analysis—HCA), and Water Quality Index (WQI), El Mountassir et al. (2020) determined the suitability of water for drinking. Their WQI shows that 6% of the samples are classified as "Good" and rest of the samples were categorized as "poor," "Very poor," and "unsuitable" for drinking. Applying the GIS-based approach and Groundwater Quality Index (GWQI), Honarbakhsh et al. (2019) investigated the suitable areas for safe drinking water at Marvdasht of Fars Province in Iran, but there is no arsenic information in the study. Kim et al. (2019) explored the groundwater productivity potential (GPP) of Okcheon city of South Korea using Boosted Regression Tree (BRT) model, Random Forest (RF) model, and the Logistic Regression (LR) model. They have identified "high quality," "moderate quality," and "low quality" areas. Applying geostatistical logistic regression models, Ayotte et al. (2006) and Yang et al. (2012) noticed the prediction of spatial probability of arsenic concentrations in groundwater in New England of the USA. Ahmadian (2013) compared the relative performance of classical geostatistical kriging methods in analysis of groundwater quality in Tehran-Karaj Plain. Our study shows the further suitable areas for arsenic-safe groundwater apart from the areas of existing safe aquifers.

6.7 Conclusion

This paper demonstrates the suitability of geostatistics in exploring the safe water pockets in shallow and deep aquifers. Mapping the proximity area of arsenic and functionalities of geostatistics allow different map data to be combined in determining suitable sites for different arsenic-safe groundwater sources. The safe water pockets can be utilized for safe drinking water option in coastal areas of Bangladesh. It is a major challenge in the coastal areas of Bangladesh to secure the potable water source to overcome the health hazards. It is necessary to minimize the risks since arsenic contamination in water potentially causes arsenicosis, saline water for hypertension, and pathogenic water for deadly diseases like cholera, typhoid, and diarrhea. Considering these health impacts, it is necessary to maintain water quality for drinking and cooking purposes. Therefore, a proper policy including safe water management can be formulated for ensuring proper utilization of arsenic-safe water and to prevent different water-related diseases. The use of safe water utilization could be an option to achieve the Sustainable Development Goal (SDG) 6 for "Clean Water and Sanitation" at some extent with exploring safe water aquifer.

Acknowledgements We would like to express our sincere thanks to the ICCO Cooperation&Kerk in Actie, Netherlands for their financial support for a research project on "Scanning and Mapping the WASH Situation in Coastal Bangladesh: Problems and Potential" (Project No.:71-02-10-013). We are grateful to MrTaritKanti Biswas and MrMahafuzur Rahman (Biplob) for their supports in collecting the relevant arsenic data from the field. We are also grateful to Mr Hussain Ahmad for his untired support in data entry operation.

Conflict of Interest The authors declare no conflict of interest.

References

Achour MH, Haroun AE, Schult CJ, Gasem KAM (2005)A new method to assess the environmental risk of a chemical process. Chem Eng Process 44(8):901–909. https://doi.org/10.1016/j.cep.2004.10.003

Adhikary PP, Dash CJ (2017) Comparison of deterministic and stochastic methods to predict spatial variation of groundwater depth. Appl Water Sci 7(1):339–348. https://doi.org/10.1007/s13201-014-0249-8

Ahmadian S (2013) Geostatistical based modelling of variations of groundwater quality during 2006 to 2009 (in Tehran-Karaj Plain). Journal of Basic Applied Science Research 3:264–272

Al-Abadi AM, Pourghasemi HR, Shahid S, Ghalib HB (2017) Spatial mapping of groundwater potential using entropy weighted linear aggregate novel approach and GIS. Arab J Sci Eng 42:1185–1199. https://doi.org/10.1007/s13369-016-2374-1

Al-Omran AM, Aly AA, Al-Wabel MI, Al-Shayaa MS, Sallam AS, Nadeem ME (2017) Geostatistical methods in evaluating spatial variability of groundwater quality in Al-Kharj Region, Saudi Arabia. Appl Water Sci 7:4013–4023. https://doi.org/10.1007/s13201-017-0552-2

Amiri V, Rezaei M, Sohrabi N (2014) Groundwater quality assessment using entropy weighted Water Quality Index (EWQI) in Lenjanat. Iran. Environmental Earth Sciences 72:3479–3490. https://doi.org/10.1007/s12665-014-3255-0

Andrade AIASS, Stigter TY (2013) The distribution of arsenic in shallow alluvial groundwater under agricultural land in central Portugal: insights from multivariate geostatistical modelling. Sci Total Environ 449:37–51. https://doi.org/10.1016/j.scitotenv.2013.01.033

Antunes IMHR, Albuquerque MTD, Seco MFM, Oliveira SF, Sanz G (2014) Uranium and Arsenic spatial distribution in the Águeda watershed groundwater. Procedia Earth and Planetary Science 8:13–17. https://doi.org/10.1016/j.proeps.2014.05.004

Ashraf M, Loftis JC, Hubbar KG (1997) Application of geostatistics to evaluate partial weather station networks. Agric Meteorol 84(3–4):255–271. https://doi.org/10.1016/S0168-1923(96)02358-1

Ayotte JD, Nolan BT, Nuckols JR, Cantor KP, Robinson GR, Baris D, Hayes L, Karagas M, Bress W, Silverman DT, Lubin JH (2006) Modeling the probability of arsenic in groundwater in New England as a tool for exposure assessment. Environ Sci Technol 40(11):3578–3585. https://doi.org/10.1021/es051972f

BBS (2011) Bangladesh Population census (2011) Zilla Statistics. GOB, Bangladesh Bureau of Statistics, Dhaka, .

Beg MK, Srivastav SK, Carranza EJM, de Smeth JB (2011) High fluoride incidence in groundwater and its potential health effects in parts of Raigarh district, Chhattisgarh. India. Current Science 100(5):750–754. www.jstor.org/stable/24075817

Belkhiri L, Mouni L (2014) Geochemical characterization of surface water and groundwater in Soummam Basin. Algeria. Nat Res Res 23(4):393–407. https://doi.org/10.1007/s11053-014-9243-y

Belkhiri L, Narany TS (2015) Using multivariate statistical analysis, geostatistical techniques and structural equation modeling to identify spatial variability of groundwater quality. Water Resour Manage 29:2073–2089. https://doi.org/10.1007/s11269-015-0929-7

Berke O (2004) Exploratory disease mapping: kriging the spatial risk function from regional count data. Int J Health Geogr 3(1):18. https://doi.org/10.1186/1476-072X-3-18

Bhuiyan MAH, Bodrud-Doza M, Islam ARMT, Rakib MA, Rahman MS, Ramanathan AL (2016) Assessment of groundwater quality of Lakshimpur district of Bangladesh using water quality indices, geostatistical methods, and multivariate analysis. Environmental Earth Sciences 75(12):1020. https://doi.org/10.1007/s12665-016-5823-y

Biswas A, Neidhardt H, Kundu AK, Halder D, Chatterjee D, Berner Z, Jacks G, Bhattacharya P (2014) Spatial, vertical and temporal variation of arsenic in shallow aquifers of the Bengal Basin: Controlling geochemical processes. Chem Geol 387:157–169. https://doi.org/10.1016/j.chemgeo.2014.08.022

Bonsor HC, MacDonald AM, Ahmed KM, Burgess WG, Basharat M, Calow RC, Dixit A, Foster SSD, Gopal K, Lapworth DJ, Moench M, Mukherjee A, Rao MS, Shamsudduha M, Smith L, Taylor RG, Tucker J, Steenbergen FV, Yadav SK, Zahid A (2017) Hydrogeological typologies of the Indo-Gangetic basin alluvial aquifer. South Asia. Hydrogeol J 25(5):1377–1406. https://doi.org/10.1007/s10040-017-1550-z

Cai L, Xu Z, Baoe P, He M, Dou L, Chen L, Zhou Y, Zhu YG (2015) Multivariate and geostatistical analyses of the spatial distribution and source of arsenic and heavy metals in the agricultural soils in Shunde, Southeast China. J Geochem Explor 148:189–195. https://doi.org/10.1016/j.gexplo.2014.09.010

CESCR (2002) General Comment No. 15: The Right to Water. (Articles 11 and 12 of the Covenant). UN Committee on Economic, Social and Cultural Rights (CESCR), (UN document: E/C.12/2002/11)

Chen J, Qian H, Wu H, Gao Y, Li X (2017) Assessment of arsenic and fluoride pollution in groundwater in Dawukou area, Northwest China, and the associated health risk for inhabitants. Environ Earth Sci 76:314. https://doi.org/10.1007/s12665-017-6629-2

Cinti D, Poncia PP, Brusca L, Tassi F, Quattrocchi F, Vasellide O (2015) Spatial distribution of arsenic, uranium and vanadium in the volcanic-sedimentary aquifers of the Vicano-Cimino Volcanic District (Central Italy). J Geochem Explor 152:123–133. https://doi.org/10.1016/j.gexplo.2015.02.008

Civita MV (2010) The combined approach when assessing and mapping groundwater vulnerability to contamination. J Water Resour Prot 2:14–28. https://doi.org/10.4236/jwarp.2010.21003

Dehghanzadeh R, HirNS SJS, Taghipour M (2015) Integrated assessment of spatial and temporal variations of groundwater quality in the Eastern Area of Urmia Salt Lake Basin using multivariate statistical analysis. Water Resour Manage 29(4):1351–1364. https://doi.org/10.1007/s11269-014-0877-7

Delbari M, Amiri M, Motlagh MB (2016) Assessing groundwater quality for irrigation using indicator kriging method. Appl Water Sci 6(4):371–381. https://doi.org/10.1007/s13201-014-0230-6

Díaz SL, Espósito ME, del Carmen BM, Amiotti NM, Schmidt ES, Sequeira ME, Paoloni JD, Nicolli HB (2016) Control factors of the spatial distribution of arsenic and other associated elements in loess soils and waters of the southern Pampa (Argentina). CATENA 140:205–216. https://doi.org/10.1016/j.catena.2016.01.013

Dokou Z, Kourgialas NN, Karatzas GP (2015) Assessing groundwater quality in Greece based on spatial and temporal analysis. Environ Monit Assess 187(774):1–18. https://doi.org/10.1007/s10661-015-4998-0

Dowling CB, Poreda RJ, Basu AR, Petters SL, Aggarwal PK (2002) Geochemical study of arsenic release mechanisms in the Bengal Basin groundwater. Water Resour Res 38:1173–1190. https://doi.org/10.1029/2001WR000968

DPHE, JICA. (2010) Situation Analysis of Arsenic Mitigation 2009. Department of Public Health Engineering and Japan International Cooperation Agency, Dhaka

Dummer TJB, Yu ZM, Nauta L, Murimboh JD, Parker L (2015) Geostatistical modelling of arsenic in drinking water wells and related toenail arsenic concentrations across Nova Scotia, Canada. Sci Total Environ 505:1248–1258. https://doi.org/10.1016/j.scitotenv.2014.02.055

El Mountassir O, Bahir M, Ouazar D, Ouhamdouch S, Chehbouni A, Ouarani M (2020) The use of GIS and water quality index to assess groundwater quality of Krimat aquifer (Essaouira; Morocco). SN Applied Sciences 2:871. https://doi.org/10.1007/s42452-020-2653-z

Elumalai V, Brindha K, Sithole B, Lakshmanan E (2017) Spatial interpolation methods and geostatistics for mapping groundwater contamination in a coastal area. Environ Sci Pollut Res 24(12):11601–11617. https://doi.org/10.1007/s11356-017-8681-6

Flanagan SV, Spayd SE, Procopio NA, Marvinney RG, Smith AE, Chillrud SN, Braman S, Zheng Y (2016) Arsenic in private well water part of 3: Socioeconomic vulnerability to exposure in Maine and New Jersey. Sci Total Environ 562:1019–1030. https://doi.org/10.1016/j.scitotenv.2016.03.217

GOB (2013) Bangladesh Water Act 2013. (Act No 14 of 2013). Article 3, Bangladesh Gazette, pp.14277–14300, 29 December 2013, Dhaka: Government of Bangladesh

Golia EE, Dimirkou A, Floras SA (2015) Spatial monitoring of arsenic and heavy metals in the Almyros area, Central Greece. Statistical approach for assessing the sources of contamination. Environ Monitor Assess 187(7):399. https://doi.org/10.1007/s10661-015-4624-1

Goovaerts P, AvRuskin G, Meliker J, Slotnick M, Jacquez G, Nriagu J (2005) Geostatistical modeling of the spatial variability of arsenic in groundwater of southeast Michigan. Water Resour Res 41:W07013. https://doi.org/10.1029/2004WR003705

Gorgij AD, Kisi O, Moghaddam AA, Taghipour A (2017) Groundwater quality ranking for drinking purposes, using the entropy method and the spatial autocorrelation index. Environmental Earth Sciences 76:269. https://doi.org/10.1007/s12665-017-6589-6

Guo H, Zhang Y, Xing L, Jia Y (2012) Spatial variation in arsenic and fluoride concentrations of shallow groundwater from the town of Shahai in the Hetao basin, Inner Mongolia. Appl Geochem 27:2187–2196. https://doi.org/10.1016/j.apgeochem.2012.01.016

Guo Q, Guo H, Yang Y, Han S, Zhang F (2014) Hydrogeochemical contrasts between low and high arsenic groundwater and its implications for arsenic mobilization in shallow aquifers of the northern Yinchuan Basin, P.R. China. J Hydrol 518:464–476. https://doi.org/10.1016/j.jhydrol.2014.06.026

Han S, Zhang F, Zhang H, An Y, Wang Y, Wu X, Wang C (2013) Spatial and temporal patterns of groundwater arsenic in shallow and deep groundwater of Yinchuan Plain, China. J Geochem Explor 135:71–78. https://doi.org/10.1016/j.gexplo.2012.11.005

Hassan MM (2005) Arsenic poisoning in Bangladesh: spatial mitigation planning with GIS and public participation. Health Policy 74(3):247–260. https://doi.org/10.1016/j.healthpol.2005.01.008

Hassan MM (2018) Arsenic in groundwater: poisoning and risk assessment. CRC Press, Boca Raton (USA). https://doi.org/10.1201/9781315117034

Hassan MM, Atkins PJ (2006) Arsenic in Bangladesh. Geogr Rev 19(4):14–17

Hassan MM, Atkins PJ (2007) Arsenic risk mapping in Bangladesh: a simulation technique of Cokriging estimation from regional count data. J Environ Sci Health Part A 42:1719–1728. https://doi.org/10.1080/10934520701564210

Hassan MM, Atkins PJ (2011) Application of geostatistics with Indicator Kriging for analyzing spatial variability of groundwater arsenic concentrations in southwest Bangladesh. J Environ Sci Health Part A 46(11):1185–1196. https://doi.org/10.1080/10934529.2011.598771

Hassan MM (2015) Scanning and mapping the WASH situation in coastal Bangladesh: problems and potential. Narrative report, Geoecological Research Team (GeRT), Dhaka

Hassan MM, Ahamed R (2017) Arsenic-safe aquifers in Coastal Bangladesh: an investigation with Ordinary Kriging Estimation. The International Archives for Photogrammetry, Remote Sensing and Spatial Information Sciences XLII-4/W5:97–105. https://doi.org/10.5194/isprs-archives-XLII-4-W5-97-2017

Honarbakhsh A, Tahmoures M, Tashayo B, Mousazadeh M, Ingram B, Ostovari Y (2019) GIS-based assessment of groundwater quality for drinking purpose in northern part of Fars province, Marvdasht. J Water Supply Res Technol 68:3. https://doi.org/10.2166/aqua.2019.119

Hossain MM, Piantanakulchai M (2013) Groundwater arsenic contamination risk prediction using GIS and classification tree method. Eng Geol 156:37–45. https://doi.org/10.1016/j.enggeo.2013.01.007

Isaaks EH, Srivastava RM (1989) An introduction to applied geostatistics. Oxford University Press, New York

Islam ARMT, Shen S, Bodrud-Doza M, Rahman MA, Das S (2017) Assessment of trace elements of groundwater and their spatial distribution in Rangpur district. Bangladesh. Arab J Geosci 10:95. https://doi.org/10.1007/s12517-017-2886-3

Javed S, Ali A, Ullah S (2017) Spatial assessment of water quality parameters in Jhelum city (Pakistan). Environ Monit Assess 189:119. https://doi.org/10.1007/s10661-017-5822-9

Johnston R, Heijnen H, Wurzel P (2001) Safe water technology. World Health Organization, Chapter VI, Geneva

Kim JC, jung HS, Lee S (2019) Spatial mapping of the groundwater potential of the Geum River Basin using ensemble models based on remote sensing images. Remote Sens 11 (2285):1–19. https://doi.org/10.3390/rs11192285

Kourgialas NN, Karatzas GP (2015) An integrated approach for the assessment of groundwater contamination risk/vulnerability using analytical and numerical tools within a GIS framework. Hydrol Sci J 60(1):111–132. https://doi.org/10.1080/02626667.2014.885653

Lark RM (2000) A comparison of some robust estimators of the variogram for use in soil survey. Eur J Soil Sci 51(1):137–157. https://doi.org/10.1046/j.1365-2389.2000.00280.x

Liu CW, Jang CS, Liao CM (2004) Evaluation of arsenic contamination potential using indicator kriging in the Yun-Lin Aquifer (Taiwan). Sci Total Environ 321:173–188. https://doi.org/10.1016/j.scitotenv.2003.09.002

Longley PA, Goodchild MF, Maguire DJ, Rhind DW (2001) geographic information systems & science. Wiley, Chichester

Ma L, Wang L, Jia Y, Yang Z (2016) Arsenic speciation in locally grown rice grains from Hunan Province, China: Spatial distribution and potential health risk. Sci Total Environ 557–558:438–444. https://doi.org/10.1016/j.scitotenv.2016.03.051

Machiwal D, Mishra A, Jha MK, Sharma A, Sisodia SS (2012) Modeling short-term spatial and temporal variability of groundwater level using geostatistics and GIS. Nat Resour Res 21(1):117–136. https://doi.org/10.1007/s11053-011-9167-8

Marko K, Al-Amri NS, Elfeki AMM (2014) Geostatistical analysis using GIS for mapping groundwater quality: case study in the recharge area of Wadi Usfan, western Saudi Arabia. Arab J Geosci 7(12):5239–5252. https://doi.org/10.1007/s12517-013-1156-2

Masoud AA (2014) Groundwater quality assessment of the shallow aquifers west of the Nile Delta (Egypt) using multivariate statistical and geostatistical techniques. J Afr Earth Sc 95:123–137. https://doi.org/10.1016/j.jafrearsci.2014.03.006

Mehrjardi RT, Jahromi MZ, Heidari A (2008) Spatial distribution of groundwater quality with geostatistics, case study: Yazd-Ardakan plain. World Appl Sci J 4(1):9–17

Mizan SA, Chatterjee A, Ahmed S (2017) Arsenic enrichment in groundwater in southern flood plain of Ganga-Son interfluves. Arab J Geosci 10:100. https://doi.org/10.1007/s12517-017-2880-9

Mohana AA, Rahman MA, Islam MR (2020) Deep and shallow tubewell water from an arsenic-contaminated area in rural Bangladesh: risk-based status. Int J Energy Water Res 4:163–179. https://doi.org/10.1007/s42108-020-00059-4

Mukherjee A, Fryar AE (2008) Deeper groundwater chemistry and geochemical modeling of the arsenic affected western Bengal basin, West Bengal, India. Appl Geochem 23:863–894. https://doi.org/10.1016/j.apgeochem.2007.07.011

Narany TS, Ramli MF, Aris AZ, Sulaiman WNA, Fakharian K (2014) Spatiotemporal variation of groundwater quality using integrated multivariate statistical and geostatistical approaches in Amol-Babol Plain. Iran. Environ Monitor Assess 186(9):5797–5815. https://doi.org/10.1007/s10661-014-3820-8

Olea RA, Raju NJ, Egozcue JJ, Powlowsky_Glahn V, Singh S, (2017) Advancements in hydro-chemistry mapping: methods and application to groundwater arsenic and iron concentrations in Varanasi, Uttar Pradesh, India. Stoch Env Res Risk Assess 2017:1–19. https://doi.org/10.1007/s00477-017-1390-3

Oliver MA, Webster R (2015) Basic steps in geostatistics: The Variogram and Kriging. Springer, Heidelberg. https://doi.org/10.1007/978-3-319-15865-5

Ou C, St-Hilaire A, Ouarda TB, Conly FM, Armstrong N, Khalil B, Proulx-McInnis S (2012) Coupling geostatistical approaches with PCA and fuzzy optimal model (FOM) for the integrated assessment of sampling locations of water quality monitoring networks (WQMNs). J Environ Monit 14(12):3118–3128. https://doi.org/10.1039/C2EM30372H

Pardo-Igúzquiza E, Chica-Olmo M, Luque-Espinar JA, Rodríguez-Galiano V (2015) Compositional cokriging formapping the probability risk of groundwater contamination by nitrates. Sci Total Environ 532:162–175. https://doi.org/10.1016/j.scitotenv.2015.06.004

Peters SC, Burkert L (2008) The occurrence and geochemistry of arsenic in groundwaters of the Newark basin of Pennsylvania. Appl Geochem 23:85–98. https://doi.org/10.1016/j.apgeochem.2007.10.008

Pi K, Wang Y, Xie X, Liu Y, Ma T, Su C (2016) Multilevel hydrogeochemical monitoring of spatial distribution of arsenic: A case study at Datong Basin, northern China. J Geochem Explor 161:16–26. https://doi.org/10.1016/j.gexplo.2015.09.002

Radloff KA, Zheng Y, Stute M, Weinman B, Bostick B, Mihajlov I, Bounds M, Rahman MM, Huq MR, Ahmed KM, Schlosser P, Geen AV (2017) Reversible adsorption and flushing of arsenic in a shallow, Holocene aquifer of Bangladesh. Appl Geochem 77:142–157. https://doi.org/10.1016/j.apgeochem.2015.11.003

Rebolledo B, Gil A, Flotats X, Sánchez JA (2016) Assessment of groundwater vulnerability to nitrates from agricultural sources using a GIS-compatible logic multicriteria model. J Environ Manage 171:70–80. https://doi.org/10.1016/j.jenvman.2016.01.041

Safe Drinking Water Foundation (2020) Human Rights. [Online] Available at: https://www.safewater.org/fact-sheets-1/2017/1/23/human-rights. Accessed on 27 June 2020

Said AA, Yurtal R (2019) Spatial groundwater quality assessment using geostatistics in Puntland, Somalia. Int J Sci Technol Res 5(1):1–9

Sengupta S, Sracek O, Jean JS, Lu HY, Wang CH, Palcsu L, Liu CC, Jen CH, Bhattacharya P (2014) Spatial variation of groundwater arsenic distribution in the Chianan Plain, SW Taiwan: role of local hydrogeological factors and geothermal Sources. J Hydrol 518:393–409. https://doi.org/10.1016/j.jhydrol.2014.03.067

Serón FJ, Badal JI, Sabadell FJ (2001) Spatial prediction procedures for regionalization and 3-D imaging of Earth structures. Phys Earth Planet Inter 123(2–4):149–168. https://doi.org/10.1016/S0031-9201(00)00207-7

Sikdar PK, Chakraborty S (2017) Numerical modelling of groundwater flow to understand the impacts of pumping on arsenic migration in the aquifer of North Bengal Plain. J Earth Syst Sci 126:29. https://doi.org/10.1007/s12040-017-0799-x

Tiwari K, Goyal R, Sarkar A (2017) Assessment of groundwater quality using geographical information system (GIS), at north-east Cairo. Egypt. Environmental Process 4:645–662. https://doi.org/10.1007/s40710-017-0257-4

Tsanis IK, Gad MA (2001) A GIS precipitation method for analysis of storm kinematics. Environ Model Softw 16(3):273–281. https://doi.org/10.1016/S1364-8152(00)00068-2

Umitsu M (1993) Late quaternary sedimentary environments and landforms in the Ganges Delta. Sediment Geol 83:177–186. https://doi.org/10.1016/0037-0738(93)90011-S

UN (2010) The human right to water and sanitation. UN General Assembly Resolution 64/292 of 28 July 2010 (https://undocs.org/A/RES/64/292)

UNECE (2012) No One Left Behind: Good Practices to ensure equitable access to water and sanitation in the Pan-European Region. New York and Geneva: UN and WHO.

UNHRC (2012) The human right to safe drinking water and sanitation. UN Human Rights Council Resolution 21/2 of September 2012 (A/HRC/21/L.1).

WHO (2004) WHO guidelines for drinking-water quality. World Health Organization, Geneva

WHO (2015) Drinking Water. Fact Sheet No 391. Geneva: World Health Organization

Wilbers GJ, Becker M, Nga LT, Sebesvari Z, Renaud FG (2014) Spatial and temporal variability of surface water pollution in the Mekong Delta. Vietnam. Sci Total Environ 485–486:653–665. https://doi.org/10.1016/j.scitotenv.2014.03.049

Yang Q, Jung HB, Marvinney RG, Culbertson CW, Zheng Y (2012) Can arsenic occurrence rates in bedrock aquifers be predicted? Environ Sci Technol 46(4):2080–2087. https://doi.org/10.1021/es203793x

Yazdanpanah N (2016) Spatiotemporal mapping of groundwater quality for irrigation using geostatistical analysis combined with a linear regression method. Model Earth Syst Environ 2:18. https://doi.org/10.1007/s40808-015-0071-9

Yeşilkanat CM, Kobya Y, Taşkin H, Çevik U (2015) Dose rate estimates and spatial interpolation maps of outdoor gamma dose rate with geostatistical methods: a case study from Artvin, Turkey. J Environ Radioact 150:132–144. https://doi.org/10.1016/j.jenvrad.2015.08.011

Yu WH, Harvey CM, Harvey CF (2003) Arsenic in groundwater in Bangladesh: a geostatistical and epidemiological framework for evaluating health effects, and potential remedies. Water Resour Res 39(6):1146–1163. https://doi.org/10.1029/2002WR001327

Chapter 7
Biodiversity Degradation
of South-Western Region in Saudi Arabia

Adel Moatamed⊙

Abstract This study aimed to monitor the degradation of biodiversity in one of the southwestern region of Saudi Arabia. The study area has a rich diversity of flora and fauna, particularly in juniper trees ecosystem. The total area covered with natural vegetation cover was 4385.6 km^2 in 1980; it decreased to 3645.5 km^2 according to the satellite image of 2019. Population growth and urban sprawl were the main factors causing the degradation of natural vegetation cover in this region.

Keywords Biodiversity · Juniper trees ecosystem · Aseer · Abha urban area · Human activities

7.1 Introduction

Biodiversity is one of the most important environmental issues, since the mid-1980s where the international interest in the biodiversity has appeared.

The Convention on Biological Diversity (CBD), adopted at the 1992 UN Conference on Environment and Development in Rio de Janeiro, Brazil, was intended to stimulate and encourage governments to counteract rapid declines in the extent and condition of natural ecosystems, the abundance of wild species and the benefits to humanity that wild nature provides. Under the Convention 190 countries pledged in 2002 to reduce significantly the rate of biodiversity loss by 2010. This objective was also incorporated into the Millennium Development Goals (UN 2009).

Recent studies underline the unprecedented scale and magnitude of biodiversity loss. The current rate of human-induced species extinction is believed to be far higher

A. Moatamed (✉)
Department of Geography, College of Humanities, King Khalid University, Abha, Saudi Arabia
e-mail: amuatmed@kku.edu.sa

Faculty of Arts, Department of Geography, Assiut University, Assiut, Egypt

Prince Sultan Bin Abdul-Aziz Center for Environment and Tourism Research and Studies, King Khalid University, Abha, Saudi Arabia

than the commonly estimated background rate of extinction (Sethi and YatishLele 2017).

The study area has a unique diversity of vegetation, wild animals and birds, it contains about 55% of endemic species, and it contains about 50% of the birds, which live in the Saudi Arabia.

Uncertainties exist in the cause–effect relationship between responses, pressures, states, and benefits. Imperfect indicators in each focal area may magnify these uncertainties. Consequently, there will be a need for an adaptive management approach to the interpretation and use of indicators in which both the responses and the effort put into them are modified iteratively according to outcomes (Sparks et al. 2011).

The Kingdom of Saudi Arabia has a great diversity of biological components of natural vegetation, wild animals, as well as distinct ecological habitats, these components varying from region to other inside the kingdom; the study area has the most richness of the biodiversity.

The author aims to assess the degradation of biodiversity, through studying the degradation of natural vegetation cover.

7.2 Material and Methods

7.2.1 Study Area

The area under investigation is extended astronomically between 17: 20 N and from 34: 40 E. It bordered on the west by the Red Sea coast, and bordered from the east and north by the borders of the Makkah area, and from the south by the Saudi-Yemeni political borders (Fig. 7.1).

The physical characteristics of the study area (Fig. 7.2) left its effects upon the climatic features so rainfall ranges from 238 to 500 mm per year. Rain falls in the spring and early summer months, the study area exposed to the southwestern wind, and the highlands are a catalyst for large rainfall.

In Saudi Arabia, there are 2250 plant species, of which 132 plant species include 837 species, of which 105 species grow in sand dunes, 90 species of saline plants, 75 species of trees, and 12 species of aquatic plants. Endemic local species, the southwestern region are the most abundant of plant content, more abundant in endemic species (Saudi wildlife Authority 2005).

The most important dense forest trees, and prevalent in the region is juniper trees, followed by the density of trees Acacieae, Acacia tortilis, and Acacia ehrenbergiana.

The most important areas of the spread of tamrix in the region of Balqarn, on the western slopes of Abha Heights there are also types of trees scattered in the valleys of Tihamat Asir as well as in the valleys of the Sarra and the plateau. As for the trees of Acacieae, and Acacia tortilis, Acacia and Salvadoraaustralis shrubs, they are frequent in Tihamaand in the highlands and plateau. However, the Acaieae trees spread in some areas of the plateau, east of Khamis Mushait.

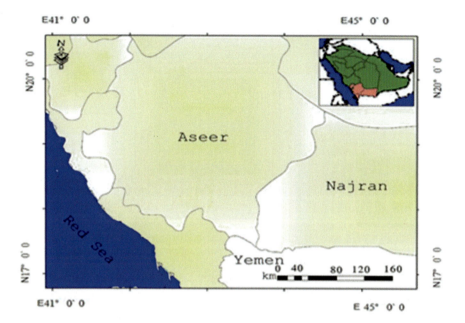

Fig. 7.1 Location of the study area

The dominant vegetation is the area in a number of species which primarily consist of juniper trees, as well as different species of broad-leaved such as domestic Neem and others. However, each sub-region within this area has its own characteristics. The region is dominated by sparse to medium density tree cover consisting primarily of juniper trees (El-Juhany and Aref 2013).

As this study relies on directed change analysis, it is a valuable technique for detecting changes that affect both land use and coverage patterns, and how to determine a reasonable identification of large changes and the direction of change.

The processes of monitoring and measuring these changes are indispensable aspects for reaching a deeper and deeper understanding of the mechanism by which the change is taking place, as well as modeling the effect of this change on the environment as well as the associated ecosystems at different levels (Muhlestein 2008).

This is remote sensing technology as a suitable source through which levels of change can be extracted with clear efficiency in both land use and its coverage pattern, during the past two decades, and there is a growing trend in the development of technologies to detect changes through the use of remote sensing technology.

The space visual analysis is one of the most important methods used to detect changes in the Earth's surface. The most important results of this cartographic method can be summarized in the following two axes:

First: The changes that occur in the Earth's organization as a result of human activities may be mainly related to economic activities or conditions, such as increasing

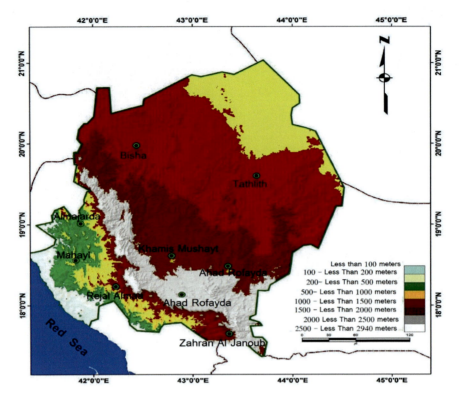

Fig. 7.2 Topography of the study area

the population associated with other changes expected in changing the patterns of use of the land and its body (Muhlestein 2008).

The second is that urban growth, especially population movements and commercial activities will transfer urban attributes to rural areas, which has negative impacts on the ecosystem (Squires 2002).

Whereas, updating the database using geographical information systems and remote sensing requires exploring recent changes, which mainly depend on the sequence and continuity of economic and urban changes and socio-economic development processes in the region (Belal and Moghanm 2011).

7.2.2 Results and Discussion

Through the last five decades, the study area exposed to intensive human activities, which led to great changes in the total area of natural vegetation and its species composition.

The total area of vegetation cover in Aseer area was 4385.6 km^2 in 1980, it decreased to 3645.5 km^2 according to the satellite image of 2019.

The maps no (Figs. 7.3 and 7.4) show the degradation in vegetation cover which was occupying nearly 44.000 km^2, but now, according to the recent satellite image 2019, it has shrunk to nearly 36.000 km^2, in other words the study area has lost nearly 750 km^2 of the green cover.

In more details we can note a distinct case of natural vegetation degradation as a direct reflect of human impact, that found in Abha urban area (the largest urban area in Asir, includes three big cities: Abha, KhamisMushayt, AhadRofida) the maps No (Fig. 7.5) shows that.

The area occupied by natural vegetation in 1970 was estimated at 754.1296 km^2, and then declined in 2010 to 648.1339 km^2, which is about 15% less than in 1970. The results of analysis of satellite image of 2019 showed that the contraction reached the greatest extent. About 47% more than it was in 1970 with an estimated area of 354.8422 km^2.

The three cities that constitute the study area (Abha, Khamis Mushayt, Ahad Rafidah) had a small area of about 28 km^2 in 1980, and then increased within five years to reach the extent of multiplication registered about 56 km^2, and continued to increase in the same way amounted to 129 km^2 in 1990 and represents the period from 1980 to 1995, which recorded significant leaps in the growth of urban mass.

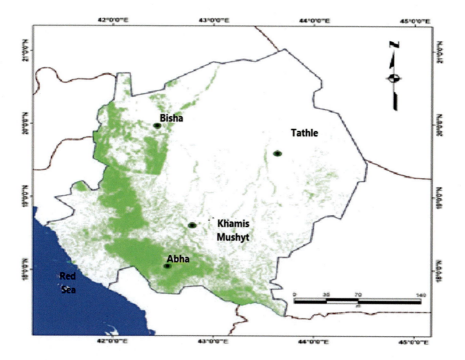

Fig. 7.3 Natural vegetation in the study area 1980

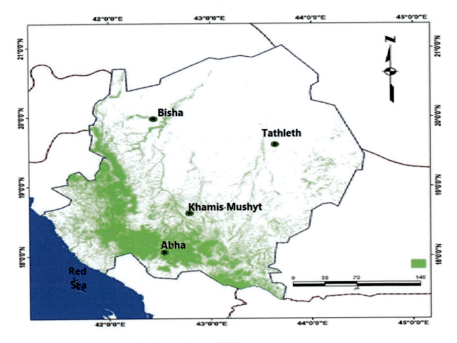

Fig. 7.4 Natural vegetation in the study area 2019

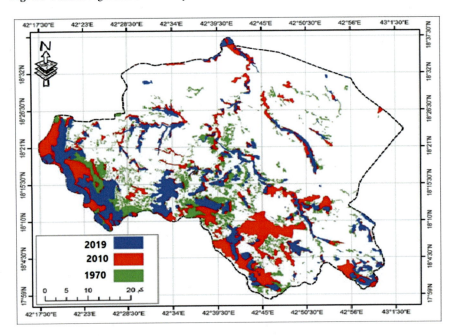

Fig. 7.5 Changes of natural vegetation cover in Abga Urban Area (1970–2019)

Figures 7.6a, b represent the growth of the built up area of Abha Urban Area in the period of 1980–2019.

It can be said that the human activities are the major responsibilities of the degradation, which recorded in the study area. Population Growth and urbanization are the most effect factors, which led to the degradation of the natural vegetation cover of the study area.

A large number of people moved from pastoral and agricultural work to service, and the transformation of these people to the lives of stability and urbanization more clearly, in addition to what represented by the abnormal increase through waves of migration to the region to work in various economic sectors as a primary source of this demographic change. Population growth requires the construction of residential and commercial areas, public facilities and new transport routes, associated with the transformation of the original use of forest land, agricultural land and fragile soils, resulting in changes in both land use patterns and cover (Muhammad Rahman 2016, p. 5).

The following figure number (Fig. 7.7) shows the numerical increase of the population of the study area from 1990 to 2019, just over a quarter of a century, indicating clear increases in the total population. The total population increased from just over half a million to 1990 to more than one million and one hundred thousand according to the latest 2019 population census.

The term urbanization generally refers to the increase in both the population and the associated construction of residential settlements, and thus includes the increase in which number and size of the cities, as well as the population movement to the urban areas (Uttara et al. 2012).

As in most areas of the Kingdom, the study area witnessed a remarkable growth in population through the past four decades. Most of this increase was accompanied by the economic leap associated with the discovery of oil.

The economic leap in the Kingdom of Saudi Arabia has been closely associated with a big change in the lifestyle and economic and social characteristics of the population.

The rate of nomads and nomadic populations living where the water and grasses has declined not know stability in its real sense. Statistical data indicate that the percentage of workers in the agricultural, pastoral, and related sectors decreased by more than 10% compared with the census of 1990 and decreased from 16 to 6% according to the results of the 2019 census.

7.3 Conclusion

According to the results of this paper it turns out how the biodiversity in the study area facing numerous of problems which leading to losing vital components of the ecosystem. The natural vegetation is one of these components which suffering from the human activities affects, the study area losing about 750 km^2 of its cover through the last five decades. The most effected factors are the population growth and the

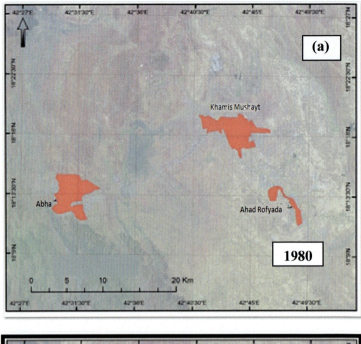

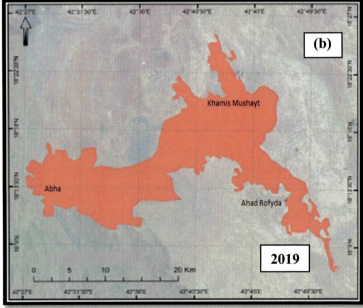

Fig. 7.6 Growth of the built up area of Abha Urban Area during the period (1980–2019). **a** In 1980, **b** In 2019

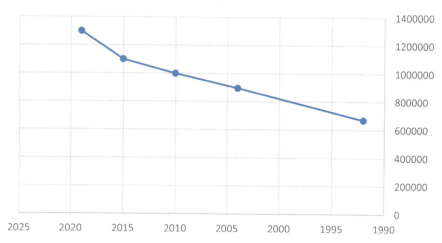

Fig. 7.7 Total population growth of the study area (1990: 2019)

urbanization. It should be noted the study area is candidate for more lose more of its bio components, so the conservation and preservation efforts must be more efficient and effective.

References

Belal AA, Moghanm FS (2011) Detecting urban growth using remote sensing and GIS techniques in Al Gharbiya governorate. Egypt J Remote Sense Space Sci 14:73–79

El-Juhany, Aref IM (2013) The present status of the natural forests in the Southwestern Saudi Arabia: 3-Asir and East Jazan Forests. World Appl Sci J 21(5):710–726. https://doi.org/10.5829/idosi.wasj.2013.21.5.2841

Muhlestein KN (2008) Land use land cover change analysis of Maverick County Texas along the US Mexico border. University of Texas at San Antonio EnvironmentalScience and Engineering PhD Program

Saudi wildlife Authority (2005) https://swa.gov.sa/En/Pages/default.aspx

Sethi P, Lele Y (2017) People, planet and progress: beyond 2015, the energy and resource institute (TREI). In: Chakrabarti D (ed). New Delhi, India

Sparks TH et al (2011) Linked indicator sets for addressing biodiversity loss. Fauna & Flora Int Oryx 45(3):411–419. https://doi.org/10.1017/S003060531100024X

Squires GD (2002) Urban Sprawl and the uneven development of metropolitan America causes, consequences, & policy responses. In: Urban Sprawl. Urban Institute Press, Washington

TurnerII BL, Skole D, Sanderson S, Fischer G, Fresco L, Leemans R (1995) Land-use and landcover changescience/research plan. IGBP report no. 35, HDP report No 7, Stockholm and Geneva

UN (2009) The millennium development goals report 2009. UN, New York

Uttara S, Nishi B ,Vanita A (2012) Impactsof urbanization on environment. Int J Res Eng Appl Sci Pakistan. http://www.euroasiapub.org.

Chapter 8
Causes and Effects of Water Logging in Dhaka City

Mallik Akram Hossain, Sanjia Mahiuddin, Arif Uddin Ahmad, and A. H. M. Monzurul Mamun

Abstract Dhaka city, the capital of Bangladesh is one of the densely populated Mega Cities in South Asian region. The city consists of two city corporations: Dhaka South City Corporation (DSCC) and Dhaka North City Corporation (DNCC), and is facing water logging every year during monsoon. The main objective of the study is to determine the causes and consequences of water logging in Dhaka City. Twenty seven vulnerable areas in the city have been identified through field visits, and reports from the national daily newspaper. To understand the reasons and consequences of water logging, intensive questionnaire survey among the respondents is conducted in 2019. Three hundred respondents have been interviewed from seven areas of the city. The findings of the research indicate that water logging is generated by both natural and man-made causes. Water logging also creates adverse social, economic, environmental, and health impact on the city dwellers. To mitigate the water logging problem, widening the drainage system, regular cleaning of the drains, and awareness of the inhabitants are urgently required to achieve the goal of a livable city.

Keywords Water logging · Monsoon · Causes · Impacts · Dhaka city

M. A. Hossain (✉) · S. Mahiuddin
Department of Geography and Environment, Jagannath University, 1100 Dhaka, Bangladesh
e-mail: mallik.a@geography.jnu.ac.bd

A. U. Ahmad
Department of Geomatics, Patuakhali Science and Technology University (PSTU), Patuakhali, Bangladesh

A. H. M. Monzurul Mamun
Geography and Environment, Pabna University of Science and Technology, Pabna, Bangladesh
e-mail: hasnat@pust.ac.bd

8.1 Introduction

In the twenty-first century, water logging emerges as a big issue globally (Rahman and Debnath 2015; Majumder et al. 2018). Water logging is the state of flooding especially in built up areas caused by heavy rainfall and responsible for creating adverse impact on life and livelihood (Towhid 2004; Islam et al. 2015). Urban storm water logging (USWL) is a global concern today for countries like USA, Canada, Europe, Australia, Philippines, Sri Lanka, Japan, and China (Li 2012; Han et al. 2006; Djordjevic et al. 2011; Akter et al. 2017). Water logging is considered as one of the severe environmental hazards in the Asian countries especially in Bangladesh, India, China, Pakistan, etc. (Sahu 2014; Sar et al. 2015). Developing countries are more vulnerable to urban flooding compared to developed countries due to their limited and mismanagement of drainage systems (Zurbrugg 2003; Haque 2013; Pervin et al. 2019). The narrow and limited capacity of drainage systems lead to urban flooding and water logging in the South Asian cities (Pervin et al. 2019). Water logging creates environmental as well as socio-economic problems, which are directly related with the changes in land use and land cover (Sar et al. 2015). During the monsoon, urban water logging has become a regular phenomenon for the city dwellers because of rapid urbanization and climate change impact, which causes disruption of social life and national economic losses every year (Akter et al. 2017; Zhang and Pan 2014). However, water logging is an environmental issue for many cities, the way of its prevention and mitigations to date have received not much attention by the city planners and policy makers (Awal 2014).

Dhaka city, a rapidly growing megacity has been suffering from many environmental problems (Dani 1962; Islam et al. 2010; Mahmud et al. 2011; Ishtiaque et al. 2014). Along with other Asian cities, Dhaka is one of the most polluted cities in the world (The Daily Star 2019). Rapid and unplanned urbanization is depleting as well as reducing the water bodies, wetlands, and flow-paths continuously, which is responsible for rainfall-flooding and drainage congestion (Chowdhury et al. 1998; Reza and Alam 2002; Kamal and Midorikawa 2004; Dewan and Nishigaki 2006; Islam et al. 2012). During monsoon, rainfall-induced water logging has been very common in Dhaka city (Islam et al. 2010; World Bank 2007; Dhaka metropolitan development plan strategic environmental assessment, prepared by SENES consultants limited in association with Techno Consult 2007). The frequency of this problem is increasing day by day especially in monsoon period due to the drainage congestion (Towhid 2004; Huq and Alam 2003). In recent years, water logging has become a big burden for the city dwellers and is creating adverse social, economic, and environmental consequences (Sar et al. 2015; Subrina and Chowdhury 2018). Water logging is not just the result of heavy rainfall and extreme climatic events; but also the outcome of the consequent changes in the built-up areas and urban environment (Hasan et al. 2018). Not only the capital city Dhaka, but also other cities are facing this hazard every year (Pervin et al. 2019). For instance, the city of Chittagong, Khulna, Sylhet, and Narayanganj are vulnerable to water logging because of unabated urbanization along with unplanned development works (Majumder et al. 2018; Islam et al. 2015;

Pervin et al. 2019; Hasan et al. 2018; Ashraf and Chowdhury 2009; Rahman et al. 2009; Khan 2017; Chakma and Chakrabartty 2017). Drainage facilities in urban areas are mostly inappropriate and also in poor condition. These causes are also responsible for urban water logging and flash-flooding (Towhid 2004; Jahan 2012; Datta et al. 2017).

Severe water logging incurs huge loss to the urban dwellers and creates negative social, physical, economic, environmental, and health impact every year especially in monsoon period (Islam et al. 2010; Mark and Chusit 2002; Mowla and Islam 2013). Contaminated storm water threatens the ecosystem by polluting the soil and water bodies. Besides, stagnant water for long time due to heavy rainfall mixed with wastewater creates bad odor and causes habitual degradation to flora and faunas alongside affecting urban health (Islam et al. 2015; Subrina and Chowdhury 2018). During rainy season, rain water becomes polluted with silt, solid waste, medical waste, various pollutants, and domestic wastes. Sometimes uncollected municipal waste is piled up on road side and the rain water washes this garbage from the roads and causes erosion and siltation (Rahman et al. 2009; Datta et al. 2017). Millions of citizens suffer immensely because infrastructures like roads, houses, schools, colleges, shops, business premises submerge underwater from 3 to 4 feet (Majumder et al. 2018; Papry and Ahmad 2015). Children and office goers are most vulnerable groups during water logged situation (Rahman et al. 2009). In urban areas, the most adverse impact of water logging is the creation of breeding sites for diseases vectors and causes various waterborne diseases to city dwellers (Towhid 2004; Rahman et al. 2009; Datta et al. 2017). During monsoon period, urban runoff mixes with sewage and causes different pollution and creates a wide range of problems associated with waterborne diseases especially in poorly drained parts of the city (Datta et al. 2017). Malaria, dengue fever, respiratory problems, eye, and skin disease are some common diseases that are caused by water logging (Mowla and Islam 2013; Alom and Khan 2014). Ground water contamination by logged water also creates adverse health impacts (Pervin et al. 2019; Phanuwan et al. 2006; Rahman et al. 2009, ten Veldhuis et al. 2010). The following section focuses on the existing drainage system of Dhaka city which is partially responsible for creating water logging.

8.2 Existing Drainage System in Dhaka City

A proper drainage system is one of the prominent preconditions for making a city livable. Good drainage pattern can play the pivotal role to reduce water logging problems and also ensure appropriate sanitation facilities for city dwellers (Datta et al. 2017; Ullah et al. 2013). On the contrary, unplanned urban development reduces the efficiency of urban drainage systems by increasing the risk of urban flooding and water logging (Pervin et al. 2019; Clemens and Veldhuis 2010). In the past, there were numerous lowlands and canals for draining storm and rain water in Dhaka city efficiently (JICA 1991). These canals were used as connecting channels of the rivers surrounded by Dhaka city, which facilitated draining rain water through canals to the

Table 8.1 Existing Drainage facilities in Dhaka City

Storm water drainage line (dia 450–3000 mm)		370 km
Box culvert		10.5 km
Open Channel (Khal)		78 km
Area under drainage facility		140 km^2
Storm water pumping station: 4	Kalyanpur	20 m^3/s
	Dholiaikhal	22 m^3/s
	Rampura	25 m^3/s
	Kamalapur	15 m^3/s

Source DWASA, 2017–2018

rivers and helped mitigate water logging in the city areas during monsoon period. According to the annual reports of DWASA (Dhaka WASA 2018, 2019), Paribag, Dhanmondi, Begunbari, Dholaikhal, Debdolai, Segunbagicha, and Arambagh canals were used for both purposes: draining out storm water and reservoirs for rain water. Over 30 years back, drainage network of Dhaka city depended on 43 major canals, many of them have now disappeared due to encroachment and development activities (Mahmud et al. 2011). The situation has drastically changed over recent decades. The road networks of the city have been developed by filling those water bodies or by making box culverts, which is not only shrinking the water carrying capacity but also decreasing ground water replenishment (Mowla and Islam 2013; Mowla 2005). In addition, the expansion of city in all directions has caused shrinkage of the natural drainage system and wetlands over the years. As a result, water logging has been a regular problem in Dhaka city (Dhaka WASA 2019). Unplanned growth of urban infrastructures is continuously depleting natural drainage systems and filling-up the water bodies of Dhaka results in water logging during rainy season (Islam et al. 2012; Mowla 2011). The Table 8.1 exhibits the drainage facilities in Dhaka city.

The drainage system of Dhaka city is taken care of by different organizations/agencies, like Dhaka Water and Sanitation Authority (DWASA), Dhaka North City Corporation (DNCC), Dhaka South City Corporation (DSCC), RAJUK, Bangladesh Water Development Board, Cantonment Board, Dhaka District Administration, Department of Housing & Settlement, Public Works Department and the Private Developers. DWASA, DNCC, DSCC, and Water Development Board are the dominant organizations. DWASA looks after the 39% drainage network of the capital city, Dhaka (Siddique 2015). On the contrary, DNCC and DSCC have responsibility to maintain around 2,500 km of surface drains and around 4,000 km of underground drains (Dhaka 2019). However, there is no one for wetlands and drainage system management. Lack of coordination and cooperation among the related organizations/agencies could be ascribed to the problems of water logging (Mahmud et al. 2011). Poorly managed and insufficient drainage system is partly responsible for blocking regular water flow during moderate or heavy rainfall and creates water logging (Jahan 2012; Datta et al. 2017). However, a limited number of studies have been conducted to find out the causes of water logging and its impacts on society, people, and environment in Dhaka city (Subrina and Chowdhury 2018; Alom and Khan 2014; Afrin et al. 2010). Considering the severity of the problems, contemporary literatures have been reviewed to understand the water logging issues related to its causes and consequences.

8.3 Materials and Methods

The study on water logging was conducted in Dhaka city consisting of DSCC and DNCC. Dhaka city is chosen for this study (Fig. 8.1) as the city faces water logging problems almost every year (Pervin et al. 2019; The Daily Star 2019; Alom and Khan 2014). This research is based on primary data-collection by questionnaire survey and field observation. Secondary data are also gathered to supplement primary data. The field survey was conducted in the period of Monsoon period (including pre-monsoon, monsoon, and post-monsoon) from March to November in 2019. Questionnaire survey was administered in 27 severely affected areas in Dhaka city by water logging from where five categories of people participated in the survey. These

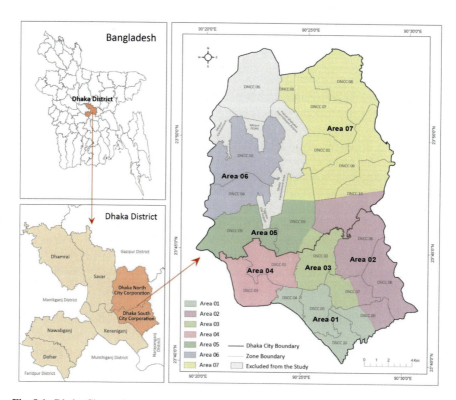

Fig. 8.1 Dhaka City study area map

include local people, CNG/bus driver, small businessmen, hawkers, and rickshaw pullers. Three hundred respondents have been interviewed in 27 spots from seven areas of the city to understand the causes and impacts of water logging. The questions for the development of questionnaire were set in accordance with the objectives stated in the research. Both open and closed questions form the basis of questionnaire that focuses on both causes and consequences of water logging. Questions related to the causes-natural and man-made and impacts of water logging (social, economic, environmental, and health impact) are set in the questionnaire. Way out of reducing water logging menace was also sought from the respondents who suffer from it. The pre-testing of questionnaire was also carried to improve the questionnaire. Feedback from the pretesting was accommodated in the final questionnaire. Collected data was processed and analyzed using simple statistics. Necessary tables and graphs are also used to project the findings of the analysis.

8.4 Causes of Water Logging

This section has tried to explain the causes of water logging in Dhaka city. The comments of the respondents through questionnaire survey have been displayed in the following section. Before moving on to dig out the causes of water logging, a short review has been given on the time and duration of water logging in the study area. The dominant season of occurring water logging is monsoon period. During questionnaire survey, about 56% respondents inform that water logging hazard occurs mainly in monsoon season while 28% state that these hazards occur post-monsoon period and only 16% inform water logging in pre-monsoon period. The following Fig. 8.2 shows the monthly distribution of water logging. In pre-monsoon season, water logging occurs mostly in the month of April and May than the month of March. During monsoon period, water logging happens due to heavy rainfall in September, August, and July than the month of June. During post-monsoon period, water logging mainly occur in October than the month of November and December (Fig. 8.2).

In terms of duration of water logging, 55.7% respondents claim that water stay less than 3 h in their area while 19.7% state that water logging continues 3–6 h; followed by 16% people saying that water stay more than 12 h. Only 9% of them opine that water logging prolong in their area from 6 to 9 and 9 to 12 h (Fig. 8.3).

The following Table 8.2 represents the duration of water logging condition in different areas of Dhaka city noticed by respondents. The findings demonstrate that water stays less than 3 h in Area 7, Area 4, Area 2, and Area 1. Likewise, water logging continues more than 3–6 h in Area 3, Area 4, Area 5, Area 1, etc. Sometimes, water stays more than 12 h in some areas for example in Area 6, Area 2 (Table 8.2).

There are many causes of water logging hazard as reported by the respondents during survey, 2019. These causes are categorized into two: natural causes and man-made causes.

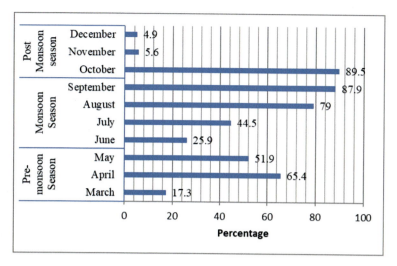

Fig. 8.2 Months of occurring water logging. *Source* Field Survey, 2019 (Multiple responses)

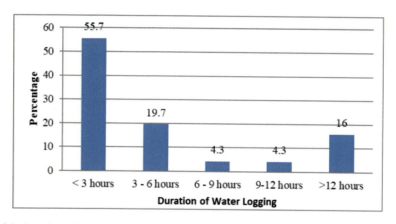

Fig. 8.3 Duration of water logging. *Source* Field Survey, 2019

8.4.1   Natural Causes of Water Logging

Approximately 99.3% respondents note that natural causes are responsible for water logging. In the period of monsoon, the city very often faces short rainfall with high intensity which creates water logging in different parts of the city, where the drainage system is inappropriate and congested. According to the questionnaire survey, 57% of the respondent has mentioned that heavy rainfall in a shorter period of time is one of the main reasons for water logging in Dhaka City. Around 42% of them claim that lowlands lead to water logging. Table 8.3 presents the area-wise distribution of natural causes of water logging.

Table 8.2 Duration of water logging condition in Dhaka City

Name of areas	<3 h		3–6 h		6–9 h		9–12 h		>12 h		Total	
	N	%	N	%	N	%	N	%	N	%	N	%
Area 1	45	51.1	25	28.4	3	3.4	4	4.5	11	12.5	88	100.0
Area 2	21	51.2	2	4.9	4	9.8	0	0.0	14	34.1	41	100.0
Area 3	17	48.6	12	34.3	1	2.9	3	8.6	2	5.7	35	100.0
Area 4	20	54.1	12	32.4	2	5.4	3	8.1	0	0.0	37	100.0
Area 5	4	44.4	3	33.3	0	0.0	2	22.2	0	0.0	9	100.0
Area 6	16	40.0	0	0.0	3	7.5	1	2.5	20	50.0	40	100.0
Area 7	44	88.0	5	10.0	0	0.0	0	0.0	1	2.0	50	100.0
Total	167	55.7	59	19.7	13	4.3	13	4.3	48	16.0	300	100.0

Source Field Survey, 2019

Table 8.3 Natural causes of water logging in areas of Dhaka City

Name of areas	Heavy rainfall		Low land		Others	
	N	%	N	%	N	%
Area 1	85	96.6	47	53.4	1	1.1
Area 2	40	97.6	32	78.0	2	4.9
Area 3	25	71.4	20	57.1	0	0.0
Area 4	34	91.9	21	56.8	0	0.0
Area 5	8	88.9	4	44.4	0	0.0
Area 6	39	97.5	35	87.5	0	0.0
Area 7	43	86.0	40	80.0	0	0.0

Source Field Survey, 2019

8.4.2 Man-Made Causes of Water Logging

As regards to anthropogenic causes, overwhelming majority (98.3%) respondents have identified that man-made causes are responsible for frequent water logging hazard in Dhaka city. Narrow drainage system (81.3%), throwing waste into drains (74.9%), City Corporation's irresponsibility (69.2%), poor waste management (55.9%), and constructional works (17.7%) are some of the major man-made causes resulting in water logging in the city as stated by the respondents during field survey (Fig. 8.4).

The research findings demonstrate that nearly 35.1% respondents of Area 4 and 27.5% of Area 6 have pointed out that construction works in rainy season creates water logged situation. Likewise, about 90% people of Area 6, 85.2% of Area 1, 84% of Area 7 believe that narrow drainage system are responsible for water logging in their areas. However, about 97.3% respondents of Area 4 inform that people throw waste into drain and it creates blockage to storm water flow. Similarly, around 86.5%

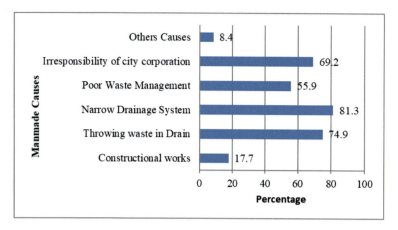

Fig. 8.4 Man-made causes of water logging. *Source* Field Survey, 2019 (Multiple responses)

people of Area 4 claim that poor waste management in their locality is causing water logging problem. Furthermore, irresponsibility of both city corporation (DNCC and DSCC) about this water logging issue are also one of the major causes of this problem as stated by 97.5% of people of Area 6 (Table 8.4). However, the findings of this study on the causes of water logging in Dhaka city have some similarities with the findings of other researchers (Towhid 2004; Subrina and Chowdhury 2018; Mowla and Islam 2013).

8.5 Impacts of Water Logging

Impacts of water logging on society, economy, environment, and health have been analyzed in this section. The adverse impact of water logging has been mentioned by the city dwellers during questionnaire survey. The major impacts as faced by the inhabitants are elaborated in the following sections.

8.5.1 Social Impacts

Water logging hazards hamper people's social life by creating different obstacles such as loss of time, interruption in daily activities, traffic problems, and damage roads, etc. About 32% respondents have said that water logging creates problem for traffic movement and destroy their working hour. In monsoon period, due to heavy rainfall many roads of Dhaka city go under knee deep water which causes damage to roads, disruption of traffic movement and creates traffic congestion in the study area. About 33% of 300 respondents claim that traffic problems is a major issue created

Table 8.4 Man-made causes of water logging in areas of Dhaka City

Name of areas	Constructional works		Throwing waste in drain		Narrow drainage system		Poor waste management		Irresponsibility of City Corporation		Others	
	N	%	N	%	N	%	N	%	N	%	N	%
Area 1	11	12.5	71	80.7	75	85.2	49	55.7	62	70.5	2	2.3
Area 2	5	12.2	33	80.5	32	78.0	23	56.1	29	70.7	7	17.1
Area 3	4	11.4	22	62.9	24	68.6	20	57.1	12	34.3	3	8.6
Area 4	13	35.1	36	97.3	30	81.1	32	86.5	27	73.0	0	0.0
Area 5	0	0.0	9	100.0	4	44.4	9	100.0	3	33.3	0	0.0
Area 6	11	27.5	23	57.5	36	90.0	23	57.5	39	97.5	4	10.0
Area 7	9	18.0	30	60.0	42	84.0	11	22.0	35	70.0	9	18.0

Source Field Survey, 2019

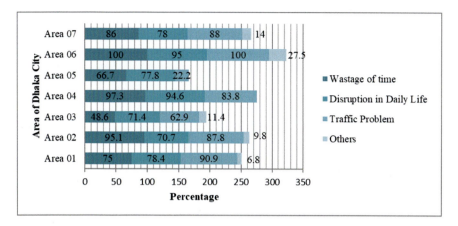

Fig. 8.5 Social impacts of water logging in the areas under Dhaka City. *Source* Field Survey, 2019

by this hazard. Disruption in daily life is another social problem caused by water logging. At the time of survey, nearly 31% inhabitants have mentioned that water logging hampers their daily life. Figure 8.5 shows the areal distribution of social problem in Dhaka city.

8.5.2 Economic Impacts

Water logging hazards have long-term effects on urban economy. Different sector of urban economy is affected by this hazard. The Fig. 8.6 shows the percentage

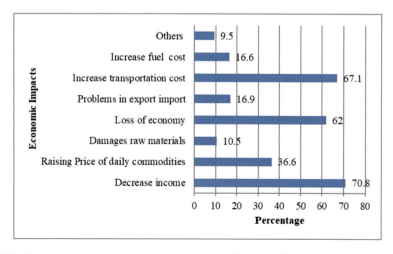

Fig. 8.6 Economic impacts of water logging. *Source* Field Survey, 2019

distribution of the economic impacts. From this figure it is evident that water logging poses severe impact on income, economy, and transport and so on. Household's, shop, and factory owners face financial losses due to the damage of their substructure, brick foundations of house, shops, institutions, etc. during water logging. About 62% people claim about economic loss is occurred by this hazard. Approximately 70.8% respondents have mentioned that water logging has a direct and indirect impact on income potential of the people of Dhaka city. Transportation cost (67%) also pop up during water logging. Around 36.6% people of respondents have noted about raising the price of daily commodities in rainy season (Fig. 8.6).

Approximately 16.9% people note that water logging hamper import–export of goods and products in the time of delivery. Fuel cost also increase in the city with transportation cost due to water logging problem. About 16.6% opine that fuel cost increases during this hazard. The Table 8.5 displays the percentage distribution of different economic impacts identified by the respondents of the study area.

8.5.3 Environmental Impacts

During questionnaire survey, respondents are asked about environmental impacts of water logging in the study area. They have mentioned different kinds of pollution: air, water, and soil which are caused by this hazard (Fig. 8.7). The logged water mixes with different kind of waste, silt, and pollutants that are washed off from roads. Consequently, the water is polluted and becomes a burden for the residents of urban area, which causes unhygienic environment resulting in environmental impacts (Towhid 2004; Subrina and Chowdhury 2018). During survey, approximately 72.6% people note that water logging is responsible for surface water pollution in their area. About 18.8% people confirm that ground water is polluted by logged water. Approximately 16.8% people claim that scarcity of drinking water is a common problem during this hazard. Ground water contamination and surface water pollution raises the problem of water scarcity of pure drinking water in the study area. About 24% people talks about soil pollution. During monsoon season, polluted logged water mixed with water bodies, and wetlands threatens aquatic life. In this regard, only 0.7% of people state about loss of aquatic life while about 74% mention about air pollution caused by this hazard.

The Table 8.6 demonstrates the percentage distribution of different environmental impacts encountered by the respondents in different areas of the study area.

8.5.4 Health Impacts

Apart from social, economic, and environmental impact, water logging is responsible for creating many health problems to the residents of affected area. Skin diseases, mosquito borne diseases, water borne diseases, etc. are common health issues created

Table 8.5 Economic impacts of water logging in the areas of Dhaka City

Name of area	Economic impacts															
	Decrease income		Raising price of daily commodities		Damages raw materials		Loss of economy		Problems in export import		Increase transportation cost		Increase fuel cost		Others	
	N	%	N	%	N	%	N	%	N	%	N	%	N	%	N	%
Area 1	60	68.2	33	37.5	9	10.2	49	55.7	13	14.8	61	69.3	11	12.5	11	12.5
Area 2	23	56.1	24	58.5	5	12.2	18	43.9	8	19.5	34	82.9	7	17.1	3	7.3
Area 3	20	57.1	11	31.4	2	5.7	17	48.6	0	0.0	15	42.9	3	8.6	2	5.7
Area 4	34	91.9	9	24.3	7	18.9	37	100.0	5	13.5	24	64.9	8	21.6	0	0.0
Area 5	7	77.8	1	11.1	0	0.0	8	88.9	4	44.4	2	22.2	1	11.1	1	11.1
Area 6	24	60.0	17	42.5	1	2.5	16	40.0	4	10.0	27	67.5	5	12.5	8	20.0
Area 7	41	82.0	13	26.0	7	14.0	38	76.0	16	32.0	35	70.0	14	28.0	3	6.0

Source Field Survey, 2019

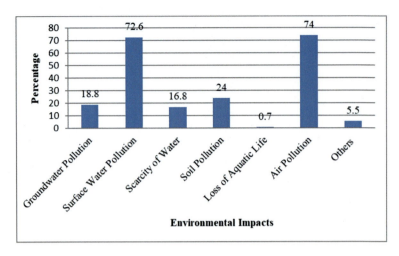

Fig. 8.7 Environmental impacts of water logging. *Source* Field Survey, 2019

by logged water and suffered by city dwellers. Contaminated logged water is suitable for spreading different types of skin diseases in the affected area. Around 42% of total sampled respondents claim that they suffer from various skin diseases owing to contact with polluted logged water on the roads, lanes, etc. Stagnant water is also the ideal breeding place for mosquitoes. Different viral diseases like dengue fever, malaria, and chikungunya spread through stagnant water. About 37% of 300 respondents have said that Dhaka city has been suffering a lot from mosquito borne diseases due to water logging. Water borne diseases emerge as another issues arisen out of water logging. For example, about 19% of the respondents have replied that stagnant rain water accelerates the incidence of water borne diseases when it is contaminated during water logging. Jaundice, typhoid, cholera, diarrhea, dysentery, etc. are common water borne diseases affected by residents of study area. The Fig. 8.8 demonstrates area wise scenario of health impact emanated from water logging.

8.6 Conclusion

Water logging has become a perennial problem for Dhaka city dwellers like other Asian cities. Over the decades, rapid and unplanned urbanization in Dhaka city has drastically reduced natural drainage system. Wetlands are being randomly filled up by the real estate companies by ignoring the directives of the government organizations. Under this circumstance, the city dweller faces tremendous suffering due to water logging every year in the monsoon season. Water logging hazard is caused in many ways. The research findings reveal that the causes of water logging are of two types: natural and man-made. Heavy rainfall, low lying areas, and other associated factors are categorized as natural causes of water logging. Almost all the

Table 8.6 Environmental impacts of water logging in the areas of DCC

Name of areas	Environmental impacts														
	Groundwater pollution		Surface water pollution		Scarcity of water		Soil pollution		Loss of aquatic life		Air pollution		Others	
	N	%	N	%	N	%	N	%	N	%	N	%	N	%
Area 1	17	19.3	61	69.3	16	18.2	16	18.2	0	0.0	59	67.0	6	6.8
Area 2	15	36.6	32	78.0	4	9.8	0	0.0	0	0.0	37	90.2	0	0.0
Area 3	0	0.0	27	77.1	2	5.7	1	2.9	0	0.0	15	42.9	2	5.7
Area 4	3	8.1	28	75.7	3	8.1	26	70.3	0	0.0	27	73.0	0	0.0
Area 5	2	22.2	7	77.8	4	44.4	1	11.1	0	0.0	1	11.1	0	0.0
Area 6	18	45.0	29	72.5	17	42.5	1	2.5	1	2.5	38	95.0	3	7.5
Area 7	0	0.0	28	56.0	3	6.0	25	50.0	1	2.0	39	78.0	5	10.0

Source Field Survey, 2019

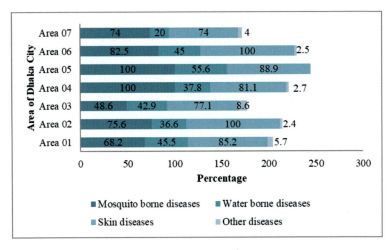

Fig. 8.8 Health impacts of water logging in the areas under Dhaka City. *Source* Field Survey, 2019

respondents (99%) claim that water logging occur due to both man-made and natural causes. Among man-made causes, narrow drainage system (81.3%), dumping waste into drains (74.9%), City Corporation's irresponsibility (69.2%), poor waste management (55.9%), and constructional works (17.7%) have been identified as some of the major man-made causes creating water logging in the city.

The findings of the research also demonstrate four major impacts: social, economic, environmental, and health. Water logging hazards hamper people's social life by creating different hassles such as loss of time, interruption in daily activities, traffic problems, damage roads, etc. Water logging completely disrupts normal life as claimed by 33% respondents. It also poses impact on various sectors of urban economy. Approximately 71% respondents have mentioned that, water logging has a direct and indirect impact on income potential of the people of Dhaka city. About 62% people report about economic loss that has been occurred by this hazard in 2019. Delivery of import–export goods is also hampered. According to the findings of this research, logged water poses negative impact on environment by polluting: air, water, and soil. Approximately 72.6% people have stated that water logging causes surface water pollution in their area. Almost 19% people complain that ground water is polluted by logged water. About one-fourth of the city dwellers think that water logging also causes soil pollution. Water logging hazard do not contaminate only water and soil, but also pollute air. About 74% people of respondents have mentioned about air pollution caused by this hazard. The most crucial impacts of water logging are the challenges of major health issues. Adverse health impacts of this hazard include skin diseases, mosquito borne diseases, and water borne diseases. Around 42% people claim that they suffer from various skin diseases due to contact with polluted logged water. Research findings show that about 37% residents have complained about mosquito menace. About 19% of the respondents have replied that stagnant rain water increases the water borne diseases for example jaundice, typhoid,

Diarrhea, and dysentery. To mitigate the water logging hazard, construction of wide drain, regular cleaning, and awareness of the people are necessary to make Dhaka city livable.

Acknowledgements This research is funded by the University Grant Commission-Jagannath University, Dhaka for the fiscal year of 2018-19. We are grateful to them for financial support.

References

Afrin M, Saika, U, Hoque MA (2010) An assessment of the causes and consequences of the water logging in the Dhaka City, The Jahangirnagar Review, Part II: Social Sciences, Vol. XXXIV, Printed in June, 2014

Akter A, Mohit SA, Chowdhury MAH (2017) Predicting urban storm water-logging for Chittagong city in Bangladesh. Int J Sustain Built Environ 6:238–249

Alom MM, Khan MZH (2014) Environmental and social impact due to urban drainage problems in Dhaka city Bangladesh. Int J Eng Adv Technol 3(6):128–132

Ashraf MA, Chowdhury MSA (2009) Drainage planning in the cities of Bangladesh: case study of drainage and water logging in Chaktai commercial Area, Chittagong. J Bangladesh Inst Plan 2:49–60

Awal MA (2014) Water logging in southwestern coastal region of Bangladesh: local adaptation and policy options. Sci Postprint 1(1):e00038 https://doi.org/10.14340/spp.2014.12A0001

Chakma M, Chakrabartty P (2017) Water logging and drainage planning in Bangladesh: a case study of Pabna Municipality, Pabna. In: International conference on planning, architecture and civil engineering, Rajshahi University of Engineering & Technology, Rajshahi, Bangladesh

Chowdhury JU, Rahman R, Bala SK, Islam MS (1998) Impact of 1998 flood on Dhaka city and performance of flood control works, Integrated Water Flow Model (IWFM), Bangladesh University of Engineering & Technology (BUET), Dhaka

Clemens FHLR, Veldhuis JAET (2010) Quantitative risk analysis of urban flooding in lowland areas (unpublished doctoral thesis). http://resolver.tudelft.nl/uuid:ef311869-db7b-408c-95ec-69d8fb7b68d2

Dani AH (1962) Dacca: a record of its changing fortunes, Ahmed Hasan Dani. Mrs. SS Dani, Dhaka

Datta RK, Alam O, Hossain MM (2017) Current status of urban drainage system and its problem im Netrokona Municipal Dhaka. JIST 22(1):165–178

Dewan AM, Nishigaki KTM (2006) Flood hazard delineation in greater Dhaka, Bangladesh using an Integrated GIS and remote sensing approach. Geo-carto Int 21:33–38

Dhaka WASA (2018) Annual Report 2016–2017, Dhaka water supply and sewerage authority. http://dwasa.org.bd/wp-content/uploads/2019/07/Annual-Report-2016-17.pdf Accessed 1 Jan 2020

Dhaka WASA (2019) Annual report 2017–2018, Dhaka water supply and sewerage authority. http://dwasa.org.bd/wp-content/uploads/2019/11/Annual-Report-2017-18.pdf Accessed 1 Jan 2020

Djordjevic S, Butler D, Gourbesville P, Mark O, Pasche E (2011) New policies to deal with climate change and other drivers impacting on resilience to flooding in urban areas: the CORFU approach. Environ Sci Policy 14:864–873

Han SQ, Xie YY, Li DM, Li PY, Sun M (2006) Risk analysis and management of urban rainstorm water logging in Tianjin. J Hydrodyn Ser B 18:552–558

Haque AKE (2013) Reducing adaptation costs to climate change through stakeholder-focused project design: the case of Khulna city in Bangladesh (No. G03520). International Institution for Environment and Development, London. https://pubs.iied.org/pdfs/G03520.pdf

Hasan MR, Hossain, MT, Khan LA, Afrin S (2018) Seasonal water logging problem in an urban area of Bangladesh: a study on Pabna pourashava. In: Paper presented at the 1st national conference on water resources engineering (NCWRE 2018), CUET, Chittagong, Bangladesh

Huq S, Alam M (2003) Flood management and vulnerability of Dhaka city. Bangladesh Center of Advance Studies (BCAS), Dhaka

Ishtiaque A, Mahmud MS, Mahmudul HR (2014) Encroachment of canals of Dhaka city, Bangladesh: an investigative approach. GeoScape 8(1):48–64. https://doi.org/10.2478/geosc-2014-0006

Islam MS, Rahman MR, Shahabuddin AKM, Ahmed R (2010) Changes in wetlands in Dhaka city: trends and physico-environmental consequences. J Life Earth Sci 5:37–42

Islam MK, Chowdhury S, Chisty KU, Rahman A (2015) Community participatory tools for water logging vulnerability assessment in chittagong city corporation area. J Bangladesh Inst Planners 8:221–231

Islam MS, Shahabuddin AKM, Kamal MM, Ahmed R (2012) Wetlands of Dhaka city: its past and present scenario. J Life Earth Sci 7:83–90. http://banglajol.info.index.php/JLES

Jahan M (2012) Impact of rural urban migration on physical and social environment: the case of Dhaka city. Int J Dev Sci 1(2):186–194

JICA (1991) Master plan on Greater Dhaka flood protection project, Flood Action Plan No. 8A, Supporting Report-I and II, Flood Plan Coordination Organization (presently WARPO), Dhaka

Kamal MM, Midorikawa S (2004) GIS-based geo-morphological mapping using remote sensing data and supplementary geo-information: a case study of the Dhaka city area, Bangladesh. Int J Appl Earth Observ Geo-Inform 6:111–125

Khan MMH (2017) Analysis of causes and impact of water logging in Khulna city of Bangladesh. In: International conference on mechanical, industrial and materials engineering 2017 (ICMIME2017), 28–30 December, 2017, RUET, Rajshahi, Bangladesh

Li C (2012) Ecohydrology and good urban design for urban storm water-logging in Beijing, China. Ecohydrol Hydrobiol 12(4):287–300. https://doi.org/10.2478/v10104-012-0029-8

Mahmud MS, Masrur A, Ishtiaque A, Haider F, Habiba U (2011) Remote sensing and GIS based spatio-temporal change analysis of wetland in Dhaka city Bangladesh. J Water Res Protect 3:781–787. https://doi.org/10.4236/jwarp.2011.311088

Majumder AK, Hossain MS, Nayeem AA (2018) Assessment of people's perception on water logging in Chittagong city corporation area, Bangladesh. Int J Multidiscipl Res Dev 5(2)

Mark O, Chusit A (2002) Modeling of urban runoff in Dhaka city, Asian Institute of Technology (AIT), Thailand

Mowla QA (2005) Eco-systems and sustainable urban design nexus: a borderless concept. In: Paper presented in the international alumni conference on technology without borders, organized by the Global IIT, Bethesda, Washington DC, USA, 20–22 May, 2005

Mowla QA (2011) Crisis in the built environment of Dhaka: an overview. In: Paper presented at proceedings of the conference on engineering research, innovation and education 2011 (CERIE 2011), 11–13 January, Sylhet, Bangladesh

Mowla QA, Islam MS (2013) Natural drainage system and water logging in Dhaka: measures to address the problems. J Bangladesh Instit Plan 6:23–33

Papry IR, Ahmad GU (2015) Drainage condition in water logged areas of central part in Chittagong city corporation. Int J Eng Sci Invent 4(1):24–29

Pervin IA, Rahman SMM, Mani N, Haque AKE, Karim H, Ganesh D (2019) Adapting to urban flooding: a case of two cities in South Asia. Water Policy, 1–27. https://doi.org/10.2166/wp.2019.174

Phanuwan C, Takizawa S, Oguma K, Katayama H, Yunika A, Ohgaki S (2006) Monitoring of human enteric viruses and coliform bacteria in waters after urban flood in Jakarta Indonesia. Water Sci Technol 54(3):203–210

Rahman KMA, Debnath SC (2015) Water logging and losses in ecosystem: A case analysis on DND embankment, Bangladesh. Int Res J Interdiscipl Multidiscipl Studies (IRJIMS) VIII(I):27–33

Rahman MM, Akteruzzaman AKM, Khan MMH, Jobber A, Rahman MM (2009) Analysis of water logging problem and its environmental effects using GIS approaches in Khulna city of Bangladesh. J Socio Res Dev 6(2):572–577

Reza MG, Alam MS (2002) Wetland transformation in the western part of Dhaka city (1963–2000). Bhugal Patrika (J Geogr) 21:23–40

Sahu AS (2014) A study on Moyna Basin water-logged areas (India) using remote sensing and GIS methods and their contemporary economic significance. Geograph J 9. https://doi.org/10.1155/2014/401324

Sar N, Chatterjee S, Adhikari MD (2015) Integrated remote sensing and GIS based spatial modeling through analytical hierarchy process (AHP) for water logging hazard, vulnerability and risk assessment in Keleghai river basin, India. Model Earth Syst Environ 1:31. https://doi.org/10.1007/s40808-015-0039-9

Siddique MAR (2015) Assessment of present status and conveyance capacity of drainage system in selected areas of Dhaka city, (M.Sc. Thesis), Department of Civil Engineering, Bangladesh University of Engineering & Technology (BUET), Dhaka

Subrina S, Chowdhury FK (2018) Urban dynamics: an undervalued issue for water logging disaster risk management in case of Dhaka city Bangladesh. Proc Eng 212:801–808

ten Veldhuis JAE, Clemens FHL, Sterk G, Berends BR (2010) Microbial risks associated with exposure to pathogens in contaminated urban flood water. Water Res 44(9):2910–2918

The Daily Star (2019) EIU Livability Index: Dhaka improves a little, Now Ranks 3rd Worst, September 14, 2019. https://www.thedailystar.net/city/liveable-city-index-2019-dhaka-now-3rd-worst-capital-eiu-1795360

Towhid K (2004) Causes and effects of water logging in Dhaka city, Bangladesh (TRITA-LWR, Master thesis), Department of Land and Water Resource Engineering, Royal Institute of Technology, Stockholm

Ullah MMMN, Hossaina MS, Shahiduzamana M, Islam MS, Islam MZ, Choudhary MSR (2013) A study on some aspects of drainage system in Rajshahi city Bangladesh. Sci J Environ Sci 2(6):118–124. https://doi.org/10.14196/sjes.v2i6.1005

World Bank (2007) Dhaka metropolitan development plan strategic environmental assessment, prepared by SENES consultants limited in association with Techno Consult International Ltd. Dhaka, Bangladesh

Zhang S, Pan B (2014) An urban storm-inundation simulation method based on GIS. J Hydrol 517:260–268

Zurbrugg C (2003) Solid waste management in developing countries. SANDEC/EAWAG, SWM introductory text on www.Sanicon.Net, https://www.eawag.ch/fileadmin/Domain1/Abteilungen/sandec/publikationen/SWM/General_Overview/Zurbruegg_2002_SWM_DC.pdf Accessed 15 Jan 2020

Chapter 9
Water Resource Development and Sustainable Initiatives of India: Present and Future

Jayant Kumar Routray

Abstract Water is the most important commodity needed by all—the animal, plant kingdoms, and above all for meeting diverse needs of the rising global population. India is a populous country. With rising population, economic and other activities, water demand for domestic and other types of consumptions is enormous. It is a challenge always for striking a balance between the demand and supply of water. This chapter attempts to make a review of water resource situation of India, its management and associated challenges, existing policies and focuses on the sustainable initiatives practiced now and a perspective to the future.

Keyword Water resource · Management issues · National water policy · Sustainable development initiatives · India

9.1 Introduction

Water is very basic and essential resource for sustaining the life on earth. Water is required every day for domestic consumption, to support all types of production systems from agricultural to nonagricultural and service activities. Earth is covered by 71% of water. Out of total amount of water, 97.5% is salt water and 2.5% is fresh water. By sources, the share of fresh water is 69% covered by glacier and ice caps, 30% by ground water, and only 0.27% is surface water. Accessible to fresh water is the key to the survival of the planet.

Above all, ground water is the most important source of water. Conservation, protection, and efficient management of water resources are the key to ensure the availability of and accessibility to water. Currently 40% of global people face water scarcity. More than 2 billion people are affected by water stress. Seventy percent of global people have safety-managed water and rest do not have. Thirty percent global people have safe sanitation. Eighty percent of wastewater goes into waterways without adequate treatment. Eighty percent of countries have laid the foundation of

J. K. Routray (✉)
Asian Institute of Technology, Bangkok, Thailand

© The Author(s), under exclusive license to Springer Nature Singapore Pte Ltd. 2022   173
N. C. Jana and R. B. Singh (eds.), *Climate, Environment and Disaster in Developing Countries*, Advances in Geographical and Environmental Sciences, https://doi.org/10.1007/978-981-16-6966-8_9

integrated water management. Seventy percent of wetlands have been lost over the last century.

Water is becoming scarce day by day to meet the rising demand due to population explosion associated with overgrown cities and metropolises and increasing production systems. With this backdrop, the Sustainable Development Goal 6 has rightly emphasized on ensuring availability and sustainable management of water. The focus of this chapter is to discuss on the water resource development and sustainable initiative of India at present and moving towards the future. This chapter has several sections including spatial and temporal variations of rainfall.

9.2 Rainfall Trends and Variability in India

Kumar et al. (2010) have made an exhaustive study on the long-term rainfall trend using a robust data set consisting of 306 stations of 135 years from 1871 to 2005. Annual average rainfall is 125 cm, contributed by South West monsoon (75% during June to September). Northeast monsoon contributes 13% during October to December and 10% is contributed by pre-monsoon cyclonic rainfall, and 2% by western disturbances.

It is less than 60 cm in Rajasthan, adjoining areas of Gujarat and Haryana, more than 400 cm in Western Coast and North East India. Rainfall variation is very wide from less than 20 to more than 400 cm. The highest rainfall in India of about 1270 cm recorded in Cherapunji in Meghalaya and the lowest rainfall of 15 cm is recorded at Jaisalmer in Rajasthan. This implies for efficient water management system for harvesting rainfall water and distribution for proper use of available water.

The sub-divisional rainfall trends show a large variability—nearly half of the subdivisions out of 30 have shown an increasing trend in annual rainfall and the rest have shown an opposite trend. The maximum increase was 2.37 mm per year and the maximum decrease was 0.76 mm per year. Among the five regions of India, a small increasing trend is observed for the North West and Peninsular India. A decreasing trend is established for North East, Central North East, and West Central India. For India as a whole, a small decreasing trend is shown in the annual rainfall.

Seasonal analysis has demonstrated that pre-monsoon rainfall has increased over 23 subdivisions and all regions, monsoon rainfall increased only over 10 subdivisions and 1 region; post-monsoon rainfall increased over 27 sub-divisions; and winter rainfall increased over 20 subdivisions and 3 regions. Pre-monsoon, post-monsoon, and winter rainfall has increased and monsoon rainfall decreased for the whole country. Many subdivisions in the South East Peninsular India have shown an increasing trend of nearly 7–9% of mean per 100 years. Chhattisgarh, Vidarbha, and East Madhya Pradesh have experienced a decreasing trend of nearly 5–10% mean per 100 years in annual rainfall (Kumar et al. 2010).

This study has indicated large spatial and temporal variation in the rainfall trends and no clear pattern was established over India. At the regional level, Prabhakar et al. (2019) have attempted a similar study for Odisha State to explain the variability of rainfall and implications on agriculture. This is a district level study and the data set was used from 1901 to 2013. Odisha receives average annual rainfall of about 1438 mm out of which 78% occurs in the monsoon season.

The range of annual rainfall varies from 1610 mm in Bhadrak district to 1140 mm in Puri district. The year 1945 is the changing point due to beginning of current warming period. The decreasing rainfall trend starts from this point. Annual rainfall is decreasing at 5% significant level for the state except three districts (Cuttack, Dhenkanal, and Jagatsinghpur) having significant positive trend. In pre-monsoon season, all districts have shown a positive trend and it is more significant in Cuttack, Gajapati, and Jajpur districts.

For monsoon season, the positive trend has been reflected in six districts (Cuttack, Deogarh, Dhenkanal, Gajapati, Ganjam, and Jagatsinghpur). Jharsuguda has shown a significant negative trend. In post-monsoon season, all districts have shown negative trend but it is more significant in Khordha district. A decreasing trend of rainfall is noticed in all districts during pre and post-monsoon periods although not so significant. The spatio-temporal variation has a strong bearing on the water resource development and management in the state.

9.3   River Basins and Water Resources of India

Chitale (1992) has given the clear picture of development of river basins in India. The description in this section is extracted from this work. In 1945, Dr. A. N. Khosla as Chairman of the Central Water Irrigation and Navigation Commission of India had propounded that planning should be on a regional basis and the drainage basins should be treated as units. In subsequent years, it was felt necessity and water was made a subject of both State and Central Government.

Damodar Valley Corporation was created in 1948 in line with the Tennessee Valley Authority of the USA as a multipurpose river valley project for protecting against flood, irrigation, generation of power, community use, industrial, and other uses. In 1956, the enactment of the River Board Act by the Parliament paved the way for establishment of the River Board for the interstate rivers. However, the states acted either through mutual negotiations, reached agreements on their shares in the river water for projects, or on the nature of interstate projects to be undertaken jointly.

Irrigation Commission of 1972 recommended that (1) basin plans should provide development possibilities of land and water resources to meet anticipated regional and local needs, (2) indicate a broad framework of various engineering works to be taken up, establish priorities of water use for different purposes, indicate priority of projects, indicate the need to earmark water for any specific future purpose, and (3) the plan should be periodically reviewed and revised in the light of changing needs and supplies. Hydrologically, Indian rivers can be classified into four groups—Himalayan Rivers, Deccan Rivers, Coastal rivers, and rivers of the inland drainage basin. The Himalayan Rivers are snowfed and have water round the year.

During the monsoon periods, these rivers become full and cross the banks due to heavy rain and cause frequent flooding. The Indus and the Ganges–Brahmaputra-Meghana system belong to the Himalayan group. These rivers with a large number of tributaries cover a drainage area of about 1 million km^2. This is the most dominant hydrological class of river valleys accounting for nearly half of India's geographical area. Generally, they have good groundwater potential. Deccan Rivers are rain fed, mostly non-perennial, and they fluctuate in volume of water.

The Narmada and the Tapti flowing westward, The Brahmani, the Mahanadi, the Godavari, the Krishna, the Pennar, and the Cauvery flowing eastward are the major rivers of the Deccan group. These rivers together cover drainage of 1.2 million km^2. Due to hilly and rocky region in major parts of these basins, they have relatively poor ground water holding capacity. The coastal rivers are mostly non-perennial, short in distance and have limited catchment area. There are a few rivers on the east coast and about 600 rivers on the west coast. The West Coast Rivers contain about 14% of India's water resources. The inland drainage basins of West Rajasthan are few and ephemeral in nature.

These rivers flow only for some distance and are lost in the desert. River basins in India are classified into three broad groups—(1) large basin with a drainage area of 20,000 km^2 and more, (2) medium river basin with a catchment area of between 2000 to 20,000 km^2, and (3) minor river basin with catchment area below 2,000 km^2. The large basins share 89%, medium basins 7.5% and minor basins 3.5% of the geographical area of India. Bay of Bengal receives 70% surface water, the Arabian Sea receives 20%, and the rest 10% drains into interior basins and lakes (Das 2009).

The tributaries of the Ganga, Brahmaputra, and Indus rivers contribute about 50% of water resources of India. The average yield per unit area of the Himalayan rivers is almost double than that of south peninsular river system. The ground water is also an important source of drinking water, irrigation, and other uses. It accounts for about 80% of domestic water requirement and more than 45% of the total irrigation in the country (Kumar et al. 2005). As per the National Water Mission of India (nwm.gov.in/?q= synopsis-water-data-india), the water resources estimated in 2011 provides the fact that India has total water resource of 1121 BCM out of which the surface water shares 690 BCM (61.5%) and ground water shares 431 BCM (38.5%). India has a live storage capacity of 253.4 BCM and projects under construction can add 51 BCM more. Per capita availability of water is 1544 Cubic Meter.

9.4 National Water Policy

Scarcity of fresh water and sustaining all life forms as a part of larger ecological system pushed forward to review, refresh, and develop the National Water Policy in 2002, although National Water Policy was adopted in 1987. The vast areas of the country are affected by floods and droughts.

Planning and implementation of water resources involve a number of complex issues—socioeconomic issues of people, environmental sustainability, resettlement, rehabilitation of the project-affected people, livestock, public health and dam safety, etc. Population rise and rapid expansion of economic activities lead to increasing demand for water for diverse purposes (domestic, agricultural, industrial, hydropower, thermal power, navigation, recreation, etc.). Production of food grains to be raised to about 350 million tons by 2025. Water quality is also another but important issue needs to be addressed. With this background, the salient features of the National Water Policy (Ministry of Water Resources, 2002) are discussed below.

Information System: A strong need arises to establish a standardized national information system with a network of data banks and databases by integrating and strengthening the existing central and state level agencies for improving the quality of data and processing capabilities. Water availability, actual use, and future demand for diverse purposes are the part of this information system.

Water Resource Planning: It should emphasize on availability of water resources and utilization to the maximum possible extent. Other possible sources like recharging ground water artificially, desalinization of brackish or sea water, rain water harvesting including rooftop rain water, inter-basin transfer, etc. need to be practiced. Water resource development should be undertaken by basin/sub-basin wise taking into account of both surface and ground water with attention on sustainable use while focusing on quantity and quality aspects. Watershed management should be practiced linking with soil conservation, conservation of and increasing forest cover. Construction of check-dams should be promoted for this purpose.

Institutional Mechanism: For planning, development, and management of the water resource on a drainage basin, multi-sectoral, multi-disciplinary, and participatory approach should be practiced involving the existing institutions from the state level to the basin and local levels in line with the existing agreements and directives of the Tribunals under the relevant laws. Water Allocation Priorities: Water allocation priority should focus on drinking water, irrigation, hydropower, ecology, agriculture, and agro industries and other industries, and navigation, etc.

Project Planning: Water development project must consider all components (settlements and population, occupation, environment, etc.) should be considered along with the application of EIA and SIA before taking the approval. Participation of beneficiaries and all stakeholders, integrated and multi-disciplinary approaches are the keys for preparing the plan. Cost estimation and benefit analysis is part of the planning. Ground Water Development: Assessment of the ground water, regulating exploitation not exceeding the recharging possibilities, environmental consequences are to be taken into account while developing projects.

Drinking Water: All irrigation and multipurpose projects must consider to provide adequate safe drinking water to urban and rural populations.

Irrigation: Irrigation planning either for the entire basin or any individual project should take into account of irrigability of land, cost-effectiveness of the project, water and land use policies, water allocation with due attention to equity and social justice, water for head and tail end farms, command area development approach, water and farm management practices, and reclamation of water logged areas.

Resettlement and Rehabilitation (R & R): Projects associated with the displacement of people in the storage area require resettlement and rehabilitation program. Careful planning and implementation of R & R should be done following the policies taking into account local condition.

Financial and Physical Sustainability: Water charges for various uses should be fixed to cover the operation and maintenance costs first and a partially the capital cost subsequently. Subsidy on water rates may be considered for the poorer and disadvantaged section of the population.

Participatory Approach to Water Resource Management: Government agencies, users, and all stakeholders, local bodies should be involved in the operation, maintenance, and management of water infrastructures and facilities at appropriate level.

Private Sector Participation: Depending on specific situation, private sector participation as water users may help in providing innovative ideas, technology, generating financial resources and introducing corporate management, and improving service efficiency, etc. Water Quality: Water quality should be monitored and treated to the acceptable level. Principle of 'polluter pays' should be strictly followed in managing polluted water.

Water Zoning and Conservation of water: Water zoning of the country is necessary for guiding, and regulating economic activities. Promotion of conservation of water should be linked with water use optimization. Measures like selective linings in the conveyance system, modernization and rehabilitation of existing system, adoption of new techniques like drip and sprinkler irrigation may be promoted, wherever possible.

Flood Control and Management: The flood prone basins should have master plan for flood control and management.

Land Erosion by Sea and River: It should be minimized and rigorously controlled by suitable cost-effective measures. Each coastal state should prepare a comprehensive coastal land management plan paying attention to the environmental and ecological impacts.

Drought prone Area Development: The needs of the drought-prone areas should be prioritized while planning water resource development projects.

Monitoring of Projects: A monitoring and evaluation system should be placed to assess the performance and socioeconomic impact of the project.

Water Sharing among the States: Necessary guidelines need to be evolved for facilitating future agreements among the basin states for water sharing. The existing Interstate Water Dispute Acts may be reviewed regularly and amended as deemed fit.

Performance Improvement: The performance of the existing water facilities should be given high priority.

Maintenance and Modernization: The structures created through massive investments should be properly maintained in good order and modernized as felt necessary. Water user associations should be encouraged to facilitate the management and maintenance of irrigation system.

Safety of Structures: Specialists at the national and state level should create proper organizational set up for ensuring the safety of storage dams and other related structures.

Science, Technology, and Training: The frontiers of knowledge need to be pushed forward in almost all areas associated with water resource development and management. Training should be an integral part of water resource development and management. It should cover the information system, sectoral planning, project planning, and project management, operation of projects and managing water distribution system.

The policies outlined above cover almost all issues associated with water resource development, areas of utilization, repairing, maintenance and operational aspects, management system, infusing new technologies and training components, etc. It looks impressive. As water is a subject jointly dealt by the State and Central Governments, Government of India Act of 1935 and revised one in 2007 has in principle given power to the states to legislate in this regard.

Despite this policy document, there are still several issues need to be addressed among which intra-basin and interstate distribution is a big challenge now and it will remain in future. Wate (2012) mentions that the major bottleneck in an effective policy formulation for water quality and management and its implementation is the current institutional setup involving many government agencies. The pollution arises from a number of diffused sources from either urban or rural. The present approach to water-related matters restricts the issue only to political boundaries, involving a number of agencies and ministries with overlapping responsibilities.

The concept of River Board is not truly reflective in solving interstate water problems. In a study by Chetan and Biswas (2019) with reference to water policy of 2002 and 2012, have very critically captioned the paper that India's National Water Policy: 'feel good' document, nothing more. This message is to look back the process of policy formulation to implementation and bring out necessary modifications to make it a practical one.

Pandit and Biswas recommends for a good consultation and drafting process is necessary condition to produce a worthy National Water Policy, it is not a sufficient condition, and does not necessarily ensure a descent outcome in terms of producing a policy document that is implementable, and which is likely to improve water management in the country. The implementation of the policy is far from satisfactory. Current policies affecting water use, management, and development are often contradictory, inefficient, or unresponsive to changing conditions (Lal 2001).

9.5   Water Resource Management Issues

The works of Jain (1999, 2019) are worth mentioning in the context of issues and challenges of water resource management. The factors behind the water-related problems are linked with (1) highly uneven pattern of rainfall and availability of water both in space and time that often leads to floods and droughts; (2) pollution of freshwater resources due to agricultural, industrial, and municipal sources; (3) highly unreliable municipal water supply with poor quality; (4) laws that give unlimited ownership of ground water to the landowner coupled with uncontrolled use of bore wells with high rate of extraction of ground water, very often exceeding the recharge; (5) less or no attention to water conservation, efficiency in use, water reuse, ground water recharge, and ecosystem sustainability; and (6) very low water prices that never discourages wastage.

Water conservation and water management measures are the needs of the day to achieve a strong and stable economic base, especially in the arid and drought-prone areas of the country (Kumar 2018). As per Kumar, some of the methods may be suggested as technical strategies to mitigate the adversaries of drought. The strategies include creation of surface storage, planning for less dependable yield, prevention of the evaporation losses from reservoirs, adjustment of water releases from the reservoir, reduction in conveyance losses, equitable distribution, maintenance of irrigation systems, better irrigation practice, irrigation schedule, cropping pattern, conjunctive use of surface and ground water, watershed development, creation of large storages, increasing small reservoirs along with major reservoirs, and transfer of water from water excess basins to water-deficit basins. With the growing population along with the expansion of economic activities, the demand for water use in various sectors is making the fresh water scarce.

The pollution of fresh water is posing a serious threat to water resources of India. India is highly vulnerable to the impacts of climate change linked with water resources due to the variability and uncertainty of water availability in meeting the seasonal water demand. Rising of temperature and extreme weather affect crop-water requirements, incidence of floods and droughts, intrusion of seawater, and water quality in rivers and lakes, etc.

To cope with the climate change issues, it is required to develop a prudent planning system with water monitoring and research program for climate sensitive areas. Adoption of new methods and techniques of planning to addressing climate uncertainty, reevaluation of technical and economic approaches for managing water resources in view of potential impacts of climate change, preparation of action plans by all states for management and operation, improving water use efficiency for meeting future needs in water scarcity regions, re-examination of engineering designs, operating rules, contingency plans, water allocation policy in different agro climatic regions, coordination between the water agencies and scientific organizations for introducing climate smart management and operational techniques, waste water reclamation, rain water harvesting, desalinization of water where needed most should be considered and recommended by scholars including Lal (2001).

9.6   Interstate River Water Dispute

Interstate river water dispute arises as most of the water resource development and planning in the country has been done following the administrative boundaries rather than adopting river basins as the hydrological units. Water conflicts have arisen as most of the river basins are shared by several states, and water demand has gone up significantly within each state for meeting domestic, industrial, and agricultural needs (Kumar 2018). Of the total waters flowing through Indian rivers, about 80% derives from interstate rivers. Each state forming part of an interstate river basin wants to develop and utilize the water flowing through its territory in a manner considered beneficial to that state in its own perception.

Development of water resource projects in one state can have adverse effects on other states (Shah 1994). There are several options to address the interstate water resource development and distribution of water among the states. Formation of Control Boards involving the concerned States and the Central government was a practice and adopted for few such projects since 1950. Despite continuous efforts by the States and the Government of India through Water Dispute Tribunals, some of the major disputes have not been resolved yet.

New cases are coming up even now of which the conflict over the Mahanadi water between Odisha and Chhattisgarh states is noteworthy, challenged through Supreme Court of India. It was mentioned by Shah (1994) that resolution of interstate water disputes is a highly complex and intricate problem. The existing practice in India of resolving the disputes through judicial tribunals has not been very successful. In the national and States' interest, such disputes need to be resolved through mutual agreements. A set of objective criteria for sharing water between states should be worked out.

The criteria may include the indicators like the proportion of respective catchment area, drainage density, amount of rainfall received and generation of surface water draining to the river system, and the potential target area to be benefitted, commitment for joint investment as per the ratio of sharing the water or the generated revenue, and commitment for the maintenance of the system and supporting associated services for common infrastructure. It should be legislated in the Parliament with an environmental management framework for implementation. The potential of harnessing more water from the river basins is very high from the years of heavy rainfall and floods.

9.7   Inter-Basin Water Transfer Through Interlinking of Rivers

This proposed large-scale project aims to manage water resources in India by linking Indian rivers by a network of reservoirs and canals to enhance irrigation and groundwater recharge, reduce frequent flooding and water shortage in India. Historically, in the nineteenth century a British engineer Arthur Cotton initiated this proposal to interlink major Indian rivers to facilitate import and export of goods. In 1970s, Dr. K. L. Rao, a dam engineer and former irrigation minister proposed 'National Water Grid' to mitigate the severe shortage of water in the South India and frequent flooding in the North India. The Brahmaputra and the Ganga basins are water surplus areas where Central and South India are water-deficit areas. He proposed that surplus water should be diverted to deficit areas. By this time, several inter-basin transfer projects were implemented and it was suggested for scaling up.

In 1980, Ministry of Water Resources published a report entitled 'National Perspectives of Water Resource Development' with two parts—the Himalayan and Peninsular components. In 1982 'National Water Development Agency (NWDA)' was set up with a committee with nominated experts and given responsibility to conduct detailed studies, surveys, and investigations of reservoirs and canals and all aspects of feasibility of interlinking Peninsular Rivers and related water resource management. NWDA produced many reports during 1982 and 2013 but no projects were developed (Wikipedia: Indian Rivers Inter-link).

Thatte (2007) in his paper has discussed in details on the historical facts behind interlinking of rivers, needs, plans, status, and prospects. Dr. K. L. Rao proposed in 1972 that 2640 km long of the Ganga-Cauvery link involving pumping over 550 m. Captain Dastur also envisaged in 1977 two canals—firstly a 4200 km long 'Himalayan Canal' at the foot of Himalayas, and secondly a 9300 km long 'Garland Canal' covering Central and Southern India. However, the proposals were found to be very expensive and technically unfeasible.

Inter-basin water transfer undoubtedly comprises a viable and immediate need based freshwater augmentation program for India and quick boost for the ongoing conventional effort (Thatte 2007). Under this situation interstate and inter-basin water resource development and management must go hand in hand by settling the interests and conflicts among the concerned states involved in the basin.

9.8   Sustainable Development of Water Resources

Meeting the scarcity of water and making the availability of clean water is very vital for ensuring healthy society. About 40% of global people face water scarcity. More than 2 billion people are affected by water stress. Only 70% of the global people had safety-managed water, rest does not have. About 39% global people had safe sanitation. Eighty percent of wastewater goes into waterways without adequate treatment. Eighty percent of countries have laid the foundation for integrated water management and 70% of natural wetlands have been lost over the last century. Therefore, Sustainable Development Goal (SDG) 6 introduced in 2015 emphasizes on access to water and sanitation for all, and other connected issues for water resource development and management for meeting various demands effectively are equally implied.

India's 600 million people are facing the water crisis. As per the Composite Water Management Index study of NITI Aayog (2019), it is estimated that the overall demand for water in India will be double, 40% of the population will have no access to drinking water by 2030. It is serious and alarming. In India, 50% of area is struggling with drought-like situations. A quarter of country's populations were affected by a severe drought in 2018.

Twenty-one major cities (Delhi, Bengaluru, Chennai, Hyderabad, and others are closer to zero ground water level (Matto 2019). Almost 90% of rural India drinking water comes from ground water and 75% of agriculture is dependent on groundwater. In the areas where overexploitation may reach the level of 'Day Zero' particularly in the states namely Tamil Nadu, Rajasthan, Uttar Pradesh, Punjab, Haryana, Karnataka, Madhya Pradesh, Andhra Pradesh, and Telangana. State governments do not assert control over the extraction of groundwater.

Further, the large volume of groundwater is becoming unfit for drinking due to pollutants such as nitrates, fluoride, arsenic, iron, salinity, and pathogens that implies for low cost purification technologies. About one fourth of India's domestic wastewater is treated only (Parvatam and Priyadarshini 2019). The above facts illustrate comprehensively in favor of sustainable development programs and initiatives. Two decades before, it is advocated by Bobba et al. (1997) that unless an integrated and sustainable approach to utilization of water resources is adopted either at a regional or a national level, problems of water scarcity, droughts, famines will continue to occur year after year in many parts of India.

9.9 Sustainable Initiatives in India

Realizing the very temporal and spatial variation of rainfall and consequent impacts in the river flow and groundwater recharge, the National Water Mission of India has the main objective to conserve water, minimize the wastage, and ensure its more equitable distribution both across and within States through integrated water resources development and management system. The Mission has five important goals—1. Comprehensive water database in public domain and assessment of impact of climate change on water resource, 2. Promotion of citizen and state action for water conservation, augmentation, and preservation, 3. Focused attention to vulnerable areas including over exploited areas, 4. Increasing water use efficiency by 20%, and 5. Promotion of basin level integrated water resources management. The impact of climate change on water resources has not been accurately quantified. Various studies indicate that climate change could contribute further intensifications of extreme weather events (National Water Mission under the National Action Plan on Climate Change 2011).

Recently in 2019, the National Institution for Transforming India (NITI) Aayog of India in association with Ministry of Jal Shakti and Ministry of Rural Development has conducted a study on 'Composite Water Management Index' to measure the performance of States in a comprehensive set of water indicators and reports relative performance in 2017–18 as well as trends from previous years (2015–16 and 2016–17). This type of benchmarking study can go a long way in creating a common frame for the progress of water in India and highlight the need for specific improvements.

States are displaying progress in water management, but the overall performance remains much below of what is required to tackle India's water challenges. High performers like Gujarat, Andhra Pradesh, Madhya Pradesh, and Himachal Pradesh have continued to demonstrate strong water management practices, but low performers are struggling to cope with water shortage. Low economic contributors have low water management scores that indicate poor water management can hamper India's economic progress. Food security is also at risk because large agricultural producers are struggling to manage their water resources effectively.

On the positive side, greater focus on water governance and increased data discipline among states is building a pathway for driving long-term success in terms of water management practices to demonstrate outcomes. As it is evidenced from 2018 data that states in India are making progress overall but have a long way to go in terms of water management if India is to afford its citizens the quality of life they deserve, support economic growth, and sustain its ecosystems on a long-term basis (CWMI 2019).

Ministry of Jal Shakti has launched several programs. The important programs are (1) Developing integrated water and crop management information system, (2) Controlling water depletion and rain water harvesting, (3) Groundwater recharging by identifying defunct, unused, abandoned tube or bore wells, (4) new initiatives namely River Development and Ganga Rejuvenation, and many others. NITI Aayog has documented a number of best water management practices in India in 2017 covering six areas under the title 'Selected Best Practices in Water Management in India'. Cases are drawn from different states on selected areas: Irrigation water management (9 cases), Drinking water management (5 cases), Urban water management (4 cases), Traditional water management (6 cases), Industrial water management (5 cases), and Watershed management (5 cases). Few cases are presented below as outlined by the NITI Aayog's CWMI Report of 2019.

1. **Mukhya Mantri Jal Swavlambhan Abhiyan (MJSA), Rajasthan**: This is a successful story based on 'Participatory and decentralized rural water management'. It was launched in 2016. It is a multi-stakeholder program that makes villages self-sufficient in water through a participatory water management approach. It involves improved water harvesting and conservation initiatives. Drones are used to identify water bodies for restoration. Village meetings are responsible for budgeting use of water resources for different uses. This has given greater power to the community members in decision meeting. The program has benefitted 11,955 villages with more than 8.8 million people, 9.3 livestock heads, and covered an area of 3.35 million ha. As a net benefit, it has reduced 56% of water supply by tankers, raised groundwater level about 1.4 m in 21 non desert districts, and additional 50,000 ha of land were made fit for cultivation in the district and 64% of installed hand-pumps had been rejuvenated.

2. **Neeru-Chettu Program, Andhra Pradesh**: This is a water mission introduced by the Andhra Pradesh to make the state drought-proof state and reduce economic inequalities through improved water conservation and management

practices. The program has a strong focus on ensuring water supply in drought-prone areas by scaling up adoption of scientific water management practices. Repair, renovation, and maintenance of irrigation assets are the key activities. The state has repaired 7,000 farm ponds, and more than 22,000 check dams. The program has enabled irrigation access to nearly 85.000 ha of land in the State.

3. **Jalyukt Shivar Abhiyan, Maharashtra**: This program was launched in 2015–16 with the mission to make Maharashtra drought free by 2019, and making 5000 villages' water scarcity free every year. The focus was on deepening and widening of streams, construction of cement and earthen stop dams, work on drainage and digging of farm ponds. This involves geo-tagging of water bodies and use of a mobile application to enable web-based monitoring. This program has led to an increase in groundwater levels of 1.5–2 m. About 11,000 villages have been declared drought free and agricultural productivity has increased by 30–50%.

4. **Mission Kaktiya, Telangana**: Program was launched in 2014 and aimed to restore over 46,000 tanks across state. The objective was to develop minor irrigation structures, promote community-based irrigation management, and effective utilization of water allocated for minor irrigation under the Godavari and Krishna river basins. More than 22,500 tanks were restored until March 2018. This has boosted the water storage capacity of water bodies and enhanced on-farm moisture retention capacity in the region. The gross irrigated area was increased by 51.5%.

5. **Sujalam Sufalam Yojana, Gujurat**: This program was introduced in May 2018 with an objective of deepening water bodies before monsoon and increasing water storage for rainwater collection. A target of 110 million cubic feet through 13,000 lakes, check dams, and reservoirs was achieved successfully.

6. **Kapil Dhara Yojana, Madhya Pradesh**: This is a project undertaken by the Madhya Pradesh State Government under the Mahatma Gandhi National Rural Employment Guarantee Programme. The purpose of this project was to develop irrigation facilities on private land of small and marginal farmers through construction of dug wells, farm ponds, and check dams, etc. Farmers without access to irrigation facilities entitled for financial support and marginalized communities were prioritized for this project. The project has contributed to improved productivity, cropping intensity, and diversity of crop production in the region and generated livelihood options.

7. **Jakhni Village, Bundelkhand, Uttar Pradesh**: Within a time period of five years, the villagers of Jakhni drought-prone village of Banda district in the Bundelkhand region without any external financial support, machinery, and resources could able to transform through water conservation such as construction of farm ponds, restoration of water bodies, collection and utilization of grey water, raising farm bunds and intensive plantation of trees. It has become water self-sufficient village and it is getting the benefits of improved production of Basmati rice and other crops. This is an excellent example for village

water budgeting model of collection of rainwater and storage within village boundaries and utilizing for life protection and economic development.

9.9.1 Conclusion

The discussions made in previous paragraphs provide a comprehensive picture of water resources development and management in India with various approaches from farmers to villagers and states. Due to temporal and spatial variation of rainfall and availability of water for different uses, various models and approaches need to be evolved and practiced in reality.

However, micro level planning approach at the village and Panchayat level should be given the highest priority for water resource development and management in addition to regular activities undertaken at the state and national levels for major projects. The bottom up and top down planning and development approach is necessary for implementation of all macro and micro level projects in the decentralized mechanism involving people (the beneficiaries) and all stakeholders to make projects effective and fruitful in solving the water scarcity problem and making the project sustainable.

People should get awareness adequately about the depletion of water resources, impacts of extreme weather, rainfall variation, and shifting of cropping seasons, available quantity and quality of water, water loss and water pollution. Continuous assessment of requirements for domestic, agricultural and allied activities, and agro-industrial production purposes along with preparation of water budgeting and distribution plan. It is essential to develop a holistic database (spatial, sectoral, socioeconomic and other attributes) along with resource mapping with focus on all types of water sources available at the village and Panchayat level.

In irrigated areas, people are highly dependent on only irrigation water release for agriculture and allied activities. They are not concerned so much for creating new sources and alternatives. In contrast to this, the rain fed areas people struggle for overcoming the water scarcity through public, community, and individual water resource projects. Agricultural land suitability analysis at the block level should be taken up by the Agricultural Department to support activities in line with the existing potential and water stress in order to optimize the available water utilization simultaneously practicing rainwater harvesting and conservation of local water resources.

The local development authorities should facilitate the villagers and local stakeholders to undertake integrated and participatory village/panchayat development planning exercise with emphasis on available natural resources, problems and issues, potential and existing policy instruments. Finally, it is the responsibility of GOs, NGOs, stakeholders, and local people to think and work together to address the water scarcity issues and move in the sustainable path in meeting the current and future water needs.

References

Chatterjee K Water resources of India. Climate Change Centre, Development Alternatives

Chetan P, Biswas AK (2019) India's national water policy: 'feel good' document, nothing more. Int J Water Resour Dev 36(6):1015–1028. https://doi.org/10.1080/07900627.2019.1576509

Chitale MA (1992) Development of India's river basins. Water Resour Dev 8(1):31–44. https://doi.org/10.1080/07900629208722531

Das B (2009) India's water resources. Availability, http//www.base.d-p-h.info/fr/fitches/dph/fiche-dph-7825.html

Ghosh BA, Singh VP, Lars B (1997) Sustainable development of water resources in India. Environ Manage 21(3):367–393

Government of India, Ministry of Water Resources (2002) National water policy, pp 1–9

Jain SK (2019) Water resources management in India–challenges and the way forward. Curr Sci 117(4):569–576

Jain SK (1999) Water resources management in India. The national Commission for Integrated Water Resources Development (NCIWRD)

Kumar CP (2018) Water resources issues and management in India. J Sci Eng Res 5(9):137–147

Kumar R, Singh RD, Sharma KD (2005) Water resources of India. Curr Sci 89(5):794–811

Kumar V, Jain SK, Singh Y (2010) Analysis of long-term rainfall trends in India. Hydrol Sci J Journal des Sciences Hydologiques 55(4):484–496. https://doi.org/10.1080/02626667.2010.481373

Lal M (2001) Climate change–implications for India's water resources. J Soc Econ Dev III(1):57–87

Matto M (2019) India's water crisis: the clock if ticking. DownToEarth June 2019

NITI Aayog (2019) Composite Water Management Index, Ministry of Jal Shakti and Ministry of Rural Development, Government of India

National Water Mission under National Action Plan on Climate Change, New Delhi (2011) vol I

Parvatam P, Priyadarshini S (2019) One day zero, India prepares for a water emergency. Nat India. https://doi.org/10.1038/nindia.2019.84

Prabhakar AK, Singh KK, Chandniha SK (2019) Assessment of regional-level long-term gridded rainfall variability over the Odisha State of India. Appl Water Sci 9(93):1–15. https://doi.org/10.1007/s13201-019-0975-z

Shah RB (1994) Interstate river water disputes: a historical review. Int J Water Resour Dev 10(2):175–189

Thatte CD (2007) Inter-basin water transfer (IBWT) for the augmentation of water resources in India: a review of needs, plans status and prospects. Water Resour Dev 23(4):709–725. https://doi.org/10.1080/07900620701488646

Wate SR (2012) An overview of policies impacting water quality and governance in India. Int J Water Resour Dev 28(2):265–279. https://doi.org/10.1080/07900627.2012.668646

Wikipedia, Indian rivers interlinking project, en.wikipedia.org/wiki/Indian_rivers_interlinking_project

Chapter 10
Mapping and Reclamation of Wastelands in Drought-Prone Purulia District of West Bengal, India Using Remote Sensing and GIS

Manoj Kumar Mahato⬤ **and Narayan Chandra Jana**⬤

Abstract Wastelands mapping and reclamation studies has been carried out in Purulia District in western most part of West Bengal, India using high-tech tools of geoinformatics. Purulia District provides vast tracts of wastelands, which covers 28.41% of total wastelands area of West Bengal and 7.51% of total geographical area of the district in 2015–16. About six categories of wastelands were identified, viz. Badland wasteland, Gravelly wasteland, Mining & Industrial wasteland, Rocky or stony wasteland, Degraded forestland, and Degraded land under plantation crop. The main objectives of the present research are to study the spatial distribution of different types of wastelands and to suggest the appropriate measures for the reclamation of various categories of wastelands. The wastelands of the Purulia have been identified and categorized through the SOI toposheets of 1:50,000 scale, SRTM DEM LISS-III, Landsat-8 OLI/TIRS C-1 Level-1, and Google Earth images by GIS software's with rigorous field survey. Based on the analysis of secondary and primary data and information the authors in the present context have given appropriate suggestions toward the reclamation of Wastelands of Purulia district.

Keywords Wastelands · Purulia District · Geoinformatics · Spatial distribution · Reclamation

10.1 Introduction

Land resource is a precious natural resource and it functions as a key for the sustenance of humankind (NCA 1976; De and Jana 1997; Ramachandra 2007; Rawat et al. 2018). Excessive exploitation of land resources results in a significant change of the landforms, which has unfavorable effect to the environment (Rawat et al. 2018). The excess population pressure, growing industrialization, rapid urbanization, and extensive agriculture have put abundant stresses on land properties, resulting into the significant reduction of agricultural land and other natural resources (Grunwald 2013).

M. K. Mahato (✉) · N. C. Jana
Department of Geography, The University of Burdwan, Bardhaman, West Bengal, India

© The Author(s), under exclusive license to Springer Nature Singapore Pte Ltd. 2022   189
N. C. Jana and R. B. Singh (eds.), *Climate, Environment and Disaster in Developing Countries*, Advances in Geographical and Environmental Sciences, https://doi.org/10.1007/978-981-16-6966-8_10

Enormous population pressure is also guiding to deforestation and natural resource degradation that has distressed environmental balance of terrestrial systems (Chandramohan and Durbude 2002; Sharma et al. 2007). Land is turning into wasteland due to various natural and anthropocentric factors like rainfall deficit, drought conditions, soil erosion, wind erosion or deposition, water logging, salinity or alkalinity of soil, floods, deforestation, and unscientific techniques of cultivation (Crosbishley and Pearce 2007; Basavarajappa and Manjunatha 2014).

Wastelands are denoted as degraded land and recently laying unutilized (NRSA 2007; Chaturvedi et al. 2014; Basavarajappa et al. 2015; Sreekala and Neelakantan, 2015) due to inherent or imposed incapability related to geographical location, environment, soil fertility, water availability and current financial constraint (NRSA2007; Basavarajappa et al. 2015). National Wastelands Development Board (NWDB 1987) defines wasteland as degraded land which can be brought under the purview of vegetative cover by reasonable effort, and which is currently underutilized; deteriorating for lack of suitable water and soil management or on account of natural causes (Rao et al. 1991; Kohli et al. 2018). National Atlas and Thematic Mapping Organization (NATMO 2010) have expressed their views regarding wasteland as "those areas which are not utilized to their full potential and whose productivity could be increased by making reasonable efforts and investment". Different sectors define the wasteland according to their land use pattern. Agriculture land lying fallow for more than two years can be termed as agricultural wasteland (NCERT 2016). Lands under the control of Revenue Department are not fit for agriculture lying barren can be termed as Revenue wasteland (Singh 2012). Similarly, grasslands and lands under the control of Forest Department, which do not have tree cover, can be termed as forest wasteland (Luna 2006; Chaturvedi et al. 2014).

Although there are different perspectives on wastelands marking but it is largely accepted that wastelands are the areas which are underutilized and which produce less than 20% of its biological productivity (Mishra et al. 2013). The wastelands have characterized degraded, fallow, uncultivated, and common land as (i) lands not obtainable for cultivation, barren, and uncultivable wastes, (ii) other appendicular land apart from fallow, permanent pastures, culturable waste and land under various trees, (iii) fallows under wastelands.

The drought-prone Purulia District in West Bengal, India provides no exceptional picture from the previously mentioned scenario. High rate of deforestation, unscientific plantation, water crisis, and lack of irrigation facilities are mainly responsible for the formation of vast tracts of wastelands in the study area. According to Wasteland Atlas of India 2019, the total wasteland in West Bengal in 2015–16 was 1654.99 km^2, of which Purulia alone has 470.19 km^2 (28.41%). Land degradation and drought condition are the serious problem of the study area, which can accelerate the amount of wasteland. These problems can be controlled by conserving land surface and ground water in the wasteland area (Rawat et al. 2018) as well as through scientific tree plantation programs (Dwivedi et al. 1997; HARSAC 2006) on wastelands in rocky and shallow deep soils. Thus, the current and appropriate information about the location and spatial arrangement of vacant or wastelands has played an emergent role for better planning of trees and treatment to exterminate the negative effects of land degradation (Contador et al. 2008; Nawar et al. 2015). Hence, there is a significant requirement for wasteland identification and reclamation in many countries around the world (Chandramohan and Durbude 2002) and also in the district of Purulia of West Bengal.

Recent developments in geographical mapping consent the researchers to work out the location aspects and distribution pattern of land use/land cover (LULC), which is shown in more accurately by using geospatial techniques (Saha et al. 1990; Sugumaran et al. 1994; Rao 1999; Singh 2006). Numerous studies have recognized the applications of RS & GIS in natural resources monitoring and management (Pramila 1994; Metternicht and Zinck 2008; Mulder et al. 2013). The satellite imageries are currently being used extensively in mapping of various land features with the help of GIS technique (Basavarajappa and Dinakar 2005), such as LULC mapping (Singh et al. 2014; Srivastava et al. 2014), LULC modeling (Singh et al. 2015; Mustak et al. 2015), groundwater (Singh et al. 2010), lake and wetlands (Thakur et al. 2012a, b), crop suitability (Mustak et al. 2013), slope estimation (Szabo et al. 2015), landscape ecology (Singh et al. 2017), urban land use dynamics, forest mapping (Singh et al. 2012), soil characterization (Paudel et al. 2015) and watershed management (Yadav et al. 2014). Further, Remote Sensing technology has proven its application in wasteland assessment and its temporal monitoring (Jain et al. 1991; Barchyn et al. 2014). Wasteland reclamation and development can be possible by RS & GIS technology and accurate field monitoring, through appropriate strategies (Breunig et al. 2008; Basavarajappa et al. 2016). The present study has been adopted to investigate different wasteland categories and their reclamation in Purulia District using applications of RS & GIS and conventional data.

10.2 Materials and Methods

10.2.1 Study Area

Bounded by the latitudes of 22°40′ N to 23°42′ N and longitudes of 85°49′ E to 86°54′ E, in the eastern fringe of the Chotonagpur plateau (Fig. 10.1), funnel shaped Purulia District is located in the western part of the West Bengal. It is surrounded by the state of Jharkhand in north, west, and south; and in the eastern part by the districts of Barddhaman, Bankura, and Jharagram of West Bengal, India. Total geographical area of the district is 6,259 km^2.

Various stratigraphic units ranging from the oldest Archaeans (Pre-Cambrian) to the younger Tertiary-Quaternary formations constitute the study area (Dunn 1929). Topographically, this area is very much diversified with dome-shaped inselbergs, spurs, escarpments, undulating upland, and erosional plain (Mahato and Jana 2019). As a part of the Chhotanagpur Granite-gneiss tract, the Purulia did not experience any severe diastrophic disturbance in its long geological history, but it could not escape the impact of orogenic forces (Dunn 1929; Singh 1969; Ray 1982; Ghosh

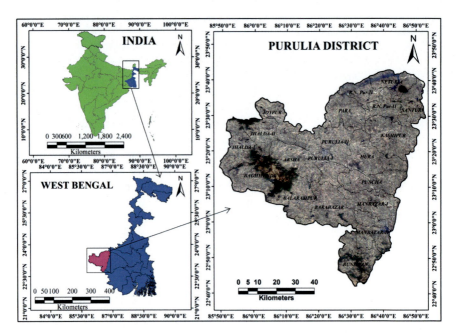

Fig. 10.1 Location of the study area

2012). The study area occupies the eastern part of the Pre-Cambrian Granite-gneiss tract (Singh 1969).

Climatologically, Purulia District is characterized by sub-tropical Monsoon type of climate with very high day temperatures during the summer months reaching up to 46 °C, whereas the winter months are plentiful cooler with lowest temperatures of up to 3 °C (Bhattacharya et al. 1985). The evaporation rate of the district is very high during the summer months due to the mean monthly average temperature of 32 °C, while the mean monthly temperature of the winter months is 13 °C. The mean long-term annual rainfall for the period of 1960–1961 to 2014–15 is 132 cm of which 80% rainfall occurs during June to September (Bhattacharya et al. 1985; Datta and Chakraborty 2015). The soil of the area is infertile laterite and red gravelly type, which is characterized by infertile, unproductive, erosion prone, lack of soil nutrients, and lower water holding capacity (NBSS and LUP 2010).

10.2.2   Lithology

In eastern fringe of Chhotanagpur gneissic complex, Purulia District generally repre-sents an Archaean complex region (Dunn 1929). The parent rock of this area is Granite-Gneiss with its varying composition (Dunn 1935). The entire study area is covered mostly by Chotonagpur granite gneisses (Dunn 1929), which include quartz biodte granite gneiss, prophyroblastic granite gneiss, massive granite composite gneiss, augen gneiss, and migmatites (Chatterjee 1946). The Chotonagpur granite gneiss holds the enclaves of metasedimentaries that include mainly the mica schists, garnetiferous sillimanite biotite schist, amphibolites, etc. (Baidya et al. 1987). The Lithology and Structural elements map of the Purulia District (Fig. 10.2) shows that most of the area (71% of the total area) is covered by Chotanagpur granite-gneiss.

Structurally the studied area is considered as one of the ancient (Pre-cambrian/Proterozoic) stable landmass of the Indian Peninsula that has not been affected by any folding movement created within the earth during later geological periods (Ball 1981). Tectonically, the study area is an old land surface, which suffered tectonic turbulences due to Gondwana drifting from Permo-Carboniferous to Jurassic period (Singh 1969).

The Pre-Cambrian, the oldest rock formed at the lowermost, which was the most dominant among the various rock types. In Permian age, lower Gondwana rock groups such as Sandstone and Shale of Kulti Formation and Coal bearing sandstone and shale of Raniganj Formation situated over the Archaeans rocks. The upper Gond-wana rocks of Triassic and Lower Cretaceous age are Sandstone, Clay and Shale of Panchet Formation, Red sandstone and Red clay of Mahadeva Formation and Clay with caliche concretion of Sijua Formation deposited unconformably over the lower

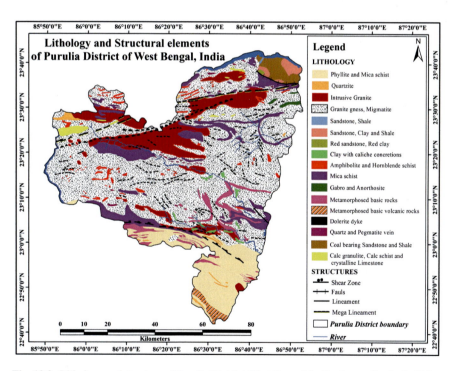

Fig. 10.2 Lithology and structure of Purulia District, West Bengal, India. *Source* Geological Map, published by Geological Survey of India, Kolkata, 2001

Gondwana in the northeastern part of the district, i.e., along the Damodar River (Geological Survey of India, Kolkata 2001) (Fig. 10.2).

10.2.3 Topographical Characteristics

Purulia district is formed steeply sloping from west to east between the Chhotanagpur plateau and the Damodar plain. In terms of topography and structure, Purulia is a fragment of the Ranchi peneplains (Singh 1978). The elevation of different parts of Purulia ranges between 78 to 699 m above the mean sea level with its great diversity of the polycyclic landscape through the undulating Archean plateau (Dunn and Dey 1942). The diversity of landforms in the district has been caused by different cycle of erosion and lithological structures. Purulia district made up a portion of the Precambrian metamorphic terrain of the southeastern Chotonagpur plateau consisting mainly of granites, gneisses, quartzite, etc. rocks, and Gondwana sediments. The main hills of the study area are Ajodhya (669 m.), Panchet (643 m.), and Joychandi (305 m.). Besides these, monadnock like residual hills, hillocks, dome-shaped inselbergs, spurs, escarpments, dissected valleys, and rocky out crops are general features. The remaining part of the district is undulating and rolling land consisting of laterite. The major lithological and geomorphic structures have exhibited striking differences in the physiography. Physiographically, this area is divided into five divisions such as (i) Hilly tract, (ii) Highly gradient rugged upland, (iii) Rugged upland, (iv) Moderately gradient rugged terrain, and (v) Low gradient rugged terrain (Fig. 10.3).

10.2.4 Drainage System

Originating from the hillocks of the Chhotanagpur plateau, several rivers flow from east to west through Purulia District. Among these Kangsabati (Kasai /Cossey/ Kansai), Kumari, Dwarakeswar, Subarnarekha, and Damodar are the important ones. The main drainage of the district is controlled by river Kasai, which drains more than three-fifth of the district's water. Apart from these main rivers, there are numerous tributaries and *Jora* (in those channels water flows only during monsoon) in the district. Most of the channels of the district are non-perennial which contain water for three to four months (Fig. 10.4). The perennial rivers like Kangshabati and Kumari are sustained water around the year but not in flowing condition. A number of Non-perennial River, *Jhor, Bandh* (artificial created water body), and a few check dam are the significant surface water bodies in the study area. All the water bodies in the study area are depend on rainwater. During pre-monsoon period, all the water bodies are dried up and drought condition prevails all over the district.

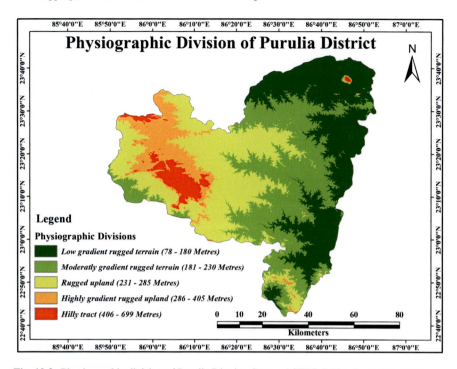

Fig. 10.3 Physiographic division of Purulia District. *Source* ASTER DEM, September 2014

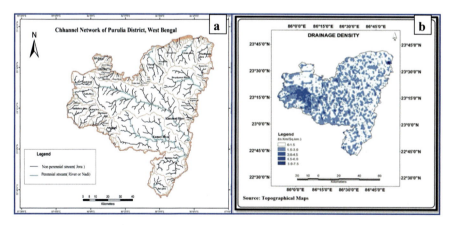

Fig. 10.4 Drainage System of Purulia District, **a** showing the channel network of Purulia District; mainstream of Kangsabati, Kumari, and Dwarakeswar river are perennial. Apart from these mainstreams, all the tributaries of these rivers are non-perennial. **b** is showing the drainage density of the district

10.2.5 Data Used and Analytical Procedures

10.2.5.1 Geospatial Data

For detailed study, the wastelands of Purulia District, West Bengal have been delineated and mapped through digital image analysis of high resolution satellite data and digital interpretation of SOI toposheets (Table 10.1), which are verified during the field visits.

Table 10.1 Geospatial data sources and analytical techniques

Analytical features	Data source	Techniques
Lithology	Geological Map, published by Geological Survey of India, Kolkata, 2001	Vector mapping for different Lithological and Structural elements
Physiography	ASTER DEM, Spatial Resolution 30 m., September 2014, USGS Earth Explorer	Vector mapping of different physiographic divisions
Drainage	ASTER DEM, Spatial Resolution 30 m., September 2014, USGS Earth Explorer	1. Raster creation for Perennial Non-perennial drainage system, 2. Drainage density raster creation by *Drainage length/area in sq.km*
Land use /Land cover (LULC)	Landsat-8 OLI/TIRS C-1 Level-1 (11 bands), Spatial Resolution- 30 m, Date of Acquisition- 20 & 29 May, 2019, USGS Earth Explorer	Land use/land covers (LULC) classification by supervised techniques
Wastelands Distribution	1. Landsat-8 OLI/TIRS C-1 Level-1 (11 bands), Spatial Resolution- 30 m, Date of Acquisition- 20 & 29 May, 2019, 2. SOI Toposheets: 73E/15, 73E/16, 73I/2, 73I/3, 73I/4, 73I/6, 73I/7, 73I/8, 73I/10, 73I/11, 73I/12, 73I/14, 73I/15, 73I/16, 73 J/1, 73 J/5, 73 J/6, 73 J/9, 73 J/10 on 1:50,000 scale, new edition 2010, Source: (NakshePortal) www.soinakshe.uk.gov.in	Overlay of Final LULC Map and Thematic Map of SOI toposheets through GIS Application for Spatial distribution of Existing Wastelands and Zonal distribution of different types of Wastelands

10.2.5.2 Reports and Records

- West Bengal District Gazetteers, Purulia, Government of West Bengal, Published by Narendra Nath Sen, State Editor, West Bengal District Gazetteers, Calcutta, 1985.
- District Census Handbook of Purulia District: 1991, 2001 and 2011.
- District Statistical Handbook, Purulia (2013, 2014 and 2015), Published by Government of West Bengal.
- Daily newspapers and periodicals for the information regarding the recent work on wastelands,
- Administrative Reports regarding policy measures at the Government level toward the reclamation of wasteland.

10.2.5.3 GIS Softwares

ArcMap, version 10.4 and Erdas Imagine, version 2013.

10.2.5.4 GPS

A handheld GPS Garmin-12 is used to record the exact locations and extent specific wasteland categories in the study area.

10.2.6 Methods

Geoinformatics techniques include SOI toposheets, Remote Sensing Satellite data, Global Positioning System (GPS), and GIS Software for mapping of forest cover, vegetation, lithology, physiography, drainage systems, and land use/land cover pattern in measuring the wasteland reclamation and management (Basavarajappa et al. 2015). With the help of satellite imageries and SOI toposheets, identification and delineation of wastelands has been done in three steps, viz. preliminary analysis, ground truth verification, and final interpretation (Fig. 10.5).

10.2.6.1 Preliminary Analysis

Visual analysis of Landsat-8 OLI/TIRS C-1 Level-1 on 1:50,000 scale geocoded data has been accepted for the mapping techniques of this study. In the first step, the base map of the study area have been prepared from SOI toposheets on 1:50,000 scale taking forest boundaries, roads, rivers, visible water bodies, district, state, and C. D.

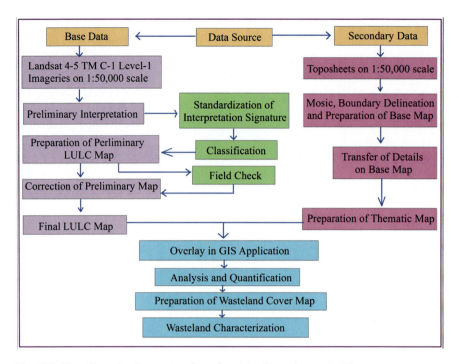

Fig. 10.5 Flow Chart showing an overview of wasteland mapping methodology

Block boundaries are taken from census handbooks of Purulia District and super-imposed on the base maps. The base map and C. D. Block boundaries have been digitized in ArcMap software. Universal Tranverse Mercator (UTM) with WGS-84 datum, 45° N Projection system have been taken to project all these maps. Satellite images of Landsat-8 OLI/TIRS C-1 Level-1 have been loaded on ArcMap software for visual interpretation. Based on the standard image interpretation keys like tone, texture, site, situation, pattern, shape, size and association, images were classified on supervised method using Maximum Likelihood Classification. By functioning the classifier panel of the ERDAS IMAGINE, training signatures of the target, i.e., Wastelands were identified. In this study, certain categories of wastelands like Rocky or stony wasteland, degraded forestland, and Gravelly wasteland are fluently delin-eated by asset of their pattern, location, and spectral separability. Besides them, Badland waste and Mining & Industrial wastelands are delineated with moderate success. However, degraded land under plantation crop could not be simply delin-eated because of its merging with fallow land, which is widely appeared throughout the district.

10.2.6.2 Ground Truth Verification

In the study area, accuracy has been verified on actual appearance of ground surface with the delineated wastelands of preliminary analysis phase. Particularly the areas where confusion in the preliminary interpretation has been thoroughly verified with the GPS recorded data.

10.2.6.3 Final Interpretation

At this stage the interpretation of the image through ground check has also been further developed. With the help of these, final interpretation is completed. Through this exercise, demarcation of wastelands as well as coverage area under particular types of wastelands has been finalized. In this way, maps have been prepared for ready to the cartographic work.

10.3 Results and Discussion

10.3.1 Land Use/Land Cover (LULC) Pattern of the Purulia District

Land use refers to multiple uses of land, which are directly related to human activities (Anderson et al. 1976) and land cover refers to natural vegetation, soil, rocks, water bodies, agricultural area, artificial land cover, and others resulting due to land transformation (NRSA 1987; Basavarajappa et al. 2016). In general, land use is of a rapidly changing nature because it takes on complex forms in the context of socio-economic, technological, and environmental changes. Land use refers to responding the advantages and deficiencies of the physical environment based on human economic potential (Bhattacharya 1965). In India, rapid industrialization, urbanization, and massive population growth have resulted in depletion of natural resources, water crisis, declining agricultural lands, and increasing wastelands (Government of India, Department of Agriculture 2016; Roy and Inamdar 2018).

Delineation of various land use/land cover categories of Purulia is made using Visual Image Interpretation Techniques on Landsat-8 OLI/TIRS C-1 Level-1 satellite images in combination with collateral data like SOI toposheets. The major classes of land use/land cover have been identified following the supervised method using the bands 2, 3, 4 & 5 of satellite imageries. In the present study, a total of seven land use/land cover classes have been taken for study which include Built-up area, Vegetation cover, Degraded forest, Agricultural land, Fallow land, Wasteland, and Water Bodies (Fig. 10.6). A total number of 145 section points were used to check the accurateness of the classification. The overall accuracy and kappa coefficient value of the classification are 96.67 and 95.87, respectively.

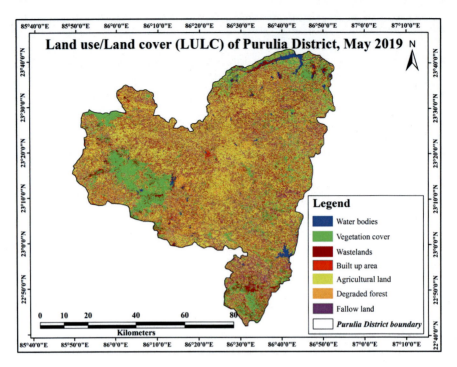

Fig. 10.6 Land use/Land cover (LULC) of Purulia District. *Source* Landsat 8 imagery, May 2019

Most of the land in Purulia is under agriculture; nearly 50.93% of the total geographical area of the district was agricultural land in 2019. Most of the agricultural land in the district is dependent on monsoon rainfall, so these lands remain vacant during the pre & post-monsoon season. Vegetation cover is 16.26% of the total area, most of which is observed in the western and southern parts of the district. The amount of wasteland is 7.54% and fallow land is 17.61% of the total geographical area. The amount of fallow land affects the seasonal diversity of the wasteland, which is varying in different years due to the nature of monsoon rainfall and irrigation condition. In addition, other land use areas are relatively less, such as 2.03% of degraded forestlands, 3.12% of water bodies, and 2.51% built-up area (Table 10.2). Irregular and scattered type of rural settlements is found all over the district, but in several parts, agglomerated urban settlements are also observed.

10.3.2 Identified Categories of Wastelands in Purulia District

Wasteland is a type of land cover that can not be used for agriculture or any other profitable purpose under the ongoing conditions of land management (Jha 1987; Khatun and Debnath 2014). In Purulia District six categories of wasteland have been

Table 10.2 Land use/Land cover pattern in Purulia District, West Bengal (May, 2019)

Land use types	Area in km^2	Percentage
Built-up area	157.1009	2.51
Water bodies	195.2808	3.12
Wastelands	471.9286	7.54
Vegetation cover	1017.7134	16.26
Degraded forest	127.4321	2.03
Agricultural land	3187.4495	50.93
Fallow land	1102.0947	17.61
Total geographical area	6259.00	100.00

Source Landsat-8 OLI/TIRS C-1 Level-1 imageries, May 2019

identified, they are badland waste, gravelly wasteland, mining & industrial wasteland, rocky or stony wasteland, degraded forestlands, and Degraded land under plantation crop (Fig. 10.7). The categories of the wastelands are identified and digitized using SOI topographical maps of 1:50,000 scale and restructured from Landsat-8 OLI/TIRS C-1 Level-1 satellite images of 1:50,000 scale through ArcMap and Erdas Imagine (NRSC/ISRO 2012).

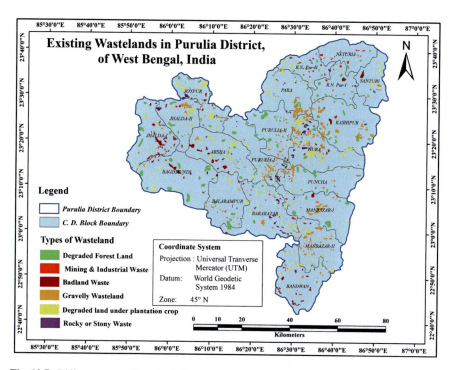

Fig. 10.7 Different categories of existing wastelands of Purulia District. *Source* SOI toposheets, Landsat-8 OLI/TIRS C-1 Level-1 satellite images and field survey

Table 10.3 Identified categories of wasteland in Purulia District in 2019

Categories of wasteland	Wasteland area in sq. km	% to Total Wastelands	% to Total Geographical area
Badland waste	11.2790	2.39	0.18
Degraded forestland	127.4207	27.00	2.03
Gravelly wasteland	113.0268	23.95	1.81
Mining & Industrial wasteland	9.0138	1.91	0.14
Degraded land under plantation crop	169.1864	35.85	2.71
Rocky or stony wasteland	42.0016	8.90	0.67
Total	471.93	100.00	7.54

Source Landsat-8 OLI/TIRS C-1 Level-1 satellite images, SOI toposheets and field survey

Being the most wasteland prone district of West Bengal, the total amount of wasteland in Purulia is 471.93 km². (in 2019) that occupies 7.54% of the total geographical area of the district. Out of total wasteland, Plain Degraded land under plantation crop alone covers 35.85% followed by degraded forestland (27.00%), gravelly wasteland (23.95%), rocky or stony wasteland (8.90%), badland waste (2.39%), and mining & industrial wasteland (1.91%). Only three categories of wastelands, i.e., Plain drought-prone, degraded forestland, and gravelly wasteland jointly cover most (86.8%) of the wasteland areas of the district (Table 10.3).

10.3.3 Spatial Distribution of Wastelands in Purulia District

The geo-database formed using ArcMap and screen digitization techniques shows the types, extension and spatial distribution of different categories of wasteland existing in the study area (Fig. 10.9). In respect to total geographical area, the intensity of wasteland is highest in Arsha C. D. Block, which is 18.01 percent to total geographical area of the block. Moreover, Hura, Joypur, and Barabazar C. D. Blocks also have high intensity of wastelands covering to 16.19%, 14.43%, and 13.40% of the total geographical area, respectively. These four C. D. Blocks cover more than 46% of the total wasteland area of the district. Jhalda-I, Arsha, Joypur, Barabazar and Hura C. D. Blocks have more than one-half of the total wasteland area in the district for the predominance of hill slope, escarpments, hillocks, intrusive granite features. The C. D. Blocks that having significant percentage sharing of wasteland are Kashipur (6.43%), Manbazar-II (5.72%), Purulia-II (4.68%), Jhalda-II (4.58%), Bandwan (3.59%), Para (3.17%), and Balarampur (3.01%). The amount of wasteland in these C. D. Blocks is high due to water crisis, prolonged drought, massive soil erosion, and extensive deforestation. Table 10.4 and Fig. 10.8 show the coverage

Table 10.4 Spatial Pattern of wastelands in the Purulia District

Sl. no	C. D. blocks	Total geographical area (TGA) (in Hectare)	Total wastelands area (in Hectare)	% to TGA	C. D. Block wise share in %
1	Jhalda-I	31,909	2841.0186	8.90	6.02
2	Jhalda-II	25,661	2161.4394	8.42	4.58
3	Arsha	37,504	6753.3183	18.01	14.31
4	Joypur	23,047	3327.1065	14.43	7.05
5	Baghmundi	42,795	1061.8425	2.48	2.25
6	Balarampur	30,088	1420.5093	4.72	3.01
7	Barabazar	41,806	5601.8091	13.40	11.87
8	Manbazar-I	40,132	1057.1232	2.63	2.24
9	Manbazar-II	28,581	2699.4396	9.44	5.72
10	Bandwan	35,125	1694.2287	4.82	3.59
11	Puncha	33,011	1000.4916	3.03	2.12
12	Hura	38,221	6187.0023	16.19	13.11
13	Kashipur	45,131	3034.5099	6.72	6.43
14	Santuri	17,969	556.8774	3.10	1.18
15	Purulia-I	32,337	991.053	3.06	2.1
16	Purulia-II	31,010	2208.6324	7.12	4.68
17	Para	31,259	1496.0181	4.78	3.17
18	Raghunathpur-I	20,182	1142.0706	5.66	2.42
19	Raghunathpur-II	19,767	1231.7373	6.23	2.61
20	Nituria	20,365	726.7722	3.59	1.54
	Purulia District	625,900	47,193.00	7.54	100.00

Source Landsat-8 OLI/TIRS C-1 Level-1 satellite imageries, SOI toposheets and field survey

area, intensity (% to TGA) and percentage sharing of wasteland according to the different C. D. Blocks of Purulia District.

N.B.: % to GTA = Total Wastelands Area / Total Geographical Area × 100; C. D. Block wise share in % = Total Wastelands Area of specific C. D. Block/grand total wasteland area (47,193.00) × 100.

Table 10.5 and Fig. 10.9 show the intensity of wasteland in the district in four grades. These grades are allocated as low (<4.00%), medium (4.01–8.00%), high (8.01–12.00%) and very high (>12.00%). Very high range wasteland found in Arsha, Hura, Joypur, and Barabazar C. D. Blocks. As the Arsha C. D. Block is located on the escarpment zone of the Ajodhya hill, the steeper slope and the faster soil erosion are responsible for the creation of wasteland. Whereas, Hura, Joypur, and Barabazar C. D. Block are existed on highly gradient rugged upland of the district. There are

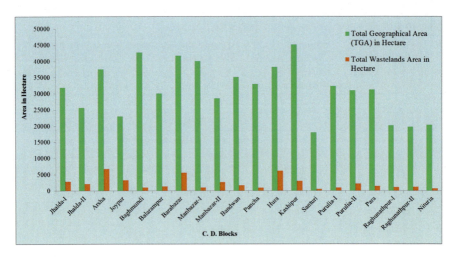

Fig. 10.8 Amount of wasteland to the total geographical area of the Purulia District. *Source* Landsat-8 OLI/TIRS C-1 Level-1 satellite imageries, SOI toposheets and field survey

Table 10.5 Wasteland intensity categorization

Range of wastelands (% to TGA)	Grades	Existing C. D. Blocks
<4.00	Low	Baghmundi, Manbazar-I, Puncha, Santuri, Purulia-I, and Nituria
4.00–8.00	Moderate	Purulia-II, Kashipur, Raghunathpur-I & II, Bandwan, Balarampur, and Para
8.01–12.00	High	Manbazar-II, Jhalda-I, and Jhalda-II
>12.00	Very high	Arsha, Hura, Joypur, and Barabazar

several small hills in these blocks and Pre-Cambrian rocks have been exposed in different places, which is the main reasons for the very high-level of wasteland.

High-grade wasteland is found in Manbazar-II and Jhalda-I & II C. D. Blocks due to the presence of highly rugged topography with numerous hillocks. Besides, the other C. D. Blocks of the district are characterized as moderate to low range wasteland because of the capacity of the land to be used is high in the moderate to low gradient rugged terrain. Overall, the main causes for uneven distribution of wasteland in Purulia District are variations in topographic complexity, water crisis, rapid soil erosion, excessive deforestation, unscientific plantation, and other anthropogenic activities.

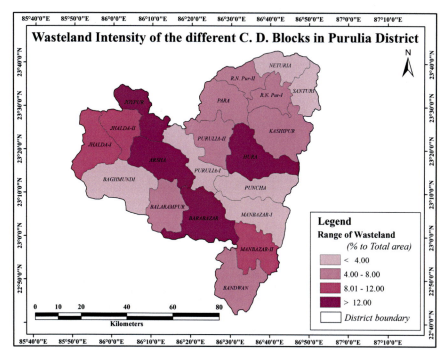

Fig. 10.9   Intensity of wasteland of the different C. D. Blocks in Purulia District. *Source* Landsat-8 OLI/TIRS C-1 Level-1 satellite imageries, SOI toposheets and field survey

10.3.4   Details of Category-Wise Wasteland Distribution

Multi-spectral data and maps show the magnitude, spatial distribution, and extent of the six categories of wastelands in Purulia District (Table 10.6). The categories of wastelands are quantified and mapped based on the characteristics of satellite imageries such as tone, color, shape, size, texture, pattern, and association (Fig. 10.10). In the Purulia District, the category- wise wasteland distribution has shown that most of the total geographical area (TGA) is covered by degraded land under plantation crop (2.71%) followed by degraded forestland (2.03%), gravelly wasteland (1.81%), rocky or stony wasteland (0.67%), badland wasteland (0.18%) and mining & industrial wasteland (0.14%), respectively (Fig. 10.11).

10.3.4.1   Degraded Land Under Plantation Crop

Degraded land under plantation crop is found sporadically scattered all over the district. The concentration of this category is found to be maximum in Arsha C. D. Block i.e., 5.52% to the total geographical area (TGA) due to the higher elevation of

Table 10.6 Category-wise wasteland distribution among the C. D. Blocks

Sl. no	C.D. blocks	Percentage to total geographical area (TGA) of the C.D. Block						Total % to TGA
		Degraded forestland	Degraded land under plantation crop	Gravelly wasteland	Rocky or stony wasteland	Badland wasteland	Mining & industrial wasteland	
1	Jhalda-I	1.84	2.00	0.60	0.66	3.80	0.00	8.90
2	Jhalda-II	2.49	2.84	1.30	1.06	0.56	0.17	8.42
3	Arsha	3.02	5.52	1.80	1.28	6.39	0.00	18.01
4	Joypur	3.80	4.13	3.48	2.70	0.15	0.17	14.43
5	Baghmundi	0.73	0.40	0.08	0.07	1.20	0.00	2.48
6	Balarampur	0.90	1.04	1.21	1.30	0.00	0.27	4.72
7	Barabazar	1.71	4.23	2.86	1.99	0.00	2.61	13.4
8	Manbazar-I	0.80	0.31	1.37	0.09	0.04	0.02	2.63
9	Manbazar-II	0.20	3.41	5.02	0.78	0.00	0.03	9.44
10	Bandwan	0.09	1.10	0.28	0.19	3.16	0.00	4.82
11	Puncha	0.70	0.51	1.41	0.41	0.00	0.00	3.03
12	Hura	2.54	5.13	4.10	3.42	0.12	0.88	16.19
13	Kashipur	1.80	1.00	3.28	0.24	0.36	0.04	6.72
14	Santuri	0.21	1.43	1.24	0.11	0.08	0.03	3.10
15	Purulia-I	0.40	0.72	1.90	0.63	0.00	0.09	3.74
16	Purulia-II	3.15	0.49	3.04	0.44	0.00	0.00	7.12
17	Para	0.32	1.96	2.13	0.27	0.00	0.10	4.78
18	Raghunathpur-I	2.14	0.17	0.24	0.65	0.32	2.14	5.66
19	Raghunathpur-II	0.50	2.33	0.00	0.27	0.00	3.13	6.23
20	Nituria	0.72	0.00	0.32	0.00	0.00	2.55	3.59
Purulia		2.03	2.71	1.81	0.67	0.18	0.14	7.54

Source Landsat-8 OLI/TIRS C-1 Level-1 satellite imageries, SOI toposheets and field survey

Ajodhya foothill, lack of ground and surface water resources, and limp & infertile soil. Besides, Hura (5.13%), Barabazar (4.23%), Joypur (4.13%), and Manbazar-II (3.14%) C. D. Blocks also have high intensity (>3% to GTA) of this category of wasteland because of water crisis and poor soil. This is followed by Jhalda-II (2.84%), Raghunathpur-II (2.33%), Jhalda-I (2.00%), Para (1.96%), Balarampur (1.04%), Bandwan (1.10%) and Santuri (1.43%) C. D. Blocks. The intensity of this category in Purulia-I & II, Puncha, Manbazar-I, Kashipur, Neturia, and Baghmundi C. D. Blocks is much less (<1.00%) because the main rivers of the district like Kangsabati, Kumari, Dwarakeswar, and Damodar are flowing along these C. D. Blocks.

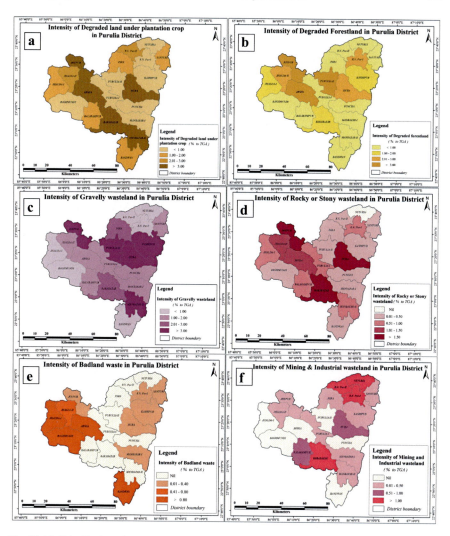

Fig. 10.10 Distribution of categories of wasteland; **a** distribution of degraded land under plantation crop among the C. D. Blocks, **b** distribution of degraded forestland among the C. D. Blocks, **c** distribution of gravelly wasteland among the C. D. Blocks, **d** distribution of rocky or stony wasteland among the C. D. Blocks, **e** distribution of badland waste among the C. D. Blocks, **f** distribution of mining & industrial wasteland among the C. D. Blocks

10.3.4.2 Degraded Forestland

The degraded forestland has covered more than 3 percent of the total geographical area in Joypur (3.80%), Purulia-II (3.15%), and Arsha (3.02%) C. D. Blocks indicating the high intensity of this category. Others C. D. Blocks having more geographical area under the degraded forestland are Hura (2.54%), Jhalda-II (2.49%),

Fig. 10.11 Different categories of wastelands in Purulia District; **a** shows the degraded land under plantation crop in Kalimati Mouza of Baghmundi C.D. Block. Soil quality is good and water holding capacity is moderate but this tract is not utilized due to lack of water. **b** Degraded forestland of Deuli Mouza of Hura C.D. Block. *Sanaijhuri* plantation for long time creates this wasteland. **c** Gravelly wasteland of Palsara Mouza of Kashipur C. D. Block. **d** Rocky wasteland of Hirbahal Mouza of Purulia-II C.D. Block. **e** *Sonaijhuri* Plantation on Maghuria Pahar (hill) in Hura C. D. Block. This type of plantation promotes soil erosion and increases rocky field. **f** Badland wasteland in Gura-hata Mouza of Arsha C.D. Block and **g** Badland wasteland of Sharangdi Mouza of Baghmundi C. D. Block are found in foothills of Ajodhya Range. Soil erosion is rapid which needs adequate plantation. **h** Heaps of industrial wastes (ash) in Durmut Mouza of Neturia Block, which converts the agricultural land into wastelands. **i** Sponge iron industry and adjoining wasteland of Kashiberia Mouza of Natundih G.P. in Raghunathpur-II C. D. Block. **j** Rock mines in rocky and gravelly waste-lands of Latpoda Mouza in Barabazar C.D. Block, which accelerates soil erosion & deforestation and creates mining wasteland. **k** Stone-chips industry of Hinjla Mouza in Barabazar C. D. Block, **l** *Panjania Granite Mine Project* (Hura Block) of WB Mineral Development & Trading Corporation Ltd. It accelerates rapid soil erosion, which is deposited on the agricultural field of lower part

Raghunathpur-I (2.14%), Jhalda-I (1.84%), and Kashipur (1.80%). These C. D. Blocks have a higher rate of economic development than the other blocks in the district, which is the main reason for the increase in their degraded forestlands. The people of these blocks always use the land profitably. In the rest of the blocks, the amount of this category of wasteland is negligible percentage (<1.00%).

10.3.4.3 Gravelly Wasteland

High concentration of gravelly wasteland is found in Hura (5.13%), Manbazar-II (5.02%), Joypur (3.48%), Kashipur (3.28%) and Purulia-II (3.04%), C. D. Blocks due to the occurrence of high-level weathered materials of granite-gneiss, migmatite, and mica-schist. Other C. D. Blocks having more area under this category are Para, Barabazar, Jhalda-II, Purulia-I, Arsha, Balarampur, Manbazar-I, and Santuri. The intensity of wasteland in these blocks is "between" 1% to 3% to the total geographical area.

10.3.4.4 Rocky or Stony Wasteland

In case of rocky or stony wasteland, Hura again tops among all the C. D. Blocks followed by Joypur, Barabazar, Balarampur, Arsha, and Jhalda-II. Their percent to total geographical area of respective blocks are 3.42%, 2.70%, 1.99%, 1.30%, 1.28% and 1.06%, respectively. In the study area, the main reason for the formation of this category is the rapid soil erosion resulting in exposure to the parent rocks on the surface soil. Rest of the C. D. Blocks in Purulia District, the amount of this category of wasteland is either negligible percent or nil.

10.3.4.5 Badland Waste

Although the badland wasteland of the district is much less than the other wastelands, it is highly intense in Arsha (6.39%), Jhalda-I (3.80%), Bandwan (3.16%), and Baghmundi (1.20) C. D. Blocks. Arsha, Jhalda-I, and Baghmundi C. D. Blocks are located in the hilly tract of Ajodhya Pahar. On the other hand, there are many hills and hillocks along the entire Bandwan C. D. Block. Therefore, the high elevation, hill slope, escarpment, and highly rugged terrain of these areas are the fundamental causes for generating badland waste. The less intensity (0.01–0.40%) of badland wasteland is seen in Raghunathpur-I, Hura, Kashipur, Manbazar-I, Santuri, and Joypur C. D. Blocks as there are some hills like Jaychandi (Raghunathpur-I), Panjania, Tilaboni (Hura), Belamu (Joypur), Senera (Santuri), Bahadurdih (Kashipur), Parasa (Manbazar-I), etc. but in the rest of the nine C. D. Blocks is nil.

10.3.4.6   Mining and Industrial wasteland

Mining wasteland refers to those lands where waste rubbish is accumulated after extraction of raw materials; land is useless after excavation of rocks, stone, sand, gravel pits, and soil (Wastelands Atlas of India, 2019). Industrial wastelands are industrial waste disposal areas (Wastelands Atlas of India, 2019) and adjoining unused lands covered by industrial ash. Mining wasteland and industrial wasteland have been shown together because small-scale industries have also developed in all the areas where mining is done in the district. The maximum area under mining & industrial wasteland is observed in Raghunathpur-II (3.13%) Block due to the presence of thermal power plants and multiple small-scale sponge iron industries. The Nituria C. D. Block has maximum wasteland in this category (2.55% out of 3.59%) because of its coal extraction and many sponge iron industries. Besides, the Raghunathpur-I (2.14%) Block has a higher intensity of this category of wasteland for the predominance of small-scale industries. On the other hand, the massive rock quarrying and predominance of the stone chips industries in the Barabazar (2.61%) C. D. Block is the main reason for the formation of this category of wasteland. Among the other C. D. Blocks, Hura, Balarampur, Joypur, Jhalda-II, Para, Purulia-I, Kashipur, Santuri, Manbazar-I & II represent less intensity (<1.00%) of this category of wasteland due to excavation of rocks, stone, gravel pits, and soil; but it is not seen in the remaining six C. D. Blocks at all.

10.4   Recommendations for Wasteland Reclamation

To meet the demands of increasing population and many other developmental activities, there is an urgent need to reduce the trend of wastelands generation and transform wastelands to their productive capacity. There are six categories of wastelands in the study area of which degraded land under plantation crop, badland, degraded forestland, and mining & industrial wastelands can be reclaimed by taking suitable management measures. Therefore, the recommended measures will be obligate for the reclamation and alteration of wastelands to the productive lands (Table 10.7).

10.5   Conclusion

This study shows the identification and monitoring of wastelands with the help of satellite imageries and SOI toposheets. GIS analysis provides accurate information and data on different categories of wasteland identification, validation, area demarcation, and mapping. The potentiality of satellite imaginary for providing accurate baseline information is currently well recognized and emerging. The study carrying out determination of wasteland at the district level can be utilized for various reclamation measures in effective manner.

Table 10.7 Reclamation strategies and suggested land use for the different categories of Wastelands in Purulia District

Categories of wasteland	Reclamation methods	Suggested land use
Degraded land under plantation crop	Increased assurance of surface water and ground water through rain water harvesting and construction of check dams, Providing irrigation facilities	Cultivated crops through suitable crop selection, Continuous cropping
Degraded forestland	Restrict illegal forest cutting, Regulate grazing activity	Plantation of native plant, Permanent plantation
Gravelly wasteland	Natural generation of grassland, Plantation of indigenous trees	Pasture development with proper channel, Natural regeneration of vegetal cover
Rocky or Stony wasteland	Regular grazing activity, Building materials	Pasture development
Badland waste	Leveling of gullies or ravines, Construction of earthen check dams, Providing diversion bunds or trench above the head	Afforestation, Agro-horticulture for food and fuel, Grassland cover
Mining & Industrial wasteland	Removal of waste materials of industries and mines for the uses of road, buildings and mine fills, Construction of infiltration wells and open wells	Sow the early successional species of plants and grasses, Use groundcovers that are consistent with growing trees

From the above analysis, it can be said that the district has the enormous potential to be developed with the conversion of a significant proportion of degraded land under plantation crop into arable lands. Apart from this, degraded forestland, gravelly wasteland, and badland waste may be used for other than agricultural activities such as livestock farming like goat rearing, piggery, poultry, etc., agro-based and cottage-based industries. The rocky or stony wasteland is expected to be the most profitable for the district to use in the brick kilns (the soil for making bricks should be brought from barren uncultivated lateritic tracts of the district) and stone chips industry.

The above recommended measures can be effective only, if the State and Central Government take the initiative along with the participation of local community. However, the role of State Government especially the Forest Department, in Purulia District is very disappointing. They have no such suitable planning for the enhancement of degraded forestlands. Acacia auriculiformis (*Sanajhuri*) is being planted in the degraded forestlands and as a results soil health is being severely deteriorated. Because, Acacia auriculiformis species has been most dangerous for soil health and forest ecosystem through reduction of soil moisture, soil fertility, and under growth. Therefore, it is necessary to plant the native tree species like Dalbergia sissoo

(*Shisham*), Butea frondosa (*Palash*), Schleichera trijuga (*Kusum*), Zizyphus xylopyra (*Kul*), Bassia latifolia (*Mohul*), Shorea robusta (*Sal*), Carissa spinarum (*Karamcha*), Terminalia arjuna (*Arjun*), etc. instead of Acacia auriculiformis species on these lands.

Acknowledgements The authors are grateful to the Department of Geography, The University of Burdwan, India for providing necessary facilities for this research. The authors wish to express sincere thanks to NRSA, Hyderabad; NWDB, New Delhi, NBSS & LUP, Kolkata, and Survey of India, NATMO, Kolkata. We are also thankful to Mr. Sumanta Kumar Baskey, Mr. Ritabrata Mukhopadhyay, Mr. Thakurdas Rajwar, and Mr. Ranjit Mahato for their impeccable help during the field survey.

References

Anderson JR, Hardy EE, Reach JT, Witmer RE (1976) A land use and land cover classification system for use with Remote Sensor data. US Geological Survey Professional Paper, 964. https://www.researchgate.net/publication/287997665

Baidya TK, Chakrabarty PS, Orubetskoy E, Kiltova VJ (1987) New geochronologic data on some granitie phases of the chhotonagpur granite gneiss complex in the northwestern Purulia district, West, Bengal. Ind Jr Earth Sci XIV(2):136–141. https://www.osti.gov/etdeweb/biblio/5121618

Ball V (1881) Geology of Manbhum and Singhbhum. Mem Geol Surv Ind 18:61–90

Barchyn TE, Martin RL, Kok JF, Hugenholtz CH (2014) Fundamental mismatches between measurements and models in aeolian sediment transport prediction: the role of small-scale variability. Aeol Res 15:245–251. https://doi.org/10.1016/j.aeolia.2014.07.002

Basavarajappa HT, Manjunatha MC (2014) Geoinformatic techniques on mapping and reclamation of wastelands in Chitradurga District, Karnataka, India. Int J Comput Eng Technol (IJCET) 5(7):99–110. https://www.researchgate.net/publication/281319668

Basavarajappa HT, Manjunatha MC, Pushpavathi KN (2015) Mapping and Reclamation of Wastelands through geomatics technique in precambrian Terrain of Mysuru District, Karnataka, India. Int J Civil Struct Eng 5(4):379–391. https://doi.org/10.6088/ijcser.2014050035

Basavarajappa HT, Pushpavathi KN, Manjunatha MC (2016) Mapping and reclamation of Wastelands in Yelanduru Taluk of Chamarajanagara District, Karnataka, India Using Geo-Informatics Technique. Int J Sci Res Sci Technol 2(3):91–98. https://www.researchgate.net/publication/303751359

Basavarajappa HT, Dinakar S (2005) Land use/land cover studies around Kollegal, Chamarajanagar district, using Remote Sensing and GIS techniques. J Indian Mineral 1:89–94. https://www.researchgate.net/publication/280066552

Bhattacharya BK, Ray P, Chakraborty BR, Sengupta S, Sen NN, Sengupta KS, Mukherji S, Maity T (1985, 2005) West Bengal District Gazetteers, Purulia, Government of West Bengal. Published by Narendra Nath Sen, State Editor, West Bengal District Gazetteers, Calcutta. www.wbpublibnet.gov.in

Breunig FM, Galvao LS, Formaggio AR (2008) Detection of sandy soil surfaces using ASTER-derived reflectance, emissivity and elevation data: Potential for the identification of land degradation. Int J Remote Sens 29(6):1833–1840. https://doi.org/10.1080/01431160701851791

Chandramohan T, Durbude DG (2002) Estimation of soil erosion potential using universal soil loss equation. J Indian Soc Remote Sens 30(4):181–190. https://link.springer.com/article/10.1007/BF03000361

Chatteijee SC (1946) Physiographic evolution of Chotonagpur. Geogr Rev (Calcutta) VIII 3–4:31–35

Chaturvedi OP, Kaushal R, Tomar JMS, Prandiyal AK, Panwar P (2014) Agroforestry for wasteland rehabilitation: mined, ravine, and degraded watershed areas. Adv Agrofor 10:233–271. https://doi.org/10.1007/978-81-322-1662-9_8

Contador JFL, Schnabel S, Gutierrez AG, Fernandez MP (2008) Mapping sensitivity to land degradation in Extremadura, SW Spain. Boletín de la Asociación de Geógrafos Españoles 53:387–390. https://doi.org/10.1002/ldr.884

Crosbishley, Pearce D (2007) Growing trees on salt-affected land, ACIAR impact Assessment series Report No. 51. www.aciar.gov.au

Datta SK, Chakraborty T (2017) District statistical handbook, Purulia, Bureau of applied economics & statistics, department of planning, statistics & programme monitoring. Government of West Bengal. http://www.purulia.gov.in/distAdmin/departments/baes/baes.html

De NK, Jana NC (1996) The land: multifaceted appraisal and management. Sri Bhumi publishing co., Calcutta

Dunn JA (1929) The Geology of North Singhbhum including parts of Ranchi and Manbhum district. Mem Geol Surv Ind vol LIV:1–66

Dunn JA (1935) Late Tertiary uplift in Singhbhum. Proc As Soc Bengal XXIX:285–288

Dunn JA, Dey AK (1942) The geology and petrology of eastern Singhbhum and surrounding areas. Mem Geol Surv India LXIX(2):281–456. https://books.google.co.in/books/about/The_Geology_and_Petrology_of_Eastern_Sin.html?id=EE1RAQAAIAAJ&redir_esc=y

Dwivedi RS, Sankar TR, Venkataratnam L (1997) The inventory and monitoring of eroded lands using remote sensing data. Int J Remote Sens 18(1):107–119. https://doi.org/10.1080/014311697219303

Ghosh S (2012) Geomorphic Land Evaluation for Sustainable Use of Land Resources in Puruliya District, West Bengal. J Lands Syst Ecol Stud 35(1):263–274

Government of India (2016) Ministry of agricuture and farmers welfare: Deparment of Agriculture, C. and W, India Annual Report 2016–17, 194. http://agriculture.gov.in/

Grunwald S (2013) The current state of digital soil mapping and what is next. Boetinger JL, Howell DW, Moore AC, Hartemink AE, Kienst-Brown S (eds) Digital soil mapping: bridging research, production and environmental applications. Springer, Heidelberg, Germany, pp 3–12. https://www.springer.com/gp/book/9789048188628

HARSAC (2006) Haryana Space Applications Centre, Wastelands Atlas of Haryana, Haryana State Remote Sensing Application Centre, Department of Science and Technology, Government of Haryana

Jain AK, Hooda RS, Nath J, Manchanda ML (1991) Mapping and monitoring of urban land use of Hisar Town, Haryana using remote sensing techniques. J Indian Soc Remote Sens 19(2):125–134. https://link.springer.com/article/10.1007/BF03008127

Jha VC (1987) Wasteland types and their effective utilization in Birbhum District. The Deccan Geographer. Secunderabad 15(2 & 3):231–242

Khatun S, Debnath GC (2014) Identification and mapping of wasteland in Birbhum District, West Bengal. Int J Adv Remote Sens GIS 3(1):713–722. http://technical.cloud-journals.com/index.php/IJARSG/article/view/Tech-313

Kohli RK, Garg VK, Dhawan A, Pandey P, Choudhary BK (2018) Applications of remote sensing and GIS in wasteland mapping. Environ Sci 6:5–31. https://www.researchgate.net/publication/323411930

Luna RK (2006) Plantation forestry in India. Offset Printers and Publishers, Dehradun, pp 361–386

Metternicht IG, Zinck AJ (2008) Remote sensing of soil salinization: impact on land management. CRC Press, Taylor & Francis Group, Boca Raton, Fla, USA, pp 257–272. http://citeseerx.ist.psu.edu/viewdoc/download?doi=10.1.1.523.5717&rep=rep1&type=pdf

Mishra CSK, Rath M (2013) Wasteland reclamation: novel approaches. Ecoscan 3:99–105. http://theecoscan.in/JournalPDF/Spl2013_v3-15%20C.%20S.%20K.%20Mishra.pdf

Mulder VL, de Bruin S, Schaepman ME (2013) Representing major soil variability at regional scale by constrained Latin Hypercube Sampling of remote sensing data. Int J Appl Earth Obs Geoinf 21(1):301–310. https://doi.org/10.1016/j.jag.2012.07.004

Mustak S, Baghmar NK, Singh SK (2015) Prediction of industrial land use using linear regression and MOLA techniques: a case study of Siltara industrial belt. Lands Environ 9(2):59–70. https://doi.org/10.21120/LE/9/2/2

National Atlas & Thematic Mapping Organisation (2010) Annual Report 2009–10. Government of India, Kolkata, (Deptt. of Science & Technology), pp 7–8

National Commission on Agriculture (1976). Report of the national commission on agriculture, ministry of agriculture and irrigation. Government of India, New Delhi, India. http://krishikosh.egranth.ac.in/handle/1/2041447

National Council of Educational Research and Training (2016) Land resources and agriculture. India: People and Economy, XII Level book chapter, 5, pp 40–44. http://ncert.nic.in/ncerts/l/leg y205.pdf

Nawar S, Buddenbaum H, Hill J (2015) Digital mapping of soil properties using multivariate statistical analysis and ASTER data in an Arid region. Remote Sens 7(2):1181–1205. https://doi.org/10.3390/rs70201181

NBSS & LUP (2010) Forest Soils of Puruliya district, West Bengal, Report No. 794, 10. https://www.nbsslup.in/kolkata.html

NRSA (1987) Manual of procedure for wasteland mapping using remote sensing techniques. National Remote Sensing Agency, Hyderabad.

NRSA (2007) Manual of national wastelands monitoring using multi-temporal satellite data, National Remote Sensing Agency, Hyderabad. https://economictimes-indiatimes.com/topic/National-Remote-Sensing-Agency-in-Hyderabad

NRSC (2005) Wastelands Atlas of India. National Remote Sensing Centre, ISRO (DOS) Govt. of India, Hyderabad, India. https://dolr.gov.in/sites/default/files/West_Bengal.pdf

NRSC (2011) Wastelands Atlas of India 2011 (Change analysis based on temporal satellite data of 2005–06 and 2008–09. National Remote Sensing Centre, ISRO (DOS) Govt. of India, Hyderabad, India. https://dolr.gov.in/sites/default/files/Wastelands_Atlas_2011.pdf

NRSC (2019) Wastelands Atlas of India 2019. National Remote Sensing Centre, ISRO (DOS) Govt. of India, Hyderabad, India. https://dolr.gov.in/sites/default/files/West%20Bengal_0.pdf

NRSC/ISRO (2010) Wastelands Atlas of India, change analysis based on multi-temporal satellite data. Version 1, pp 1–16. https://dolr.gov.in/sites/default/files/west%20Bengal.pdf

NWDB (1987) National Wasteland Development Board, Description, classification, identification and mapping of wastelands. Ministry of Environment and Forests, Government of India, New Delhi

Paudel D, Thakur JK, Singh SK, Srivastava PK (2015) Soil characterization based on land cover heterogeneity over a tropical landscape: an integrated approach using earth observation data sets. Geocarto Int 30(2):218–241. https://doi.org/10.1080/10106049.2014.905639

Pramila R (1994) Assessment of soil degradation Hazards in Jalor and Ahor Tehsil of Jalor district (Western Rajasthan) by remote sensing. J Indian Soc Remote Sens 22(3):169–181. https://link.springer.com/article/0.1007/BF03024778

Ramachandra TV (2007) Comparative assessment of techniques for bio-resource monitoring using GIS and remote sensing. ICFAI J Environ Sci 1(2):2–8. https://www.researchgate.net/publication/256453918_Comparative_assessment_of_techniques_for_bioresource_monitoring_using_GIS_and_Remote_sensing/link/00b49522afb183a525000000/download

Rao DP (1999) Remote sensing application for land use and urban planning: retrospective and perspective. In: Proceedings of the ISRS national symposium on remote sensing application for natural resources retrospective and perspective, Bangalore, India, pp 287–297

Rao DP, Gautam NC, Sahai B (1991) IRS-1A application for Wasteland mapping. Current Sci 61(25):3–4. https://www.researchgate.net/publication/323411930_Applications_of_Remote_Sensing_and_GIS_in_Wasteland_mapping

Rawat KS, Mishra SV, Singh SK (2018) Integration of earth observation data and spatial approach to delineate and manage Aeolian Sand-affected Wasteland in highly productive lands of Haryana, India. Int J Geophys 2847504:1–7. https://doi.org/10.1155/2018/2847504

Ray S (1982) Some aspects of geoecological adjustment in the upper Kangsabati basin‖, West Bengal. Ind Jr Lands Syst, Calcutta, 5(1–2):25–27. https://shodhganga.inflibnet.ac.in

Roy A, Inamdar AB (2018) Multi-temporal Land Use Land Cover (LULC) change analysis of a dry semi-arid river basin in western India following a robust multi-sensor satellite image calibration strategy. Heliyon 5:e01478. https://doi.org/10.1016/j.heliyon.2019.e01478

Saha SK, Kudrat M, Bhan SK (1990) Digital processing of Landsat TM data for wasteland mapping in parts of Aligarh District (Uttar Pradesh), India. Int J Remote Sens 11(3):485–492. https://doi.org/10.1080/01431169008955034

Sharma A, Moorti T, Chauhan SK (2007) A study on the estimation of wasteland and proposed strategy for their regeneration in Western Himalayas; agricultural situation in India 47 CSSRI–2007. Annual Report 2006–07

Singh AN (2006) Geospatial database for sustainable reclamation of degraded lands. In: Proceedings of the abstracts of the ISPRS TC-IV international symposium on geo spatial databases for sustainable development, Goa, India, 151–152

Singh RP (1969) Geomorphological evolution of Chotanagpur highlands. Nat Geog Soc Ind Varabasi 5:4–29. https://search.library.wisc.edu/catalog/999615257002121

Singh S (1978) Physiographic regions, landforms and erosion surfaces of the Ranchi Plateau. Nat Geograph 12 (l):43–65

Singh S (2012) Common lands made 'Wastelands' making of the 'Wastelands' into Common Lands. Ecol Secur 7(5):14–33. https://dlc.dlib.indiana.edu/dlc/bitstream/handle/10535/9074/SINGH_0217.doc.pdf?sequence=1&isAllowed=y#:~:text=In%20the%20Panjab%2C%20the%20areas,for%20the%20extension%20of%20cultivation

Singh SK, Kewat SK, Aier B, Kanduri VP, Ahirwar S (2012) Plant community characteristics and soil status in different land use systems at Dimapur, Nagaland, India. Forest Res Papers 73(4):305–312. https://doi.org/10.2478/v10111-012-0029-x

Singh SK, Mustak S, Srivastava PK, Szab´o S, Islam T (2015) Predicting spatial and decadal LULC changes through cellular automata markov chain models using earth observation datasets and geo-information. Environ Process 2(1): 61–78. https://link.springer.com/article/ https://doi.org/10.1007/s40710-015-0062-x

Singh SK, Pandey AC, Singh D (2014) Land use fragmentation analysis using remote sensing and Fragstats. In: Srivastava PK, Mukherjee S, Islam T, Gupta M (eds) Remote sensing applications in environmental research. Springer International Publishing, Cham, Switzerland, 9, pp 151–176. https://link.springer.com/chapter/10.1007/978-3-319-05906-8_9

Singh SK, Singh CK, Mukherjee S (2010) Impact of land use and land-cover change on groundwater quality in the Lower Shiwalik hills: a remote sensing and GIS based approach. Cent Eur J Geosci 2(2):124–131. https://link.springer.com/article/10.2478/v10085-010-0003-x

Singh SK, Srivastava PK, Szabo SG, Petropoulos P, Gupta M, Islam T (2017) Landscape transform and spatial metrics for mapping spatiotemporal land cover dynamics using Earth Observation data sets. Geocarto Int 32(2):113–127. https://doi.org/10.1080/10106049.2015.1130084

Sreekala S, Neelakantan R (2015) Mapping of wasteland and change detection using geospatial data - a case study from Annavasal block of Pudukkottai District, Tamilnadu, India. Int J Innov Sci Eng Technol 2(12):707–714. http://ijiset.com/vol2/v2s12/IJISET_V2_I12_77.pdf

Srivastava PK, Singh SK, Gupta M, Thakur JK, Mukherjee S (2014) Appraisal of land use/land cover of mangrove forest ecosystem musing support vector machine. Environ Earth Sci 71(5):2245–2255. https://doi.org/10.1007/s12665-013-2628-0

Sugumaran R, Sandhya G, Rao KS, Jadhav RN, Kimothi MM (1994) Potential of satellite data in delineation of wastelands and correlation with ground information. J Indian Soc Remote Sens 22(2):113–118. https://link.springer.com/article/10.1007%2FBF03023881

Szabo G, Singh SK, Szabo S (2015) Slope angle and aspect as influencing factors on the accuracy of the SRTM and the ASTER GDEM databases. Phys Chem Earth 83(84):137–145. https://doi.org/10.1016/j.pce.2015.06.003

Thakur JK, Srivastava PK, Pratihast AK, Singh SK (2012). Estimation of evapotranspiration from wetlands using geospatial and hydro-meteorological data in Geospatial Techniques for Managing

Natural Resources. Thakur JK, Singh SK, Ramanathan AM, Prasad BK, Gossel W (eds). Springer and Capital, pp 53–67. https://www.researchgate.net/_publication/271905531

Thakur JK, Srivastava PK, Singh SK (2012) Ecological monitoring of wetlands in semi-arid Konya closed basin, Turkey. Reg Environ Change 12(1):133–144. https://link.springer.com/article/10.1007/s10113-011-0241-x

Xiaoyan L, Wang Z, Kaishan S, Bai Z, Dianwei L, Zhixing G (2007) Assessment for salinized wasteland expansion and land use change using GIS and remote sensing in the west part of Northeast China. Environ Model Assess 131(1–3):421–437. https://doi.org/10.1007/s10661-006-9487-z

Yadav SK, Singh SK, Gupta M, Srivastava PK (2014) Morphometric analysis of Upper Tons basin from Northern Foreland of Peninsular India using CARTOSAT satellite and GIS. Geocarto Int 29(8):895–914. https://doi.org/10.1080/10106049.2013.868043

Chapter 11
Soil, Water Salinization and Its Impact on Household Food Insecurity in the Indian Sundarbans

Nabanita Mukherjee◉ and **Giyasuddin Siddique**◉

Abstract The present study is an attempt to comprehend the effects of soil and water salinity on household food security. Excess salinization of soil and water exacerbates significant long-term environmental risks in a coastal and deltaic environment especially in the densely populated tropical deltas. This study focuses on the coastal regions of the Indian Sundarbans directly confronting the Bay of Bengal based on data collected from the household Survey 2017–2020; soil and water survey during Pre and Post Monsoon period of 2019 to investigate into the effects of soil and water salinity on human life and ecological sustainability of the deltas. The result reveals significant effects of soil and water salinization on climate sensitive subsistence economy. Growth of crop and plant gets affected by increased soil salinity which further leads to soil sterility and poor seed germination. Water salinization leads to the toxicity of specific ions which directly affects those households dependent on fresh water ponds for irrigation and fishing activities. The findings clearly suggest policy intervention to deal with enhanced soil and water salinity in context of changing climate.

Keywords Soil and water salinization · Household food security · Tropical delta · Changing climate · Ecological sustainability

11.1 Introduction

Climate change in deltaic areas has been responsible for change in pattern of weather; rise of sea level; greater occurrences of extreme events; higher storm surges consequent upon inundation of coastal land; saline water intrusion and contamination of soils and water; loss of wetlands, cropland, biodiversity and finally degradation of coastal livelihood and economy (Dasgupta et al. 2015a; Chen and Mueller 2018; Okur and Orcen 2020). About 600 million people worldwide dwell in the low lying coasts that are potentially threatened by the negative effects of salinization (CIESIN 2010;

N. Mukherjee (✉) · G. Siddique
Department of Geography, The University of Burdwan, Bardhaman, India

N. C. Jana and R. B. Singh (eds.), *Climate, Environment and Disaster in Developing Countries*, Advances in Geographical and Environmental Sciences, https://doi.org/10.1007/978-981-16-6966-8_11

Wheeler 2011). In the twenty-first Century, the anticipated rate of global temperature escalation and sea level rise would be 1.4–5.8 °C and 1.8–5.9 mm/year, respectively, following the global climate change models (Dasgupta et al. 2015a; Teh and Koh 2016). Hence, the situation may lead to severe vulnerability of one billion population (Hansen and Sato 2012; Brecht et al. 2012; Dasgupta et al. 2015a) overall.

Agriculture and Climate change has complex and intricate interrelations where minimum to abrupt changes in climatic condition affects food availability and accessibility to a greater extent. Following the Food and Agriculture Organization (FAO) statistics, there would be decline of major food crops (20–30% in rice, 5–50% in wheat and 20–45% in maize yields), if the current green house gas emissions remain unrestrained (Arora 2019). Decline in crop yield with respect to rising population has also been notified by World Food Programme (WFP) report (World Food Programme 2018) which questions the future food security. Ground water salinization as a consequence of changing climate (Jahan 2018) directly alters the availability of fresh water for irrigation (Haque and Reza 2017). Absorption of saline water by soil turns it sterile and unproductive and deteriorates 20–30% net agricultural land (Islam et al. 2019). Even salinization and alkalanization pose weakening effect on soil for which they become prone to further water and wind erosion (Jahan 2018; Okur and Orcen 2020). Climate change also impinges on ground water recharge and evapotranspiration patterns that exacerbate groundwater depletion (Colombani et al. 2016) which is considered as main source of potable water for 1.5–2 billion people across the world. Saline water incursion into surface and subsurface water sources probes threat to the chemical composition of natural water resources, aquatic balance, present and future water security (Teh and Koh 2016; Rahman et al. 2018). It also causes emergence of grave health hazards (Okur and Orcen 2020), e.g., hypertension (Scheelbeek et al. 2016, 2017), pre-eclampsia in pregnant women (Khan et al. 2011, 2014), anemia of women and children (Panda et al. 2016), cardio vascular diseases (Talukder et al. 2016) and infant mortality (Dasgupta et al. 2015b; Islam et al. 2019).

Sea level rise, higher storm and tidal surges, fiercer cyclones and consequent coastal erosion, and breaching of mud embankment are the main reasons behind saline water intrusion in the Sundarbans (Mukherjee and Siddique 2019; Sahana et al. 2020). A host of geographers, scientists, and environmentalists have already discussed the effects of such intrusion on soil, and water salinization in the active deltaic ecosystem. Some of the researchers have analyzed pH, salinity and other physical–chemical characteristics of soil by laboratory testing (Halder and Debnath 2014; Ataullah et al. 2017) while others have used Remote Sensing and Geographical Information System techniques to unravel the land use change, spatial extent, and nature of land salinization (Das et al. 2016; Ramteke et al. 2017). Similarly, climate change induced salinization of water resources- surface water, ground water (Dhar 2011; Banerjee 2013; Akter et al. 2018; Islam et al. 2019) has also been discussed in several research articles. Effect of climate change on climate sensitive subsistence systems—agriculture (Murshid 2012; Debnath 2013; Mondal et al. 2015; Hasnat et al. 2016; Clarke et al. 2018) and fishing (Ali 1999; Chand et al. 2012; Bhattacharjee et al. 2013) are also highlighted by numerous researchers. Building of adaptive capacities by sustainable agriculture, aquaculture practices, and water management to develop resilience are illustrated in some of the literatures (Sanchez Triana et al. 2016; Ghosh and Chattopadhyay 2017). But most of the investigations are based on the Bangladesh part of Sundarbans and in case of Indian part, the areal stretch is quite limited. Thus finding the need of more study, the present work fills the research gap of constant monitoring of vulnerable locations in Indian Sundarbans. The work has been carried out to emphasize the impact of climate change induced soil and water salinization on household food insecurity in six ocean confronting Blocks such as Sagar, Namkhana, Patharpratima, Kultali, Basanti, and Gosaba of the South 24 Parganas District that are highly prone to climatic perturbations. Such studies may be crucial for understanding multidimensional effects of salinity and planning for appropriate adaptation to achieve food and water security in vulnerable coastal regions.

11.2   The Study Area

The study area includes 12 *mouzas* from six Blocks located in the southern part of South 24 Parganas District, and south western part of the Indian Sundarbans. Following the District Disaster Management Plan (2018–19) of West Bengal, the six Blocks of Sagar, Namkhana, Patharpratima, Kultali, Basanti, and Gosaba are under the severe category of multi hazard prone area. From each of Blocks two *mouzas* have been selected for current analysis such as Ghoramara and Dhablat *mouzas* from Sagar Block; Iswaripur, Baliara *mouzas* from Namkhana Block; Gobardhanpur, Purba Sripatinagar *mouzas* from Patharpratima Block; Bhubaneshwari, Kisorimohanpur *mouzas* from Kultali Block; Birinchibari, Lot 126 *mouzas* from Basanti and Kumirmari, Hetalbari *mouzas* from Gosaba Block. These *mouzas* have reported higher coastal erosion that are identified through Govt. reports (District Disaster Management Plan 2018–19), expert judgment and questionnaire survey. The area under study is encircled by Hooghly River in the west and Hariabhanga River in east, Bay of Bengal lies at the southern part. The other most important rivers in the study area include Gabtala, Baratala, Saptamukhi, Hetania Doania, Thakuran, Matla, Bidyadhari, Gomor, Garal, Bangaduani, Gosaba, etc. Purba Medinipur District lies in the west and North 24 Parganas lies in the north east corner. The study area shares international boundary with Bangladesh. The formation of the area dates back to mid-Holocene period between 5000 and 2500 yr BP. The temperature during summer and winter remains 34 °C in and 11 °C, respectively, and relative humidity stays 70–80%. Most of the rainfall occurs during monsoon. The average elevation is 2 m above the sea level (Sahana and Sajjad 2019) (Fig. 11.1).

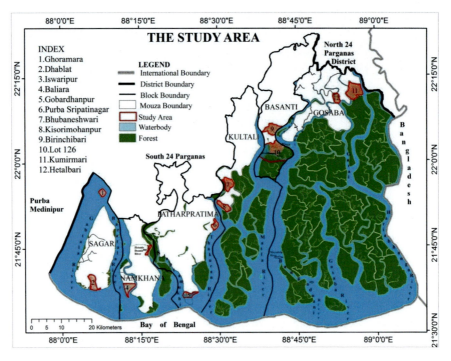

Fig. 11.1 The study area

11.3 Database and Methodology

Soil samples have been collected randomly from 12 locations during Pre and Post Monsoon season of the year 2019 to observe spatial and seasonal variation of pH and salinity levels of the studied *mouzas*. These are obtained from a depth of 25 cm to ensure the exclusion of top exposed layer. Then clean plastic containers are used to procure them so that the physico-chemical properties are retained. Similarly surface water samples have been derived randomly from 12 ponds located in marginal areas from a depth of 0.5 m during Pre and Post Monsoon season of 2019. Three samples were mixed together for making one water sample for a particular pond. Samples were collected in 100 ml cleaned plastic bottles. The water quality parameters have been selected by rigorous review of literature (Mitra et al. 2009; Dasgupta et al. 2014; Zaman et al. 2015; Mondal et al. 2016). The parameters considered are pH, salinity, Total Dissolved Solid (TDS), Conductivity, Dissolved Oxygen (DO), Temperature, and Total Hardness; they are analyzed on the basis of Bureau of Indian Standard (BIS (Bureau of Indian Standards) 10500 2012) quality. The soil and water samples are brought to the Department of Geography, The University of Burdwan for laboratory testing and further investigation. Figure 11.2a and b shows the soil and water sample collection sites, the location has been obtained from GPS survey. A sample size of 886 households from 12 *mouzas* have been selected by random sampling technique from

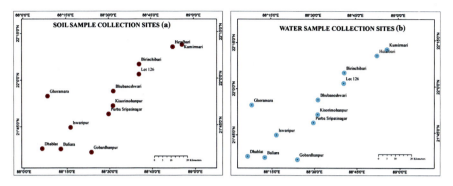

Fig. 11.2 Soil sample collection sites (**a**), Water sample collection sites (**b**)

six Blocks (Ghoramara—68, Dhablat—82, Iswaripur—64, Baliara—74, Gobard-hanpur—58, Purba Sripatinagar—70, Bhubaneshwari—72, Kisorimohanpur—72, Birinchibari—72, Lot 126—75, Kumirmari—92, Hetalbari—87 Households) for primary household survey to understand the impact of soil and water salinization on household food insecurity (National Statistical Service Organization 2018). The survey took place since January 2017 to March 2020. The extent of flooded and saline areas have been identified by Modified Normalized Difference Water Index (MNDWI) and Normalized Difference Salinity Index (NDSI) indices (Landsat 8 OLI, 2019 image), respectively, and correlation among them has been established to prove that flooding and water logging accentuates the problem of salinization in the study area. MNDWI provides the most accurate detection of flooded areas compared to other commonly used band ratio indices and has the most stable threshold (Ji et al. 2009; Ogilvie et al. 2015; Chen and Mueller 2018). NDSI is also proved to be very effectual for soil salinity mapping (Allbed and Kumar 2013). ArcGIS software has been used for map making while PAST (PAleontological STatistics Sofware Package) and Excel have been used for preparing graphs.

11.4   Results and Discussion

11.4.1   Soil Characteristics

Generally soil of the area is deep fine textured and heavily structured with slow drainage characteristics. Figure 11.3a describes that there has been ascend of soil pH values and drop of salinity in most of the *mouzas* from Pre to Post Monsoon season. Iswaripur and Baliara *mouza* from Namkhana and Lot 126 *mouza* from Basanti record just the opposite condition. There has been embankment breaching and saline water intrusion in these parts during the Post Monsoon field survey which is the main reason behind such changes. Typically, salinity remains lower in Post

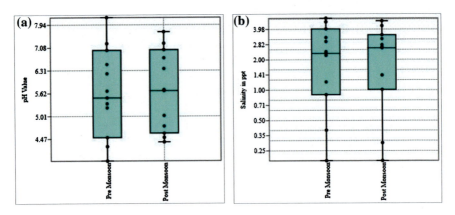

Fig. 11.3 Soil pH characteristics (**a**) and Salinity characteristics (**b**)

Monsoon period due to additional fresh water from monsoon rainfall that removes the soluble salts. Again Pre Monsoon higher evapotranspiration draws out the soil moisture through capillary action which leaves the salt deposits on the top soil, known as salt encrustations (Nath and De 1999; Halder and Debnath 2014). Reverse to salinity condition, soil pH remains low in Pre Monsoon and it gets escalated with fresh water inflow during Monsoon season (Biswas et al. 2017; Yadav et al. 2019). The pH of soil ranging from 5.5 to 7 is considered as most appropriate condition for crop and plant growth. In the study area, pH value above 5.5 have been found in Kumirmari (5.6), Hetalbari (6.29), Bhubaneshwari (6.6), Kisorimohanpur (7), Birinchibari (7.23), and Lot 126 (7.95) while all the other *mouzas* reported lower than that, lowest values have been found in Gobardhanpur (4.2), Purba Sripatinagar (4.45) and Ghoramara (4.6) *mouzas*. The most common crops that grow in the study area include *Oriza sativa, Lycopersicon esculentum, Cucumis sativus, Spinacia oleracea, Brassica oleracea capitata, Saccharum officinarum, Solanum tuberosum*, etc. and their growth is affected with >3 ppt salinity (FAO Corporate Document Repository 1994; Halder and Debnath 2014). In the study area, lowest salinity level has been found in Kultali Block (average 0.65) while highest has been found in Patharpratima (average 4.7). Most of *the mouzas* recorded 1–3 ppt salinity level. Studies describe that in low lying flat terrain, accumulated salts move very slowly and remain in the soil for longer periods, ranging from seasons to decades to (McNeil and Cox 2007) even 100,000s years (Herczeg et al. 2001; Cañedo-Argüellesa et al. 2013), this causes salinity retention in soils. In some of the *mouzas,* pH and salinity beyond standard limit has been found which is detrimental to crop growth.

11.4.2 Water Quality Analysis

Figure 11.4 illustrates that in the study area, water quality parameters like pH, DO, temperature have been escalated while salinity, TDS, Conductivity, and Total Hardness have fallen from Pre to Post Monsoon season. In most of the studied ponds, pH value of water is within the standard limit (6.5–8.5) while Birinchibari (5), Ghoramara (5.1), and Purba Sripatinagar (5.62) indicate acidification, especially in the Pre Monsoon period. The permissible limit of salinity, TDS and Conductivity is 0.5 ppt, 0.5 ppt, and 300 μ mhos/cm, respectively. Here Salinity, TDS (except Purba Sripatinagar) levels are above the standard limit while Conductivity levels are excessively higher during Pre Monsoon in Ghoramara (334 μ mhos/cm), Dhablat (548 μ mhos/cm) and Kumirmari (611 μ mhos/cm) *mouzas*. Conductivity is a substitute measure of Salinity (Corwin and Yemoto 2017) and it assesses the concentration of ionized constituents (Huq and Alam 2005) in the water. TDS determines solid

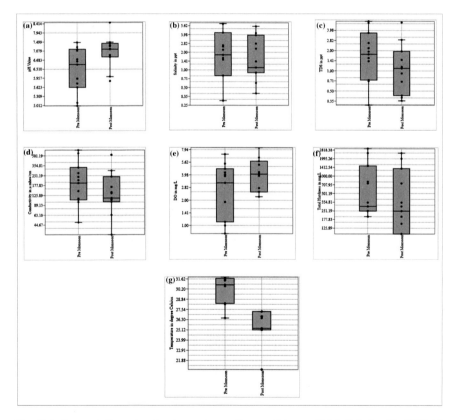

Fig. 11.4 Box and Jitter plot showing water quality condition—pH (**a**), Salinity (**b**), TDS (**c**), Conductivity (**d**), DO (**e**), Total Hardness (**f**), and Temperature (**g**)

materials (calcium, magnesium, sodium, potassium, carbonates, bicarbonates, chlorides, sulfate, phosphate, silica, and iron) dissolved in the water (Ahamed et al. 2015). Greater concentration of Salinity and TDS is lethal for human consumption as well as harmful for the survival of fresh water fishes; though different swimming animals have different tolerance level of salinity and temperature (Dubey et al. 2016) conditions. Higher TDS level causes paralysis of the tongue, lips, face; irritability, and dizziness. It also harms fishes and aquatic plants (Chang 2005). DO is dissolved oxygen available in the water for living organisms and its standard level varies from 4 to 6 mg/l. *Mouzas* like Ghoramara, Dhablat, Baliara, Gobardhanpur, and Bhubaneshwari report very lower DO levels especially during Pre Monsoon period, lower oxygen availability affects organisms thriving in the water bodies. Pre Monsoon temperature in the study area varies from 26.5 to 31.9 °C while Post Monsoon temperature varies from 21 to 27.3 °C. Higher temperature induces growth and death rate of aquatic organisms very fast irrespective of DO level; it undermines photosynthesis of plants and solubility of oxygen in the water (Mondal et al. 2016). Total hardness condition determines dissolved Ca^{2+} and Mg^{2+} substance in the water reflected as $CaCO_3$ (Li et al. 2013) and among the studied ponds, hard water conditions (Standard 300 mg/L) have been found almost everywhere except in Kultali and Gosaba Blocks. Elevated temperature and evaporation condition lowers the water level during Pre Monsoon season which makes sodium ions and other solids present in the water more concentrated, contributing to elevated salinity, TDS, hardness, and conductivity. DO and pH levels drop owing to fresh water addition during monsoon rainfall. Gastro enteric disease, skin itching, fever, diarrhea, dengue, malaria, and women reproductive tract diseases are main saline water-related health hazards in the area.

11.4.3 Impact of Soil and Water Salinization on Food Insecurity of Households

The main problems responsible to food insecurity in the region include soil, water salinization; higher acidity of soil; non-scientific nutrient management; lack of fresh water for irrigation; low crop diversification; lack of alternate livelihood options; low productivity of freshwater fish, and under utilization/improper use of brackish water (NAIP 2011). Figure 11.5 shows the household perception on the effect of soil and water salinization on food security of the people inhabited. Most of the households of the study area are of the opinion that they are partially supported by their family farm as monocropping with an importance of *Aman* rice farming is practiced here. They have to buy most of their food items directly from the market. Cultivation of rice and vegetables are mostly for the sustenance of their family. Households which do not possess their own land are found to be highly food insecure, especially during the occurrence of natural hazard. Super Cyclone *Amphan* on May, 2020 with amplitude of 180 km/h has devastated the Sundarbans and severely affected the food and

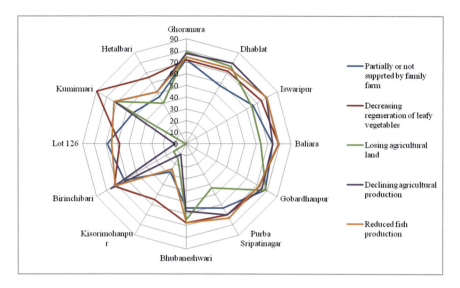

Fig. 11.5 Household perception (percentage of households) on effect of soil and water salinization on food security

water security of the marginal villagers. There are fewer households which opined that water logging is a very common phenomenon here and thus they prefer integrated farming during the time. Mainly rice, fishes, dyke horticulture, legumes, and vegetables have been found to be most desired form of integrated farming, but very less number of individuals are involved in it owing to lack of capability and financial crisis. Increased salinity level in the study area favors shrimp farming, thus salt water is deliberately retained in the ponds and in most cases rice fields and cultivable lands are converted into shrimp farms. This phenomenon instigates groundwater salinity, soil degradation, lower yield, mangrove destruction, water pollution, sedimentation, disease outbreaks, loss of biodiversity, and destruction of local ecosystem (Vineis et al. 2011; Islam Didar-Ul and Bhuiyan 2016; Johnson et al. 2016). An interrelationship between MNDWI and NDSI indices has been established in Fig. 11.6. Karl Pearson's product moment correlation statistics is 0.32 with 0.0003 p value which signifies that there is positively significant correlation between extent of flooded areas and salinity values. Premature reclamation and unsystematic construction of mud embankments have caused higher siltation of river beds for which accretion has been accentuated effectively and the inland areas have remained at a level lower than the creek channels. The most obvious effect has been inundation of inland area at every count of embankment breaching. Soil salinization, poor drainage system, improper irrigational facilities are the major reasons behind the compulsion of practicing monocropping in the area (on an average 80% land is mono cropped). Very few farmers can afford deep tube well for desirable irrigation and mechanization of farm equipments during Pre and Post Monsoon season have been the main reasons behind minimum crop diversification in the area. There is greater dependency on

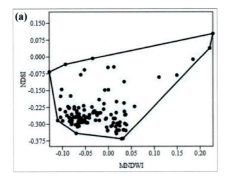 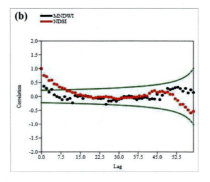

Fig. 11.6 Correlation between MNDWI and NDSI values (**a**), Lag (**b**)

rainwater harvesting as a source of fresh water and pond waters are mostly used for irrigation during summer at an individual household level. Water is lifted by motor pumps in most cases. Ground water lifting is a very costly due to higher salinity, very few households lift groundwater by shallow pumps during dry spell. Following the Census (2011) statistics, except Birinchibari (22.78% unirrigated), all the other *mouzas* have > 70% cultivated land unirrigated. > 2ppt salinity of water is detrimental for irrigational purpose but few of the ponds surveyed have recorded salinity above this limit, hence there is sheer need of salt cleansing of ponds so that irrigation can be provided all around the year. Here people favor the use of HYV seeds for higher yield in spite of higher cost of production, as the salinity level markedly increased during post *Aila* (a cyclone that wrecked havoc in the Sundarbans in May 2009 with a speed above 120 km/h) period. Demand on pesticides and insecticides has been increased at par with the use of HYV seeds but still people prefer HYV for their benefit. Traditional variety of rice is generally preferred for rainfed agriculture and HYVs are used mostly during summer. Sometimes villagers follow combination of both the varieties for better yield. Traditional variety of cultivated rice species are: *Bahurupi, Balam, Dudhersar, Rupsal, Valuki, Patnai* in Sagar Block; *Dudhersar, Lal Dhan, Paloi, Patnai, Sabita Patnai, Hamilton* in Namkhana Block; *Dudhersar, Gobindabhog, Malabati, Patnai* in Patharpratima Block; *Dudhersar, Gobindobhog, Marishal, Sonachur,* in Kultali Block; *Chinekamini, Dudhersar, Gheus, Gobind-abhog, Gopalbhog, Kalomota, Khejurchori, Lilabati* in Basanti and Gosaba Block. The HYV varieties of rice includes: *Keralasundari* in Sagar; *Anushree, Barsha, JP72, Maharaj, Pankaj, Pratiksha, Ranjit, Sabita, Shyamali* in Namkhana Block; *Pratiksha, pankaj, Masuri* in Patharpratima Block; *Pankaj CR 1017, CR 1036, Barshalakshmi, Lal Shankar, Suvashshyamali, Santoshi, Lal Swarno* in Kultali Block; *CR1017, C5, Ganga-Kaberi, N-Sankar, Pankaj, Ranjit, Pratiksha, Masuri* in Basanti, and Gosaba Block. *Dudhersar* and *Gobindabhog* variety are mostly used among traditional species as the financial benefit is more from them owing to better taste and market value. Not all the households interviewed possessed land for cultivation; many of them are engaged as agricultural laborers. Following their perception (having farming experience of 5–50 years), there has been decrease of production of crops and green leafy vegetables during post *Aila* period. Before *Aila* damage, the production of rice was 10–12 packets per *bigha* [1/3 acre] of land but it decreased to 5–8 packets currently. The green leafy vegetables with decreased trend of regeneration include *Alternanthera sp, Alternanthera phyloxeroides, Amaranthus sp, Basella sp, Saouda maaritiama and Polygonium plebium* etc. Along with the agricultural problems already discussed, land loss owing to coastal erosion plays most important role. Agricultural production suffered a setback in the study area also due to frequent land loss and embankment breaching incidents. Land loss has been found to be high in most of the *mouzas* except Kisorimohanpur, Birinchibari and Lot 126. Kisorimohanpur of Kultali Block and Lot 126 of Basanti Block reported that though production decreased for 3–4 years after *Aila*, but now agricultural production has been increasing owing to washing out of soil nutrients by monsoon rain. High degree of decline in crop yield, plant growth, and coastal land loss have been reported by Ghoramara, Dhablat, Iswaripur, Baliara, Gobardhanpur, and Kumirmari *mouzas*. The household study has also informed reduction in the availability of fresh water fishes in the ponds and rivers. The water sample study conducted already informed us the level of contamination of water in the surveyed ponds. Higher Salinity, Conductivity, TDS, Total Hardness, and lower DO have been recorded in many of the ponds (marginal area) which affects thriving of fresh water fishes. People having bigger ponds culture *Oreochromis niloticus, Clarias batrachus, Oxygaster clupeoides* fishes but in general, fishing occupation is quite threatened owing to water pollution. Commercial fishing in rivers with boats and trawlers are not also cost effective for small fishermen caused by shifting of spawning grounds of fishes from warm saline waters; diminishing quantity and quality of fish; use of age old tools and population pressure (overfishing). Previously they used to earn Rs. 7000/month but now they get only Rs. 5000/month. Both fresh and brackish water fishes like *Cyprinus rubrofuscus, Clarias batrachus, Heteropneustes fossilis, Oreochromis niloticus, Punticus sophore, Wallago attu, Mystus tengara, Channa striata, Notopterus notopterus, Tenualosa toil* are rapidly decreasing in number. Crab collection is another important activity performed by the locals. There is a greater demand of hard shelled male crab species and the price depends upon size, weight of the species (Price Rs. 100 for <100 gm; Rs. 250–300 for >100 gm; Rs. 900 for >400 gm) and seasonality (peak season is from November to January). Female crabs are more preferred sometimes as they lay eggs and tastes better, a female crab of 250 gm weight costs higher than male ones. Overexploitation of these swimming animals has also affected this supplementary occupation. There has been a rapid transformation of agricultural lands into aquaculture farms and these farms are also converted to brick fields for higher financial gain during post *Aila* period. Field investigation also has reported that the newly accreted portions called the *char* lands are getting rapidly transformed into aquaculture ponds, it does not only harm the ecosystem but also the future protection of remote Islands from higher cyclonic or storm surges. Forest Rights Act (2006) is not implemented here hence these illegal transformation of lands for aquaculture practices are on the rise under the interference of local elite class people and corrupted bureaucrats. In Basanti Block (Lot 126 *mouza*) longer stretches of aquaculture farms can be observed at the cost of agricultural lands.

11.5   Conclusion and Policy Implications

Cultivation of salt tolerant species in the context of growing concerns of climate change proves to be a great adaptation strategy in the Sundarbans. Farmers need to be encouraged to rediscover and apply traditional knowledge to select salt tolerant varieties, though they provide lesser yield, but supports environmental sustainability. These species deal with irregular rainfall; saline and waterlogged conditions pretty well, along with that, they offer cost effectiveness; better taste; minimum use of chemical fertilizers, pesticides, and insecticides; more yield with the extent of time; better thatch material and lesser need of man power. Ghosh and Chattopadhyay (2017) in their research study have estimated that the HYV seeds generate productivity of 6.2 quintal/acre while traditional variety provides 3.4 quintal/acre. But the collective advantages seem to be much alluring than the backdrop, especially in the era of climate change induced soil and water salinization. Govt. of West Bengal (Agricultural Department; Sundarban Affairs Department) and NGOs (Sundarban Social Development Centre, Tagore Society for Rural Development, Society for Environment and Development) are already providing seeds of both HYV and traditional variety of species (*Swarna Masuri, Pankaj, Dudhersar, Hamilton, CR 1017, Pratiksha, R. Masuri, C5, Lunishree, Sabita*) to farmers but they must propagate the salt tolerant varieties more (*Nona bokhr, Talmugur, Lal getu, Sada getu, Nona khirish, Hamilton, Matla*) due to the benefits already mentioned. Agricultural program of Sundarban Affairs Department (Sundarban Affairs Department 2018) also incorporates distribution of fruit plants among school students; supply of oil seeds, pulses, cotton, and other agricultural inputs during Rabi season for double cropping to promote additional income generation. Extension of this program is necessary as only 20% agricultural land remains multi cropped here. The West Bengal Biodiversity Board encountered some traditional varieties of rice while preparing People's Biodiversity Register; hence the Board has been pleading the farmers from their end to opt these for long-term food security. But better awareness generation, provision of facilities (multi cropping, integrated farming, land-reshaping, drainage improvement, low cost mechanization, capacity building, soil health card), and encouragement on greater interaction between farmers are needed to ensure all these. Government organizations must enhance the procurement price of the traditional species in the market to persuade farmers in selecting traditional varieties over HYV's. Equally, greater institutional support in terms of provision of crop insurance; cold storage facilities; generation of seed banks (traditional salt tolerant species), and soil testing measures utmost necessities so that farmers can select the crop more suitable for their land.

Jal Dharo Jal Bharo Abhiyan (2011) launched by the Govt. of West Bengal has initiated solar power operated minor irrigation schemes using sprinklers, tube wells, pumps, dug wells to increase the area under irrigation and also has promoted rainwater harvesting in water bodies. The program was later converged with MNREGA. *Jalatirtha* scheme (2015) announced by the Govt. of West Bengal resolved many saline water-related issues. Community Area Development and Water Management Programme launched by the Govt. of India in 1974–75 have reduced the gap between irrigation potential and actual irrigated area. But all these schemes are much more extensive and successful in Birbhum, Puruliya, Bankura, Jhargram, and Paschim Medinipore District. After their accomplishment, they are extended in the Sundarban parts of North and South 24 Parganas Districts recently. *Pradhan Mantri Krishi Sinchayee Yojana* (2015) by the Government of India also provides irrigation facilities in the coastal fringes of West Bengal. But major extension of these programs in the Sundarbans is the need of the time to facilitate sustainable development. Irrigation of cultivated land has been satisfactory only in Birinchibari *mouza* in the study area. Thus solemn improvement in this sector is a basic requirement of agricultural development. Here lies the necessity of improvement of pond water as it is mostly used during the summer and winter period for irrigation at individual HH level. Absence of canal irrigation makes people greatly dependent on rainwater harvesting for fresh water, and it is proved that rainwater harvesting is a cheap method for cultivation in dry period and all sorts of household activities. The Block Development Offices and PHE (Public Health Engineering) Department, West Bengal (entrusted to provide safe and quality water) must collect water samples from time to time and test them in Government laboratories to monitor physico-chemical parameters of the water. Low cost technique of desalinization of water, e.g., mixture of rice husk ash and cementing materials to separate sodium chloride and use of solar still water cones needs to be arranged. Govt. initiative is required as in most cases farmers are unable to afford salt cleansing due to sheer poverty and income instability. Fisheries Division of Sundarban Affairs Department distribute fingerlings, lime, fish feed for composite culture and train fisher folks to fruitfully utilize their ponds. Sometimes pH measurement facility and technical advices like prevention of baby crab, fish catching to foster their growth, and restrictions on *meen* (shrimp fry) farming to safeguard mud embankments, are extended by the Department. Panchayat and Rural Development Department, West Bengal train fewer self help groups on fish cultivation and improvements on its sale in market through *Meen Mitras* (Govt. representative who help in fishing). Finally, it can be concluded that extensive awareness needs to be generated by the Govt. and non Govt. authorities through household visits, training programs, camps, group discussions, mass meetings on using traditional rice variety, creating local seed banks of salt tolerant varieties, rain water harvesting and guiding aquaculture practices (both in brackish and fresh water). Central and State Govt. schemes (*Pradhan Mantri Kisan Samman Nidhi, Prime Minister Fasal Bima Yojna, Bangla Shasya Bima*) for economic assistance to poor farmers need to be endorsed by the Government officials with frequent visits in the remote islands to guarantee their implementation. Local schools and colleges may come forward to collaborate with these agencies to make people aware of sustainable crop yields by active campaigning.

Acknowledgements   The first author is thankful to UGC, New Delhi, India for providing fellowship to carry out the research work. Authors are thankful to the Soil laboratory technicians of the Department of Geography, the University of Burdwan, for extending their help in conducting a part of research analysis. We are also grateful to the survey respondents for giving us their valuable time.

References

Ahamed AJ, Loganathan K, Jayakumar R (2015) Hydrochemical characteristics and quality assessment of groundwater in Amaravathi river basin of Karur district, Tamil Nadu, South India. Sustain Water Resourc Manag. https://doi.org/10.1007/s40899-015-0026-3

Akter S, Ahmed KR, Marandi A, Schüth C (2018) Possible factors for increasing water salinity in an embanked coastal island in the southwest Bengal Delta of Bangladesh. Sci Total Environ. https://doi.org/10.1016/j.scitotenv.2020.136668

Ali MY (1999) Fish resources vulnerability and adaptation to climate change in Bangladesh. In: Huq S, Karim Z, Asaduzzaman M, Mahtab F (eds) Vulnerability and adaptation to climate change for Bangladesh. Springer, Dordrecht

Allbed A, Kumar L (2013) Soil salinity mapping and monitoring in arid and semi-arid regions using remote sensing technology: a review. Adv Remote Sens 2:373–385. https://doi.org/10.4236/ars.2013.24040

Arora NK (2019) Impact of climate change on agriculture production and its sustainable solutions. Environ Sustain 2:95–96. https://doi.org/10.1007/s42398-019-00078-w

Ataullah Md, Chowdhury Md MR, Hoque S, Ahmed A (2017) Physico-chemical properties of soils and ecological zonations of soil habitats of Sundarbans of Bangladesh. Int J Pure Appl Res1(1):80–93

Banerjee K (2013) Decadal change in the surface water salinity profile of indian sundarbans: a potential indicator of climate change. J Marine Sci Res Dev. https://doi.org/10.4172/2155-9910.S11-002

Bhattacharjee D, Samanta B, Danda A, Bhadury P (2013) Impact of climate change in the sundarban aquatic ecosystems: phytoplankton as proxies. In: Sundaresan J, Sreekesh S, Ramanathan A, Sonnenschein L, Boojh R (eds) Climate change and island and coastal vulnerability. Springer, Dordrecht

BIS (Bureau of Indian Standards) 10500 (2012) Specification for drinking water. Indian Standards Institution, New Delhi

Biswas S, Zaman S, Mitra A (2017) Soil characteristics of indian sundarbans: the designated world heritage site. Sci J Biomed Eng Biomed Sci 1(2):053–059

Brecht H, Dasgupta S, Laplante B, Murray S, Wheeler D (2012) Sea-level rise and storm surges: high stakes for a small number of developing countries. J Environ Dev 21:120–138

Cañedo-Argüellesa M, Kefford BF, Piscart C, Prat N, Schäfer RB, Schulz CJ (2013) Salinisation of rivers: an urgent ecological issue. Environ Pollut 173:157–167

Census (2011) District Census Handbook. South 24 Parganas District, West Bengal, Government of India

Chand BK, Trivedi RK, Dubey SK, Beg MM (2012) Aquaculture in changing climate of Sundarban. National Initiative on Climate Resilient Agriculture, West Bengal University of Animal & Fishery Sciences

Chang H (2005) Spatial and temporal variations of water quality in the Han River and its tributaries, Seoul, Korea, 1993–2002. Water Air Soil Pollut 161:267–284

Chen J, Mueller V (2018) Coastal climate change, soil salinity and human migration in Bangladesh. Nature Clim Change 8:981–985. https://doi.org/10.1038/s41558-018-0313-8

CIESIN (2010) Low-elevation coastal zone (LECZ) rural-urban estimates. http://sedac.ciesin.columbia.edu/gpw/lecz.jsp

Clarke D, Lázár AN, Saleh AFM, Jahiruddin M (2018) Prospects for agriculture under climate change and soil salinisation. In: Nicholls R, Hutton C, Adger W, Hanson S, Rahman M, Salehin M (eds) Ecosystem services for well-being in deltas. Palgrave Macmillan, Cham

Colombani N, Osti A, Volta G et al (2016) Impact of climate change on salinization of coastal water resources. Water Resour Manage 30:2483–2496. https://doi.org/10.1007/s11269-016-1292-z

Corwin D, Yemoto K (2017) Salinity: electrical conductivity and total dissolved solid. Methods Soil Anal. https://doi.org/10.2136/msa2015.0039

Das S, Roy Choudhury M, Das S, Nagarajan M (2016) Earth observation and geospatial techniques for soil salinity and land capability assessment over Sundarban Bay of Bengal Coast, India. Geodesy Cartogr 65(2):163–192

Dasgupta S, Hossain MM, Huq M et al (2015a) Climate change and soil salinity: the case of coastal Bangladesh. Ambio 44:815–826. https://doi.org/10.1007/s13280-015-0681-5

Dasgupta S, Huq M, Wheeler D (2015b) Drinking water salinity and infant mortality in Coastal Bangladesh Development Research Group. Environment and Energy Team, Research Support Team, World Bank Group, Washington, DC

Dasgupta S, Kamal FA, Khan ZH, Choudhury S, Nishat A (2014) River salinity and climate change evidence from Coastal Bangladesh. The World Bank, Development Research Group, Environment and Energy Team

Debnath A (2013) Condition of agricultural productivity of Gosaba C.D. Block, South 24 Parganas, West Bengal, India after Severe Cyclone Aila. Int J Sci Res Publ 3(7):97–100

Dhar S (2011) Impact of climate change on the salinity situation of the Piyali River Sundarbans, India. J Water Resourc Protect. https://doi.org/10.4236/jwarp.2011.37059

District Disaster Management Plan (2018–19) South 24 Parganas District, Government of West Bengal

Dubey SK, Trivedi RK, Chand BK, Mandal B, Rout SK (2016) Farmers' perceptions of climate change, impacts on freshwater aquaculture and adaptation strategies in climatic change hotspots: a case of the Indian Sundarban delta. Environ Dev. https://doi.org/10.1016/j.envdev.2016.12.002

FAO Corporate Document Repository (1994) Water quality for agriculture in FAO Irrigation and Drainage Papers. Version 29, Agriculture and consumer protection, Food and Agriculture Organization of the United Nations

Ghosh S, Chattopadhyay KS (2017) A study on indigenous rice varieties in Sundarban Delta and their role in ensuring local food security in the face of climate change threats. Agro-Economic Research Centre, Visva-Bharati, Santiniketan West Bengal

Halder A, Debnath A (2014) Assessment of climate induced soil salinity conditions of Gosaba Island, West Bengal and Its Influence on Local Livelihood. In: Singh M et al (eds) Climate change and biodiversity. Proceedings of IGU Rohtak conference, Vol. 1, Advances in geographical and environmental sciences, Springer, Japan. https://doi.org/10.1007/978-4-431-54838-6_3

Hansen JE, Sato M (2012) Paleoclimate implications for human-made climate change. Springer, Vienna

Haque MZ, Reza MI (2017) Salinity intrusion affecting the ecological integrity of Sundarbans Mangrove Forests, Bangladesh. Int J Conservat Sci 132–141

Hasnat M A, Hossain N, Muhibbullah M, Sarwar MDI, Shormin T (2016) Impacts of Climate change on agriculture and changing adaptive strategies in the Coastal Area of Lakshmipur District, Bangladesh. Curr World Environ. https://doi.org/10.12944/CWE.11.3.03

Herczeg AL, Dogramaci SS, Leaney FWJ (2001) Origin of dissolved salts in a large, semi-arid groundwater system: Murray Basin, Australia. Marine Freshwater Res 52(1):41–52

Huq SMI, Alam MD (2005) A handbook on analysis of soil, plant and water. BACERDU, University of Dhaka, Bangladesh

Islam Didar-Ul SM, Bhuiyan MAH (2016) Impact scenarios of shrimp farming in coastal region of Bangladesh: an approach of an ecological model for sustainable management. Aquacult Int 24:1163–1190. https://doi.org/10.1007/s10499-016-9978-z

Islam MA, Hoque MA, Ahmed KM et al (2019) Impact of climate change and land use on groundwater salinization in Southern Bangladesh—implications for other Asian Deltas. Environ Manage 64:640–649. https://doi.org/10.1007/s00267-019-01220-4

Jahan A (2018) The effect of Salinity in the flora and fauna of the Sundarbans and the impacts on local livelihood. Master thesis in Sustainable Development, Department of Earth Science, Uppsala University

Ji L, Zhang L, Wylie B (2009) Analysis of dynamic thresholds for the normalized difference water index. Photogramm Eng Remote Sens 75:1307–1317

Johnson AF, Hutton CW, Hornby D et al (2016) Is shrimp farming a successful adaptation to salinity intrusion? A geospatial associative analysis of poverty in the populous Ganges–Brahmaputra–Meghna Delta of Bangladesh. Sustain Sci 11:423–439. https://doi.org/10.1007/s11625-016-0356-6

Khan AE, Ireson A, Kovats S, Mojumder SK, Khusru A, Rahman A, Vineis P (2011) Drinking water salinity and maternal health in coastal Bangladesh: implications of climate change. Environ Health Perspect 119:1328–1332. https://doi.org/10.1289/ehp.1002804

Khan AE, Scheelbeek PFD, Shilpi AB, Chan Q, Mojumder SK, Rahman A, Haines A, Vineis P (2014) Salinity in drinking water and the risk of (pre) eclampsia and gestational hypertension in coastal Bangladesh: a case-control study. PLoS One. https://doi.org/10.1371/journal.pone.0108715

Li P, Wu J, Qian H (2013) Assessment of groundwater quality for irrigation purposes and identification of hydrogeochemical evolution mechanisms in Pengyang County, China. Environ Earth Sci 69:2211–2225

McNeil HV, Cox ME (2007) Defining the climatic signal in stream salinity trends using the interdecadal Pacific Oscillation and its rate of change. Hydrol Earth Syst Sci 11(4):1295–1307

Mitra A, Gangopadhyay A, Dube A, Schmidt ACK, Banerjee K (2009) Observed changes in water mass properties in the Indian Sundarbans (northwestern Bay of Bengal) during 1980–2007. Curr Sci 97(10):25

Mondal MS, Saleh AFM, Razzaque Akanda MA, Biswas SK AZ Md, Zaman Moslehuddin S, Lazar AN, Clarke D (2015) Simulating yield response of rice to salinity stress with the AquaCrop model. Environ Sc: Process Impacts 17(6):1118–1126

Mondal I, Bandyopadhyay J, Paul AK (2016) Water quality modeling for seasonal fluctuation of Ichamati river. Model Earth Syst Environ. https://doi.org/10.1007/s40808-016-0153-3

Mukherjee N, Siddique G (2019) Assessment of climatic variability risks with application of livelihood vulnerability indices. Environ Dev Sustain. https://doi.org/10.1007/s10668-019-00415-3

Murshid SM (2012) Impact of sea level rise on agriculture using groundwater in Bangladesh. M.Sc thesis CoMEM programme, University of Southampton. https://repository.tudelft.nl/islandora/object/uuid:e484b9b8-e1d1-40b1-99ee-5c7a274ef500/?collection=research

NAIP (2011) Strategies for sustainable management of degraded coastal land and water for enhancing livelihood security of farming communities, Central Soil Salinity Research Institute (CSSRI), Regional Research Station (RRS), Canning Town

Nath D, De DK (1999) Water and soil characteristics of different estuaries in Sundarbans. In: Guha Bakshi DN, Sanyal P, Naskar KR (eds) Sundarbans Mangal. Naya Prokash, Kolkata, pp 128–130

National Statistical Service Organization (2018) Statistics, Australian Bureau of Statistics

Ogilvie A et al (2015) Decadal monitoring of the Niger inner delta food dynamics using MODIS optical data. J Hydrol (amst) 523:368–383

Okur B, Orcen A (2020) Soil salinization and climate change. In: Prasad NMV (eds) Climate change and soil interactions. Elsevier, pp 331–350. https://doi.org/10.1016/B978-0-12-818032-7.00012-6

Panda S, Sadhu C, Pramanik G, Pahari S, Hossain J (2016) Concerning public health situation of under-nutrition in children and anemia in women in Indian Sundarbans delta: a community based cross-sectional investigation. BMC Nutrition 2:1–10. https://doi.org/10.1186/s40795-016-0105-3

Rahman Md S, Di L, Yu EG, Tang J, Lin L, Zhang C, Yu Z, Gaigalas J (2018) Impact of climate change on soil salinity: a remote sensing based investigation in Coastal Bangladesh. In: The seventh international conference on agro-geoinformatics, August, Hangzhou, China

Ramteke IK, Reddy GPO, Sen TK, Singh SK, Chaterjee S, Rajankar PB, Das SN (2017) Land use/land cover change dynamics in coastal ecosystem of Sundarban Delta, West Bengal: a case study of Bali Island. Int J Sci Environ Technol 6(6):3565–3577

Sahana M, Sajjad H (2019) Vulnerability to storm surge flood using remote sensing and GIS techniques: a study on Sundarban Biosphere Reserve, India. Remote Sensing Applications: Society and Environment. https://doi.org/10.1016/j.rsase.2018.10.008

Sahana M, Rehman S, Ahmed R et al (2020) Analyzing climate variability and its effects in Sundarban Biosphere Reserve, India: reaffirmation from local communities. Environ Dev Sustain. https://doi.org/10.1007/s10668-020-00682-5

Sanchez Triana E, Leonard O, Paul T (2016) Managing water-related risks in the West Bengal Sundarbans: policy alternatives and institutions. Int J Water Resourc Dev. https://doi.org/10.1080/07900627.2016.1202099

Scheelbeek PFD, Khan AE, Mojumder S, Elliott P, Vineis P (2016) Drinking water sodium and elevated blood pressure of healthy pregnant women in salinity-affected coastal areas. Hypertension. https://doi.org/10.1161/HYPERTENSIONAHA.116.07743

Scheelbeek PFD, Chowdhury MAH, Haines A, Alam DS, Hoque MA, Butler AP, Khan AE, Mojumder SK, Blangiardo MAG, Elliot P, Vineis P (2017) Drinking water salinity and raised blood pressure: evidence from a cohort study in coastal Bangladesh. Environ Health Perspect. https://doi.org/10.1289/EHP659

Sundarban Affairs Department (2018) Agricultural Extension and Fishery Programme, Government of West Bengal

Talukder MdRR, Rutherford S, Phung D, Islam MdZ, Chu C (2016) The effect of drinking water salinity on blood pressure in young adults of coastal Bangladesh. Environ Pollut 214:248–254

Teh SY, Koh HL (2016) Climate change and soil salinization: impact on agriculture, water and food security. Int J Agricult Forestry Plantation 2:1–9

Vineis P, Chan Q, Khan A (2011) Climate change impacts on water salinity and health. J Epidemiol Global Health 1:5–10

Wheeler D (2011) Quantifying vulnerability to climate change: Implications for adaptation assistance. Center for Global Development, Working Paper No. 240, Washington

World Food Programme (2018) 2018 Global Report on Food Crises. Italy, Rome

Yadav GK, Jagdhani AD, Sawale DD, Krishnapriya MK, Yadav RS (2019) Seasonal variation of soil chemical characteristics at agriculture technical school, Manjri Farm, Pune, India. Int J Curr Microbiol App Sci 8(10):2623–2628

Zaman S, Agarwal S, Mitra A, Amin G, Prmanick P, Mitra A (2015) Impact of Aila on the dissolved oxygen level in the Indian sundarbans region. J Energy Environ Carbon Credits 5(3):1–4

Chapter 12
Impact of Water Reservoir and Irrigation Canals on Land Use and Land Cover Changes in Upper Kumari River Basin, West Bengal, India

Piya Bhattacharjee and **Debasish Das**

Abstract It is observed that irrigation has many beneficial impacts on agriculture. However, there are many adverse environmental impacts that have been noticed as well. The objective of the investigation is to assess such types of impact. A research has been carried out in the upper catchment area of Kumari River, Purulia, West Bengal. Two reservoirs, i.e., Kumari and Hanumata and related canals are investigated. Kumari irrigation scheme covers 3255 ha during Kharif season and 320 ha for Rabi season. Hanumata irrigation scheme covers 1265 ha during Kharif season and 460 ha during Rabi season. Settlements have been evacuated due to construction of the reservoirs. Geographically, it is a part of the Chota Nagpur Plateau, lies between 22°55′ and 23°15′N latitude and 86°08′ and 86°38′ E longitude. Application of spatial technologies aided by field work is done for the study. Kumari Reservoir has got two main irrigation canals which flow in the NNE-SSE direction. These two main canals (location is N 23°09′30.4″ and E 86°17′23.8″ and elevation is 228 m) have been further re-directed in the E-W directions. Moisture content of the agricultural field here is 2.3–2.5% (dry season) and up to 8% (syn-monsoon), and pH is 7.2–8.0.

Keywords Reservoir · Irrigation canal · Land use · Land cover · River basin

12.1 Introduction

It is known that human activities have a tremendous impact on natural environment and they are the main cause for natural degradation. The expansion of dam construction pertaining to water resource management has become an important aspect in many parts of the world including India. Irrigated agriculture makes production more stable than rain-fed agriculture. Higher agricultural productivity creates job opportunities. Dam construction and some related irrigation practices sometimes have adverse impacts on environment such as water logging, soil salinization, ecological damage, and adverse socio-economic impact.

P. Bhattacharjee · D. Das (✉)
Department of Environmental Science, University of Kalyani, Kalyani 741235, West Bengal, India

© The Author(s), under exclusive license to Springer Nature Singapore Pte Ltd. 2022
N. C. Jana and R. B. Singh (eds.), *Climate, Environment and Disaster in Developing Countries*, Advances in Geographical and Environmental Sciences, https://doi.org/10.1007/978-981-16-6966-8_12

Fig. 12.1 Kumari reservoir

The specific objective was to assess the land use land cover changes before and after the construction of reservoir and irrigation canals by using multi-temporal google earth images. Dams have long been recognized for electricity generation, flood protection, and making water available for irrigation. The damming of a reservoir has got dramatic consequences both upstream and downstream of the dam. Prior to dam construction, most natural rivers have a flow rate that varies widely throughout the year in response to varying climatic conditions. It has been observed that the Kumari Reservoir (Fig. 12.1) has got two main irrigation canals (Fig. 12.2) which flow in the NNE-SSE direction. These two main canals (location of the starting point is N23°09′30.4″ and E 86°17′23.8″ and elevation is 228 m) have been further re-directed in the E-W directions. Many irrigated agricultural plots (Fig. 12.3) have been identified on both sides of the canals. Moisture content of the agricultural field here is 2.3–2.5% (dry season) and up to 8% (syn-monsoon) and pH is 7.2–8.0. The agricultural fields which lack recent irrigation show dry soil condition.

12.2 Materials and Methods

Goggle earth images of the year 1973, 1992, 2008 (pre and post monsoon season) were consulted and subsequently thematic layers were developed for further analysis in GIS environment (Burrough 1990; Burrough and Mc Donnel 1998).

Fig. 12.2 Irrigation canal

Fig. 12.3 Agricultural plots adjacent to irrigation canal

12.3 Study Area

The district of Purulia is located at the border of Jharkhand and West Bengal. It was carved out of the former district of Manbhum of Bihar in 1956 and added to West Bengal. Geographically, it is a part of the Chota Nagpur Plateau consisting of "succession of rolling uplands with intervening hollows" and infertile lateritic soil. It lies between 22°55′ and 23°15′ N latitude and 86°08′ and 86°38′ E longitude. The district has an area of 6259.0 km^2 and in 2011 the total population stood at 29.30 lakhs (Census Handbook 2011). Like the topography of the district, its population structure has several points of dissimilarity with that of the rest of West Bengal. It has the highest percentage of scheduled tribes 19.58% (West Bengal 5.72%). The bulk of its population consists of Hinduized and semi-Hinduized communities who still preserve some of the lifestyle of their tribal forefathers. Economically, the district is one of the poorest among the sixteen districts of West Bengal. Only 45.01% people are landowning and 33.37% landless, there being 20% increase of landless people from 1961 to 1971. The district was once covered by thick sal (*Shorea robusta*) forest populated by rich fauna, but now this forest is almost lost. Only occasional trees and shrubs attest to the past environment. The upper catchment area of the Kumari River Basin is situated in the western part of Purulia district, W.B., of eastern India bounded by latitude 22°55′ to 23°15′N and longitude 86°8′ to 86°38′E. It is actually a regenerated landmass which includes variety of landscape units (Sing 1969). Tributaries of Kumari River are the main drainage in the area. In spite of moderate average annual rainfall (1400 mm) (Mishra 2012) groundwater recharge is inadequate due to crystalline nature of the country rock and uneven relief of the terrain (Mukherji and Das 1989).

Hot summer and cold winter characterize the climate of the district. The average annual rainfall in the district is 1363 mm. May is generally the hottest month with a mean daily maximum temperature of 40.3 °C and a mean daily minimum of 27.2 °C. January is the coldest month with the mean daily maximum temperature at 25.5 °C and the mean daily minimum at 12.8 °C. Purulia district belongs to Agro-eco Sub region (AESR) of Chota Nagpur Plateau.

Natural Vegetation of the district consists of trees, shrubs, and grasses. The major tree species are *Shorea robusta* (Sal), *Butea frondosa* (Palash), *Terminalis arjuna* (Arjun), *Bombax malabaricum* (Shimul), etc. The important shrubs and herbs are *Flacontia nemontchi* (Boinchi), *Jatropa gossypifolia* (Lal bharenda), etc. The few grasses found in the field are *Saccharum munja* (Sar*), Solanum niagram* (Kakmachi), etc.

12.4   Physiography and Landforms of the Area

The study area is an erosional landscape developed on Precambrian basement rock. Granite and meta sedimentaries are the man lithologies. Among the two physiographic divisions, high land is composed of denudational and structural hills. The rocks of the area are traversed by several sets of joints among which NNE-SSW master joints are prominent which show (dip > 50°). Low lying region is undulating and forms the peneplain region. The thickness of the soil cover varies from 3 to 6 m.

The soils of the study area are broadly categorized into residual type which has been derived from weathering of Precambrian granite gneiss/composite gneiss.

Residual soil has got an extensive areal extent. Alluvial soils are present in the valleys of major streams/rivers. Latosols are present in the upland areas where as in the valleys reddish clay loam or simply reddish clay are common.

12.5   Geological Setting

The study area forms a part of the south eastern extension of the Chota Nagpur plateau of eastern India, which is developed on the Precambrian granite basement rock. Geologically the area constitutes a part of Chota Nagpur gneissic complex, which has been cut across by two shear zones. Lithologically the area was composed of metasedimentaries and granitic clan of rocks. It is actually a regenerated landmass (Singh 1969), which includes a variety of landscape units. Peneplained Precambrian mountain is the most conspicuous feature of the area. Many erosional landforms, like inselberg, bornhardts, and undulating valleys are the evidences of action of erosional process which worked unhindered for a vast period of time. The area must have been suffered by several cycle of erosion aided by geotectonic upliftment (Mukherji and Das 1989).

12.6   Drainage and Groundwater Condition of the Area

Observed drainage pattern is rectangular to angulate as per toposheet study, the higher order stream like Kumari River and adjoining rivers show the evidence of structural control. Joint pattern has influenced the main stream course. The main stream shows a base flow during dry months where as they show a bankful stage during monsoon. As the rocks are generally impervious, groundwater accumulation mainly takes place in the weathered residuum. Many water wells have been developed in this region. Wells in the upper catchment area show almost near-surface water table condition. Whereas wells situated downstream after the reservoir show water table at higher depth.

12.7   Reservoir and Irrigation Canals

Kumari River is the main drainage system of the study area. A reservoir has been constructed in the upper catchment area of Kumari River. With a catchment area of 94.72 km^2, length of the main canal is 37 km and the subsidiary canal's length is 24 km. Canals have been constructed to supply irrigation water into the agricultural field. Cultivated land is seen adjacent to the irrigation canals. Canals are constructed toward E-W alignment. And, in some places canals are constructed branching out from the main canal. In this area targeted area under Kumari irrigation scheme is 3642

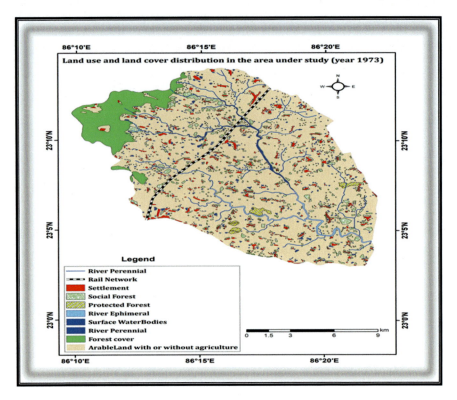

Fig. 12.4 Map with multiple themes of the year 1973. *Data Source* SOI map

ha (Kharif) and 432 ha (Rabi crop). However, 3255 ha (Kharif) has been achieved and 320 ha (Rabi crop) has been achieved.

When a comparative exercise is made among the multi temporal thematic maps, it is observed that major changes have been taken place in between irrigated land and non-irrigated land. That means, the construction of water reservoir and subsequent laying out of irrigation canals have wreaked havoc toward the changes in agricultural productivity. Non-agricultural land use has been converted to agricultural land use. Social forestry area has been changed with temporal variations and they are shown as follows (Figs. 12.4, 12.5, 12.6, 12.7, 12.8 and 12.9).

12.8 Environmental Impact of the Reservoir and Canals

Reservoir construction is very much successful process to enhance the total development and management system of water resource utilization (Devic 2014). Irrigation through constructed canal is actually an art of channelizing water flow for a specific purpose like supporting water flow for agriculture. This purpose is usually to water

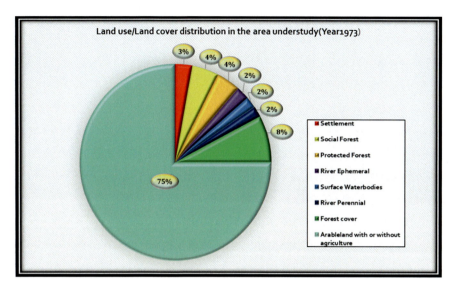

Fig. 12.5 Land use land cover distribution in the area under study, 1973

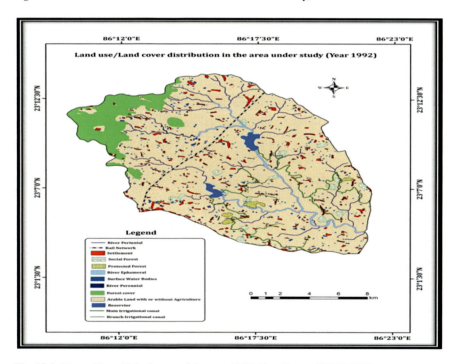

Fig. 12.6 Map with multiple themes of the year 1992. *Data Source* IRS IB 1992

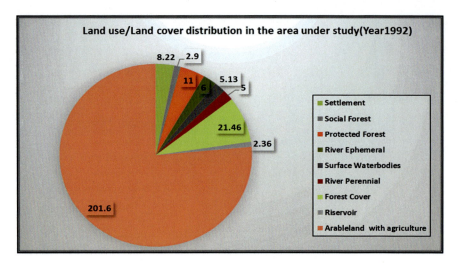

Fig. 12.7 Land use land cover distribution in the area under study, 1992

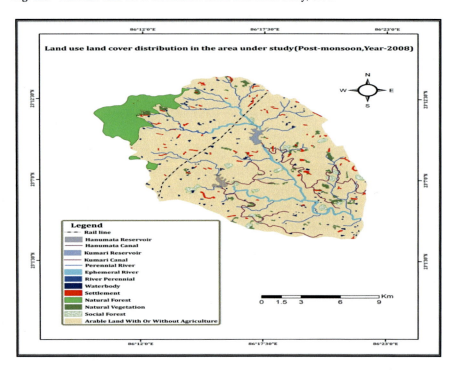

Fig. 12.8 Map with multiple themes of the year 2008. *Data Source* Google Earth images

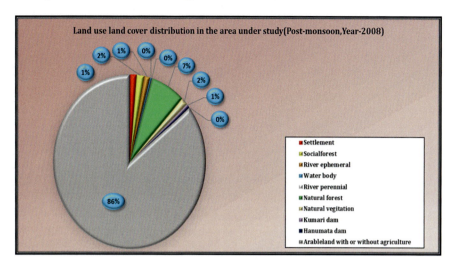

Fig. 12.9 Land use land cover distribution in the area under study, 2008

agricultural crops, to maintain landscapes, or to provide much-needed water during a drought.

However, this exercise has got a direct impact on the surrounding environment. This environmental impact includes: increased groundwater level in the irrigated areas, decreased downstream water flow, and increased evaporation loss. This article takes a closer look at the environmental impact of irrigation (https://en.m.wikipedia.org/wiki/environmental_impact_of_irrigation).

12.9 Conclusion

It is evident from the present investigation that Kumari River is a very vital surface water resource for local and adjoining community. This modification has been accompanied by laying out of irrigation canals and which ultimately created some modifications in land use practice. Reservoir construction will ultimately create some adverse environmental impact like submergence of settlement and habitation. Water logging may create soil salinity as well. Further expansion/changes in social forestry and in other thematic land uses is clearly discernible in remote sensing and GIS data base. The present exercise proves the effective synergism of remote sensing and GIS application in land use studies.

Acknowledgements The present investigation is a part of Ph.D. research work of Piya Bhattacharjee under the supervision of Debasish Das. This research work was funded by SVMS scholarship of W.B. state government, India.

References

Burrough PA (1990) Principles of GIS for land resources assessment. Oxford Science Publication

Burrough PA, Mc Donnel RA (1998) Principles of GIS. Oxford University Press, Oxford Census Handbook 2011

Devic G (2014) Environmental impact of reservoir. Environmental Indicators (2015). ISBN:978-94-017-9498-5

Mishra S (2012) Climate change adaptation in arid region of West Bengal, Climate change Policy paper III, WWF India

Mukherji AL, Das D (1989) A study on the development of basons and their hydrogeomorphic features in and around Ajodhya Plateau, eastern India. In: Proceedings of international symposium intermontane basins: geology and resources, Chiang Mai, Thailand, pp 409–417

Sing RP (1969) Geomorphological evolution of Choto Nagpur highlands, India. National Geographical Society of India, Varanasi, p 127

Web link: https://en.m.wikipedia.org/wiki/environmental_impact_of_irrigation

Chapter 13
Spatial Pattern of Arsenic Contamination in Floodplain Aquifers, Western Bank of Bhagirathi River, Lower Ganges Delta, West Bengal, India

Sunam Chatterjee

Abstract Arsenic (As) concentration above World Health Organization (WHO) recommended value is reported from either side of Bhagirathi River in lower Ganges delta. Millions of people are living at risk. The study area is located on the western bank of the Bhagirathi River. The database is collected from Public Health Engineering Department (PHED). The study focuses on spatial mapping, depth-wise distribution, association with geomorphology, and seasonal variation of As concentration. It is found that about 48.40% of tube wells (N = 1535) have arsenic concentrations above the WHO limit, i.e., >0.01 mg/L. Vertical distribution of arsenic contamination shows 52.77% of samples (N = 1535) within 40–80 m depth are contaminated. At deeper depths (>100 m) the count of contaminated tube well are few; though the magnitude of As concentration is extremely erratic. Seasonal variations study finds that about 70% of samples have to exceed As concentration in the pre-monsoon season. The contaminated tube wells (>0.01 mg/L) are mostly occurring within the meander belt of the Bhagirathi River and have been associated with different floodplain features. The lithological study reveals the Holocene aquifers consisting of grey-coloured younger alluvial deposits of *Katwa* formation and *Present-day* deposits are found contaminated; while, Pleistocene aquifers are almost safe.

Keywords Arsenic · Holocene aquifer · Quaternary · Seasonal variation · Lithology · Bhagirathi floodplain

13.1 Introduction

Arsenic (As) contamination in groundwater is a worldwide problem (Herath et al. 2016). About 200 million people are exposed to arsenic contamination through drinking water (George et al. 2014). Arsenic in tube well water has been reported from different countries such as India, Pakistan, Bangladesh, Nepal, China, Vietnam,

S. Chatterjee (✉)
Independent Researcher, Burdwan city, West Bengal, India

© The Author(s), under exclusive license to Springer Nature Singapore Pte Ltd. 2022   245
N. C. Jana and R. B. Singh (eds.), *Climate, Environment and Disaster in Developing Countries*, Advances in Geographical and Environmental Sciences, https://doi.org/10.1007/978-981-16-6966-8_13

Cambodia, Ethiopia, Chili, Argentina, Italy, etc. (Nickson et al. 2007; Chakraborti et al. 2013; Ali et al. 2019; Shahid et al. 2018; Chakraborty et al. 2015; Mueller 2017; Guo et al. 2014; Winkel et al. 2011; Agusa et al. 2007; Merola et al. 2014; Steinmaus et al. 2013; Francisca and Perez 2009; Rotiroti et al. 2014). Excessive arsenic has already been detected in the floodplains of the Ganges–Brahmaputra, Mekong, and Red rivers (Edmunds et al. 2015). Millions of people are affected by this groundwater hazard (Chakraborti et al. 2018). Arsenic enters the human body mainly through drinking water and food, which leads to chronic health problems (Saha et al. 1999). As a result, many carcinogenic and non-carcinogenic symptoms will appear among the people exposed to it (Hong et al. 2014).

In India, groundwater arsenic contamination has been linked up with many river systems, viz. Ganges, Tista, Brahmaputra, etc. (Verma et al. 2019; Bhattacharyya and Mukherjee 2009). Occurrence of arsenic in tube well water has been found in several parts of the middle and lower Ganga plain (Shah 2008, 2017; Bhattacharya et al. 1997; Singh 2015). But the condition is severe in the deltaic part (Chakraborty et al. 2015). About 45% of the total population is directly or indirectly affected by arsenic contamination in groundwater (Stüben et al. 2003). Bhagirathi Hooghly sub basin in the lower Gangetic delta is a vast floodplain made up of river-borne sediments descending from the Himalayas and plateaus (Allison 1998). Arsenic naturally exists in these sediments (Purkait 2018; Acharyya and Shah 2007). Almost the entire area of the Bhagirathi Hooghly floodplain is affected by arsenic in groundwater. The districts lying on the east bank of this river are in the grip of terrible arsenic contamination. Many studies have shown the presence of arsenic in the groundwater on the east bank of the Bhagirathi River and its ill effects (Mukherjee 2010a; Sankar et al. 2014; Hazra 2010; Chowdhury et al. 2000; Rahman et al. 2003; Mukherjee and Taraknath 2015; GuhaMazumder 2003; Shrivastava et al. 2014). On the other hand, some scattered research papers have found arsenic contamination in the floodplain on the west bank of the Bhagirathi River (Acharyya and Shah 2007; Mukherjee 2010b). The most important issue in arsenic research is the identification of arsenic contaminated tube wells and the provision of safe drinking water for the communities affected by arsenic exposure (Smith et al. 2000). Before starting work on this goal, a clear idea about the distribution; its spatial association with surface and subsurface elements and variability pattern is known to be required. Modern researchers believe that the spatial heterogeneity in the distribution of tube well As content remains poorly understood phenomena (Acharyya et al. 2000; Islam et al. 2004). Keeping this in mind, in the present research, emphasis has been laid on the distribution pattern, depth-wise variation, mapping, seasonal variation of As contaminated tube well, and spatial association of arsenic in the tube well water of the two most arsenic contaminated blocks in Burdwan district on the west bank of the Bhagirathi River.

13.2 Study Area

The area of the present study is located in the western bank of the Bhagirathi River (Fig. 13.1). Geographically, it is a floodplain tract situated in the juxtaposition between Bhagirathi River and Damodar fan formation. The plain is formed in the Pleistocene–Holocene time period and is characterized by the presence of extensive clay layers with varying thickness, which is underlain by silt, sand, and gravel (Acharyya and Shah 2007; Mukherjee 2010b; Mukherjee et al. 2009). Shifting and incision of Bhagirathi River sculpture the area and leaves behind its imprints although the region.

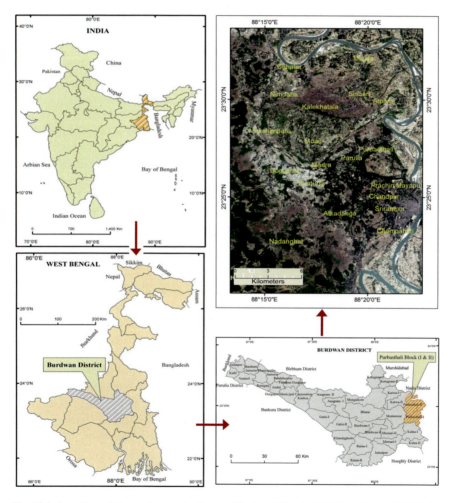

Fig. 13.1 Location of the study area in the west bank of Bhagirathi River, Purbasthali blocks, Bardhaman district, West Bengal, India

Administratively, Purbasthali I and II blocks of Burdwan district are the focus area of this research. It includes about 340.91 km^2 area and carries a total population of about 419,332 as per the 2011 census. Climate of the area is warm and subtropical humid monsoon type. The summer time average temperature is about 32 °centigrade and in winter it is about 18 °centigrade. The annual average precipitation is about 141.7 cm (Indian Metrology Department). The area is rich in groundwater resource potential (Purkait 2018). The aquifers are transmissive and multilayered (Chatterjee et al. 2018). Their condition varies from unconfined to semi confined in nature (Mukherjee 2010a). The surface morphology of the area is endowed with several hydro-geomorphic features (Fig. 13.1).

13.3   Regional Setting

Bhagirathi River floodplain genealogically belongs to the Ganges delta (Rudra 2014). It is a vast plain of alluvial origin. The upper part of the delta is influenced by the fluvial process and the lower portion is dominated by tidal deposits (Mukhopadhyay and Ghosh 2010). The Himalayan Rivers deposit a thick pile of sediments to form this upper segment of the fluvio-deltaic complex. The western bank of the Bhagirathi River is naturally delimited by the Bhagirathi River in the east and terraced Damodar fan formation in the west. The landscape bears the imprint of multiple cycles of erosion and deposition (Mukhopadhyay and Ghosh 2010). It is apparently a monotonous plain with elevations ranging 5–25 m above mean sea level. The master slope is in north–south direction allowing the Bhagirathi River to flow in that direction (Biswas 2010). Another drainage artery Khari River traverses the region from west to east following the gently dipping Damodar fan, facing eastward. The overall drainage is dendritic with subparallel disposition of distributaries (Fig. 13.2). The area is ornamented with several scars of channel migration, ox-bow lake, relict channels, fragmented narrow meandering levees, back swamps, floodplain marshes, inter levee depressions, etc. (Mukherjee and Taraknath. 2015). The regional set up map (Fig. 13.2) shows four major Quaternary units in the western part of the Bhagirathi River. These are present-day deposits of Bhagirathi, *Katwa* formation, *Chinsurah* formation, and *Sijua* formation (Modified after GSI, 2014). The detailed arrangement of Quaternary formation is summarized in Table 13.1. The present-day deposits are the recent ones and *Sijua* formation is the oldest (Mukhopadhyay and Ghosh 2010). The Holocene deposit of younger alluvium dominates in the first three units and is represented by grey alluvial sediments rich in carbonaceous matter. The *Sijua* formation is characterized by iron-coated oxidized sand and caliches (Acharyya and Shah 2007; Mukherjee 2010c).

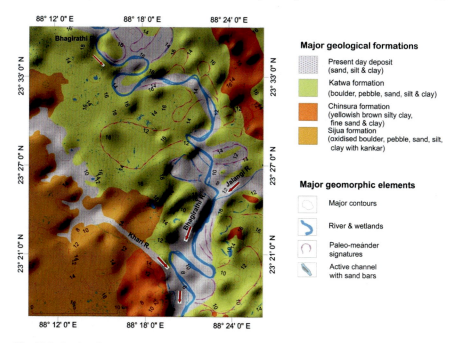

Fig. 13.2 Regional setup map of quaternary formations and major geomorphic elements, purbasthali area,west bank of Bhagirathi river, Bardhaman district, West Bengal, India. *Source* Modified after 79A Quadrant geology map, Krishnanagar, GSI (2014)

Table 13.1 General lithostratigraphic succession of the study area, western bank of Bhagirathi River, Purbasthali blocks, Bardhaman district, West Bengal, India (modified after 79A Quadrant geology map, Krishna Nagar, GSI, 2014)

Lithology	Depositional environment	Formation	Age
Sand, silt and clay	Deltaic, formation with extensive peat layers	Recent sediments	Holocene
Boulder, pebble, sand, silt and clay	Eustatic change; sandy to coarse sandy deposition and formation of floodplain swamps	Bhagirathi - Ganga Formation	Holocene
		Katwa Formation	
Yellowish brown silty clay and fine sandy clay		Chinsura Formation	Holocene
Oxidized boulder, pebble, sand, silt, clay with kankar	Sub areal oxidized older alluvial deposition	Sijua Formation	Upper pleistocene to lower holocene

Source Quadrant Geology map, (Krishnanagar Quadrant (79A), 2014)

13.4 Materials and Methods

In the present study, a handful of tube wells (N = 1535) have been used to study the pattern of arsenic contamination in Purbasthali Blocks. Data on arsenic levels in tube well water were collected from the Public Health Engineering Department (PHED) water testing laboratory, Nadanghat, Burdwan. The project is divided into several sections and specific methods are followed to achieve the desired objectives.

13.4.1 Distribution of As Contaminated Tube Wells

Distribution of contaminated tube wells is done by classifying the village-wise tube wells into two major groups, i.e., safe and contaminated. In this regard, the World Health Organization (WHO) approved tolerance level is taken into consideration. For this work, the total number of tube wells in each village exceeding WHO-approved tolerance level are identified as contaminated. The village-wise distribution of contaminated tube wells percentage value is calculated. Finally, the village-wise percentage values of contaminated tube wells are classified into seven groups and a thematic map showing the distribution of contaminated tube wells is created in ArcGIS 10.3.

13.4.2 Mapping of As Concentration

Mapping of arsenic concentration in groundwater is an important task of present work. For this purpose, all the location of the tube wells collected by handheld GPS (Garmin etrex 30) and the respective arsenic concentration values of each tube wells are imported to ArcGIS. Then an interpolated surface map is created from the coordinate-based tube well arsenic concentration data. Interpolation is a popular technique for mapping in GIS. In this case, the *spline* interpolation technique has been used. Incidentally, the *spline* forms a piecewise polynomial regression and forms an interpolated curve that passes through each sample point resulting in an ideal Interpolated surface. This will help the researcher to identify the zone of severe arsenic concentration and the pattern of groundwater arsenic concentration in the study area.

13.4.3 Depth-Wise Distribution of As

The depth-wise distribution of arsenic contamination has also been seriously considered in the present work. The tube well depths (N = 1535) are classified into six categories following a 20 m interval and are graphically plotted to show the relation between arsenic concentration and tube well depth. In a separate diagram depth class-wise average arsenic concentration and percentage of tube wells are represented. This will allow the readers to understand the spatial distribution of arsenic contamination properly.

13.4.4   Floodplain Morphology and As Contaminated Tube Well

Groundwater arsenic contamination has a close connection with the geomorphic elements of floodplain in the parts of the moribund delta of the Bhagirathi flood-plain (Mukherjee and Taraknath 2015). For analyzing the association between flood-plain morphology and arsenic contaminated tube wells, the overlay technique is adopted. The geomorphology map has been prepared by assimilating the survey of India topographical sheets (73A/2; 73A/3; 73A/6; 73A/7) and Shuttle Rader Topo-graphic Mission (SRTM) elevation data, and the coordinate locations of the tube wells with arsenic concentration above WHO and Bureau of India (BI) tolerance level are overlaid in a GIS environment and presented in the figure.

13.4.5   Litho-Stratigraphy and As Contaminated Tube Well

Ten boreholes data are collected from the office of the Agri Irrigation Department, Purbasthali I and II Blocks to understand the relationship between stratigraphy and tube well arsenic concentration. Boreholes are selected from each geological forma-tion. In this regard, four boreholes were selected from the *Katwa* formation where the maximum count of contaminated tube wells is found and three boreholes are selected from both the Present-day deposits and *Sijua* formation. The locations of the boreholes are mapped in the borehole site map Fig. 13.3. These borehole data are then plotted and mapped on a specific scale using AutoCAD software, and finally the As concentration values are plotted vertically along with borehole data with their corresponding depth. In this study, the tube well-wise As concentration of the corre-sponding villages of borehole locations are classified into five groups, viz. Arsenic Free (AF) having As concentration below WHO tolerance limit, i.e., ≤ 0.01 mg/L; Low Arsenic (LA) with As concentration ranging between 0.011 and 0.029 mg/L; Moderate Arsenic (MA) with As value 0.03–0.049 mg/L; High Arsenic (HA) with As concentration between 0.05 and 0.07 mg/L, and Very High Arsenic (VHA) with As magnitude ranging between >0.07 mg/L.

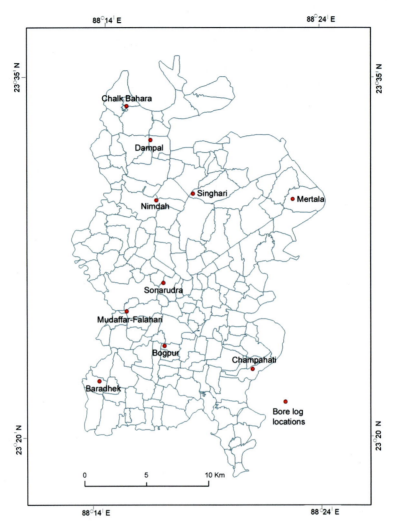

Fig. 13.3 Borehole site map showing the litholog location (based on Agri irrigation department information) in the west bank of Bhagirathi River, Purbasthali blocks, Bardhaman district, West Bengal, India

13.4.6 Seasonal Variation of As Concentration

The seasonal variability of arsenic concentration in tube well water is done by conducting a case study in *Mandra* village of Purbasthali I block. *Mandra* is chosen depending on its accessibility from the PHED laboratory. Total twenty (20) tube wells water samples are collected from *Mandra* in both pre and post- monsoon period of 2012. The sample sites are indicated in Fig. 13.4. The water samples are collected

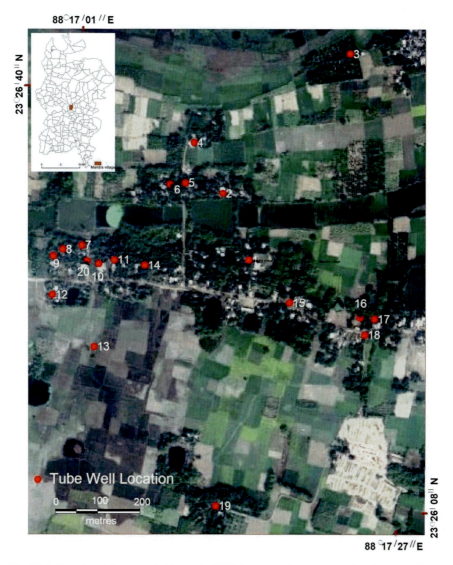

Fig. 13.4 Sample site locations (collected by GPS) for seasonal variation study in Mandra village, N = 20, west bank of Bhagirathi River, Purbasthali blocks, Burdwan district, West Bengal, India

with one drop of dilute nitric acid (1:1) AR grade and brought to the laboratory for testing. UV spectrophotometer is used to measure the level of As concentration. Finally, the seasonal variation in tube well As concentration is graphically presented.

13.5 Results and Discussion

13.5.1 Distribution of As Contaminated Tube Wells

The people of Purbasthali Blocks (I and II) are using tube wells for their daily needs and drinking purposes Plate 13.1. This tube well water contains dissolved arsenic ranging from 0.006 to 0.618 mg/L. For identifying arsenic contamination in tube well water at village level, the WHO threshold value is considered as the cut off and village-wise percentage of tube wells exceeding WHO tolerance level are calculated; a map based on the percentage data is prepared (Fig. 13.5). This map shows that villages located in the present and earlier meander belt of the Bhagirathi floodplain contain a significant count of tube wells with arsenic content ≥ 0.011 mg/L. It is found that most of the villages along the river, especially in the north, east-central, and east of the study area are contaminated. On the other hand, arsenic levels in the water of tube wells in most of the villages in the south and southwest of the region are within the prescribed tolerance level of WHO. The information presented in this

Plate 13.1 People of *Mandra* are using tube Well water for their daily needs, west bank of Bhagirathi River, Purbasthali 1, Bardhaman district, West Bengal, India

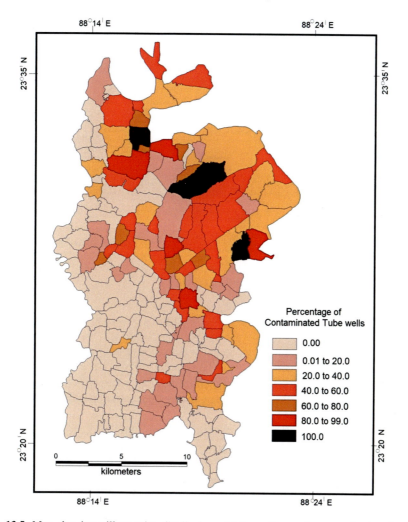

Fig. 13.5 Map showing village wise distribution of tube well arsenic contamination (in %) following WHO tolerance level, Purbasthali blocks, Bardhaman district West Bengal India

map is more thoroughly documented in Table 13.2. It illustrates contaminated tube wells (in %) and risk population and their potential regional extent. The table also depicts that 72% of the total population is living in fully or partially contaminated villages which occupy about 66% of the study area. About 11.80% of the population lives in villages where 60% or more tube wells contain arsenic levels ≥0.011 mg/L. It occupies more than 13% of the administrative area of study.

Table 13.2 Count of arsenic contaminated (following WHO tolerance level) tube wells and distribution of risk population & area in % values, western bank of Bhagirathi River, Purbasthali blocks, Bardhaman district, West Bengal, India

Tube well contamination (In %)	% of risk population	Cumulative % of risk population	% of risk area	Cumulative % of risk area
100	2.2	2.2	2.8	2.8
99.99–80.00	3.9	6.1	4.75	7.55
79.99–60.00	5.7	11.8	5.6	13.15
59.99–40.00	16.7	28.5	15.8	28.95
39.99–20.00	14.7	43.2	14.8	43.75
19.99–0.01	28.8	72	22	65.75
0	28	100	34.25	100

Source PHED water testing report (2005–2006), Census data (2011) and author's calculation

13.5.2 Mapping of As Concentration

The study of arsenic concentration in tube well water throughout the study area has been done based on PHED water samples of N = 1535 tube wells. It is found that 24.04% of tube wells have arsenic concentration above 0.05 mg/L and 48.34% of tube wells have arsenic levels above WHO tolerance limit, i.e., >0.01 mg/L (Table 13.3). The individual tube well arsenic concentration values are classified into six distinct classes based on the magnitude of arsenic concentration and represented in a concise manner in Table 13.3. Table 13.3 depicts the maximum (24.30%) tube wells arsenic concentration value, which ranges between >0.01 mg/L and 0.05 mg/L. It reports further that about 13.49% of tube wells (N = 1535) of arsenic concentration is above 0.101 mg/L.

Determining the level of arsenic in tube well water is a costly and time-consuming process. Therefore, mapping of arsenic concentration based on existing data is very important in identifying the spatial pattern of arsenic pollution in tube wells; which is a prerequisite for the management and rehabilitation purpose. An interpolated surface map of arsenic concentration has been created in GIS environment based on existing water testing data (N = 1535) and tube well coordinate locations. It illustrates the pattern of arsenic concentration in the tube well water of the entire study area. In

Table 13.3 Arsenic concentration in tube wells (mg/L) and their account in percentage value, N = 1535, west bank of Bhagirathi River, Purbasthali blocks, Bardhaman district, West Bengal, India

Arsenic (mg/l)	≤ 0.01	0.011–0.050	0.051–0.100	0.101–0.200	0.201–0.500	≥0.501
Percentage of tubewell	51.66	24.30	10.55	10.10	3.19	0.20

Source PHED water testing report (2005–2006)

this context, it should be noted that the level of dissolved arsenic in the sampled tube well is ranging between 0.006 and 0.618 mg/L. The interpolated surface map identified eleven chromatic zones separated by ten isolines of arsenic concentration at 0.01 mg/L intervals ranging from 0.01 to 0.1 mg / L. The maximum arsenic concentration is shown in deep red colour (Fig. 13.6). A pseudopodia arrangement is seen in the eastern and north-central part of the study area. The figure further depicts that high arsenic concentration patches (>0.100 mg/L) are surrounded in some cases by relatively low concentration contours and in some cases by reverse condition also prevails. It clearly shows that the distribution of arsenic in the area is quite varied.

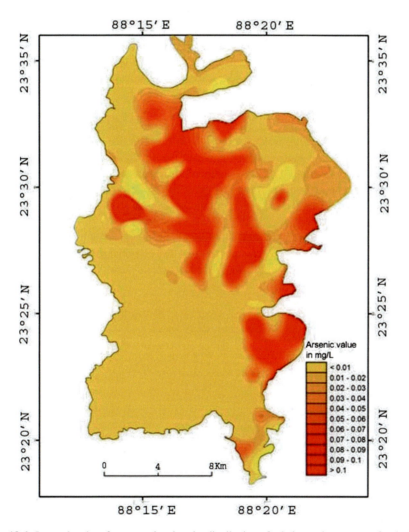

Fig. 13.6 Interpolated surface map showing the distribution of tubal arsenic concentration in the west bank of Bhagirathi River, purbasthali blocks, Bardhaman district, West Bengal, India

The variation in tube well arsenic concentration needs to be checked by other spatial elements like floodplain geomorphology and lithological sequences associations to explain the diverse formations.

13.5.3 Depth Wise Distribution of Arsenic

Distribution of arsenic concentration varies significantly according to the aquifer depth. The sampled (N = 1535) tube wells depth varies from 06 to 123 m. It is seen that most of the tube wells of the study area (about 52.77%) are buried at a depth of 40 to 80 m (Table 13.4).

The depth-wise distribution of arsenic contaminated (>0.01 mg/L) tube wells and the amount of dissolved arsenic at different aquifers is clearly shown in Table 13.4. It shows that tube wells with a depth of less than 60 m have more than 50 percent of their sample is above WHO recommended limit (i.e., (>0.01 mg/L). However, with increasing depth, a trend of decreasing percentage of tube wells with arsenic concentration value exceeding >0.01 mg/L are found. The magnitude of arsenic concentration is alarming at all depth classes. But, the maximum value is seen at 20–39.9 m depth. For better understanding of the depth-wise magnitude of arsenic concentration, a separate diagram is prepared in Fig. 13.7. It represents the depth-wise average arsenic concentration and percentage of tube wells board. The figure shows that the 20–59.9 m depth zone contains an average arsenic concentration above 0.04 mg/L and carries about 42.22% (N = 1535) tube wells.

Finally, scatter plots of tube well-wise dissolved arsenic value with corresponding aquifer depth are presented graphically to show the relationship of arsenic concentration with increasing depth (Fig. 13.8). There lies a slight negative correlation between arsenic concentration and depth.

Table 13.4 Depth wise distribution of arsenic contaminated tube wells in percentage (following WHO standards) and maximum arsenic concentration in the tube well water, west bank of Bhagirathi River, Purbasthali blocks, Bardhaman district, West Bengal, India

Tube well depth class	Total no of tube wells	Contaminated tube wells (in %)	Safe tube wells	Maximum As concentration (mg/L)
<20	104	55.77	44.23	0.249
20.0–39.9	290	72.41	27.59	0.618
40.0–59.9	358	51.68	48.32	0.348
60.0–79.9	452	39.16	60.84	0.572
80.0–99.9	285	35.44	64.56	0.500
≥100	46	23.91	76.09	0.073
Total	1535	48.34	51.66	0.618

Source PHED water testing report (2005–2006)

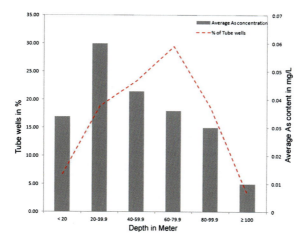

Fig. 13.7 Depth wise distribution of of tube Wells (in %) and average arsenic concentration (mg/L) along death zones in the west bank of Bhagirathi river, purbasthali blocks, Bardhaman district, West Bengal, India

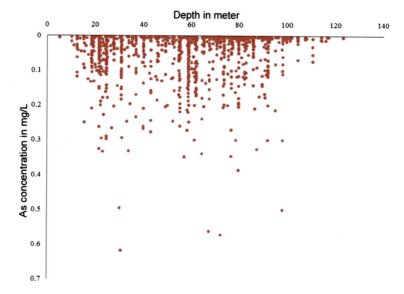

Fig. 13.8 Scatter diagram showing the relation between Tube well depth (meter) and arsenic concentration (mg/L), N = 1535, west bank of Bhagirathi river, Purbasthali blocks Bardhaman district, West Bengal, India

13.5.4   Geomorphic Control

Geomorphologically the area has been divided into three distinct zones carrying different fluvio geomorphic landform assemblage and relief features (Fig. 13.2). These are presently active meander belt, earlier meander belt of Bhagirathi flood-plain, and Damodar fan toe. The present-day active meander belt lies along the active channel of Bhagirathi and consists of present channel, point bars, ox-bow lakes, avulsions, and present-day floodplain. The earlier meander belt of Bhagirathi River is a relatively older geomorphic surface occupying the intermediate zone between present-day meander belt and Damodar fan toe. This zone is ornamented with flood-plains, paleochannels, levees, meander scar, back swamps, floodplain marshes, etc. The rest lies in the south-west corner of the study area. It contains previous spill channels of Damodar River, low-lying swamps, floodplain, and active channel of Khari River.

Quaternary litho-stratigraphy map also finds close association with geomorphic units. The regional quaternary lithology map of the area depicts four distinct zones. These are present-day deposits, *Katwa* formation, *Chinsurah* formation, and *Sijua* formation. It shows that the present-day active meander belt is represented by *Present-day deposits* of Bhagirathi River consisting of sand, silt, and clay. The earlier meander belt of Bhagirathi is characterized by uniform lithology of fining upward sequence of sandy alluvial deposits deposited during the Holocene period. The lithology varies from boulder, pebble, sand, silt, and clay and is locally designated as *Katwa* forma-tion. The *Chinsurah* formation is totally absent in Purbasthali Blocks (I and II) and no tube well samples are present in this unit, thus little attention is paid to it. It consists of yellowish-brown sand, silt, and clay deposits of alluvial origin. The *Sjua* formation lying in the south-western part of the area is overlain with older deltaic deposits of Damodar River and its spill channels. It contains oxidized coarser sandy deposits with kankar. The tube wells with arsenic concentration above 0.01 mg/L are overlaid on the fluvial geomorphology map (Fig. 13.9). The figure depicts that the maximum percentage of tube wells with arsenic concentration above 0.01 mg/L are occurring in the earlier meander belt of Bhagirathi River and has been associated with floodplain levees, meander scar, paleochannels, back swamps, etc. The overlay analysis shows that among the 742 tube wells (48.40% of N = 1535) that exceed the WHO tolerance limit (i.e. >0.01 mg/L), almost 83.00% of tune wells are occur-ring in this geomorphic unit. The present-day meander belt contains only 15.63% of contaminated tube wells. On the contrary, the Damodar fan toe associated with *Sijua* formation is almost safe and contains only 1.35% of contaminated tube wells.

13.5.5   Stratigraphic Control

Three sets of lithologs are selected from three major Quaternary litho-units of the study area in 3:4:3 manner as stated in the methodology section; i.e., present-day

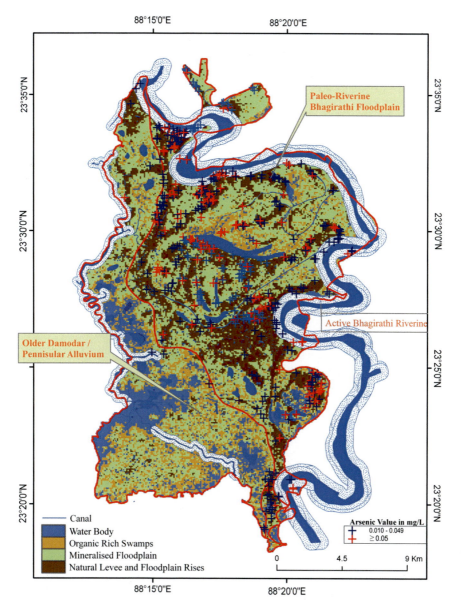

Fig. 13.9 Floodplain geomorphology map showing spatial associations between tube well with arsenic concentration >0.01 mg/L and geomorphic elements in the west bank of Bhagirathi river, Purbasthali blocks, West Bengal, India

younger alluvial deposits (3), earlier meander belt represented by Katwa formation, and Sijua formation overlain with older alluvial deposits. The sediment colouration varies between grey, brown, and yellow in tone. The western and south-western part of the area is dominated by brown sediments. This brown sediment is mostly deposited by the Damodar River and its spill channels and is rich in illite, siderite, and iron coated oxyhydroxide (Mukherjee 2010b). On the contrary, the fluviatile sequence of the present-day Bhagirathi meander belt consists of yellow to grey sand and alternate arrangement of sandy channel fills and sticky over bank deposits with the presence of kaolinite and semi decomposed organic remains (Mukherjee 2010b). In Katwa formation, the lithologs are dominated by intermittent overlapping of brown and grey sediments with irregular channel fills.

The lithologs along with neighbourhood tube well arsenic concentration value plotting in respective aquifers provide basic information about the stratigraphic control and help in exploring the pattern of arsenic contamination and its spatial association in shallow aquifers.

In the previous section, it is observed that *Katwa* formation consists of the maximum percentage of contaminated tube wells. From *Katwa* formation, four lithologs are chosen to correlate with tube well arsenic concentration and underlying aquifers (Fig. 13.10). These are *Champahati, Sonarudra, Nimdah,* and *Singhari* mouza. The overall lithology has a fining upward sequence. The borehole depth is ranging from 100 to 125 m below ground level. Three to five aquifers are identified. They consist of brown and grey sediments. The aquifers with grey sediments are frequently associated with dissolved arsenic >0.01 mg/L. The neighbourhood

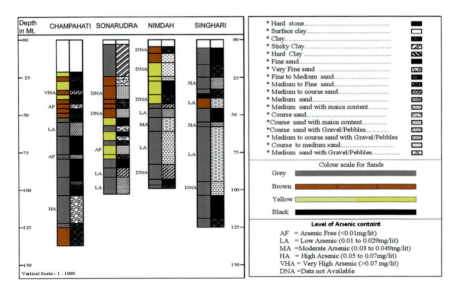

Fig. 13.10 Quaternary lithologs and allocation of neighbourhood tube wells arsenic concentration value in *Katwa* formation: associational study Indo Western bank of Bhagirathi River, purbasthali blocks, Bardhaman district, West Bengal, India

tube wells' arsenic concentration value ranges between 0.006 and 0.274 mg/L. The lithology of *Sonarudra* has typical thick sticky clay capping at the top and characterized by low arsenic concentration in the neighbourhood tube wells. In case of the other three lithologs, the clay capping is relatively thin and the arsenic safe (≤0.010 mg/L) contaminated (>0.01 mg/L) aquifers are found alternatively with increasing depth up to 100 m.

The lithologs from *Present-day* deposits also have fining upward sequence of sand silt and clay but the brown sand coated with iron oxyhydroxide is totally absent here. Borehole depth ranges from 92 to 116 m. Grey sediments dominate in all three lithologs from chalk bahara, dompal, and mertola (Fig. 13.11). The colour tone of sediments varies from yellow to grey. Arsenic concentration in adjoining tube well ranges between 0.006 and 0.055 mg/L. The alternate arrangement of safe and contaminated aquifers also prevails here.

The lithologs from *Sijua* formation depth range between 106 and 125 m and are dominated by sandy alluvial deposits (Fig. 13.12). The aquifers are found at relatively deeper depths. Sediment character varies between oxidized older alluviums to recent time channel deposits of Damodar spill channels. The aquifers are mostly of Pleistocene origin. It is important to mention that here all the neighbouring tube wells' arsenic concentration value is restricted within WHO permissible limit, i.e. ≤0.01 mg/L and is totally safe in respect of arsenic contamination.

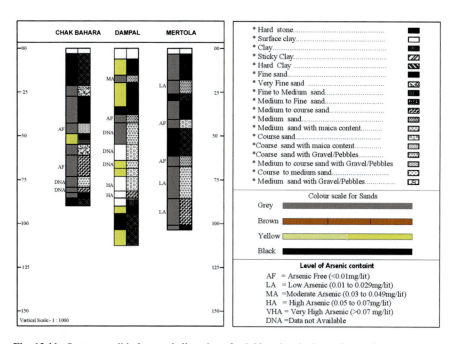

Fig. 13.11 Quaternary lithologs and allocation of neighbourhood tube wells arsenic concentration value in recent formation of *Present-day* deposits of Bhagirathi River (west bank), purbasthali blocks, Bardhaman district, West Bengal, India

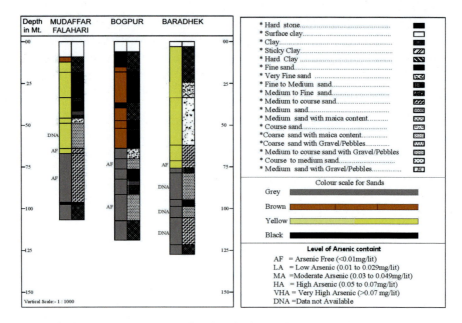

Fig. 13.12 Quaternary lithologs and allocation of neighbourhood tube wells arsenic concentration value in *Sijua* formation, west bank of Bhagirathi River, purbasthali blocks, Bardhaman district, West Bengal, India

13.5.6   Seasonal Variation of As Concentration: A Case Study

Seasonal variation of arsenic concentration is an important aspect of groundwater arsenic research (Thundiyil et al. 2007). The seasonal variation study, based on twenty tube wells of Mandra village, depicts that in 25% of samples, the post-monsoon arsenic concentration exceeds the pre-monsoonal value (Table 13.5). The maximum concentration is found in sample 4 with post-monsoon As concentration value of 0.052 mg/L which is 0.031 mg/L higher than the respective pre-monsoonal value (Table 13.5). In 70% of samples, the pre-monsoonal As concentration overrides the post-monsoonal value (Table 13.5). The maximum variation is found in sample 18; where pre-monsoonal value (0.101 mg/L) is 0.047 mg/L higher than the respective post-monsoonal amount (Table 13.5). Another interesting fact is in 5% samples both pre-monsoon and post-monsoon As concentration did not vary at all and the magnitude of concentration is also as high as 0.152 mg/L Table 13.5. The graphical presentation of pre-monsoon and post-monsoon As concentrations are represented in the line graph (Fig. 13.13).

Table 13.5 Seasonal variation of arsenic concentration (in mg/L): a case study on *Mantra* village, N = 20, west bank of Bhagirathi River, Purbasthali I block, Bardhaman district, West Bengal, India

Samples	Pre-monsoon As concentration (mg/L)	Post-monsoon As concentration (mg/L)	Samples	Pre-monsoon As concentration (mg/L)	Post-monsoon As concentration (mg/L)
Sp 1	0.012	0.014	Sp 11	0.024	0.013
Sp 2	0.016	0.008	Sp 12	0.005	0.007
Sp 3	0.010	0.004	Sp 13	0.009	0.003
Sp 4	0.021	0.052	Sp 14	0.020	0.011
Sp 5	0.020	0.025	Sp 15	0.059	0.044
Sp 6	0.010	0.003	Sp 16	0.030	0.018
Sp 7	0.042	0.030	Sp 17	0.152	0.152
Sp 8	0.013	0.010	Sp 18	0.101	0.054
Sp 9	0.006	0.007	Sp 19	0.074	0.062
Sp 10	0.012	0.003	Sp 20	0.011	0.008

Source Primary sampling survey and laboratory testing of water samples at PHED lab. Nadanghat, Bardhaman district

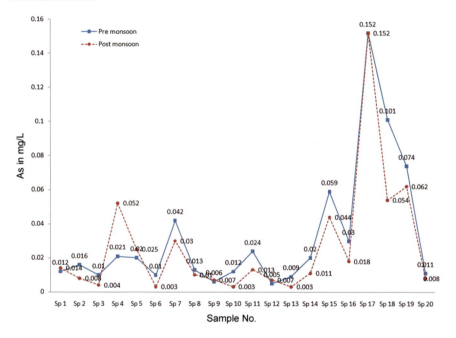

Fig. 13.13 Line graph showing pre-monsoon and post-monsoonal arsenic concentration in the tube Wells of *Mandra* village, N = 20, west bank of Bhagirathi River, purbasthali I block, Bardhaman district, West Bengal, India

13.5.7   Severity and Impact of As Contamination

The spread of arsenic toxicity and its risk are depicted well in Table 13.6. It is seen that more than 50% population at ten gram panchayats out of seventeen (58.82%), Purbasthali blocks (I and II) are at risk (Table 13.6). Four gram panchayat is found with a cent percent population living at risk (Table 13.6). Arsenic in tube well water and its prolonged exposure (more than six months) may cause different carcinogenic and non-carcinogenic health impacts (Mazumder and Dasgupta 2011; Mazumder 2015). The glimpses of the environmental pandemic are represented in Plate 13.2. The most common manifestation among the victims is skin lesions of different grades (Mazumder 2000). Besides this, arsenic often triggers cardiovascular diseases (Wu et al. 1989), diabetes (Rahman and Tondel 1998), pulmonary problems (Mazumder 2000; Ondine and Ehrenstein 2005). The extreme effect of arsenic contamination leads to cancers of skin, lungs, kidney, liver, etc. (Chen et al. 1992; Smith et al. 1998). It also causes several socio-economic consequences on the victims and population

Table 13.6 Gram Panchayat wise distribution of maximum arsenic concentration in the tube well water (mg/L), distribution of risk population (in %) and its spatial extent (% of GP area) in Purbasthali blocks located in the western bank of Bhagirathi River, Bardhaman district, West Bengal, India

Block name	Gram Panchayat	Total population	Unsafe (In %)	Total area (sq km)	Unsafe (In %)	Maximum As concentration (mg/L)
PURBASTHAL- I	Bogpur	28,404	0.00	33.74	0.00	0.016
	Nasaratpur	45,164	58.45	12.28	37.14	0.386
	Jahannagar	20,287	28.68	15.57	21.29	0.199
	Dogachia	24,221	5.40	28.28	3.30	0.158
	Samudragarh	32,207	68.27	14.60	57.38	0.333
	Srirampur	29,473	100.00	10.86	100.00	0.618
	Nadanghat	27,641	0.00	33.11	0.00	0.006
PURBASTHALI–II	Kalekhantola-I	30,373	100.00	25.48	100.00	0.348
	Kalekhantola-II	17,918	68.12	10.65	64.62	0.572
	Majida	31,462	53.90	32.26	38.70	0.495
	Mertola	10,495	84.93	18.25	79.52	0.247
	Mukshimpara	24,720	19.48	27.31	21.73	0.195
	Nimdaho	32,802	80.03	27.17	83.01	0.248
	Patuli	13,381	65.02	1076	67.65	0.249
	Pilla	24,990	31.80	21.42	38.10	0.184
	Purbasthali	20,351	100.00	10.51	100.00	0.339
	Jhaudanga	5443	100.00	08.68	100.00	0.078
Total		419,332	72.00	340.91	65.75	0.618

Source PHED water testing report and Census data (2011)

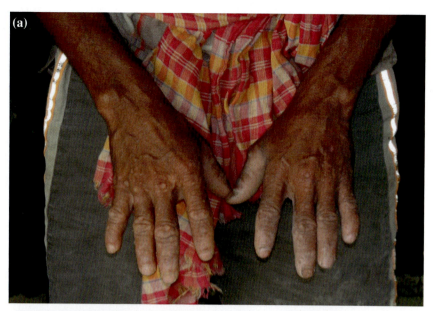

Plate 13.2 Health impacts on the people associated with chronic arsenic exposure in the western bank of Bhagirathi River, Purbasthali blocks, Bardhaman district, West Bengal, India. **a**. Keratosis skin manifestation due to consumption of As laden tube well water. **b**. Skin cancer: a virulent carcinogenic manifestation

at risk (Hassan et al. 2005). Social isolation often hampers conjugal life and induces morbidity among the victims (Mandal 1998; Chowdhury et al. 2006). The poor people are sometimes burdened up with the treatment of arsenicosis symptoms and ill effects (Sunam et al. 2010). They lose all their savings and assets in fighting against the silent killer. They are suffering from dietary deficiencies, lack of alternative safe sources, and increasing financial constraints (Sunam et al. 2010). All these make the situation potentially grave.

13.6   Conclusions

Tube well arsenic concentrations above 0.01 mg/L are occurring along the present and past meander belt of Bhagirathi River affecting 72% of the population of that area. The villages of Purbasthali Blocks lying in the western boundary line and south-western corner are safe. The distribution of arsenic in tube wells is extremely heterogeneous in nature with large differences in the magnitude of As concentration. The vertical distribution of arsenic along tube well depth shows that more than 50% of shallow tube wells, buried below 60 m ground level are found to be contaminated following WHO tolerance limit, i.e., 0.01 mg/L. Another interesting finding is the amount of dissolved arsenic concentration, which is erratic in nature, with more than 0.201 mg/L As concentration found at all aquifers up to hundred meters depth. The seasonal variation study reveals that pre-monsoonal concentration of As in tube wells is higher than the post-monsoonal value, though some reverse scene is also present in 25% of samples. This changing As concentration might have a significant influence among the risk population. The spatial association study explores that both the contaminated tube wells (>0.01 mg/L) and tube wells with higher As concentration value used to occur in the Bhagirathi floodplain specifically restricted within the earlier meander belt of the river. The late Quaternary *Katwa* formation lithologically built up by younger deltaic deposits represents this floodplain unit. The curvilinear floodplain elements like paleochannels meander scars, relict channels, back swamp, earlier levees, etc. have a close association with contaminated tube wells. The stratigraphic correlation finds that younger alluvial deposits with grey sands deposited during the Holocene period are typically contaminated with arsenic; In contrast, the Pleistocene aquifers of *Sijua* formation are almost safe.

References

Acharyya SK, Shah BA (2007) Arsenic-contaminated groundwater from parts of Damodar fan-delta and west of Bhagirathi River, West Bengal, India: influence of fluvial geomorphology and Quaternary morphostratigraphy. Environ Geol 52:489–501. https://doi.org/10.1007/s00254-006-0482-z

Acharyya S, Lahiri S, Raymahashay B et al (2000) Arsenic toxicity of groundwater in parts of the Bengal basin in India and Bangladesh: the role of quaternary stratigraphy and Holocene sea-level fluctuation. Environ Geol 39:1127–1137. https://doi.org/10.1007/s002540000107

Agusa T, Kubota R, Kunito T, Minh TB, Trang PTK, Chamnan C, Tanabe S (2007) Arsenic pollution in groundwater of Vietnam and Cambodia: a review. Biomed Res Trace Elements 18(1):35–47. https://www.researchgate.net/deref/253A%252F%252Fdx.doi.org%252 F10.11299%252Fbrte.18.35

Ali W, Rasool A, Junaid M, Zhang H (2019) A comprehensive review on current status, mechanism, and possible sources of arsenic contamination in groundwater: a global perspective with prominence of Pakistan scenario. Environ Geochem Health 41(2):737–760. https://doi.org/10. 1007/s10653-018-0169-x

Allison M (1998) Geologic framework and environmental status of the ganges-Brahmaputra Delta. J Coast Res 14(3), 827–836. Retrieved June 21, 2020, from www.jstor.org/stable/4298836

Bhattacharya P, Chatterjee D, Jacks G (1997) Occurrence of arsenic-contaminatedgroundwater in alluvial aquifers from delta plains, Eastern India: options for safe drinking water supply. Int J Water Resour Dev 13(1):79–92. https://doi.org/10.1080/07900629749944

Bhattacharyya D, Mukherjee PK (2009) Contamination of shallow aquifers by arsenic in upper reaches of Tista river at Siliguri-Jalpaiguri area of West Bengal India. Environ Geol 57:1687–1692. https://doi.org/10.1007/s00254-008-1450-6

Biswas B (2010) Geomorphic controls of arsenic in ground water in Purbasthali I & II blocks of Burdwan District, West Bengal, India. Int J Environ Sci 1(4):419–439. http://www.environmentp ortal.in/files/Arsenic%20in%20Ground%20Water.pdf

Chakraborti D, Rahman MM, Das B et al (2013) Groundwater arsenic contamination in Ganga–Meghna–Brahmaputra plain, its health effects and an approach for mitigation. Environ Earth Sci 70:1993–2008. https://doi.org/10.1007/s12665-013-2699-y

Chakraborti D, Singh SK, Rahman MM, Dutta RN, Mukherjee SC, Pati S, Kar PB (2018) Groundwater arsenic contamination in the Ganga River Basin: a future health danger. Int J Environ Res Public Health 15(2):180. https://doi.org/10.3390/ijerph15020180

Chakraborty M, Mukherjee A, Ahmed KM (2015) A Review of Groundwater Arsenic in the Bengal Basin, Bangladesh and India: from Source to Sink. Curr Pollution Rep 1:220–247. https://doi. org/10.1007/s40726-015-0022-0

Chatterjee D, Mazumder M, Barman S, Adhikari J, Kundu A, Mukherjee A, Chatterjee D (2018) Arsenic in groundwater: distribution and geochemistry in Nadia District, West Bengal, India. In: Mukherjee A (ed) Groundwater of South Asia. Springer Hydrogeology. Springer, Singapore, https://doi.org/10.1007/978-981-10-3889-1_22

Chen CJ, Chen CW, Wu MM, Kuo TL (1992) Cancer potential in liver, lung, bladder and kidney due to ingested inorganic arsenic in drinking water. Br J Cancer 66(5):888–892. https://doi.org/ 10.1038/bjc.1992.380

Chowdhury UK, Biswas BK, Chowdhury TR, Samanta G, Mandal BK, Basu GC, Roy S (2000) Groundwater arsenic contamination in Bangladesh and West Bengal India. Environ Health Perspect 108(5):393–397. https://doi.org/10.1289/ehp.00108393

Chowdhury MAI, Uddin MT, Ahmed MF, Ali MA, Rasul, SMA, Hoque MA, Islam MS (2006) Collapse of socio-economic base of Bangladesh by arsenic contamination in groundwater. https:// agris.fao.org/agris-search/search.do?recordID=AV20120134090

Edmunds WM, Ahmed KM, Whitehead PG (2015) A review of arsenic and its impacts in groundwater of the Ganges–Brahmaputra–Meghna deltaBangladesh. Environ Sci Process Impacts 17(6):1032–1046. https://doi.org/10.1039/C4EM00673A

Francisca FM, Perez MEC (2009) Assessment of natural arsenic in groundwater in Cordoba ProvinceArgentina. Environ Geochem Health 31(6):673. https://doi.org/10.1007/s10653-008-9245-y

George CM, Sima L, Arias M, Mihalic J, Cabrera LZ, Danz D, Gilman RH (2014) Arsenic exposure in drinking water: an unrecognized health threat in Peru. Bull World Health Organ 92(8): 565–572. ISSN 0042–9686. https://doi.org/10.2471/BLT.13.128496

GuhaMazumder DN (2003) Chronic arsenic toxicity: clinical features, epidemiology, and treatment: experience in West Bengal. J Environ SciHealth Part a: Toxic/hazardous SubsEnviron Eng 38:141–163. https://doi.org/10.1081/ESE-120016886

Guo H, Wen D, Liu Z, Jia Y, Guo Q (2014) A review of high arsenic groundwater in Mainland and Taiwan, China: distribution, characteristics and geochemical processes. Appl Geochem 41:196–217. https://doi.org/10.1016/j.apgeochem.2013.12.016

Hassan MM, Atkins PJ, Dunn CE (2005) Social implications of arsenic poisoning in Bangladesh. Soc Sci Med 61(10):2201–2211. https://doi.org/10.1016/j.socscimed.2005.04.021

Hazra DH (2010). Occurrence of arsenic in groundwater and its relation to aquifer sediment characteristics in Murshidabad district, West Bengal. Indian J Geosci 61–76

Herath I, Vithanage M, Bundschuh J et al (2016) Natural arsenic in global groundwaters: distribution and geochemical triggers for mobilization. Curr Pollution Rep 2:68–89. https://doi.org/10.1007/s40726-016-0028-2

Hong YS, Song KH, Chung JY (2014) Health effects of chronic arsenic exposure. J Prevent Med Public Health 47(5):245. https://doi.org/10.3961/jpmph.14.035

Islam F, Gault A, Boothman C et al (2004) Role of metal-reducing bacteria in arsenic release from Bengal delta sediments. Nature 430:68–71. https://doi.org/10.1038/nature02638

Mandal BK (1998) Status of arsenic problem in two blocks out of sixty in eight groundwater arsenic affected districts of West Bengal, India. Doctoral dissertation, PhD thesis, Jadavpur University

Mazumder DG (2000) Diagnosis and treatment of chronic arsenic poisoning. In: United Nations synthesis report on arsenic in drinking water. https://www.who.int/water_sanitation_health/dwq/arsenicun4.pdf

Mazumder DG (2015) Health effects chronic arsenic toxicity. In Handbook of arsenic toxicology. Academic Press, pp. 137–177. https://doi.org/10.1016/B978-0-12-418688-0.00006-X

Mazumder DG, Dasgupta UB (2011) Chronic arsenic toxicity: studies in West Bengal, India. Kaohsiung J Med Sci 27(9):360–370. https://doi.org/10.1016/j.kjms.2011.05.003

Merola RB, Kravchenko J, Rango T, Vengosh A (2014) Arsenic exposure of rural populations from the Rift Valley of Ethiopia as monitored by keratin in toenails. J Eposure Sci Environ Epidemiol 24(2):121–126. https://doi.org/10.1038/jes.2013.77

Mueller B (2017) Arsenic in groundwater in the southern lowlands of Nepal and its mitigation options: a review. Environ Rev 25(3):296–305. https://doi.org/10.1139/er-2016-0068

Mukherjee A, Fryar AE, Thomas WA (2009) Geologic, geomorphic and hydrologic framework and evolution of the Bengal basin, India and Bangladesh. J Asian Earth Sci 34(3):227–244. https://doi.org/10.1016/j.jseaes.2008.05.011

Mukherjee P (2010) Studies on distribution pattern of arsenic in groundwater and sediments from Nadia and 24-Paraganas (N) districts, West Bengal. Indian J Geosci 64(1–4):91–98. https://www.researchgate.net/publication/321490216

Mukherjee P (2010) Search for groundwater arsenic in Pleistocene sequence of the Damodar River flood plain, West Bengal. https://www.researchgate.net/publication/275965647

Mukherjee P (2010) Characterisation of "orange-sand" vis-à-vis "grey-sand" from the arsenic-affected areas of West Bengal. https://www.researchgate.net/publication/275965501

Mukherjee PK, Pal T, Chattopadhyay S (2015) Role of geomorphic elements on distribution of arsenic in groundwater—A case study in parts of Murshidabad and Nadia districts, West Bengal. https://doi.org/10.13140/RG.2.1.4601.8080

Mukhopadhyay DK, Ghosh G (2010) Arsenic pollution of groundwater in parts of West Bengal-A case study. Indian J Geosci 64(1–4):41–48

Nickson R, Sengupta C, Mitra P, Dave SN, Banerjee AK, Bhattacharya A, Kumar M (2007) Current knowledge on the distribution of arsenic in groundwater in five states of India. J Environ Sci Health Part A 42(12):1707–1718. https://doi.org/10.1080/10934520701564194

Von Ehrenstein OS, Mazumder DG, Yuan Y, Samanta S, Balmes J, Sil A, Smith AH (2005) Decrements in lung function related to arsenic in drinking water in West Bengal, India. Am J Epidemiol 162(6):533–541 https://doi.org/10.1093/aje/kwi236

Purkait B (2018) Arsenic in Bengal Delta. Geograph Rev India 80:95–121. https://www.researchg
ate.net/publication/330934694

Rahman MM, Mandal BK, Chowdhury TR, Sengupta MK, Chowdhury UK, Lodh D, Chakraborti D
(2003) Arsenic groundwater contamination and sufferings of people in North 24-Parganas, one of
the nine arsenic affected districts of West Bengal, India. J Environ Sci Health Part A 38(1):25–59.
https://doi.org/10.1081/ESE-120016658

Rahman M, Tondel M, Ahmad SA, Axelson O (1998) Diabetes Mellitus associated with arsenic
exposure in Bangladesh. Am J Epidemiol 148(2):198 203 https://doi.org/10.1093/oxfordjournals.
aje.a009624

Rotiroti M, Sacchi E, Fumagalli L, Bonomi T (2014) Origin of arsenic in groundwater from the
multilayer aquifer in Cremona (northern Italy). Environ Sci Technol 48(10):5395–5403. https://
doi.org/10.1021/es405805v

Rudra K (2014) Changing river courses in the western part of the Ganga-Brahmaputra delta.
Geomorphology 227:87–100. https://doi.org/10.1016/j.geomorph.2014.05.013

Saha JC, Dikshit AK, Bandyopadhyay M, Saha KC (1999) A review of arsenic poisoning and its
effects on human health. Crit Rev Environ Sci Technol 29(3):281–313. https://doi.org/10.1080/
10643389991259227

Sankar MS, Vega MA, Defoe PP, Kibria MG, Ford S, Telfeyan K, Datta S (2014) Elevated arsenic
and manganese in groundwaters of Murshidabad, West Bengal, India. Sci Total Environ 488
https://doi.org/10.1016/j.scitotenv.2014.02.077

Shah BA (2008) Role of Quaternary stratigraphy on arsenic-contaminated groundwater from parts
of Middle Ganga Plain, UP–Bihar India. Environ Geol 53:1553–1561. https://doi.org/10.1007/
s00254-007-0766-y

Shah BA (2017) Groundwater arsenic contamination from parts of the Ghaghara Basin, India:
influence of fluvial geomorphology and quaternary morphostratigraphy. Appl Water Sci 7:2587–
2595. https://doi.org/10.1007/s13201-016-0459-3

Shahid M, Niazi NK, Dumat C et al (2018) A meta-analysis of the distribution, sources and health
risks of arsenic-contaminated groundwater in Pakistan. Environ Pollut 242(Pt A):307–319. https://
doi.org/10.1016/j.envpol.2018.06.083

Shrivastava A, Barla A, Yadav H, Bose S (2014) Arsenic contamination in shallow groundwater
and agricultural soil of Chakdaha block, West Bengal India. Front Environ Sci 2:50. https://doi.
org/10.3389/fenvs.2014.00050

Singh SK (2015) Groundwater arsenic contamination in the middle-gangetic plain, Bihar (India):
the danger arrived. Int Res J Environ Sci 4:70–76. https://www.researchgate.net/publication/277
012220

Smith AH, Goycolea M, Haque R, Biggs ML (1998) Marked increase in bladder and lung cancer
mortality in a region of Northern Chile due to arsenic in drinking water. Am J Epidemiol
147(7):660–669. https://doi.org/10.1093/oxfordjournals.aje.a009507

Smith AH, Lingas EO, Rahman M (2000) Contamination of drinking-water by arsenic in
Bangladesh: a public health emergency. Bull World Health Organ 78:1093–1103. https://www.
scielosp.org/article/bwho/2000.v78n9/1093-1103/en/

Steinmaus CM, Ferreccio C, Romo JA, Yuan Y, Cortes S, Marshall G, Smith AH (2013) Drinking
water arsenic in northern Chile: high cancer risks 40 years after exposure cessation. Cancer
Epidemiol Prevent Biomark 22(4):623–630. https://doi.org/10.1158/1055-9965.EPI-12-1190Pu
blishedApril2013

Stüben D, Berner Z, Chandrasekharam D, Karmakar J (2003) Arsenic enrichment in groundwater
of West Bengal, India: geochemical evidence for mobilization of as under reducing conditions.
Appl Geochem 18(9):1417–1434. https://doi.org/10.1016/S0883-2927(03)00060-X

Sunam C, Mousumi D, Prakasam C, Biplab B (2010) Socio-economic implications of arsenic
toxicity in Purbasthali I and II blocks of Burdwan district. Environ Ecol 28(4A):2426–2432

Thundiyil JG, Yuan Y, Smith AH, Steinmaus C (2007) Seasonal variation of arsenic concentration
in wells in Nevada. Environ Res 104(3):367–373. https://doi.org/10.1016/j.envres.2007.02.007

Verma S, Mukherjee A, Mahanta C, Choudhury R, Badoni RP, Joshi G (2019) Arsenic fate in the Brahmaputra river basin aquifers: controls of geogenic processes, provenance and water-rock interactions. Appl Geochem 107:171–186. https://doi.org/10.1016/j.apgeochem.2019.06.004

Winkel LH, Trang PTK, Lan VM, Stengel C, Amini M, Ha NT, Berg M (2011) Arsenic pollution of groundwater in Vietnam exacerbated by deep aquifer exploitation for more than a century. Proc Natl Acad Sci 108(4):1246–1251. https://doi.org/10.1073/pnas.1011915108

Wu MM, Kuo TL, Hwang YH, Chen CJ (1989) Dose-response relation between arsenic concentration in well water and mortality from cancers and vascular diseases. Am J Epidemiol 130(6): 1123–1132. https://doi.org/10.1093/oxfordjournals.aje.a115439

Chapter 14
Spatial Pattern of Groundwater Depletion, Its Access and Adaptive Agricultural Strategies in Barddhaman District, West Bengal, India

Biswajit Ghosh and Namita Chakma

Abstract Quantification of spatio-temporal distribution of natural groundwater recharge with respect to different geohydrological settings is important for efficient management of this resource. Based on secondary data the present work intends to study the spatial pattern of groundwater level and natural groundwater recharge by using the soil moisture balance approach in five selected Community Development Blocks of Barddhaman district, West Bengal with different geohydrological settings. It also documents groundwater depletion, resultant changes in irrigation technology and adaptation approaches taken by farmers to minimise groundwater exhaustion based on primary data collected from two 'semi-critical' blocks and 22 villages. It has been found that rainfall in monsoon period is the key factor for the recharge of groundwater and the amount of annual recharge varies from 10 to 20% except in Jamuria. Deepening of bore well due to groundwater depletion started in 1980s and still continuing. Temporal pattern of change to submersible pump from centrifugal one represents a S-shaped curve. Water saving methods are limited within a very small part of the total operational holding. Promotion of water saving technologies and practices may reduce groundwater depletion and improve access to the small holders.

Keywords Groundwater recharge · Groundwater depletion · Soil moisture balance · Geo-hydrology · Bore well

14.1 Introduction

Increasing population and demand of food are accelerating the stretch of agricultural dimension which in turn are creating severe stress on availability of water (Singh 2011). But, nearly two fold increase will be required in food production by mid of this

B. Ghosh (✉)
Khorod Amina High School, Satgachia, Barddhaman, West Bengal, India

N. Chakma
Department of Geography, The University of Burdwan, Barddhaman, West Bengal 713 104, India

© The Author(s), under exclusive license to Springer Nature Singapore Pte Ltd. 2022   273
N. C. Jana and R. B. Singh (eds.), *Climate, Environment and Disaster in Developing Countries*, Advances in Geographical and Environmental Sciences, https://doi.org/10.1007/978-981-16-6966-8_14

century (FAO 2015). This will indeed create additional pressure on already stressed water resource due to which ecological supply system and ecological balance of water will get destroyed (Huang et al. 2018; Gleeson et al. 2012; Steenbergen et al. 2015). Therefore, estimation and conservation of this renewable resource need a careful attention at this juncture for sustenance of human being.

Renewability groundwater is determined by its recharge rate that is used to estimate groundwater resources varying over time and space (Kumar and Singh 2011; Kumar 2012). Apparently, quantification of spatial and temporal rate of recharge of groundwater should be the fundamental pre-requisite for sustainable utilisation of this important natural resource particularly in those areas where groundwater is in high demand as the key natural resource to economic development (Kumar 2003).

This study attempts to estimate a comprehensive and holistic picture of the recharge of groundwater and spatial pattern of distribution of groundwater level in association with geohydrological condition and climatic parameters of Barddhaman district, West Bengal. On the other side it also documents groundwater depletion, resultant changes in water exhausting irrigation methods and farmers tendency towards water saving farming strategies in the selected study area.

14.2 The Study Area

Barddhaman district holds a prime position in rice production in West Bengal, India and therefore, precisely known as 'rice bowl' of the state. In the district about three-fifth of the population is engaged in agricultural and allied activities (Census of India 2011). The hydrogeological setting of Barddhaman district has created abundant groundwater resources. But intensive uses of groundwater for irrigation purposes and conventional agricultural practices particularly, in last two decades have caused a significant reduction of these natural resources. As per a report of the State Water Investigation Directorate, five C.D. blocks in the eastern part of the district namely, Bhatar, Ketugram-I, Mangalkot, Memari-II and Manteswar are at 'semi-critical stage' (Ray and Shekhar 2009) (based on the ratio of withdrawal and recharge of groundwater and its long term trend).

Therefore, present study has selected five C.D. blocks on the basis of their different level of dependency on groundwater resources. Manteswar and Memari-II have a high percentage (66.64 and 80.74%) of net shown area irrigated by groundwater whereas, three blocks Katwa-I, Katwa-II and Jamuria have different geological settings and low coverage of net shown area irrigated by groundwater (32.66%, 27.01%, and 0.2% respectively). Such selection will help to estimate the pattern of groundwater resources relative to geohydrological settings in selected blocks (Fig. 14.1).

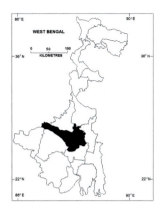
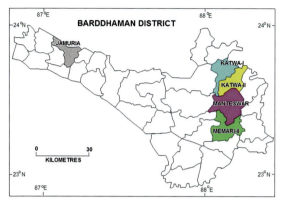

Fig. 14.1 Location of study area

14.3   Objectives of the Study

The objectives of the present study are as follows:

i.   To estimate recharge of groundwater and spatial pattern of seasonal distribution of groundwater level relative to climatic parameters and hydrogeological condition.
ii. To evaluate groundwater depletion, resultant changes in irrigation technique and adaptation to water saving practices to manage this resource.

14.4   Materials and Methods

The study area being located in alluvial plain, administrative blocks are selected as groundwater assessment unit following the methodology of Central Ground Water Board (CGWB 2006) Depths to groundwater level data (1990–2015) have been collected from Groundwater Yearbook of West Bengal and SWID office, Kolkata. Data about groundwater monitoring wells covering five C.D. blocks have been collected for Post-Monsoon (January), Pre-Monsoon (April), Monsoon (August) and Post-Monsoon (November) periods in a year. Long-term water level fluctuation has been analysed by comparing the decadal mean water level data with that of the year next to the decade. Rate and trend of declination has been shown by hydrographs of the groundwater monitoring wells in selected C.D. blocks.

Soil moisture balance is a significant method to estimate the recharge rate of groundwater (Ruston et al. 2006; Silva and Rushton 2007).To assess the recharge rate of groundwater different climatic data have been collected from secondary sources of WBADMIP (Government of West Bengal 2015). The date ranges from January, 2014 to December, 2015. Recharge rate has been assessed using Soil Moisture Balance method (Kumar 2003) and the formula is as follows:

$$G_r = P_1 - E_a + S_c - R_o \qquad (14.1)$$

where, G_r = recharge; P_1 = precipitation; E_a = actual evapotranspiration; S_c = change in soil water storage and R_o = runoff.

Recharge co-efficient is calculated as total amount of recharge relative to total rainfall (Kumar 2014).

Finally existing agricultural practices are evaluated based on primary data selecting two C.D. blocks of Manteswar and Memari-II which at present are experiencing 'semi-critical' condition. For the collection of primary data one village has been selected from each of 22 g Panchayats of the two selected blocks. The farming households were the ultimate stage of sampling. 20 farmers from every selected village were taken on random method for collecting data based on personal interview selecting one person in a family having enough experience and information on farming.

Temporal comparison of Recharge rate is conducted by 'approximate test' using normal distribution. The hypotheses have been tested using the following formula:

$$Z = (p - p0)/\text{SE of } p \qquad (14.2)$$

where, p and $p0$ are the proportions of samples from the population, and SE is the standard error of estimate (Das 2010).

14.5 Geology and Hydrological Settings of the Selected C.D. Blocks, Barddhaman District

Among the selected five blocks, Manteswar, Memari-II, Katwa-I and Katwa-II are located in the western sub-Basin part of the Bengal Basin, the largest fluvio-Deltaic to the shallow marine sedimentary basin on Earth today. The western sub-Basin represents a stratigraphy ranging from Lower Cretaceous to the most recent Alluvium (Sengupta 1966).

The study area except Jamuria C.D. block is a part of Bengal Basin originated by a combined effect of sedimentation, plate tectonic and movement of marine water in different geological period of time. The area concerned got sediments by the river Bhagirathi-Hugli and its tributaries and the source of sediment was the peninsular Indian shield (Roy and Chatterjee 2015). Deposition of Himalayan sediments in the Holocene period changed the geometry of the area to a greater extent (Alam et al. 2003).

The surface geology of the selected five blocks (Fig. 14.2) is classified into three (Ray and Shekhar 2009) as follows:

(a) *Carboniferous-Cretaceous Sedimentaries*: Most of the areas of Jamuria block which is a part of the Chhotonagpur plateau is covered by this formation. This

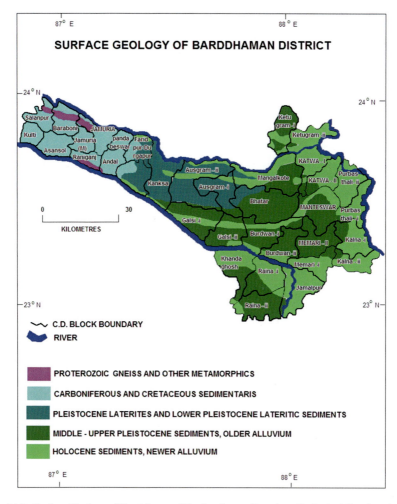

Fig. 14.2 Surface Geology of Barddhaman District. *Source* Based on Geological Quadrangle map 73M and 79A, GSI (1999)

is called as the Raniganj formation with the lithology of sandstone-coal-shale. Another formation called supra-Panchet with the lithology of coarse grained ferruginous sandstone-red siltstone can be noted in a very little portion in the eastern part of this block.

(b) *Upper Pleistocene to lower Holocene sediments*: It consists of older alluvium. Most of the areas of Manteswar and Memari-II blocks are covered by this formation with the lithology of oxidised boulder-pebble-sand-silt–clay with *kankar*. The lithological formation also contains caliches and minor ferruginous concretion with the development of Pedocal soil.

(c) *Holocene sediments*: It consists of newer alluviam deposited recently by the
     river Bhagirathi-Hugli and its tributaries. This sediment covers the entire part
     of Katwa-I and Katwa-II blocks which are nearer to Bhagirathi-Hugli and Ajay
     River. Same deposition can be seen on both sides of Khari River in Manteswar
     block. This formation is characterised by unoxidised sand with subordinate silt
     and clay. It is marked by abandoned channels, swamps and water logging.

   Hydrogeological units of the area can be identified as follows:

(i) *Porous hydrogeological unit*: Manteswar, Memari-II, Katwa-I and Katwa-II
     blocks are consist of unconsolidated alluvium with the order of clay-silt-sand-
     gravel carried by the Bhagirathi-Hugli river system. These sediments were
     deposited in Mio-Pliocene age and Quaternary age. Due to the presence of
     sand the hydrogeological unit is very porous to cause easy movement and
     occurrence of groundwater within it (Ray and Shekhar 2009).
(ii) *Fissured hydrogeological unit*: Hydrogeological settings in Jamuria C.D. block
     is characterised by Archaean to Proterozoic gneisses and schists with fractures,
     Joints and other fissures form the secondary openings in the formation.

14.6   Results and Discussion

Depth to water level and its long-term trend are the two parameters studied to depict
the tendency of ground water regime (2000–2015). The increasing trend of depth to
water level is more in the area of older alluvium zone where agricultural activity is
too dominant to cause limitless withdrawal of groundwater for irrigation purposes.

14.6.1   Pre-monsoon Groundwater Level

Fluctuation of groundwater level between April, 2010 and April, 2011 (pre-monsoon)
is within the range of 0.5–2.5 m. Out of analysed wells (Fig. 14.3), 77% wells have
reported a falling and 23% wells have rised. 54% wells belonging to 0–3 m falling
category, which indicates that regional fluctuation of the area is mainly restricted
within 3 m. The highest pre-monsoon rise of water level was recorded at Mandalpur
(6.7 m) in Jamuria and highest pre-monsoon fall was recorded at Paharhati (5.61 m) in
Memari-II C.D. block. The map depicting depth to groundwater observed during pre-
monsoon season (for the period 2011) is presented in Fig. 14.4 and (for pre-monsoon
decadal mean 2002–2011) in Fig. 14.5.

   Almost the entire parts of Katwa-I C.D. block are occupied by newer alluvium
yet we can identify a spatial variation in term of groundwater level. In this block very
low depth to groundwater level (<5 m bgl) is found in the area situated in proximity
to Hugly River and the depth gradually increases with increasing distance away from
this river. Similar picture can be seen in Katwa-II, Manteswar and Memari-II C.D.

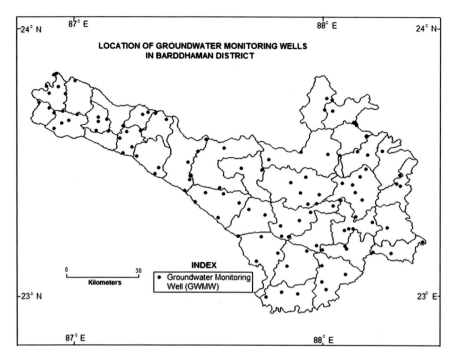

Fig. 14.3 Location of groundwater monitoring wells. *Source* Based on Groundwater Yearbook of West Bengal (Central Ground Water Board 2008, 2009, 2010, 2011, 2012, 2013, 2014, 2015, 2016)

block. At the south-western part of Manteswar the surface geology is covered by older alluvium where the depth is very high (20–40 m bgl). Such very high depth is found in north-eastern part of Memari-II C.D. block but is low at its south-western part. In Jamuria which is entirely covered by hard rock the groundwater level remains within 5–10 mbgl (bellow ground level). Higher depth (10–20 m bgl) is found at very few places.

14.6.2 Post-monsoon Groundwater Level

Fluctuation of groundwater level between November, 2010 and November, 2011 (post-monsoon) is generally found within the range of 1–5 m (76.9%). The percentage of GWMW in fluctuation category of 0–2 m and 5–10 m are 53.8% and 46.2%, respectively. Annual post-monsoon falls were found in all the wells except Katwa town (1.28 m raise). The highest quantum of fall was observed at Paharhati (7.67 m) closely followed by Damodarpur (7.03%). Most of the wells show falling of groundwater level due to lack of rainfall during 2011. The map depicting the status of

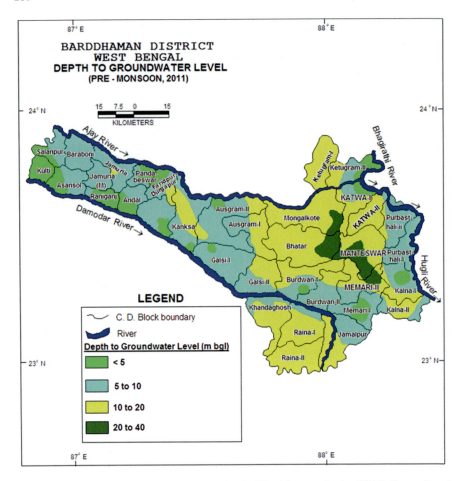

Fig. 14.4 Pre-monsoon depth to groundwater level of Barddhaman district (2011). *Source* Based on data collected from Groundwater Yearbook of West Bengal (Central Ground Water Board 2008, 2009, 2010, 2011, 2012, 2013, 2014, 2015, 2016)

groundwater level observed during post-monsoon season (for the period 2011) is presented in Fig. 14.6 and (for post-monsoon decadal mean 2002–2011) in Fig. 14.7.

Being post-monsoon season (November) all the wells in the area have raised by 0–8 m with respect to the pre-monsoon season. As a result we can see a difference in spatiality of the distribution of piezometric surface in the area. The highest quantum of rise has been recorded in Khaitan of Katwa-I C.D block due to which almost all parts of this block is covered by piezometric surface ranging from <5 m bgl and 5–10 m bgl where in pre-monsoon season it was 10–20 m bgl. The same phenomena can be found in Katwa-II also. In Manteswar C.D. block the area of the zone with piezometric surface 20–40 m bgl in south-western part has decreased than pre-monsoon season due to which the area of the zone with piezometric surface 10–20 m bgl has increased.

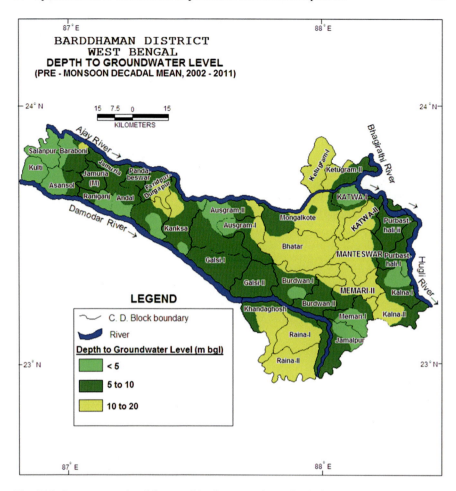

Fig. 14.5 Pre-monsoon decadal mean of depth to groundwater level of Barddhaman district (2002–2011). *Source* Based on data collected from Groundwater Yearbook of West Bengal (Central Ground Water Board 2008, 2009, 2010, 2011, 2012, 2013, 2014, 2015, 2016)

Same situation can be noticed at the northern part of Memari-II C.D. block. In Jamuria the piezometric surface ranges from 5–10 m bgl and 10–20 m bgl at few places. Here the piezometric surface has risen by 2–5 m than pre-monsoon season.

14.6.3 Pre-monsoon Long-Term Tendency of Groundwater Level

Long-term trend is analysed comparing the mean decadal fluctuation of groundwater level with that of the year next to the decade. For example pre-monsoon long-term

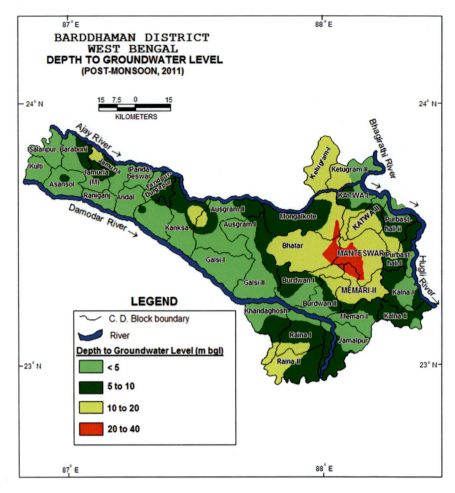

Fig. 14.6 Post-monsoon depth to groundwater level of Barddhaman district (2011). *Source* Based on data collected from Groundwater Yearbook of West Bengal (Central Ground Water Board 2008, 2009, 2010, 2011, 2012, 2013, 2014, 2015, 2016)

trend can be shown by comparing the decadal mean water level data (April, 1990–April, 1999) with that of April, 2000. In the study area, out of analysed wells 11.1% have shown rise where as 88.9% have shown fall. The rise and fall of water level is generally confined within 0–2 m (11.1% rise and 55.6% fall). Of remaining wells, 11.1% wells show 2–4 m fall and 22.2% wells show >4 m fall.

Pre-monsoon long-term status of the mean of water level data (April, 2000–April, 2009) and that of April, 2010 also represents a falling trend for all analysed wells. The fall of water level is found within 2–4 m (46.2%). Of remaining wells, 30.8% wells show 0–2 m fall and 23.0% wells show >4 m fall. Again, pre-monsoon water level for the period of 2010 to 2015 show a falling trend 92.3% wells.

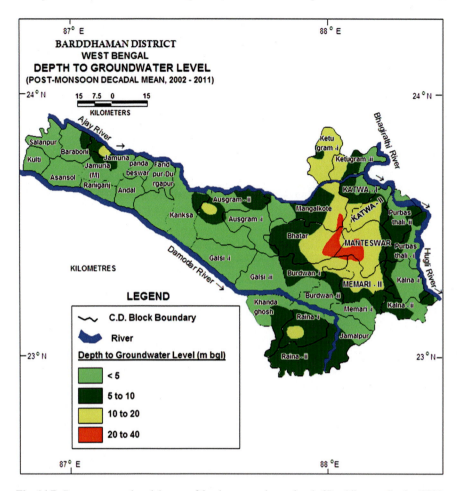

Fig. 14.7 Post-monsoon decadal mean of depth to groundwater level of Barddhaman district (2002–2011). *Source* Based on data collected from Groundwater Yearbook of West Bengal (Central Ground Water Board 2008, 2009, 2010, 2011, 2012, 2013, 2014, 2015, 2016)

14.6.4 Post-Monsoon Long-Term Tendency of Groundwater Level

The month of November is post-monsoon season before which the aquifers are adequately recharged by monsoon rainfall. By comparing the decadal mean of water level data (November, 1990–November, 1999) with that of November, 2000 we have find out that out of analysed wells 33.3% have shown rise where as 66.7% have shown fall. The rise and fall of water level is limited within 0–2 m (22.2% rise and 33.3% fall). Of remaining wells, 22.2% wells show 2–4 m fall and 11.1% wells show >4 m fall.

Similarly, Post-monsoon long-term status of mean of water level data (November, 2000–November, 2009) and that of November, 2010 represents fall of 61.5% wells. Again post-monsoon water level for the period of 2010 to 2015 shows a fall of 92.3% wells.

Hence, long-term trend of groundwater level for both pre-monsoon and post-monsoon season indicate the increasing distance of it from the surface with time which is represented more clearly by positive values of slopes in hydrographs of some selected wells (Fig. 14.8).

14.6.5   Estimation of Groundwater Recharge

Recharge of groundwater is a combined effect of temperature, rainfall, surface runoff, change in soil moisture storage and evapotranspiration. Amount of recharge in Manteswar, Memari-II, Jamuria, Katwa-I and Katwa-II is 14.42%, 16.86%, 24.27%, 14.45% and 13.31% of the total rainfall, respectively, in the year of 2014. Due to lack of rainfall, loss of soil moisture caused by evapotranspiration and withdrawal of groundwater; the months of January, March, April, October, November and December experience discharge of groundwater. Discharge is maximum in October and November months which are caused by lack of rain and relatively high temperature resulting in continuous evapotranspiration from moist soil immediately after rainy season than in winter season. High amount of rainfall by summer depression in the month of May and by south-east monsoon in the month of June to September; groundwater is recharged significantly in all the C.D. blocks. (Figs. 14.9 and 14.10).

According to Kumar (2014) the recharge co-efficient (ratio between water volume of recharge and rainfall) of an alluvial track is 0.10 to 0.20 (Kumar 2014). In the studied area the recharge co-efficient varies from 0.133 to 0.169 in 2014 and that in 2015 is 0.142 to 0.167 (Table 14.1). The study area consists of alluvial surface geology and therefore, the recharge rate is normal according to the norm of groundwater recharge in alluvial area. Though the area is normally recharged but Manteswar and Memari-II blocs are experiencing semi-critical condition in terms of groundwater storage. Agricultural irrigation to a significant extent may be responsible for it.

14.6.6   Groundwater Depletion and Investments in Irrigation Purpose

To a large extent intensive operation of agricultural activities in Manteswar and Memari- II C.D. blocks is responsible for groundwater depletion. High yielding varieties of paddy were started to cultivate in the late 1980s that caused shifting of large operational holding under this crop. Rice being the more water consuming and dominant crop (more than 93% of gross cropped area) for both *kharif* and *rabi*

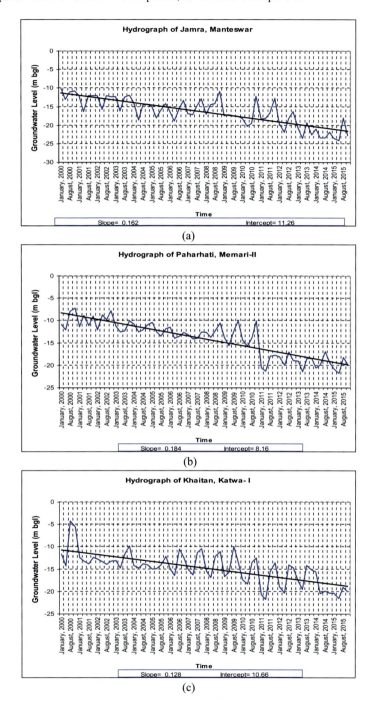

Fig. 14.8 a–e Hydrographs of some selected wells in the study area showing long-term trend of groundwater level

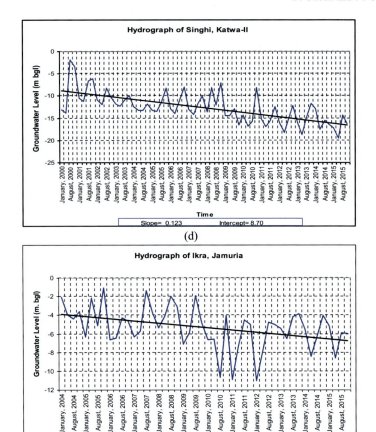

(d)

(e)

Fig. 14.8 (continued)

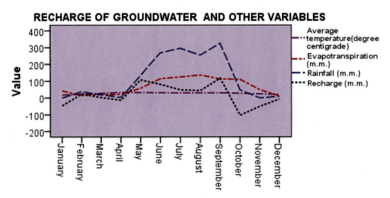

Fig. 14.9 Combined effect of climatic variables on groundwater recharge

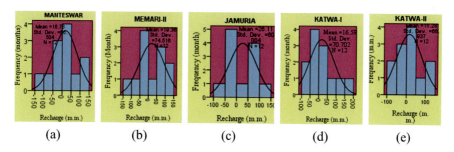

Fig. 14.10   a–e Recharge of groundwater in the study area, 2014

season in Manteswar the demand for irrigation water increased rapidly and threatened its sustainable supply in the long-run. In Memari-II the operational holding under rice and potato is more than 60% and 30% of the gross cropped area, respectively, which has also resulted in a gap between demand and supply of water for irrigation purposes (Figs. 14.11 and 14.12). The demand–supply break in agricultural water consumption caused depletion of groundwater with increasing depth to groundwater level.

If we look at the temporal change of the cropping pattern we will find that before the early 1900s more than 55% of operational holding were being used for rice cultivation only in kharif (rainy) season and were left unused for other time of the year. But, with the development of intensive irrigation system rice occupied maximum operational holding in both wet and dry season. In Memari-II 37% of operational holding are used for potato cultivation after cultivating rice in kharif season. Farmers also cultivate mustard, til, pulses, vegetables etc. but the percentage of cultivated area under these crops is very low. So the lesser degree of crop diversification and increasing specialisation of crops consuming more water is the main reason of groundwater depletion in the study area.

Increasing depth to groundwater level adversely influenced the farming community. Diesel engine fitted with centrifugal pump, firstly installed for groundwater withdrawal during the 1970s, were needed to be replaced by submersible pump due to fall of groundwater level and resultant bore well deepening. During the last three decades, all the farmers have replaced their centrifugal pump by submersible pumps (Fig. 14.13). Such changes necessitated more frequency of investments from the farmers. Farmers deepened their bore wells by 5–10 feet for multiple years based on the extent of fall of groundwater level. Large and medium farmers were the first to experience this shift and small farmers followed them for their poor financial condition. That happened at different times based on the pattern of depth to groundwater level. Therefore the respective frequency curve is representing a zigzag line with a normal distribution (Fig. 14.13).

The cumulative frequency of adoption of submersible pump has formed S-shaped curve (Fig. 14.14) showing an agreement with the Roger's thesis (Roser 1962). Firstly, the curve is flat due to lack of awareness and resultant minimum adoption rate that is increased later by growing awareness to make the curve exponential. The

Table 14.1 Result of z test between the year of 2014 and 2015

Climatic factors	Calculated |z| value						Tabulated |z| value at 5% level of significance	Remarks
	Manteswar	Memari-II	Jamuria	Katwa-I	Katwa-II			
Temperature	0.29	0.29	1.31	0.29	0.29		1.64	No difference
No. of rainy day	0.27	0.33	0.006	−0.07	0.53			No difference
Surface runoff	9.1	10.69	3.85	9.15	7.68			Significantly different
Soil moisture	24.58	10.69	43.77	27.09	31.36			Significantly different
Evapotranspiration	1.01	2.73	6.86	1.72	2.16			Significantly different
Rainfall	10.39	7.98	8.95	8.58	8.58			Significantly different
Recharge of groundwater	1.86	1.65	4.63	2.72	2.73			Significantly different

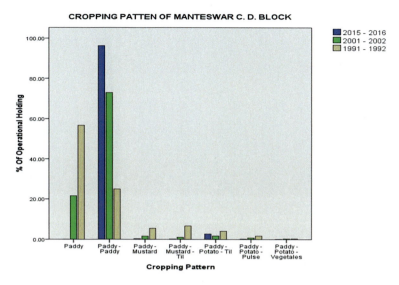

Fig. 14.11 Changing cropping pattern in Manteswar C.D. block. *Source* Based on field survey, 2014–2016

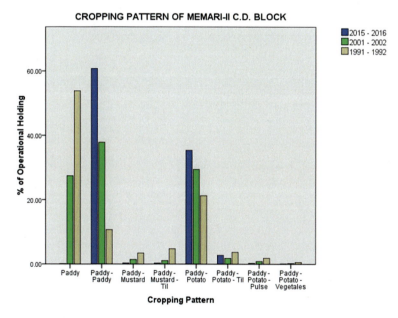

Fig. 14.12 Changing cropping pattern in Memari-II C.D. block. *Source* Based on field survey, 2014–2016

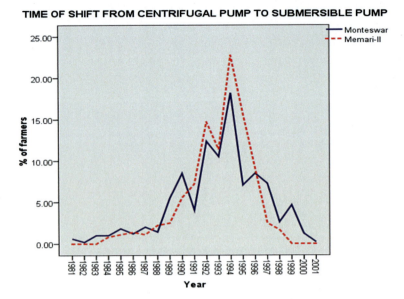

Fig. 14.13 Curves showing frequency of change to submersible from centrifugal pump. *Source* Based on field survey, 2014–2016

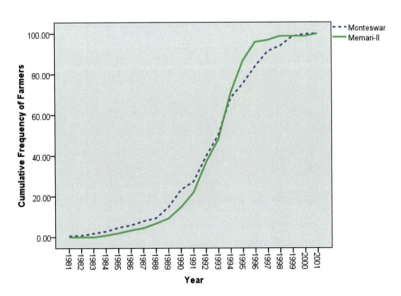

Fig. 14.14 Curves showing cumulative frequency of change to submersible from centrifugal pump. *Source* Based on field survey, 2014–2016

curve again becomes flat due to the completion of the adoption. Hence, the curve becomes S-shaped.

14.6.7 Sources of Investment and Access to Groundwater

The large farmers did not face any difficulties due to increased agricultural investment for their own saving. Medium and small farmers suffered for it. Small farmers were forced to borrow a significant proportion of their total investments from the money lenders of their villages with a very high interest rate than institutional sources. Therefore, increased financial pressure forced the small farmers to share their submersible pump with others that decreased their access to groundwater. Besides, due to financial differences small farmers were unable to participate with the large farmers in the informal water market developed in the agricultural scenario. Large farmers were selling water to neighbour farmers at cheaper rate than small farmers. Hence, the unexpected function of informal water market caused the enlargement of operational holding of large farmers and shrinking that of small farmers.

In some places, large farmers monopolistically have occupied this informal water market. Illegal 'hooking' of power has become highly profitable to them. They sell water at cheaper rate and small and medium farmers who are afraid of illegal 'hooking' for legal compensation, can not compete with them.

14.6.8 Adaptation Strategies

Farmers are practising different strategies in response to declining groundwater resources. About 1.98% farmers have reduced the area of operational holding under paddy to increase water productivity of crops. But this fact is negatively related to the size of operational holding. Small farmers were reducing their operational holding under paddy to replace it by vegetables cultivation where most of the large farmers were using it for permanent plantation. About 20.02% of the farmers are showing interest in the cultivation of *Basmati* rice in response to the decline in water table. Late transplantation of Basmati causes less use of irrigation water than other varieties of paddy. Again, Basmati is more profitable due to its very high market value. Therefore Basmati cultivation is beneficial from both economic and ecological view point. Large farmers are cultivating Basmati more than small farmers. About 15% of the farmers sow paddy seed directly on the main field using an instrument called drum seeder. 3.72% farmers have selected a water saving strategy, System of Rice Intensification (SRI) method, which increases water use efficiency using intermittent irrigation (Ghosh and Chakma 2015). But, SRI was mostly limited within very large and educated farmers. Very few farmers (1.58%) reported an increase in area under more water productive crops like onion, wheat and vegetables that require less water than paddy (Ghosh and Chakma 2019) but, the proportion of land under these water efficient crops are very negligible.

14.7   Conclusion

Manteswar, Memari-II, Katwa-I and Katwa-II are identical in term of geomorphic characteristics. There is a small difference in geological and meteorological perspectives which create minor differences in recharge rates of groundwater. Groundwater is normally recharged according to the actual norm of recharge of an alluvial track with respect to the total amount of rainfall. But, in Jamuria due to totally different geological and geohydrological settings recharge rate is more than other blocks. All blocks are representing a declining trend of depth to groundwater level. Irrespective of this fact Manteswar and Memari-II are showing semi-critical state in terms of groundwater condition due to high rate of fluctuation of groundwater level between pre-monsoon and post-monsoon period whereas, Katwa-I and Katwa-II are in safe condition (Ray and Shekhar 2009).

Withdrawal of groundwater for agricultural irrigation and the existing water consuming cropping pattern is largely responsible for groundwater depletion particularly in Manteswar and Memari-II blocks. Water consuming crops are occupying a maximum proportion of operational holding and very few area of holding under water efficient crops have made the crop production system water exploitative in nature. Groundwater depletion is causing deepening of bore wells and dependence on more water exploitative irrigation techniques that has multiplied the cost of farming decreasing small farmers' access to groundwater. The cumulative frequency distribution curve representing replacement of centrifugal pump by submersible pumps is 'S-shaped'. Informal water market in agricultural scenario is increasing groundwater depletion and the inequality between large and small farmers.

More emphasis is needed to be given on water efficient agricultural practices to improve water productivity of crops. Agro-methodological transformations are required for developing a balanced cropping patterns based on priority-based selection of crops and water saving strategies. Suitable policies are required for smallholders to recover their access groundwater as well as to stop dominancy of large farmers in informal water market.

References

Alam M, Alam MM, Curray JR, Chowdhury MLR, Gani MR (2003) An overview of the sedimentary geology of the Bengal Basin in relation to the regional tectonic framework and basinfill history. Sediment Geol 155:179–208

CGWB (2006) Report on ground water resources of West Bengal, based on GEC, 1997, Methodology

Census of India (2011) District census handbook, Barddhaman. Village and Town Directory, West Bengal: Directorate of Census Operations

Central Ground Water Board (2008) Ground water year book, 2007–08, of West Bengal, Central Ground Water Board, Eastern Region

Central Ground Water Board (2009) Ground water year book, 2008–09, of West Bengal, Central Ground Water Board, Eastern Region

Central Ground Water Board (2010) Ground water year book, 2009–10, of West Bengal, Central Ground Water Board, Eastern Region

Central Ground Water Board (2011) Ground water year book, 2010–11, of West Bengal, Central Ground Water Board, Eastern Region

Central Ground Water Board (2012) Ground water year book, 2011–12, of West Bengal, Central Ground Water Board, Eastern Region

Central Ground Water Board (2013) Ground water year book, 2012–13, of West Bengal, Central Ground Water Board, Eastern Region

Central Ground Water Board (2014) Ground water year book, 2013–14, of West Bengal, Central Ground Water Board, Eastern Region

Central Ground Water Board (2015) Ground water year book, 2014–15, of West Bengal, Central Ground Water Board, Eastern Region

Central Ground Water Board (2016) Ground water year book, 2015–16, of West Bengal, Central Ground Water Board, Eastern Region

Das NG (2010) Statistical method combined edition, vol I & II. Tata McGraw Hill Education Private Limited, New Delhi

FAO (2015) The state of food insecurity in the world-2012. www.fao.org/docrep/016/.../i3027e.pdf, Retrieved on February 2, 2015

GSI (1999) Geology & Mineral Resources of the States of India, Part-I, West Bengal, Geological Survey of India, Miscellaneous Publication No 30, 1999

Geological Quadrangle map 73M and 79A, Geological Survey of India, Kolkata

Ghosh B, Chakma N (2019) Composite indicator of land, water and energy for measuring agricultural sustainability at micro level, Barddhaman District, West Bengal, India. Ecol Ind 102:21–32

Ghosh B, Chakma N (2015) Impacts of rice intensification system on two C. D. blocks of Barddhaman district, West Bengal. Curr Sci 109(2):342–346.

Gleeson T, Wada Y, Bierkens MFP, Van Beek LPH (2012) Water balance of global aquifers revealed by groundwater footprint. Nature 488(7410):197–200

Government of West Bengal (2015) WEB GIS map, accelerated development of minor irrigation project. Retrieved from 2015-06-08 to 2016-02-01, from http://122.176.66.254/GISWEB/map1.htm

Huang H, Han Y, Jia D (2018) Impact of climate change on the blue water footprint of agriculture on a regional scale. Wat Supply 19(1):52–59

Kumar CP (2014) Groundwater data requirement and analysis. Scientific essay, ISBN (Book): 978-3-656-75553-1, GRIN Publishing GmbH, Munich, Germany

Kumar CP, Singh S (2011) Impact of climate change on water resources and national water mission of India. In: Proceedings of national seminar on global warming and its impact on water resources, vol 1, 14 January 2011, Kolkata, pp 1–9

Kumar CP (2012) Climate change and impact on groundwater resources. Int J Eng Sci 1: 43–60

Kumar CP (2003) Estimation of ground water recharge using soil moisture balance approach, national institute of hydrology. Technical Report No TR-142, p 66

Ray A, Shekhar S (2009) Ground water issues and development strategies in West Bengal. Bhujal News 24(1):1–17

Roser M (1962) Diffusion of innovations. The Free Press, New York

Roy AB, Chatterjee A (2015) Tectonic framework and evolutionary history of the Bengal Basin in the Indian subcontinent, current Science, 109, 271–279

Ruston KR, Eilers VHM, Carter RC (2006) Improved soil moisture balance methodology for recharge estimation. J Hydrol 318(1–4):379–399

Sengupta S (1966) Geological and Geophysical studies in the western part of Bengal Basin, India. Am Assoc Pet Geol Bull 50:1001–1017

Silva CSD, Rushton KR (2007) Groundwater recharge estimation using improved soil moisture balance methodology for a tropical climate with distinct dry seasons. Hydrol Sci J 52(5):1051–1067

Singh K (2011) Groundwater depletion in Punjab: measurement and countering strategies. Indian J Agric Econ 66:583–589

Steenbergen FV, Kaisarani AB, Khan NU, Gohar MH (2015) A case of groundwater depletion in Balochistan, Pakistan: enter into void. J Hydrol Reg Stud 4:36–47

Chapter 15
Detection of Land Use/Land Cover Changes of Irga Watershed in the North–Eastern Fringe of Chota Nagpur Plateau, Jharkhand, India

Ratan Pal and **Narayan Chandra Jana**

Abstract The land use/land cover (LULC) change is a prodigious physio-socio-economic concern as it is one of the important factors which is responsible for soil erosion, climate change, interruption of bio-geo-chemical and hydrological cycle, depletion of water quality, loss of biodiversity, ecological balance, etc. Continuous population growth demands more food and shelter, which caused excess pressure on land forcing to change the existing pattern of land use, and ultimately the land cover. The present study attempts a quantitative evaluation of the LULC changes in the Irga Watershed for the last three decades. Six major LULC categories are identified viz. Built-up area, Vegetation cover, Agricultural land, Waterbody, River, and Barren land. The investigation reviled that the agricultural land, vegetation cover, and built-up area are the major LULC categories, occupying around 90% of the total geographical land of the study area. The findings also suggest that the agricultural land and the built-up areas are constantly expanding while the area under vegetation cover and barren land decreasing incessantly. The anthropogenic activities are found as the main culprit for LULC changes.

Keywords LULC · Physico-socio-economic concern · Geospatial technology · Land use categories · Kappa coefficient

15.1 Introduction

Land cover has been changing since the first appearance of human beings on the earth to manage the natural environment. It is intensified from the Neolithic Revolution or the First Agricultural Revolution—the period of transformation of human societies from hunting and gathering to farming, making an increasing possibility for food of the growing population (Bocquet-Appel 2011). The availability of food helped to develop the settled communities of human beings. It permitted them to observe and experiment on plants and animals that led to the domestication of plants and

R. Pal (✉) · N. C. Jana
Department of Geography, The University of Burdwan, Bardhaman, West Bengal 713104, India

© The Author(s), under exclusive license to Springer Nature Singapore Pte Ltd. 2022
N. C. Jana and R. B. Singh (eds.), *Climate, Environment and Disaster in Developing Countries*, Advances in Geographical and Environmental Sciences, https://doi.org/10.1007/978-981-16-6966-8_15

animals (Pollard et al. 2015). However, the Second Agricultural Revolution (late seventeenth to late nineteenth centuries), the Industrial Revolution (late eighteenth century) and finally the third Agricultural Revolution or the Green Revolution (1960s) have intensively changed the pattern and intensity of land use (Martinez et al. 2009; Kuchay and Ramachandra 2016). As a result, degradation of the natural environment is an inevitable outcome we are facing today. In a study, Judson (1965) estimated that there is an increase of River-Born sediment from 9.9 billion t yr-1 before the domestication of plants to 26.5 billion t yr-1 in the 1960s (Singh and Phadke 2006). According to FAO estimation (FAO 1993), the demands for food, fibre, and products from bio-energy are growing at the rate of 3.7% per annum in developing countries.

Since the historic past, most of the natural resources have been heavily used and depleted. More than half of the world's landscape has been modified by the anthropogenic activities (Goldewijk et al. 2011). The multi-faceted impact of the modification of land use and land cover includes climate change (Fan et al. 2014; Nayak and Mandal 2012; Fu and Weng 2016; Li et al. 2018a; Gogoi et al. 2019), and its abnormal behaviour (Halder et al. 2016), interruption of the bio-geo-chemical cycle (Pouyat et al. 2007), and the hydrological cycle (Petchprayoon et al. 2010; Li et al. 2018b; Arulbalaji and Maya 2019; Garg et al. 2019), changes in soil properties (Biro et al. 2011; Roy and Sreekesh 2016), exhaustion of soil nutrients (Xiao-Yin et al. 2015), landscape change (Herold et al. 2002; Fichera et al. 2012; Jaafari et al. 2015) and fragmentation (Nagendra et al. 2004; Nurwanda et al. 2014), decrease of water quality (Zhang et al. 2009), intensified soil erosion, (Sharma et al. 2011; Paiboonvorachat and Oyana 2011; Gashaw et al. 2019; Kidane et al. 2019), loss of biodiversity (Dupouey et al. 2002; Jiang et al. 2003; Nelson et al. 2010; Sharma et al. 2018), ecological balance (Rodriguez-Echeverry et al. 2018; Tolessa et al. 2017), etc. Around 30–40% of soil organic carbon (SOC) is lost due to the alteration of forest cover and grasslands into agricultural lands (FAO 2019). In general, modification of LULC has an overall impact on almost every natural parameter (Zhang 2001; Lambin et al. 2003; Li et al. 2017; Nath et al. 2018), and it is a barrier to sustainable development (Sui and Ming 2002; Dewan and Yamaguchi 2009).

Human beings are directly dependent on land to meet their basic needs like food, clothes, and shelter, and luxury amenities as well. The economic and developmental activities are drastically modifying the global landscape. In the name of development, relentless stress of the flourishing population on the land changed the original natural settings of land use and land cover (Weinzettel et al. 2013). For continuing the growth and development in various sectors of the economy, issues that emerged due to population growth need to be addressed (Ramakrishna 1998). Although India has only 2.4% of terrestrial land area, it serves 17% of the world population and 11% of the world's livestock. The pressure on the Indian land is almost 4–6 times more than the global average (Roy and Murthy 2009). In India, to support the bursting population, farmers are practising intensive agriculture and annually take back around 2000 to 2500 × 106 m^3 of water (Roy and Roy 2010).

From the beginning of the twenty-first century, land use/cover change has become the major issue at the global as well as national and regional level because of the concern over issues like global warming and climate change (Yang et al. 2017).

Economically developing and demographically rapid progressing country like India is more sensitive to LULC changes because of its overdependence on primary activities especially on agriculture. High population pressure forces to extract maximum output from the limited land. This led to maximum pressure on land and hence degradation of land is more critical than the other population-adjusted countries (Roy and Roy 2010). Therefore, reasonable knowledge about changing patterns of land use/land cover and its future trends is the foremost criteria to understand and assess environmental consequences of LULCCs and for land use planning and effective management of natural resources (Giri et al. 2005).

According to the India state of forest report, 2011, (https://cdn.s3waas.gov.in/, p. 33), Giridih district has 17.41% area under forest, while Jharkhand has 28.82%. A somewhat different estimation was made by Abushnaf et al. (2015) that the Giridih district has 26.37% area under forest and agricultural land covers about 45%, while 5.34 and 10.78% area covered by built-up and wasteland, respectively. Sharma, et al. (Sharma et al. 2011) found that cropland and wasteland are the two dominant land use categories, together constituting more than 80% of the total geographical area of the Maithon reservoir catchment area. The objective of the present study is to analyse the temporal changes in land use and land cover of the Irga Watershed.

15.2 Study Area

According to Soil and Land Use Survey of India, 2017, the Irga Watershed is a sub-watershed of No. 2A2I6 (Soil and Land Use Survey of India 2017). The watershed located in the Giridih district of Jharkhand and covers an area of 479.5 km2. The latitudinal extension is from 24° 10′ 50″ N to 24° 28′ 20″ N latitude and longitudinal from 85° 52′ 50″ E to 86° 9′ 15″ E (Fig. 15.1).

The climate of the study area may be described as the tropical sub-humid dry bioclimatic type (Velayutham et al. 1999), received an average rainfall of 1130 mm, mainly in the rainy season (June–September) due to monsoonal effect (Kumar et al. 2016). The annual average temperature is 25 °C (Roy 2018), though it ranges from a minimum of 10 °C to a maximum of 47 °C (Kumar et al. 2016). The study area lies over the Koderma Plateau, which is a part of the Eastern Plateau physiographic division of India (Velayutham et al. 1999; Roy 2018).

15.3 Data Used and Methodology

Application of remote sensing (RS) paired with Geographical Information System (GIS) is the most important method for quantification of LULC changes, because of its digital format, geo-referencing procedures, repetitive nature, cost-effective, time saving, and reliable mapping can help to detect the land use changes by storing the information as images (Singh 1989; Chen et al. 2005; Rahaman et al. 2011). The

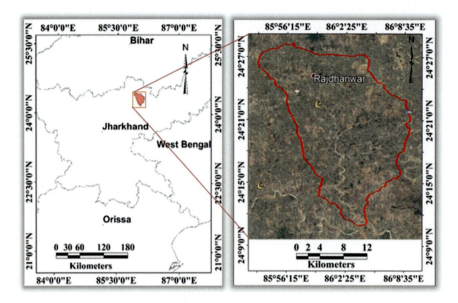

Fig. 15.1 Location of the study area

processing and interpretation of satellite images using software (GIS technique) is very handy and helpful to detect land use/cover changes (Lu et al. 2011; Corner et al. 2014).

In this study, four satellite images of Landsat series (Table 15.1) were downloaded from the United States Geological Survey (http://earthexplorer.usgs.gov/) which are processed by using ERDAS IMAGINE (V. 9.2) and Arc GIS (V. 10.4). Detail information regarding satellite data is given in Table 15.1. Four topographical maps (G45T15, G45T16, G45U3 & G45U4) of 1:50,000 scale have been downloaded from http://soinakshe.uk.gov.in/. The study area is demarcated by mosaicking and digitizing of these topographical maps (Fig. 15.2).

Table 15.1 Satellite images used for LULC classification

Sl. no	Spacecraft	Acquisition date	Path/row	Sensor	Image type	Source
1	Landsat 5	January 28, 1990	140/43	TM	Level-1 Geo TIFF	United States Geological Survey (http://earthexplore rUsgs.gov/)
2	Landsat 5	January 24, 2000	140/43	TM	Level-1 Geo TIFF	
3	Landsat 7	January 27, 2010	140/43	ETM +	Level-1 Geo TIFF	
4	Landsat 8	January 31, 2020	140/43	OIL/TIRS	Level-1 Geo TIFF	

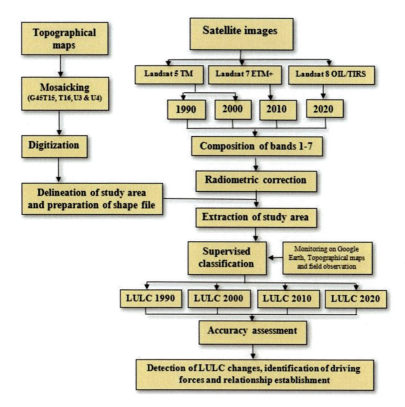

Fig. 15.2 Detail methodological flow chart

The supervised classification (maximum likelihood classifier) of Landsat images of 1990, 2000, 2010, and 2020 have been done by using Arc GIS (V. 10.4) for the classification of LULC into six different classes (Table 15.2). For this, along with processing and interpretation of satellite images, monitoring on the google earth, topographical maps, and GPS survey have also been done to increase the accuracy of the work.

An accuracy assessment has been done to validate the results by using the error matrix. Overall accuracy, producer's accuracy, user's accuracy, and the Kappa coefficient are the parameters calculated from the error matrix (Agyemang et al. 2011). The overall accuracy of the matrix is calculated by dividing the total number of correctly classified pixels (total of diagonal values) by the total number of training pixels (Congalton 1991). The producer's accuracy is the ratio of the total number of correctly classified pixels in each category and the total number of reference pixels in that category (column total). It represents how well a certain area has been classified. It considers the error of omission, the proportion of observed features on the reality that are not classified on the map (Banko 1998). The user's accuracy is derived by dividing the total number of correctly classified pixels in each category by the total

Table 15.2 Description of different LULC categories

Sl. no	LULC category	Description
1	Built-up area	All the residential areas (urban, semi-urban, village, etc.), man-made structures, industrial and commercial areas, mining areas, and transportation infrastructures
2	Vegetation cover	Dense forest, mixed forest, deciduous forest, and the area under scrubs
3	Agricultural land	Cultivated and croplands, pasture, and agricultural land with current fallow
4	Waterbody	All the stagnant inland waterbody including marshes, swamps, artificial lakes/dams, tanks, and ponds
5	River	Area under flowing water through channels including sand bars on the riverbed (Rills and gullies are not properly incorporated)
6	Barren land	Sandy and rocky surface which is not presently in cultivatable condition

number of classified pixels in that category (row total). It represents how well the ground reality is depicted on the map, e.g. reliability of the map (Banko 1998). It is inversely proportional to the error of commission. This kind of error informs about the proportion of wrong classification, which is mistakenly included in a certain category (Story and Congalton 1986). The results are often multiplied by 100 to convert into the percentage for better understanding.

Kappa coefficient, developed by Cohen (1960), is an important and common method used for the validation of thematic maps prepared from remotely sensed data in the GIS environment since the early 1980s (Congalton and Mead 1983; Congalton et al. 1983). It compares map information with the ground reality. Unlike overall accuracy, the Kappa coefficient considers non-diagonal elements as well (Rosenfield and Fitzpatrick-Lins 1986). It can be calculated by using the following formula (Bishop et al. 1975).

$$K = \frac{N \sum_{i=1}^{r} x_{ii} - \sum_{i=1}^{r} (x_{i+})(x_{+i})}{N^2 - \sum_{i=1}^{r} (x_{i+})(x_{+i})} \tag{15.1}$$

where, K = Kappa coefficient, N = total number of observations, r = number of rows and columns in error matrix, x_{ii} = observation in row I and column i, x_{i+} = marginal total of row i, and x_{+i} = marginal total of column i.

15.4 Results and Discussion

15.4.1 Analysis of LULC Change

Four satellite images of ten years interval have been analysed for the detection of decadal changes of different LULC categories. For minimizing classification error, all four satellite images are collected in the month of January, and the cloud coverage is set to less than 10%. The study area has been divided into six major LULC categories, namely Built-up areas, Vegetation cover, Agricultural land, Waterbody, River, and Barren land (Table 15.2).

Prepared maps (Fig. 15.3) of LULC indicate that agricultural land is the dominant category for all the years. It occupied 48.85% in 1990 which is increased to 52.91, 55.66, and 57.11% in 2000, 2010, and 2020, respectively (Table 15.3 and Fig. 15.4). To continue the supply of increasing demand for food, shelter and work, the area under vegetation cover and barren land is continuously converted into agricultural lands and built-up areas, while the area under waterbody and river remains more or less constant. This is why the area under vegetation cover and barren land is continuously decreased (Table 15.4). Vegetation cover, the second dominant category, covered an area of 34.86% in 1990, which is reduced to 29.78% and 25.97% in 2000 and 2010 respectively, and in 2020 it occupied only 22.53% of the total geographical land of the study area. The share under the built-up area is increased from 5.32 to 7.19%, to 8.83%, and finally to 11.08% over the selected years in ascending order. Another important category is barren land which is reduced its share from 6.9% in 1990 to 5.21% in 2020. Waterbody and river are not as significant as other LULC categories in terms of area coverage and the percentage of change, they together constitute only 4.07% of the total geographical land in 2020 (Table 15.3).

15.4.2 Decadal Change

The built-up area is the category which has the highest positive decadal growth rate of 35.16%, 22.80%, and 25.48%, respectively, over the last three decades from 1990 to 2020. The annual growth rate was 3.516%, 2.28%, and 2.548%, respectively, during the same period, while the overall average yearly changing rate from 1990 to 2020 was 3.609%. The vegetation cover has the highest negative decadal changing rate. It was -14.57% (-1.457% per year) during 1990–2000, which is reduced to -12.78% (-1.279% per year) from 2000–2010, and again increased to -13.25% (-1.325% per year) in the last decade, e.g. from 2010 to 2020. (Table 15.4 and Fig. 15.5). The agricultural land has a positive changing rate with a declining trend because agricultural land already occupied a huge area, its little scope of conversion from other categories, and even expansion of built-up area on agricultural land. Its decadal growth rate was 8.31% (0.831% per year), 5.12% (0.512% per year), and 2.61% (0.261% per year) for the last three decades. The decadal changing rate of

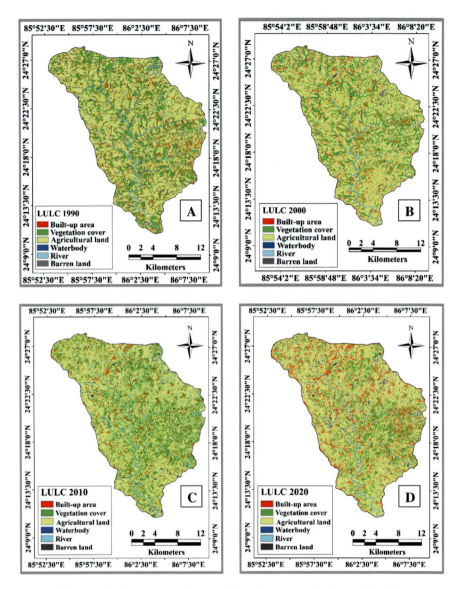

Fig. 15.3 Land use land cover maps of the Irga Watershed (A-1990, B-2000, C-2010, D-2020)

barren land is considerably highly negative, which was −11.59, −9.5, and −5.63% because the actual geographical area under this category is low, which is 33.09 km^2 in 1990 and it is reduced to 24.98 km^2 in 2020, a reduction of only 8.11 km^2 during the last thirty years. Waterbody and river may be considered as more or less constant, but one thing is noticeable that waterbody had a negative change rate during the 2000s but in the 2020s its trend showing a slightly positive and hence it is the only category

Table 15.3 Area under different LULC categories

LULC category	1990		2000		2010		2020	
	Area in km²	Area in %	Area in km²	Area in %	Area in km²	Area in %	Area in km²	Area in %
Built-up area	25.51	5.32	34.48	7.19	42.34	8.83	53.13	11.08
Vegetation cover	167.15	34.86	142.80	29.78	124.53	25.97	108.03	22.53
Agricultural land	234.24	48.85	253.70	52.91	266.89	55.66	273.85	57.11
Waterbody	7.10	1.48	6.80	1.42	6.80	1.42	7.00	1.46
River	12.41	2.59	12.47	2.60	12.47	2.60	12.51	2.61
Barren land	33.09	6.90	29.25	6.10	26.47	5.52	24.98	5.21
Total	**479.50**	**100**	**479.50**	**100**	**479.50**	**100**	**479.50**	**100**

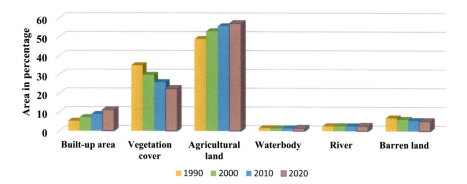

Fig. 15.4 Proportional coverage of different LULC categories from 1990 to 2020

Table 15.4 Decadal and yearly changing rate of different LULC categories

LULC category	1990–2000		2000–2010		2010–2020		1990–2020	
	Decadal change (%)	Yearly change (%)	Decadal change (%)	Yearly change (%)	Decadal change (%)	Yearly change (%)	Overall change (%)	Yearly change (%)
Built-up area	35.16	3.516	22.8	2.28	25.48	2.548	108.27	3.609
Vegetation cover	−14.57	−1.457	−12.79	−1.279	−13.25	−1.325	−35.37	−1.179
Agricultural land	8.31	0.831	5.12	0.512	2.61	0.261	15.91	0.564
Waterbody	−4.23	−0.423	0	0	2.94	0.294	−1.4	−0.047
River	0.48	0.048	0	0	0.32	0.032	0.81	0.027
Barren land	−11.59	−1.159	−9.5	−0.95	−5.63	−0.563	−24.51	−0.817

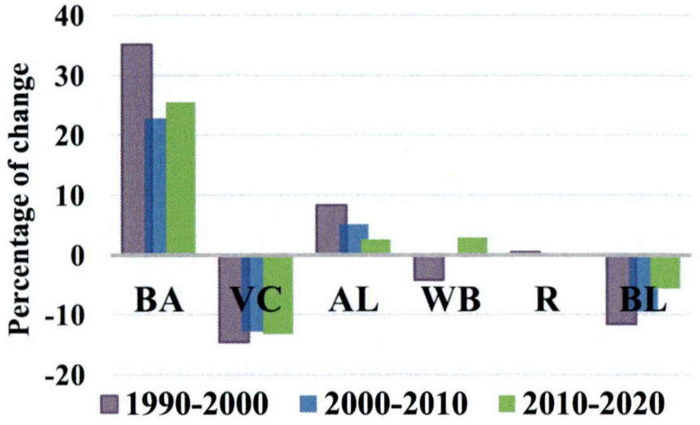

Fig. 15.5 Decadal changing
rate of different land use land
cover categories
[BA—Built-up area,
VC—Vegetation cover,
AL—Agricultural land,
WB—Waterbody, R—River,
BL-Barren land]

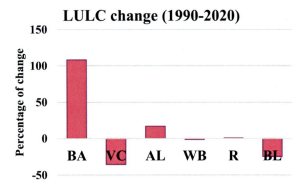

Fig. 15.6 Average changing
rate of different LULC
categories from 1990 to 2020
[BA—Built-up area,
VC—Vegetation cover,
AL—Agricultural land,
WB—Waterbody, R—River,
BL—Barren land]

which has the dual character. On an average, from 1990 to 2020, there is about 108.27
and 15.91% increase in the built-up area and agricultural land, respectively, while
vegetation cover and barren land face negative change of −35.37% and −24.51%,
respectively. The change in area under waterbody and river is negligible (Fig. 15.6).

15.4.3 Identification of Drivers

There are two types of driving forces to change LULC, namely physical forces and
anthropogenic forces. Natural hazards like volcanism, tsunami, extreme earthquake,
etc., can change LULC drastically in any area. In the present study, there is no
such physical incidence in the selected period. In fact, the contribution of physical
phenomena to change LULC is a passive component compared to anthropogenic
factors. Actions taken by human beings are the prime driving force to change LULC
in the present study. Farmers are racticing intensive agriculture intending to earn
maximum output from the land without taking proper care of land. As the population
is increasing, the built-up area and the area under agricultural land is increasing
accordingly because of the collective demands for food and work. The combined
effect of these is seen on decreasing vegetation coverage.

15.5   Accuracy Assessment

The United States Geological Survey (USGS) reported that accuracy assessment of prepared maps from remotely sensed data is an important step for the validation of results because the accuracy of the maps largely depends on the ability of the interpreters/researchers. USGS defines the accuracy of spatial data as the closeness of results from processed data to the true values or the values accepted as being true (USGS 1990). Productive utilization of this kind of generated data is only possible if the quality and accuracy are known (Congalton and Green 1999; Adam et al. 2013). In such assessment, prepared map data is compared with ground truth data. Without validation, the value of the map may be reduced as it may create confusion in the mind of users.

In this study, the Stratified Random Sampling technic has been applied for the selection of a total of 225 training pixels. For the three dominant LULC categories (built-up area, vegetation cover, and agricultural land) which constitute around 90% of the total geographical land of the study area, considering 50 training pixels per class while 25 each for the other three categories (waterbody, river, and barren land).

Anderson et al. (1976) suggest the threshold of interpretation accuracy of a produced LULC map should be 85%, while Carletta (1996) considers 80% is the minimum to accept the map. From Tables 15.5, 15.6, 15.7, and 15.8, it is noticeable that the overall accuracy of the maps is 86, 88.45, 92, and 93.34%, while the Kappa coefficient is 0.831, 0.858, 0.902, and 0.918 in the year 1990, 2000, 2010, and 2020, respectively. The producer accuracy and user accuracy for waterbody and river is 100% in the respective years with the only exception in 1990, which has 96.15% producer accuracy for river and 96% user accuracy for waterbody. The producer

Table 15.5   Confusion matrix of different LULC categories for the year 1990

LULC category	Built-up area	Vegetation cover	Agriculturalland	Waterbody	River	Barren land	Total	User accuracy (%)
Built-up area	**42**	1	3	0	0	4	50	**84**
Vegetation cover	2	**44**	4	0	0	0	50	**88**
Agricultural land	1	7	**40**	0	0	2	50	**80**
Waterbody	0	0	0	**24**	1	0	25	**96**
River	0	0	0	0	**25**	0	25	**100**
Barren land	3	1	2	0	0	**19**	25	**76**
Total	48	53	49	24	26	25	**225**	
Producer accuracy (%)	**87.50**	**83.02**	**81.63**	**100.0**	**96.15**	**76.00**	**Overall accuracy 86%**	
							Kappa coefficient 0.831	

Table 15.6 Confusion matrix of different LULC categories for the year 2000

LULC category	Built-up area	Vegetation cover	Agricultural land	Waterbody	River	Barren land	Total	User accuracy (%)
Built-up area	**41**	3	4	0	0	2	50	**82.00**
Vegetation cover	2	**45**	3	0	0	0	50	**90.00**
Agricultural land	3	3	**42**	0	0	2	50	**84.00**
Waterbody	0	0	0	**25**	0	0	25	**100.00**
River	0	0	0	0	**25**	0	25	**100.00**
Barren land	2	0	2	0	0	**21**	25	**84.00**
Total	48	51	51	25	25	25	**225**	
Producer accuracy (%)	**85.42**	**88.24**	**82.35**	**100.0**	**100.0**	**85.00**	**Overall accuracy 88.45%**	
							Kappa coefficient 0.858	

Table 15.7 Confusion matrix of different LULC categories for the year 2010

LULC category	Built-up area	Vegetation cover	Agricultural land	Water-body	River	Barren land	Total	User accuracy (%)
Built-up area	**47**	1	2	0	0	0	50	**94.00**
Vegetation cover	0	**44**	4	0	0	2	50	**88.00**
Agricultural land	1	3	**43**	0	0	3	50	**86.00**
Waterbody	0	0	0	**25**	0	0	25	**100.00**
River	0	0	0	0	**25**	0	25	**100.00**
Barren land	1	0	1	0	0	**23**	25	**92.00**
Total	49	48	50	25	25	28	**225**	
Producer accuracy (%)	**95.92**	**91.67**	**86.00**	**100.0**	**100.0**	**82.14**	**Overall accuracy 92%**	
							Kappa coefficient 0.902	

Table 15.8 Confusion matrix of different LULC categories for the year 2020

LULC category	Built-up area	Vegetation cover	Agricultural land	Waterbody	River	Barren land	Total	User accuracy (%)
Built-up area	45	2	2	0	0	1	50	90
Vegetation cover	0	48	2	0	0	0	50	96
Agricultural land	1	3	44	0	0	2	50	88
Waterbody	0	0	0	25	0	0	25	100
River	0	0	0	0	25	0	25	100
Barren land	1	0	1	0	0	23	25	92
Total	47	53	49	25	25	26	225	
Producer accuracy (%)	95.74	90.57	89.79	100.0	100.0	88.46	Overall accuracy 93.34%	
							Kappa coefficient 0.918	

accuracy is found lowest for barren land in all the years except the year 2000, it was lowest for agricultural land (82.35%). The user accuracy is lowest in barren land (76%) for 1990, built-up area for 2000 (82%), and agricultural land for 2010 (86%) and 2020 (88%).

15.6 Conclusion

The present study has examined the decadal change of different LULC categories from 1990 to 2020 in the Irga Watershed, Jharkhand. The investigation revealed that the area is suffering due to decreasing area under forest, on an average rate of 1.179% per year for the last three decades (1990–2020). This rate was 1.325% in the last decade (2010–2020). The principal cause for changing the LULC pattern is the increasing rate of population and built-up area and associated economic and developmental activities. In the last three decades (1990–2020), the area under built-up has been increased more than 100% at an average annual increasing rate of 3.609% per year. This rate was 2.548% in the last decade (2010–2020). If this continues, it will negatively effect ecological balance, biodiversity, global heat balance, and will trigger local climate change. Along with the development of cost-effective technologies to mitigate land degradation, there is a need to increase awareness with sound scientific evidence on the importance of natural balance among the people should be effective. The impact of LULCCs on various aspects is a further scope of study in the study area.

Acknowledgements We would like to thank the Survey of India (SOI) and the United States Geological Survey (USGS) for supplying necessary data and information, and the Department of Geography, the University of Burdwan for providing a comfortable environment to carry out this research.

References

Abushnaf AMA, Pandey RK, Lal D, Kumar M (2015) Land use/land cover mapping of Giridih District of Jharkhand by using remote sensing & GIS. Int J Modern Eng Res (IJMER) 5(11):2249–6645

Adam AHM, Elhag AMH, Salih AM (2013) Accuracy assessment of land use & land cover classification (LU/LC), case study of Shomadi area Renk County-Upper Nile State, South Sudan. Int J Sci Res Publ 3(5):2250–3153

Agyemang TK, Heblinski J, Schmieder K, Sajadyan H, Vardanyan L (2011) Accuracy assessment of supervised classification of submersed macrophytes: the case of the Gavaraget region of Lake Sevan, Armenia. Hydrobiologia 661:85–96. https://doi.org/10.1007/s10750-010-0465-7

Anderson JR, Hardy EE, Roach JT, Witmer RE (1976) A land use and land cover classification system for use with remote sensor data. U.S. Geological Survey Professional Paper, No. 964. USGS, Washington, DC

Arulbalaji P, Maya K (2019) Effects of Land use dynamics on hydrological response of watershed: a case study of Chittar Watershed, Vamanapuram River Basin, Thiruvananthapuram District, Kerala, India. Water Conservat Sci Eng. https://doi.org/10.1007/s41101-019-00066-5

Banko G (1998) A review of assessing the accuracy of classifications of remotely sensed data and of methods including remote sensing data in forest inventory. International Institute for Applied Systems Analysis (IIASA), INTERIM REPORT IR-98-081

Biro K, Pradhan B, Buchroithner M, Makeschin F (2011) Land use/land cover change analysis and its impact on soil properties in the northern part of Gadarif Region, Sudan. Land Degradat Dev. doi:https://doi.org/10.1002/ldr.1116

Bishop Y, Fienberg S, Holland P (1975) Discrete multivariate analysis—theory and practice. MIT Press, Cambridge

Bocquet-Appel J-P (2011) When the World's population took off: the springboard of the neolithic demographic transition. Science 333(6042):560–561. https://doi.org/10.1126/science.1208880

Carletta J (1996) Assessing agreement on classification tasks: the kappa statistic. Comput Linguist 22:249–254

Chen X, Vierling L, Deering D (2005) A simple and effective radiometric correction method to improve landscape change detection across sensors and across time. Remote Sens Environ 98(1):63–79

Cohen J (1960) A coefficient of agreement for nominal scales. Educ Psychol Measur 20(1):37–40

Congalton R (1991) A review of assessing the accuracy of classifications of remotely sensed data. Remote Sens Environ 37:35–46

Congalton R, Green K (1999) Assessing the accuracy of remotely sensed data, principles and practices. Lewis Publishers, New York

Congalton R, Mead R (1983) A quantitative method to test for consistency and correctness of photointerpretation. Photogramm Eng Remote Sens 49(1):69–74

Congalton R, Oderwald R, Mead R (1983) Assessing Landsat classification accuracy using discrete multivariate statistical techniques. Photogramm Eng Remote Sens 49(12):1671–1678

Corner RJ, Dewan AM, Chakma S (2014) Monitoring and prediction of land use and land cover (LULC) change. In: Dewan A, Corner R (eds) Dhaka megacity, geospatial perspectives on urbanization, environment and health. Springer, Netherlands, Dordrecht, pp 75–97. https://doi.org/10.1007/978-94-007-6735-5

Dewan AM, Yamaguchi Y (2009) Land use and land cover change in Greater Dhaka, Bangladesh: using remote sensing to promote sustainable urbanization. Appl Geogr 29:390–401

Dupouey JL, Dambrine E, Laffite JD, Moares C (2002) Irreversible Impact of past land use on forest soils and biodiversity. Ecology 83(11):2978. https://doi.org/10.2307/3071833

FAO (2019) Soil erosion: the greatest challenge to sustainable soil management. Rome. Licence: CC BY-NC-SA 3.0 IGO. ISBN 978-92-5-131426-5

FAO (1993) The state of food and agriculture. FAO Agriculture Series, no 26, Rome. ISBN 92-5-103360-9

Fan X, Ma Z, Yang Q, Han Y, Mahmood R (2014) Land use/land cover changes and regional climate over the Loess Plateau during 2001–2009. Part II: interrelationship from observations. Climatic Change 129(3–4):441–455. https://doi.org/10.1007/s10584-014-1068-5

Fichera CR, Modica G, Pollino M (2012) Land Cover classification and change-detection analysis using multi-temporal remote sensed imagery and landscape metrics. Eur J Remote Sen 45:1–18

Fu P, Weng Q (2016) A time series analysis of urbanization induced land use and land cover change and its impact on land surface temperature with Landsat imagery. Remote Sens Environ 175:205–214

Garg V, Nikam BR, Thakur PK, Aggarwal SP, Gupta PK, Srivastav SK (2019) Human-induced land use land cover change and its impact on hydrology. HydroResearch 48–56. https://doi.org/10.1016/j.hydres.2019.06.001

Gashaw T, Tulu T, Argaw M, Worqlul AW (2019) Modelling the impacts of land use–land cover changes on soil erosion and sediment yield in the Andassa watershed, upper Blue Nile basin, Ethiopia. Environ Earth Sci 78:679. https://doi.org/10.1007/s12665-019-8726-x

Giri C, Zhu Z, Reed B (2005) A comparative analysis of the global land cover 2000 and MODIS land cover data sets. Remote Sens Environ 94:123–132

Gogoi PP, Vinoj V, Swain D, Roberts G, Dash J, Tripathy S (2019) Land use and land cover change effect on surface temperature over Eastern India. Sci Rep. https://doi.org/10.1038/s41598-019-45213-z

Goldewijk K, Beusen A, Drecht G, Vos M (2011) The HYDE 3.1 spatially explicit database of Human-induced global land-change over the past 12,000 years. Global Ecol Biogeogr 20(1):73–86

Halder S, Saha SK, Dirmeyer PA, Chase TN, Goswami BN (2016) Investigating the impact of land-use land-cover change on Indian summer monsoon daily rainfall and temperature during 1951–2005 using a regional climate model. Hydrol Earth Syst Sci 20:1765–1784

Herold M, Scepan J, Clarke KC (2002) The use of remote sensing and landscape metrics to describe structures and changes in urban land uses. Environ Plan 34:1443–1458

Jaafari S, Sakieh Y, Shabani AA, Danehkar A, Nazarisamani A (2015) Landscape change assessment of reservation areas using remote sensing and landscape metrics (Case study: Jajroud reservation, Iran). Environ Dev Sustain 17:1–17

Jiang Y, Kang M, Gao Q, He L, Xiong M, Jia Z, Jin Z (2003) Impact of land use on plant biodiversity and measures for biodiversity conservation in the Loess Plateau in China—a case study in a hilly-gully region of the Northern Loess Plateau. Biodivers Conserv 12:2121–2133. https://doi.org/10.1023/A:1024194532292

Judson S (1965) Physical geology, 3rd edn. Prentice Hall, NJ, USA, pp 143–144

Kidane M, Bezie A, Kesete N, Tolessa T (2019) The impact of land use and land cover (LULC) dynamics on soil erosion and sediment yield in Ethiopia. Heliyon. https://doi.org/10.1016/j.heliyon.2019.e02981

Kuchay SA, Ramachandra TV (2016) Land use land cover change analysis of Uttara Kannada. Imp J Interdisc Res (IJIR) 2(4)

Kumar M, Denis DM, Suryavanshi S (2016) Long-term climatic trend analysis of Giridih district, Jharkhand (India) using statistical approach. Model Earth Syst Environ 2:116

Lambin EF, Geist HJ, Lepers E (2003) Dynamics of land-use and land-cover change in tropical regions. Ann Rev Env Resour 28:205–241

Li H, Xiao P, Feng X, Yang Y, Wang L, Zhang W, Wang X, Feng W, Chang X (2017) Using land long-term data records to map land cover changes in China Over 1981–2010. IEEE J Sel Top Appl Earth Observat Remote Sens 10:1372–1389. https://doi.org/10.1109/JSTARS.2016.2645203

Li J, Zheng X, Zhang C, Chen Y (2018a) Impact of land-use and land-cover change on meteorology in the Beijing–Tianjin–Hebei Region from 1990 to 2010. Sustainability 10:176. https://doi.org/10.3390/su10010176

Li P, Li H, Yang G, Zhang Q, Diao Y (2018b) Assessing the hydrologic impacts of land use change in the Taihu Lake Basin of China from 1985 to 2010, 10:1512. https://doi.org/10.3390/w10111512

Lu D, Moran E, Hetrick S, Li G (2011) Land use and land cover detection. In: Weng Q (ed) Advances in environmental remote sensing. CRS Press, Boca Raton, pp 273–291

Martinez ML, Perez-Maqueo O, Vazquez G, Castillo-Campos G, Garcia-Franco J, Mehltreter K, Landgrave R (2009) Effects of land use change on biodiversity and ecosystem services in tropical montane cloud forests of Mexico. For Ecol Manage 258(9):1856–1863

Nagendra H, Munroe DK, Southworth J (2004) From pattern to process: landscape fragmentation and the analysis of land use/land cover change. Agric Ecosyst Environ 101:111–115

Nath B, Niu Z, Singh RP (2018) Land use and land cover changes, and environment and risk evaluation of Dujiangyan City (SW China) using remote sensing and GIS techniques. Sustainability 10:4631

Nayak S, Mandal M (2012) Impact of land-use and land-cover changes on temperature trends over Western India. Curr Sci 102:1166–1173

Nelson E, Sander H, Hawthorne P, Conte M, Ennaanay D, Wolny S, Manson S, Polasky S (2010) Projecting global land-use change and its effect on ecosystem service provision and biodiversity with simple models. PLoS One 5(12):e14327. https://doi.org/10.1371/journal.pone.0014327

Nurwanda A, Zain AFM, Rustiadi E (2014) Analysis of land cover changes and landscape fragmentation in Batanghari Regency, Jambi Province. In: Proceedings of the social and behavioural sciences, international conference, 3–4 November 2015. Intelligent Planning Towards Smart Cities, CITIES, Surabaya, Indonesia

Paiboonvorachat C, Oyana TJ (2011) Land-cover changes and potential impacts on soil erosion in the Nan watershed Thailand. Int J Remote Sens 32(21):6587–6609. https://doi.org/10.1080/01431161.2010.512935

Petchprayoon P, Blanken PD, Ekkawatpanit C, Husseinc K (2010) Hydrological impacts of land use/land cover change in a large river basin in central–northern Thailand. Int J Climatol 30:1917–1930. https://doi.org/10.1002/joc.2131

Pollard E, Rosenberg C, Tigor R (2015) Worlds together, worlds apart, concise. W.W. Norton & Company, New York

Pouyat RV, Pataki DE, Belt KT, Groffman PM, Hom J, Band LE (2007) Effects of urban land-use change on biogeochemical cycles. Global Change, The IGBP series, pp 45–58. https://doi.org/10.1007/978-3-540-32730-1_5

Rahaman A, Kumar S, Fazal S, Siddiqui MA (2011) Assessment of land use/land cover change in the North–West District of Delhi using remote sensing and GIS techniques. J Indian Soc Remote Sens. https://doi.org/10.1007/s12524-011-0165-4

Ramakrishna PS (1998) Sustainable development, climate change and tropical rain forest landscape. Clim Change 39(2–3):583–600

Rodriguez-Echeverry J, Echeverria C, Oyarzun C, Morales L (2018) Impact of land-use change on biodiversity and ecosystem services in the Chilean temperate forests. Landsc Ecol 33(3):439–453. https://doi.org/10.1007/s10980-018-0612-5

Rosenfield G, Fitzpatrick-Lins K (1986) A coefficient of agreement as a measure of thematic classification accuracy. Photogramm Eng Remote Sens 52(2):223–227

Roy P (2018) Application of USLE in a GIS environment to estimate soil erosion in the Irga watershed, Jharkhand, India. Phys Geogr. https://doi.org/10.1080/02723646.2018.1550301

Roy PS, Murthy MSR (2009) Efficient land use planning and policies using geospatial inputs: an Indian experience. In: Denman AC, Penrod OM (eds) Land use policy. Nova Science Publishers, Inc.

Roy PS, Roy A (2010) Land use and land cover change in India: a remote sensing & GIS perspective. J Indian Inst Sci 90(4):489–502

Roy P, Sreekesh S (2016) Effect of land cover on soil particle size and organic carbon in the plough layer. In Raju NJ (ed) Geostatistical and geospatial approaches for the characterization of natural resources in the environment. Capital Publishing Company, pp 385–390. https://doi.org/10.1007/978-3-319-18663-4_59

Sharma A, Tiwari KN, Bhadoria PBS (2011) Effect of land use land cover change on soil erosion potential in an agricultural watershed. Environ Monit Assess 173:789–801. https://doi.org/10.1007/s10661-010-1423-6

Sharma R, Nehren U, Rahman S, Meyer M, Rimal B, Aria Seta G, Baral H (2018) Modelling land use and land cover changes and their effects on biodiversity in Central Kalimantan, Indobesia. Land 7(2):57. https://doi.org/10.3390/land7020057

Singh A (1989) Review article, digital change detection techniques using remotely sensed data. Int J Remote Sens 10(6):989–1003. https://doi.org/10.1080/01431168908903939

Singh R, Phadke VS (2006) Assessing soil loss by water erosion in Jamni River Basin, Bundelkhand region, India, adopting Universal Soil Loss Equation using GIS. Current Science

Soil and Land Use Survey of India (2017) In micro-watershed Atlas of Jharkhand, District-Giridih, Department of Agriculture, Cooperation and Farmers Welfare Ministry of Agriculture and Farmers Welfare. https://slusi.dacnet.nic.in/dmwai/JHARKHAND/Districts/GIRIDIH.html. Accessed 23 March 2020

Story M, Congalton RG (1986) Accuracy assessment: a user's perspective. Photogramm Eng Remote Sens 52(3):397–399

Sui LY, Ming CB (2002) The study framework of land use/cover change based on sustainable development in China. Geogr Res 21:324–330

Tolessa T, Senbeta F, Kidane M (2017) The impact of land use/land cover change on ecosystem services in the central highlands of Ethiopia. Ecosyst Serv 23:47–54. https://doi.org/10.1016/j.ecoser.2016.11.010

USGS (1990) The spatial data transfer standard. United States Geological Survey, Draft

Velayutham M, Mandal DK, Mandal C, Sehgal J (1999) Agro-ecological subregions of India for planning and development (Technical Publication No. 35). National Bureau of Soil Survey and Land Use Planning, Nagpur

Weinzettel J, Fertwich E, Peters G, Steen-Olsen K, Galli A (2013) Affluence drives the global displacement of land use. Glob Environ Chang 23(2):433–438

Xiao-Yin N, Yan-Hua W, Hao Y, Jia-Wen Z, Jun Z, Mei-Na X, Shan-Shan W, Biao X (2015) Effect of land use on soil erosion and nutrients in Dianchi Lake Watershed, China. Pedosphere 25(1):103–111. ISSN 1002-0160/CN 32-1315/P

Yang Y, Zhang S, Liu Y, Xing X, De Sherbinin A (2017) Analyzing historical land use changes using a historical land use reconstruction model: a case study in Zhenlai County, North-Eastern China. Sci Rep 7:41275. https://doi.org/10.1038/srep41275

Zhang M (2001) Progress of land science centered on land use/land cover change. Adv Geogr 20:297–304

Zhang Y, Wang Y, Wang Y, Xi H (2009) Investigating the impacts of land use/land cover (LULC) change in the pearl river delta region on water quality in the pearl river estuary and Hong Kong's coast. Remote Sens 1(4):1055–1064

Chapter 16
An Analytical Study on Interplay Between Physiographic Condition and Land Use Land Cover Dynamicity

Swetasree Nagⓘ**, Malabika Biswas Roy**ⓘ**, and Pankaj Kumar Roy**ⓘ

Abstract Land use land cover dynamicity over a region primarily depends on numerous physiographical factors like relief, slope, drainage, etc. A rhythmic correlation between these factors will promote sustainable land use land cover practices within a region. Likewise in a low relief area having a good drainage cover permits the development of settlement and agricultural activity. But sometimes an unnatural behaviour between these parameters may hamper the land utilisation activities. In this work, authors have tried to draw attention to such kind of behaviour among the topographical elements of the Kuya River Basin which makes the topography more erosive and complex in nature. This paper also enumerates how the land use land cover practices of the study area can be influenced by the topographical complexity. To explore such a topographic complex zone, a weighted overlay analysis has been carried out using six major morphometric parameters. A multiple correspondence analysis (MCA) has also been executed which proves that human beings cannot overcome the adverseness of nature as they were well accustomed with the physiographical complexity by utilising the land in a sustainable way.

Keywords Topographic complexity · Multiple correspondence analyses · Land use land cover dynamicity · Sustainable management

16.1 Introduction

Physiographic conditions play a major role determining the land utilisation pattern of the region. Depending upon the nature of relief and slope the drainage network develops. As 75% of the Indian population has been depending on the agricultural activity, areas having high water resource potentiality are occupied first. Under such circumstances settlement and communication network starts to develop from the

S. Nag · P. Kumar Roy
School of Water Resources Engineering, Jadavpur University, Kolkata, West Bengal, India

M. Biswas Roy (✉)
Department of Geography, Women's College, Kolkata, West Bengal, India

© The Author(s), under exclusive license to Springer Nature Singapore Pte Ltd. 2022
N. C. Jana and R. B. Singh (eds.), *Climate, Environment and Disaster in Developing Countries*, Advances in Geographical and Environmental Sciences, https://doi.org/10.1007/978-981-16-6966-8_16

313

areas having good drainage condition. Therefore, the overall geomorphology of an area may influence the development of the economic condition. Such scenarios are very much observable in the hilly terrain region, where due to the difficulties of physical landscape, the scope of the developmental activity is less prominent. Slope is the basic element affecting the land use pattern in those areas. Prus and Budz (2015) studied the impact of land use structure, which strictly varies with the physiographic location with selected land slope intervals, where with increase in the land slope, agricultural land reduces significantly. Kumar and Malik (2014) discussed the challenges of water management in a hilly terrain area of Uttarakhand. Overall the impact of physiography has an influence on the spatial distribution of rainfall, temperature and also the vegetation cover of any region. Depending upon the nature of physical landscape the land use and land cover also varies significantly. So many studies have been conducted regarding the changes in land use land cover trend analysis (Butt et al. 2015; Mondal et al. 2015; Rawat and Kumar 2015). In a recent decade, the local environmental issue related to land use land cover change has become a global concern because of large-scale conversion of forests, farmlands and expansion of urban areas worldwide. Such change leads to significant alteration of local and regional climatic characteristics as well as impose adverse effects on the biotic environment of any region. The growing population is the emerging factor resulting in unscientific and unplanned changes in land use land cover in any region. Depending upon the human need, people are altering the land resources like they encroach the vegetated land to expend their agricultural activities (Reis 2008). They are also responsible for the alteration of natural flow paths, channel depth, reducing channel storage capacity which in effect leads to unnatural flash flood around the world (Rogger et al. 2017). Debnath et al. (2017) conducted a study on channel migration pattern which directly and indirectly affects land use modification. Haque and Basak (2017) analysed the temporal changes in land use land cover pattern of Bangladesh where within the time span of 20 years vegetation cover has been rapidly decreasing and converted into build-up areas. Such kind of development in socio-economic activities has enhanced the rapid modification of the land resources which creates various natural disasters, deforestation, biodiversity loss, etc. Lambing et al. (2001) reviewed the causes of land use land cover changes and stated that not only the growing population, economic development of any region is also responsible for the modification of land resources. Granados et al. (2001) conducted a study on Morelia city of Mexico where it has been found that the transformation of the city into urban areas was taking place in a much unplanned way, without considering the landscape types or even the geological condition having seismic risk of that area. Deng et al. (2008) also studied that due to urban encroachment and rapid economic development, maximum changes have been observed in the cropland areas causing a drastically changes in land use pattern within a short time span. Large-scale forest degradation and increase of hill slope modification for crop production are also related to unscientific alteration of land use land cover of a particular area. So, proper management and monitoring of land resources considering the physical landscape and other soil and geological characteristics of a region is necessary for better environmental management and planning purposes in future.

An extensive agricultural activity has been very much dominant over the Kuya River Basin area, which mainly started after the completion of the Mayurakshi canal irrigation project in 1985. Such changes in land use land cover including the growth of urbanisation and industrialisation are affecting the natural ecosystem of the basin. Under such circumstances, an attempt has been made to assess the quantitative changes in land use land cover within the Kuya River Basin of West Bengal during the study period of 18 years. The physiographical attributes of the study area have also been analysed. Finally, a statistical technique, i.e. Multiple Correspondence Analysis (MCA), has been applied here to find out the impact of physiography with the changes of land use land cover pattern of the basin area.

16.2   Materials and Methods

16.2.1   Study Area

Kuya River, an important tributary of Mayurakshi River, has been selected for this study which passes through a varying topography of Chota Nagpur Plateau to the lower alluvial tract of the Bengal (Fig. 16.1). Kopai and Brakeswar (Twin Rivers) are the main channels of Kuya River, together have made the main Kuya channel. The river originates from a large pond of Khajuri village of Jharkhand. It flows towards south east direction over Birbhum and Murshidabad districts of West Bengal and then it has joined the Babla River near Nalghosha village of Murshidabad districts. Ultimately it merges with the main Mayurakshi River. The areal extension of the river basin area is from 23° 26′ 18″ North to 23° 56′ 30″ North latitude and 87° 13′ East to 88° 9′ 30″ East longitude covering the area of around 1527 km^2.

16.2.2   Data Used

A number of topographical maps of Survey of India having 1: 50,000 scales covering the study area have been used to delineate the river basin area. All these maps were geographically rectified assigning Universal Transverse Mercator (UTM), WGS 1984 datum and 45 N zone projection system and mosaicked with the help of Arc GIS software system. Digital elevation model (DEM) of Satellite Radar Topography Mission (SRTM) has been used to study the elevation profile of the region. On the other hand, two Landsat satellite imageries of ETM + and OLI/TIRS sensors provided by U.S. Geological Survey's (USGS) Earth explorer (http://earthexplorer. usgs.gov/) has been also used for image classification for the period of 2000 and 2018. Specifications about the used datasets have been mentioned in Table 16.1.

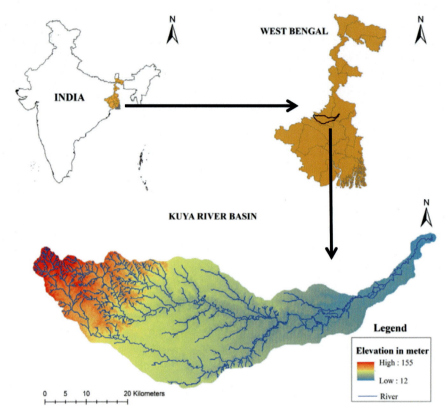

Fig. 16.1 Location map of the study area

16.2.3   Extraction of Morphometric Attributes

The morphometric analysis of the river basin broadly describes the quantitative nature of that region in all aspects like areal, linear and relief. All these parameters are highly influenced by the landform making process, transmission of water and sediment throughout the basin, discharge pattern, etc. In this study, the six major morphometric parameters from linear, relief and areal aspect have been chosen which are very much influenced by the physiography of the particular area. These are Basin Relief (Bh), Basin Slope (S), Drainage Density (Dd), Ruggedness Index (Rn), Dissection Index (DI) and Length of Overland Flow (Lo). All the parameters have been calculated using standard formulae as mentioned in Table 16.2. The major databases that have been used for morphometric analysis are SRTM DEM having 30 m resolution and Topographical maps (scale 1:50,000) provided by Survey of India, Government of India.

Table 16.1 Details of the used datasets

Data type	Details			Source
Topographical maps (1:50,000)	73 M/1, 73 M/10, 73 M/13, 73 M/14, 73 M/5, 73 M/9, 76 M/6, 79A/1			Survey of India (SOI)
SRTM DEM	Resolution—30 m			USGS earth explorer
Satellite images	Landsat 7	DOA—29/03/2000	Path/row—139/43,139/44	Resolution—30 m
	Landsat 8	DOA—07/03/2018		

Table 16.2 Morphometric parameters with formulae

Sl no	Morphometric parameters	Formulas	References
1	Basin Relief (Bh)	Vertical distance between the lowest and highest points of basin	Schumm (1956)
2	Basin Slope (S)	S = H/Lb Where H = basin relief (m) and Lb = Basin length (km)	Miller (1953)
3	Drainage Density (Dd)	Dd = L/A Where, L = Total length of stream (km), A = Area of basin (km^2)	Horton (1945)
4	Ruggedness Number (Rn)	Rn = H*Dd Where, H = Basin relief (m), Dd = Drainage density	Schumm (1956)
5	Dissection Index (DI)	DI = H/Ra Where, H = basin relief (m) and Ra = Absolute relief (m)	DovNir (1957)
6	Length of overland flow (Lo)	Lo = 1/2 D Km Where, D = Drainage density (Km/Km^2)	Horton (1945)

16.2.4 Preparation of Topographic Complexity Map

The dissimilarities among the morphometric attributes promote the topographic complexity of a region. As the universal relationship shows that in a hilly terrain area, the drainage density has always remained higher due to the nature of relief and slope of the particular region. But, in some cases, such relation will become inversed and the interplay between the physiographic attributes does not go hand in hand. To understand such topographical behaviour of the Kuya River Basin, all major morphometric parameters like basin relief, slope, drainage density, ruggedness index, dissection index and overland flow have been combined in a GIS platform. All these maps have been reclassified into five classes ranging from 1 to 5 indicating very low, low, moderate, high and very high classes and using raster calculator tool of Arc GIS software the thematic layer of Topographical Complexity map has been produced. The algorithm which has been applied to generate topographic complexity map is mentioned in Eq. (16.1).

$$\text{Topographic Complexity} = \text{Reclass}\big[\text{Relative Relief} + \text{Slope} + \text{Drainage Density}$$
$$+ \text{Ruggedness Index} + \text{Dissection Index} + \text{Overland Flow}\big]$$
$$(16.1)$$

16.2.5 Land Use Land Cover Classification

2000 and 2018 Landsat satellite imageries have been used to classify the land use land cover (LULC) map of the study area. The imageries had been downloaded from U.S. Geological Survey's (USGS) data centre. Image processing and supervised classification of LULC has been done in ERDAS Imagine software applying maximum likelihood algorithm method. Maximum likelihood algorithm is one of the most effective supervised classification methods. It calculates the spectral distance between the pixels and classifies the digital image by assigning per pixel signatures from corresponding satellite imagery. Nine LULC classes have been identified in this study area, viz. (i) Open Forest (ii) Dense Forest (iii) Exposer (iv) Waste Land (v) River/Water Bodies (vi) Agricultural Land (vii) Agricultural Fallow Land (viii) Cultivable Wet Land (ix) Settlement.

In case of land use land cover change detection analysis, performance of accuracy assessment is very much essential for individual classifications. In this work, to remove mixed pixel error and to improve the classification accuracy, a non-parametric Kappa (k) statistical test has been performed here. This statistic has been introduced by Jacob Cohen in the year of 1960. It is a measurement of agreement between predefined producers and user assigned ratings. The classified images of LULC maps for the period of 2000 and 2018 of Kuya Basin are given in Fig. 16.4. The overall classification accuracy was 95 and 96% and kappa statistics were 0.96 and 0.94, respectively, for the classification of 2000 and 2018 images. In this present study, the maximum level of accuracy has been achieved for the said two periods, as the kappa statistics is above 0.9 for both cases (Lea and Curtis 2010). It is calculated by the following formula (Cohen 1960)

$$k = \frac{P_o - P_e}{1 - P_e} \qquad (16.2)$$

where, P_o = relative observed agreement, P_e = hypothetical probability of chance agreement.

16.2.6 Multiple Correspondence Analyses (MCA)

MCA is a statistical technique which is an extension of the simple correspondence analysis for summarising a data table containing two or more than two categorical variables (IodiceD'Enza et al. 2012). This is a powerful technique to analyse and represents large and complex datasets. There are various statistical techniques that are very well-known to others like regression analysis which mathematically measures the mean cause effect relationship among the variable but the multiple correspondence analysis is formally designed to explore the relationship among the categorical variables graphically. Husson and Josse (2014) specifically define that the major objectives of MCA is to assess the similarities and associations between the variables. Tekleab et al. (2014) also analysed the association between malaria RDT result with different socio-economic and demographic variables in the African countries. In this work, with the help of SPSS software, Multiple Correspondence Analysis (MCA) has been used to examine the interrelationship between the temporal changes of land use land cover and the physiographic complexity of the Kuya River Basin area.

16.3 Results and Discussion

16.3.1 Morphometric Analysis

Basin Relief (Bh). The relief measurements of a drainage basin is an important aspect as it reflects the characteristics of major landform making processes and the hydrological nature of the region. A relief map of the study area has been prepared using 30 m resolution of SRTM DEM which exhibits the Kuya River runs through a varying topography from its source to mouth (Fig. 16.2a). It has been also observed that the maximum relief (150 m) is mainly dominated over the plateau area of Chota Nagpur region whereas, in the river mouth area which is mainly covered by the alluvial tract of Bengal plain, the relief is lower (20 m) compared to the source of the river.

Basin Slope (S). There is a strong co-relation observed between the relief and slope of the study area. The general slope of the basin decreases towards the eastern direction (Fig. 16.2b). The digital elevation model of the study area shows the area having steep slope is mainly concentrated over the zone of high relief area and moderate to low slope is identified in the river mouth or the plain land of the basin.

Drainage Density (Dd). To quantify the river basin properties of a particular region, drainage density is an important parameter. It shows the drainage properties with relation to the physiography of the region. The study of drainage density also reflects the nature of hydrological responses to the basin hydrology, its runoff potentiality depending upon the soil and vegetation cover over the region (Melton 1957; Chorley et al. 1957; Nag 1998; Ozdemir and Bird 2009). In this work, the drainage network of the basin area has been extracted from topographical sheets (scale 1:50,000) given by Survey of India to calculate the drainage density of the study area. It represents the density of the drainage network per km^2 of that basin area. The map depicts (Fig. 16.2c) that the higher drainage density of Kuya River Basin is mainly concentrated in lower part of the region, indicating the maximum inundated as well as low lying relief condition of the basin area.

Ruggedness Index (Rn). Ruggedness index is another parameter which shows the ratio between basin relief and drainage density of a particular area. It is affected

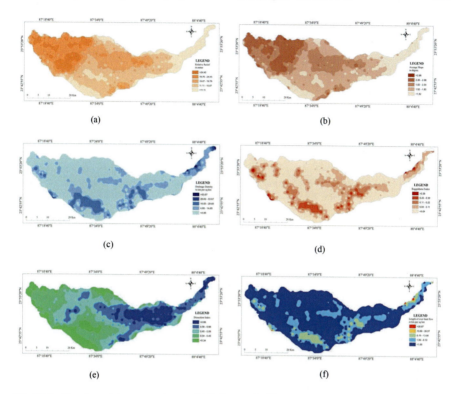

Fig. 16.2 Morphometric parameters. **a** Relief map. **b** Slope map. **c** Drainage density map. **d** Ruggedness index map. **e** Dissection index map. **f** Overland flow map

by the topography, lithological condition and gradient of that region. As shown in Fig. 16.2d, the map depicts that the high value of ruggedness index coincides with the highland exposers of the area dominated by high relief, steep slope as well as in the high stream frequency area. The rest of the part of the basin area implies less rugged topography such that low stream frequency is prevailing over there.

Dissection Index (DI). Dissection index generally measures the degree of vertical erosion which affects the whole physiography of the basin area. The value of DI ranges from 0 (complete absence of dissection) to 1 (vertical cliff). Here, the high DI value of Kuya River Basin significantly indicates highly dissected drainage condition as shown in Fig. 16.2e, where the area is mainly dominated by 1st order stream and lower Di value is mainly concentrated within the low relief zone where the stream power index is almost neglected.

Overland Flow Index (Lo). The overland flow indicates the length of channel before it gets concentrated into the definite stream. The length of overland flow is equal to half of the reciprocal drainage density of the particular basin (Horton 1945). Here in the Kuya River Basin area, as shown in Fig. 16.2f, high accumulation of water towards the river mouth has been observed where the length of the river also increases. Depending upon the soil characteristics and the underlain structure of any region, the rate of accumulation of water varies significantly. As the water saturation is high, the percolation rate decreases resultant to flood, which is very much observed in the river mouth of the Kuya Basin.

16.3.2 Topographic Complexity Map

To explore the degree of asymmetry between the morphometric parameters, topographic complexity map has been generated as shown in Fig. 16.3. The map represents five distinct classes of complexity zone varying from very high to very low having numerals values ranging from 16 to 22. The higher value of topographic complexity denotes high asymmetry in the morphometric attributes which promotes that the nature of physiography and the drainage properties of that particular place do not follow their conventional correlations. Whereas the lower value of complexity which ranges from 6 to 8 defines comparatively well-adjusted physiography as all the parameters imply each other in such a way that a more or less symmetric nature of correlation is maintained among them which further defines the land towards stability of the region. As depicted from Fig. 16.3, the very high topographic complex zone has been observed in the lower reach of the Kuya Basin where the Kopai and Kuya Rivers merge with each other and the very low complex zone prevails in the middle part of the river basin.

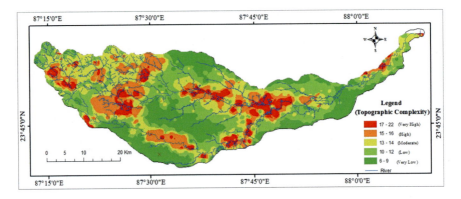

Fig. 16.3 Topographic complexity map of Kuya River Basin

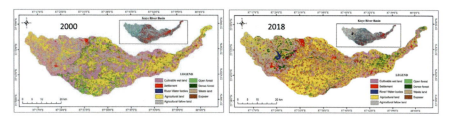

Fig. 16.4 Land use and land cover map of Kuya river basin

16.3.3   Land Use Land Cover Scenario

The LULC map of the study area has been prepared using Landsat digital data for the period of 2000 and 2018 (Fig. 16.4). The classification has been done by measuring the pixel numbers and multiplying by the spatial resolution of the remote data, i.e. 30 m. The areal statistical reports of land use land cover analysis were summarised in Table 16.3. The data reveals there has been a marked change observed in each LULC class of the Kuya River Basin during the study period of 18 years. In the year 2000, about 2.02% (61.81 km^2) area of the Kuya River Basin was under open forest, 3.18% (97.21 km^2) under dense forest, 3.05% (93.12 km^2) under waste land, 0.98% (30.06 km^2) under settlement, 0.34% (10.42 km^2) under exposer, 18.34% (560.25 km^2) under agricultural land, 14.36% (438.71 km^2) under cultivable wet land, 0.31% (9.43 km^2) under water bodies and 7.42% (226.55 km^2) area under agricultural fallow

Table 16.3 Area distribution under land use/land covers classes

Land use and land cover	2000		2018		2000–2018	
	Area in km^2	Area in %	Area in km^2	Area in %	Change in area (in km^2)	Change in area (in %)
Open forest	61.81	2.02	40.78	1.34	−21.02	−0.69
Dense forest	97.21	3.18	118.98	3.90	21.77	0.71
Waste land	93.12	3.05	0.42	0.01	−92.70	−3.03
Settlement	30.06	0.98	124.56	4.08	94.50	3.09
Exposer	10.42	0.34	43.38	1.42	32.96	1.08
Agricultural land	560.25	18.34	712.95	23.35	152.70	5.01
Cultivable wet land	438.71	14.36	286.24	9.37	−152.50	−4.99
River/water bodies	9.43	0.31	20.38	0.67	10.95	0.36
Agricultural fallow land	226.55	7.42	179.27	5.87	−47.27	−1.55
Total	1527	100	1527	100		

land. Whereas during the year 2018, the area under these land categories was found about 1.34% (40.78 km^2) under open forest, 3.90% (118.98 km^2) under dense forest, 0.01% (0.42 km^2) under waste land, 4.08% (124.56 km^2) under settlement, 1.42% (43.38 km^2) under exposer, 23.35% (712.95 km^2) under agricultural land, 9.37% (286.24 km^2) under cultivable wet land, 0.67% (20.38 km^2) under water bodies and about 5.87% (179.27 km^2) under agricultural fallow land.

16.3.4 Temporal Change Analysis of Land Use Land Cover Scenario

In the LULC detection analysis, the post classification method is much important. This method reveals both positive and negative changes, i.e. areal commission (gain) and areal omission (loss) in land use land cover pattern of the region. From Table 16.3, it has been observed that within the time span of 18 years (2000–2018) the agricultural land has raised up by 152.70 km^2 area, which has gained about 5.01% of area from other land use land cover classes registered in the year 2000. Likewise the settlement of the study area has encroached 94.50 km^2 areas from the existing area in 2000, showing an increasing behaviour as it has been gaining about 3.09% of area from other land use land cover classes. Such an expanding nature of land use land cover classes has also been noticed in exposer (32.96 km^2), dense forest (21.77 km^2) and water bodies (10.95 km^2) in the study area. On the other hand, for the rest of the land use land cover classes, i.e. open forest, waste land, cultivable wet land and agricultural fallow land, a distinct decreasing nature has been observed from the previous year (year 2000) land use land cover classification, i.e. by 21.02 km^2, 92.70 km^2, and 152.50 km^2 and 47.27 km^2, respectively. From this account, it can be said that about 0.69, 3.03, 4.99 and 1.55% area has been lost from the existing area in the year 2000.

From the above discussion of LULC change analysis, it can be said that the human interference has been increased significantly over the Kuya River Basin area. During the studied time span of 18 years, the aerial extent of settlement and agricultural land has been changed by gaining areas from other land use land cover classes. Besides this, the area of dense forest has also been changed in an increasing manner which reflects the role of joint forest management over the study area. In case of other land use land cover classes, mainly cultivable wet land, waste land and fallow land have changed their area through areal loss.

A LULC conversion matrix has also been prepared including different combinations of "from-to" classes of each land use land cover features through time, shown in Table 16.4. During the study period of 18 years, the reconfiguration of land use land cover class, i.e. from one class to another class has been produced by Arc GIS software. The result indicates that maximum positive changes have been noticed from each class to agricultural land. Such expansion of agricultural activity reinforces the development of settlement over the region. Also the negative changes in some land

Table 16.4 Conversion table of land use land cover classes from the year 2000 to 2018

LULC classes (2000) (area in km²)	LULC classes (2018) (area in km²)									
	Open forest	Dense forest	Waste land	Exposer	Settlement	Agricultural land	Cultivable Wet land	River/Water bodies	Agricultural fallow land	Total (2000)
Open forest	**3.58**	8.88	0.006	4.13	2.98	29.74	10.75	0.56	1.10	61.81
Dense forest	4.20	**12.02**	0.007	4.84	9.56	48.62	15.24	0.77	1.87	97.21
Waste land	2.93	9.53	**0.13**	4.52	10.84	28.23	16.13	4.39	16.31	93.12
Exposer	0.32	1.01	0.0009	**0.69**	0.87	5.01	0.62	1.86	0.03	10.42
Settlement	0.03	2.24	0.008	1.91	**12.67**	6.71	4.55	0.75	0.39	30.06
Agricultural land	12.46	50.13	0.07	11.99	31.34	**341.04**	76.59	3.57	32.50	560.25
Cultivable wet land	14.00	27.41	0.050	13.23	42.30	179.96	**130.91**	1.55	28.93	438.71
River/Water bodies	0.29	0.65	0.032	0.14	0.27	4.29	0.23	**3.41**	0.11	9.43
Agricultural fallow land	2.16	7.03	0.12	3.07	12.52	68.89	31.08	3.51	**97.99**	226.55
Total (2018)	40.78	118.98	0.42	43.38	124.56	712.95	286.24	20.38	179.27	1527

use land cover classes like waste land, open forest, that have been shifted to the major change in land use land cover viz. agricultural land of the river basin area. So the major impact of such land use land cover shift was subjected to agricultural land and settlement, which in turn creates more stress on the overall land resources of the region.

16.3.5 Correspondence Analysis Between the Physiography with Land Use Land Cover for the Period of 2000 and 2018

The results of MCA for the period of 2000 and 2018 produced an output as a correspondence Tables 16.5 and 16.6 with row column plots of the variables (Fig. 16.5a and b). It has been observed from the graphical representation that within the high complexity zone of physiography as indicated by class 5 and 4, the association of waste land and agricultural fallow land were high, whereas the other classes mainly settlement, agricultural land and water bodies has been concentrated within the low to moderately complex zone of the river basin area.

However an interesting observation has been made from the output of 2018 as shown in Fig. 16.5b. Within a time span of 18 years the association of agricultural land has been shifted from moderately complex zone to lower topographic complexity zone in the study area. From such observation, it can be said that the existence of normal correlation among the physical factors may allow more developmental activities to utilise the land resources. In a similar way, local people of the Kuya River Basin adopted their land use practice depending upon the nature of physiographical complexity so as to maintain the land suitability and enhance the land productivity as well.

From these two statistical analyses, it can also be stated that mainly the non-productive land use practices such as waste land, agricultural fallow land, etc., are highly associated within the zone of very high topographical dissimilarities whereas in a low topographic dissimilarities area people are utilising the land resources by practising it in a productive manner viz. agricultural activity, development of settlement, etc.

16.4 Conclusions

Physiographic conditions of any region have a control on land use practices of that area. Based on the physical characteristics like relief, drainage, slope of the region people are adopting their land utilisation pattern. But sometimes because of the occurrence of dissimilar nature among the physical factors, utilisation of land resources

Table 16.5 Correspondence table for the year 2000

Complexity classes	Name Open forest	Dense forest	Waste land	Exposer	Cultivable wet land	Settlement	River/water bodies	Agricultural fallow land	Agricultural land	Active margin
1	14.07	18.44	5.20	1.75	66.26	3.57	0.99	7.10	65.00	182.39
2	12.80	22.89	9.96	2.34	78.97	5.78	1.95	18.62	117.50	270.97
3	17.67	33.17	26.50	3.81	134.19	9.75	3.78	80.99	193.39	503.24
4	12.29	17.46	40.13	2.08	120.88	9.02	2.01	96.58	143.73	444.18
5	4.41	4.89	10.94	0.43	37.15	1.84	0.57	22.56	36.13	118.91
Active margin	61.42	96.85	92.74	10.40	437.45	29.95	9.30	225.84	555.75	1519.68

Table 16.6 Correspondence table for the year 2018

Complexity classes	Name									
	Open forest	Dense forest	Waste land	Exposer	Cultivable wet land	Settlement	River/water bodies	Agricultural fallow land	Agricultural land	Active margin
1	4.21	7.94	0.02	2.65	30.23	14.19	1.63	15.77	106.12	182.74
2	7.84	18.58	0.02	5.75	48.95	19.66	2.59	31.45	136.21	271.05
3	13.84	37.04	0.10	13.53	96.94	40.51	5.37	133.43	162.85	503.60
4	10.81	41.10	0.23	17.02	81.13	39.94	9.23	153.61	91.27	444.34
5	3.79	12.43	0.06	4.36	28.13	10.10	1.51	33.50	25.07	118.94
Active margin	40.49	117.1	0.42	43.31	285.37	124.40	20.32	367.76	521.52	1520.66

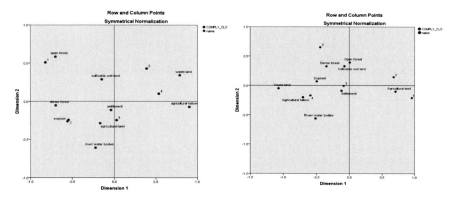

Fig. 16.5 Multiple correspondence anaysis. **a** Variable plot for the year 2000. **b** Variable plot for the year 2018

cannot be done in a productive way. The major findings of this work have been mentioned as follows:

- This paper enumerated the influence of physiography on land use land cover practices over the Kuya River Basin area using remote sensing and GIS technique.
- It has been represented how the dissimilar relationship among the morphometric parameters makes the physiography complex in nature.
- Finally it has been proved that within the study area the productive land utilisation has been mainly associated with the lower complex topography and as the complexity increases the major non-productive land use like dense forest, fallow land, waste land, exposers are becoming dominant, which ultimately denotes how the complexities in physical factors can modify the land utilisation pattern of the Kuya River Basin area.
- Such kind of research can be helpful for the society for sustainable management of land resources. There is a need to consider the natural limitation of the area before modifying the land utilisation pattern of the respective area, which will reduce the environmental hazard as well as sustain the natural resources for future use.

Acknowledgements   Authors express their sincere thankfulness to the Survey of India, Government of India for providing relevant topographical maps of the study area. Authors also acknowledge the computer laboratory of School of Water Resources Engineering, Jadavpur University for providing the necessary facilities to carry out the present study.

References

Butt A, Shabbir R, Ahmad SS, Aziz N (2015) Land use change mapping and analysis using remote sensing and GIS: a case study of Simly watershed, Islamabad, Pakistan. Egypt J Remote Sens Space Sci 18:251–259. https://doi.org/10.1016/j.ejrs.2015.07.003

Chorley RJ, Donald EGM, Pogorzelski HA (1957) New standard for estimating drainage basin shape. Am J Sci 255:138–141. https://doi.org/10.2475/ajs.255.2.138

Cohen J (1960) A coefficient of agreement for nominal scales. Educ Psychol Measur 20(1):37–46. https://doi.org/10.1177/001316446002000104

Debnath J, DasN, Ahmed I, Bhowmik M (2017) Channel migration and its impact on land use/land cover using RS and GIS: a study on Khowai River of Tripura, North–East India. Egypt J Remote Sens Space Sci 20. https://doi.org/10.1016/j.ejrs.2017.01.009

Deng J, Wang K, Deng Y, Qi G et al (2008) PCA-based land-use change detection and analysis using multitemporal and multisensor satellite data. Int J Remote Sens 29:4823–4838. https://doi.org/10.1080/01431160801950162

DovNir (1957) The ratio of relative and absolute altitudes of Mt. Carmel, a contribution to the problem of relief analysis and relief classification. Geogr Rev USA 47:564–569

Granados E, Gerardo B, Mendoza M, Duhau E (2001) Predicting land-cover and land-use change in the urban fringe: a case in Morelia city, Mexico. Landsc Urban Plan 55:271–285. https://doi.org/10.1016/S0169-2046(01)00160-8

Haque M, Basak R (2017) Land cover change detection using GIS and remote sensing techniques: a spatio-temporal study on TanguarHaor, Sunamganj, Bangladesh. Egypt J Remote Sens Space Sci 20:251–263. https://doi.org/10.1016/j.ejrs.2016.12.003

Horton RE (1945) Erosional development of streams and their drainage basins; hydrophysical approach to quantitative morphology. Geol Soc Am Bull 56(3):275

Husson F and Josse J (2014) Multiple correspondence analysis, the visualization and verbalization of data, Chapter: Multiple correspondence analysis 161–184

IodiceD'Enza A, Enza D, Greenacre M (2012) Multiple correspondence analysis for the quantification and visualization of large categorical data sets. Adv Stat Methods Anal Large Data-Sets. https://doi.org/10.1007/978-3-642-21037-2_41

Kumar A, Malik A (2014) Water resource management in hilly areas: challenges and opportunities. National Seminar on Natural Resource Management and Environmental Concerns

Lambin EF, Turner BL, Geist HJ, Agbola SB, Angelsen A, Bruce JW et al (2001) The causes of land-use and land-cover change: moving beyond the myths. Glob Environ Chang 11(4):261–269. https://doi.org/10.1016/S0959-3780(01)00007-3

Lea C, Curtis AC (2010) Thematic accuracy assessment procedures: National Park Service Vegetation Inventory, version 2.0. National Resource Report NPS/2010/NRR-2010/204, National Park Service, Fort Collins, Colorado, USA

Melton MA (1957) An analysis of the relations among elements of climate, surface properties, and geomorphology. Department of Geology, Columbia University, New York

Miller VC (1953) A quantitative geomorphic study of drainage basin characteristics in the Clinch Mountain area,Varginia and Tennessee. Columbia University, Department of Geology, ONR, Geography Branch, New York

Mondal S, Sharma N, Kappas M, Garg PK (2015) Critical assessment of land use land cover dynamics using multi-temporal satellite images. Environments 2:61–90. https://doi.org/10.3390/environments2010061

Nag SK (1998) Morphometric analysis using remote sensing techniques in the Chaka Sub-basin, Purulia District, West Bengal. J Indian Soc Remote Sens 26:69–76. https://doi.org/10.1007/BF03007341

Ozdemir H, Bird D (2009) Evaluation of morphometric parameters of drainage networks derived from topographic maps and DEM in point of floods. Environ Geol 56:1405–1415. https://doi.org/10.1007/s00254-008-1235-y

Prus B, Budz L (2015) Analysis of relief impact on land use structure on an example of Nowy Targ district. Infrastruct Ecol Rural Areas 483–497. https://doi.org/10.14597/infraeco.2015.2.2.039

Rawat JS, Kumar M (2015) Monitoring land use/cover change using remote sensing and GIS techniques: a case study of Hawalbagh block, district Almora, Uttarakhand, India. Egypt J Remote Sens Space Sci 18:77–84. https://doi.org/10.1016/j.ejrs.2015.02.002

Reis S (2008) Analyzing land use/land cover changes using remote sensing and GIS in Rize, North–East Turkey. Sensors 8:6188–6202. https://doi.org/10.3390/s8106188

Rogger M, Agnoletti M, Alaoui A, Bathurst JC, Bodner G, Borga M et al (2017) Land use change impacts on floods at the catchment scale: challenges and opportunities for future research. Water Resour Res 53(7):5209–5219. https://doi.org/10.1002/2017WR020723

Schumm SA (1956) The evolution of drainage systems and slopes in bad lands at Perth, Amboi, New Jersey. Geol Soc Am Bull 67(5):597–646

Tekleab S, Mohamed Y, Uhlenbrook S,Wenninger J (2014) Hydrologic responses to land cover change: the case of Jedeb mesoscale catchment, Abay/Upper Blue Nile Basin, Ethiopia. Hydrol Process 28. https://doi.org/10.1002/hyp.9998

Chapter 17
Assessment of Land Use and Land Cover Change Dynamics Using Remote Sensing and GIS Techniques in Most Effected Parts of Rajpur-Sonarpur Municipality

Bijay Halder⊙**, Papiya Banik, and Jatisankar Bandyopadhyay**⊙

Abstract Earth surface processes are affected by climate change, environmental degradation, urbanization and a huge amount of greenhouse gas (GHG) emission. Rapid urbanization process is hammering the environment process. Huge population breakdowns the natural process like vegetation degradation, land scarcity, ground water shortage and construction area. Now the rural–urban fringe areas are developing their urbanization due to land scarcity problem and that's the reason for land use and land cover change. Some parts of Rajpur-Sonarpur municipality have huge amount of population pressure and those areas are facing land degradation. Remote sensing and GIS technology were building their platform for various types of investigation using satellite imageries. Multi-temporal Landsat OLI data was used to calculate the total LU/LC change in Rajpur-Sonarpur municipality. The supervised classification technique was used along with maximum likelihood method for detecting the areal change in the year of 2014 and 2019. The result shows that the build-up areas were mostly increased in last 5 years around 193.595 ha area due to population pressure and the total vegetation area lost was 166.244 ha. Many open spaces and agricultural land were converted into build-up area. The overloading population is the main factor for land degradation.

Keywords Remote sensing and GIS · Land use and land cover · Satellite imageries · Change detection · Rajpur-Sonarpur municipality

B. Halder (✉) · J. Bandyopadhyay
Department of Remote Sensing and GIS, Vidyasagar University, Midnapore, India

P. Banik
Department of Geography, University of Calcutta, Kolkata, India

J. Bandyopadhyay
Centre for Environmental Studies, Vidyasagar University, Midnapore, India

17.1   Introduction

Earth surface is shielded by the terrestrial and also the water body zone. Rapid climate change as well as the urbanization process, environmental change and greenhouse gas (GHG) were transforming the land degradation over the earth. Population density is one of the influencing aspects for the land use and land cover (LULC) change. Due to population pressure on a particular area, it causes land insufficiency and environmental degradation. The LULC changes work on an important aspect in the study of global transformed situation for the total changes of the important principles to decision-making for the ecosystem management and environmental planning for future interaction universally (Dwivedi et al. 2005; Fan et al. 2008; Zhao et al. 2004). In twenty-first century, land cover dynamics is the main dramatic implication of the global concern (Meshesha et al. 2016). Land use change is the change dynamics of physical and also the biological management of the forest land change into the farming land, land scarcity, vegetation degradation and agricultural area loss (Prakasam 2010; Shiferaw and Singh 2011).

Many parts of the world, mostly the developing countries, population was depending on agricultural land and that was affected in land use change (Meshesha et al. 2016). Most of the area has restricted study for long period land use trend detection (Goldewijk and Ramankutty 2004). Last 50 years in east Africa, agricultural land extended into marginal land (Yitaferu 2007). Controlling aspects were socio-economic conditions of the society, population pressure, and physiographic feature and also the land category has resulted in land use transformation (Meshesha et al. 2016). Several parts of the world communication between many anthropogenic and natural factors cause land use change and the utilization of this nature of resource by the human population in the time and space (Dai and Khorram 1999; Owojori and Xie 2005).

To calculate approximately land use and land cover dynamics, geospatial technologies and geographic information system tools are used (Carlson and Sanchez-Azofeifa 1999; Dezso et al. 2005; Guerschman et al. 2003). Remote sensing is a more actual, easy and time-consuming method. It assists in recognizing the actual transformation in the earth lithosphere via change detection technique (Dai and Khorram 1999). Change/accuracy detection is not the only way to detect the changes on the earth surface with the help of satellite images. There are numerous ways or methods for identifying the changes on earth surface. But the change detection technique is the maximum appropriate or effective technique for the identification of earth surface transformation. This technique is implemented with two steps; pre-classification and post-classification. In the phase of pre-classification, nominal changes are acknowledged, and in the post-classification phase, changes are identified appropriately with the feature of different attributes. Massive land use changes are originated in the southernmost wards of the study zone. Rest of the wards have also experienced the degradation of the green zone due to urban expansion.

The remote sensing procedure is the easiest method to estimating the earth surface dynamics on a specific area in different time ages. Using remote sensing and GIS techniques, satellite imagery was an excellent data source for studying landscape modified situation of the earth. Land use pattern dimensions, such as quantity of land use, size and shape of the patches specify more about the covered area (Forman 1995). Remote sensing methods of high-resolution images of different years are used to recognize earth dynamic change. It can help in regional planning and is also useful for environment planning for public. Remotely sensed changed detection is based on artificial neural network (Dai and Khorram 1999). A new technique for multispectral image classification using different tanning algorithms was used to distinguish transformation on pixel by pixel basis in the real-time application. The accuracy assessment is not only the method for detecting the LU/LC change analysis and check the earth surface transformation phenomenon. In this case, so many methods or algorithms have been developed to detecting earth surface change more accurately.

The total lithosphere is called land use and land cover area. Vegetation and water bodies are identified as the land cover area (Forman 1995). In this time, the development of technologies, increase of urban facilities, and change in anthropological existence are the causes of the change in human demands. Many flats, complexes, roads, and other urban facilities are built to fulfil these human demands; these are known as land use area. The systematic gradual variations of earth and other temporal and spatial fluctuations of the earth are easily identified by the researchers.

17.2 Study Area

A ward of Rajpur-Sonarpur Municipality area was designated as study zone in South 24 Parganas district of West Bengal, India. Rajpur-Sonarpur municipality was found in the year 1980. Now this municipality have 35 wards. Total population is 4, 70,000 (2011 census). This actual study area is covered by 7.412 km^2. Study area is situated in 88°24'17.81″ E–88°25'55.14″ E and 22°25'47.87″N–22°27'42.35″ N. The elevation range in this study area was 9 m or 30 ft. Average temperature was 26.29 °C and average rainfall was 137.08 mm. Main rainfall occurrs throughout May–October (Fig. 17.1).

Earlier the State Highway 1 (also known as E. M. Bypass) scheme, Rajpur-Sonarpur was linked to Kolkata by N.S.C Bose Road (Netaji Subhash Chandra Bose Road) and Garia Main Road. This is a vision of distant past. Afterward, the construction of State Highway 1 (SH-1) and the Eastern Metropolitan Bypass is connecting the place with heart of Kolkata easily. Other State Highway is Diamond Harbour Main Road. The study area was selected because, after the expansion of southern metro position, this urban development was rapidly improved and affected the LULC pattern. Due to urban expansion, population pressure, and development, the cities areas are mostly increased health facilities, transportation accessibility, communication system, and many others urban amenities. Rajpur-Sonarpur municipality also increased the urban amenities center for the overall development of the study area.

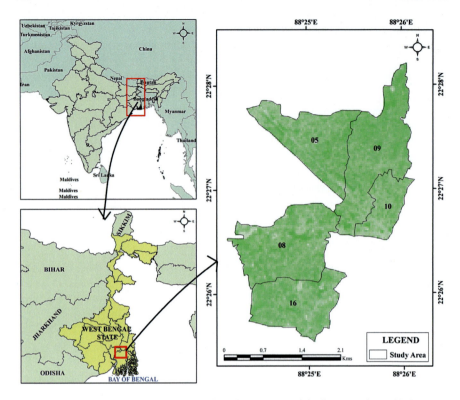

Fig. 17.1 The location map of this study (Rajpur-Sonarpur Municipality) area situated in South 24 Parganas, West Bengal

17.3 Materials and Methods

17.3.1 Data Used

In the data for study area, LULC change is calculated using Landsat image. Different year images were used to identify the real areal transformation situation for earth surface (https://earthexplorer.usgs.gov/). Images of years 2014 and 2019 were taken for LULC change detection. Image information is showing in Table 17.1.

17.3.2 Pre-processing and Classification

Pre-processing of the satellite image is an important portion to distinguish the changes on the earth surface. To improve the image quality, data was enhanced using histogram equalization (Meshesha et al. 2016). Data are pre-processed in ERDAS

Table 17.1 Details of satellite data used

Data	Sensor	Year of acquisition	Bands/colour	Resolution (m)	Date	Source
Landsat 8	OLI	2014	Multi-spectral	30	16/11/2014	https://earthe xplorer.usgs. gov/
Landsat 8	OLI	2019	Multi-spectral	30	14/11/2019	https://earthe xplorer.usgs. gov/

IMAGINE and existing Area of Interest (AOI). Two satellite imageries are taken for colour composite and it assistances to draw 75 training sites for 6 major LULC classes (Fig. 17.3). We got the spectral signatures from the Landsat image. To recognize the land use transformation, initially satellite image was geo-referenced, transformed and enhanced. Different colour composites were used for documentation of different types of LULC classes. The infrared colour composite was applied for identifying the vegetation covered area, and the field visits that make appropriate LULC classification (Fig. 17.2).

Supervised classification is used for LULC classification to identify the LULC change of two different years. Different size, shape, texture, tone and colour are helping to identify different LULC classes like water body, settlement, wetland, vegetation, open space and agricultural land. Maximum likelihood classifier decision rule is used for supervised classification (Table 17.2). ArcGIS software is used for thematic map generation.

17.3.3 Post-classification

The post-classification phase is used to improve classification accuracy (Cheruto et al. 2016). Using different categories of satellite data, medium spatial resolution of Landsat data mixed pixels are a common problem (Lu and Weng 2005), mainly the urban areas are heterogeneous combinations including residential area, roads, railway, soil, trees, water and grassland (Jensen et al. 2007). Visual interpretation is a significant portion of LULC change analysis. After image classification, the geographical area for each LULC classes was calculated for change in land use type within two different years (Fig. 17.3). Also, the total percentages were calculated in ERDAS IMAGINE 14.

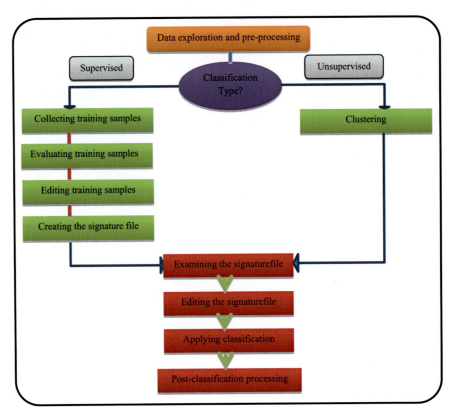

Fig. 17.2 Satellite Image classification techniques. Both Supervised and Unsupervised techniques are measured in this figure. *Source* ArcGIS website

17.3.4 Accuracy Assessment

Accuracy assessment is an important part of every individual classification of satellite image for change detection documentation (Owojori and Xie 2005). The standard accuracy assessment technique was conduct by 'error matrix'. There was a square matrix calculation of rows and columns in LULC classes. One set of data was automatically nominated by random point and one data was reference data, which was ground truth data, aerial photograph, provisionally tested maps or other data. For the accuracy assessment of land cover identification from satellite data, stratified random method was used to represent different types of land cover classes of the region. The accuracy was to detect by many point, based on different types of ground data and visual interpretation. The assessment between classification results and references data was accepted statistically using error matrices (Rosenfield and Fitzpatrick-Lins 1986).

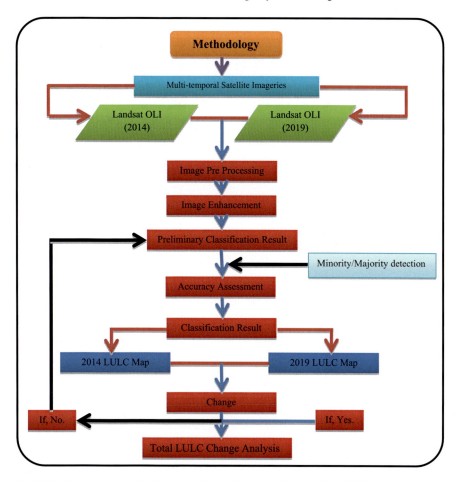

Fig. 17.3 The detailed methodology for Rajpur-Sonarpur Municipality LULC change detection using satellite images and field survey

Table 17.2 Land use and land cover class name and the description of the class area

Sl. no	Class name	Description
1	Water body	Open water, lakes, ponds, canal or nala
2	Vegetation cover	Forest, grass, green cover area
3	Build-up area	Residential area, commercial, transportations, urban fringe area, roads, industrial belt
4	Open space	Fallow lands, play ground
5	Agricultural land	Crop land, cultivated land

17.3.5 Kappa Coefficient

Kappa coefficient is not only a diagonal component but also the component in the confusion matrix (Shiferaw and Singh 2011). The kappa values are a matric that can compare with observed accuracy and expected accuracy or random change. This value is not only the calculating factor of a single classifier but also the evaluate classifiers. The kappa does not take in the degree of disagreement between observers and all disagreement is treated equally as total disagreement (Cohen 1968).

$$K = (p_o - p_e)/(1 - p_e) \qquad (17.1)$$

K means Kappa value, p_o is considered as a total accuracy and p_e is the random accuracy of the classified image. According to Altman (1991), kappa value is most important for accuracy assessment.

17.4   Results and Discussion

17.4.1 Land Use and Land Cover

LULC change dynamics consists of five classes, namely Vegetation, Build-up Area, Water Bodies, Open Space and Agricultural Land. Due to calculating the change, there was decreased vegetation area and other land use areas and increased in build-up area.

Build-up Area Residential area is a frequently effected in land use change. Due to population growth, build-up area was capturing the other area, which was open space, particularly vegetation cover area and agricultural land. In this change of 2 years (years 2014–2019), build-up area was more increased about 193.595 (ha) or 26.12%.

Vegetation Cover basically in Rajpur-Sonarpur area is not any forest cover area. But in this location, mo vegetation cover areas are situated. The southern part of this area was showing more vegetated area. Total change of this vegetation area was 22.43% that was 166.244 (ha) areas. When urbanization procedure moved frontward, the pressure was increased in open space, agricultural land and vegetation. In this area, due to metro expansion and other capability improved, vegetation covered area was rapidly decreased (Fig. 17.4).

Water Bodies. Due to this 2 years' LULC change detection, water bodies were decreased in total area 2.816 (ha) or 0.37(%). In 2014, water body was about 48.4722 (ha) or 6.54% in total area. But after change detected in 2019, the area was decreased by 6.16%. Due to infrastructural development of this region, there was an extreme demand of land for residential area.

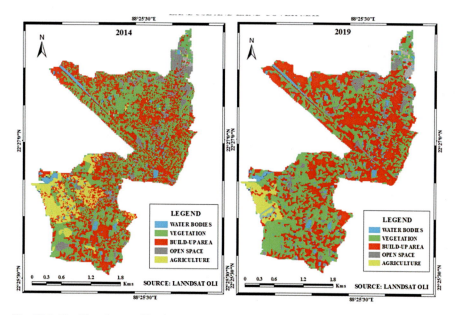

Fig. 17.4 Total Land use and land cover change dynamic over the most populated part of Rajpur-Sonarpur Municipality area

Agricultural land. The economic constancy in Rural–Urban fringe or rural area mostly dependent in there Agricultural Land. Mans are habited to change their productivity including agricultural area and other thinks in technology. But in this era, they were lost their land property. Because in urban region is better sustainable rather than rural region. So in this case, Rajpur-Sonarpur Municipality region now growing to urbanized and their economy was increased, that the reason for agricultural land converted into residential area.

Open Space. Another important land class was open space; it was decreased in 15.009 (ha) or 2.03%. Open space is a better solution to build a residential area or industries. In this study area, the north-east part this entire region was very good location of Open Space. But it the sense of in 2019, Open space was decreased.

In the map basically agriculture originate mainly in the ward no. 8 and few of the area of the wards in 2014, but 2019 the agricultural land reductions very high rate and appear into basically ward no. 8. The map demonstrated that the decrease of land cover resulted as the growth of the pollution, temperature, decrease of ponds and water bodies decrease of ecosystem, also increase of settlement resulted as the increase of population density, lack of area for habitat, etc. (Table 17.3). From the map basically out of five elements settlement is increases and it increases very high amount. Due to rapid growth of population, growth of urban facilities, extension of metropolitan city Kolkata, migration of people from rural area, also due extension people settled to the sub-urban area (Figs. 17.4 and 17.5).

Table 17.3 Total land use and land cover change calculation over the study area

LULC classes	Land use land cover area (ha) and total % area				Land use land cover area (ha) and total % area	
	2014		2019		2014–2019	
	Area (ha)	Area (%)	Area (ha)	Area (%)	Area (ha)	Area (%)
Water bodies	48.4722	6.54	45.6585	6.16	−2.816	−0.37
Vegetation area	381.033	51.41	214.779	28.98	−166.244	−22.43
Open space	69.558	9.385	54.579	7.36	−15.009	−2.03
Agricultural land	11.6363	1.57	2.088	0.28	−9.5613	−1.29
Build-up area	230.5	31.1	424.095	57.22	193.595	26.12
Total	741.1995	100	741.1995	100	–	–

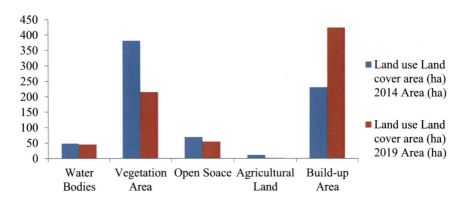

Fig. 17.5 Total LULC change during 2014–2019 in the Rajpur-Sonarpur study area

17.4.2 Estimation of Vegetation Degradation

The Normalized Difference Vegetation Index (NDVI) can identify or measure the vegetation quality. In the map, vegetation quality identifies the degradation or decreases of vegetation of five selected wards of Rajpur-Sonarpur Municipality area. The Landsat satellite imagery was used to create the NDVI map.

The Normalized Difference Vegetation Index (NDVI) is calculated by (Eq. 17.2)

$$NDVI = (NIR - Red)/(NIR + Red) \qquad (17.2)$$

Basically, NDVI was calculated in Healthy vegetation sheltered area; where the vegetation area was very high, the NIR (Near-Infrared) value was extra penetrating rather than non-vegetated area. In 2014, the vegetation area percentage of these wards 51.41% and in 2019 the percentage of vegetation remains decreases, the percentage

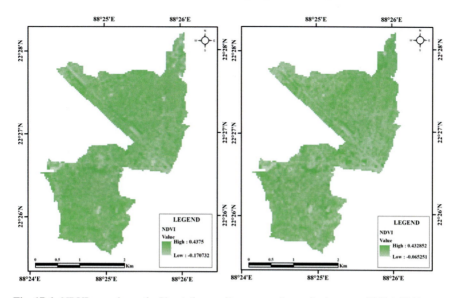

Fig. 17.6 NDVI map shows the Vegetation quality or area change in the year of 2014–2019

28.98%. The vegetation decreases means increase of settlement and other essential factors in this area. Especially the vegetation degraded mostly in the 5, 9, 10 wards.

But on the other two wards the degradation frequency is less than the other three. In these three wards, the vegetation degradation indicate that the other facility recover in those three wards, which resulted as the decrease of vegetation. It also resulted as the increase of settlement (Fig. 17.6).

17.4.3 Urban Extension in Last Five years

Urbanization is effect on the environment and over the earth surface. In some recent era, due to population pressure a huge amount of area covering with the concrete. Urban extension is directly hammering the environmental process and breakdown the natural process. Population pressure is grip up the other land cover area like open space, agricultural land and sometime water body also. At the present time, rural–urban springe and also the rural area are captured by the population. Rajpur-Sonarpur municipality area is located in South 24 Parganas district of West Bengal. In some decades, this area facing a huge amount of population pressure and the results are rapid urbanization, land degradation and land use and land cover change. In last 5 years this area facing the land degradation, agricultural land losses and rapid urbanization. North-west parts of this area facing huge build up area change. Due to urban extension, transport accessibility, less land value and better facility breakdown the natural process. 193.595 Ha area increased in last five years (Fig. 17.7). Similarly

Fig. 17.7 Urban extension
in Rajpur-Sonarpur
municipality

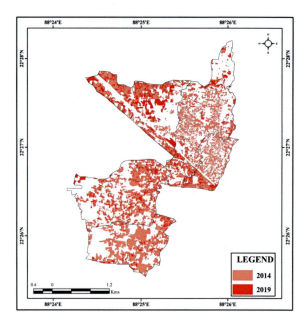

the vegetation, agricultural land and open spaces are degraded due to these types of
urbanization presses as well as land use and land cover change dynamic. Essay to
accessible the Kolkata area for hospitality, job and many urban amenities.

17.4.4 Reason Behind the Land Use and Land Cover Change

Numerous social and natural factors are accountable for land use and land cover
change dynamics. The population concentration was the most controlling factor for
LULC change. The human influence like, build residential area, commercial, industrial area also the constructional build-up. Fast population growth is the increasing
factor for the LULC change dynamics. In rural–urban fringe area, due to some
decades urban facility was rapidly increased and site by site it was affected by the
land use and land cover change dynamics. The main reason is that, nowadays most
of the family sifted for nuclear family and they need their housing area. Population
pressure was reduced the open space, vegetation area and it causes a huge amount
of land use and land cover change, increases land subsidence.

17.4.5 *Significance of Land Use and Land Cover Change Dynamics*

Land use and land cover transformation may not be the essential outcome in land degradation and soil erosion. If the land use and land cover class fast transformation into farmland, barren land and fertile soil is more massive erosion and degradation, without vegetation area. In this classified image of LULC change detection in different LULC classes between 2014 and 2019 identifying the open space, water bodies, agricultural land and vegetation area were rehabilitated into settlement area. Due to those types of land use and land cover change, this area facing a enormous amount of land degradation, soil erosion, land subsidence, and also unavailability of groundwater.

In Ethiopia, the rapid development of agricultural land into a steeper slope has aggravated for soil erosion and land degradation. The alteration in land use and land cover class significantly affected in surface runoff, soil erosion, sedimentation, land degradation, flood, drought, migration, biodiversity change and decrease of agricultural productivity. In this study area, massive land use and land cover change due to metro expansion were mostly affected in the complete the environment.

17.5 Conclusion

The key determination of the study is to discover the urban growth or urban development of a peripheral part of Kolkata, basically, classify the urban expansion of Kolkata to its peripheral area. Rajpur-Sonarpur is the peripheral part of Kolkata, which is situated in South 24 Parganas. From the survey, it was originate that the south most part of Kolkata or the Garia zone of Rajpur-Sonarpur Municipality area highly influenced by Kolkata and a massive amount of development found at this area. The urban development also growth due to population growth of theses municipality wards. The inhabitant's growth makes the scarcity of land, high land value, and building of new flats. It also reasons for the degradation of open spaces. It is very significant that increase the urban facilities to give a better life of the people of the wards, and limit the horizontal expansion, also increase the vertical expansion. This help to reduce the land scarcity problems, open spaces and vegetation degradation problems, etc.

Also additional problem is bad road condition; low accessibility of buses, to reduce this kind of problems municipality should emphasis on the road repairing to avoid the accidents of road. In these areas, there is no govt. hospital found, also the govt. health centres are not sufficient. Government should take some action to solve the problem. This helps the lower class people to get a good health facility for free or very low cost. Electricity available at all most all of the household, but the power cut makes a big problem. Government should take care about the matter and to reduce the problem. Because some time the students are suffers for this at their exam time. Government schools are found in this area. But the condition of these schools are not good, to improve the infrastructural developments schools, colleges, banks, commercial sector development is also important. Infrastructural development not refers to develop building, construction of flats, but it refers the overall development of any area.

References

Altman DG (1991) Mathematics for kappa. Pr Stat Med Res 1991:406–407

Carlson TN, Sanchez-Azofeifa GA (1999) Satellite remote sensing of land use changes in and around San Jose, Costa Rica. Remote Sens Environ 70(3):247–256

Cheruto MC, Kauti MK, Kisangau DP, Kariuki PC (2016) Assessment of land use and land cover change using GIS and remote sensing techniques: a case study of Makueni County, Kenya. http://repository.seku.ac.ke/handle/123456789/3062

Cohen J (1968) Weighted kappa: nominal scale agreement provision for scaled disagreement or partial credit. Psychol Bull 70(4):213. https://doi.org/10.1037/h0026256

Dai XL, Khorram S (1999) Remotely sensed change detection based on artificial neural networks. Photogramm Eng Remote Sens 65:1187–1194

De Wit AJW, Clevers JGPW (2004) Efficiency and accuracy of per-field classification for operational crop mapping. Int J Remote Sens 25(20):4091–4112. https://doi.org/10.1080/01431160310001619580

Dezso Z, Bartholy J, Pongracz R, Barcza Z (2005) Analysis of land-use/land-cover change in the Carpathian region based on remote sensing techniques. Phys Chem Earth Parts A/B/C 30(1–3):109–115. https://www.sciencedirect.com/science/article/pii/S1474706504002013

Dwivedi RS, Sreenivas K, Ramana KV (2005) Cover: land-use/land-cover change analysis in part of Ethiopia using Landsat Thematic Mapper data. Int J Remote Sens 26(7):1285–1287. https://doi.org/10.1080/01431160512331337763

Fan F, Wang Y, Wang Z (2008) Temporal and spatial change detecting (1998–2003) and predicting of land use and land cover in Core corridor of Pearl River Delta (China) by using TM and ETM+ images. Environ Monit Assess 137(1–3):127. https://doi.org/10.1007/s10661-007-9734-y

Forman RT (1995) Some general principles of landscape and regional ecology. Landsc Ecol 10(3):133–142. https://doi.org/10.1007/BF00133027

Goldewijk KK, Ramankutty N (2004) Land cover change over the last three centuries due to human activities: The availability of new global data sets. GeoJournal 61(4):335–344. https://doi.org/10.1007/s10708-004-5050-z

Guerschman JP, Paruelo JM, Bella CD, Giallorenzi MC, Pacin F (2003) Land cover classification in the Argentine Pampas using multi-temporal Landsat TM data. Int J Remote Sens 24(17):3381–3402. https://doi.org/10.1080/0143116021000021288

Jensen R, Mausel P, Dias N, Gonser R, Yang C, Everitt J, Fletcher R (2007) Spectral analysis of coastal vegetation and land cover using AISA+ hyperspectral data. Geocarto Int 22(1):17–28. https://doi.org/10.1080/10106040701204354

Lu D, Weng Q (2005) Urban classification using full spectral information of Landsat ETM+ imagery in Marion County, Indiana. Photogramm Eng Remote Sens 71(11):1275–1284. https://doi.org/10.14358/PERS.71.11.1275

Meshesha TW, Tripathi SK, Khare D (2016) Analyses of land use and land cover change dynamics using GIS and remote sensing during 1984 and 2015 in the Beressa Watershed Northern Central Highland of Ethiopia. Model Earth Syst Environ 2(4):1–12. https://doi.org/10.1007/s40808-016-0233-4

Owojori A, Xie H (2005, March) Landsat image-based LULC changes of San Antonio, Texas using advanced atmospheric correction and object-oriented image analysis approaches. In: 5th international symposium on remote sensing of urban areas, Tempe, AZ

Prakasam C (2010) Land use and land cover change detection through remote sensing approach: a case study of Kodaikanaltaluk, Tamil Nadu. Int J Geomat Geosci 1(2):150

Rosenfield GH, Fitzpatrick-Lins K (1986) A coefficient of agreement as a measure of thematic classification accuracy. Photogramm Eng Remote Sens 52(2):223–227. https://pubs.er.usgs.gov/publication/70014667

Shiferaw A, Singh KL (2011) Evaluating the land use and land cover dynamics in Borena Woreda South Wollo Highlands, Ethiopia. Ethiop J Bus Econ (The) 2(1)

Yitaferu B (2007) Land degradation and options for sustainable land management in the Lake Tana Basin (LTB), Amhara Region, Ethiopia (Doctoral dissertation, University of Bern). http://197.156.72.153:8080/xmlui/handle/123456789/722

Zhao GX, Lin G, Warner T (2004) Using Thematic Mapper data for change detection and sustainable use of cultivated land: a case study in the Yellow River delta, China. Int J Remote Sens 25(13):2509–2522. https://doi.org/10.1080/01431160310001619571

Part III
Disaster

Chapter 18
Disaster Risk Reduction in the Changing Scenario

Tapati Banerjee⊙

Abstract Present World is suffering from huge loss of people and property due to regular disaster events along with the prevailing worldwide pandemic situation in health sector that skyrocketed the phenomena. UNDRR warned that global average of annual loss will be increased up to US $415 billion by 2030 as climate-related natural hazards are increasing in number and intensity. This leads to adopt a strategic operational management for risk reduction addressing pre-disaster prevention at national level and post-disaster rapid recovery according to local need. Therefore, a globally accepted framework viz. *PEOPLES* Framework is discussed considering multi-scale dimension and linking climate change scenario with geospatial solution.

Keywords Disaster · Risk management · Risk reduction · Geospatial solution · *PEOPLES* framework

18.1 Introduction

Today in the changing scenario, the whole world is suffering in a worst manner out of various types of disasters like natural or anthropological that affects in the physical, social, economic and even in health sector. Disaster always has a negative interaction with living entities and it never follows any political, social, or geographical boundaries, neither confined to physical components. The present world is facing a lot of challenges in their day-to-day activities. The report of United Nations of Disaster Risk Reduction (UNDRR, erstwhile UNISDR) of 2015 reveals that since 1980 nearly 1.6 billion people lost their life in disasters.

UNDRR also warned that global average of annual loss will be increased up to US $415 billion by 2030 as the climate-related natural hazards are increasing in number and intensity giving rise to new vulnerabilities with differential spatial and socio-economic impacts on communities. The prevailing worldwide pandemic

T. Banerjee (✉)
Department of Science and Technology, Government of India, National Atlas and Thematic
Mapping Organisation, Kolkata, West Bengal, India
e-mail: dir.natmo@nic.in

© The Author(s), under exclusive license to Springer Nature Singapore Pte Ltd. 2022   351
N. C. Jana and R. B. Singh (eds.), *Climate, Environment and Disaster
in Developing Countries*, Advances in Geographical and Environmental
Sciences, https://doi.org/10.1007/978-981-16-6966-8_18

situation in health sector skyrocketed this warning. The study on disaster's impact on development together with the development increases or decreases the risk of disaster is accepted worldwide. Various terms and terminologies are widely associated with the study on disaster.

The published terminologies of the United Nations International Strategy for Disaster Reduction (UNISDR 2009), a secretariat of the International Strategy for Disaster Reduction (ISDR) are globally accepted. The UNDDR (erstwhile UNISDR) (UNISDR 2009) defines disaster as: "A serious disruption of the functioning of a community or a society involving widespread human, material, or environmental losses and impacts which exceeds the ability of the affected community to cope using only its own resources." Disaster is a combination of peril and unprotected phenomena positioned toward vulnerable font to prevent potential negative consequences.

18.2   Disaster Risk

Disaster may be responsible for loss of life, damage to property, negative impact on natural environment leading toward disruption and environmental degradation followed by instability of mental health, economic and social well-being, etc. In this context different disaster-related terminologies like risk, disaster risk, hazards, vulnerability, etc., are very much relevant in present scenario. In general, the concept of risk is about the integration of the likelihood of the events to be occurred with its negative consequences that are involved in our day-to-day lives.

Disaster risk is the potential catastrophic damages which may occur in any time and any moment. Hazard is a dangerous phenomenon which causes loss of life leading to disease, damage, etc. Hazard may be single or sequential or any combination with respect to its origin and sequels. Basically, hazard is of three types: Natural hazards (viz. earthquakes, floods, heat waves, droughts, cyclones, outbreaks of epidemic diseases, etc.); Technological hazards (viz. industrial accidents, nuclear explosions, etc.); and Environmental hazards (viz. land degradation, loss of biodiversity, land, water, and air pollution; El Niño; ozone depletion, etc.).

The hazard is characterized by hazard identification, nature, intensity, extent, predictability, and manageability. On the contrary, vulnerability is a poor set of conditions which arise from inadequate protection of physical, economic, social, political, and overall environmental factors. Coping capacity of an individual or a system is the ability to manage the disaster situation by using available resources. Resilience means the ability to "spring back from" a disaster. In general the concept of risk is about the integration of the likelihood of any type of unnatural event to be happened with its negative consequences damaging our day-to-day lives.

Disaster risk is the probable adversity fatalities that may occur in any moment. Disaster risk (R) is a well-defined function of Hazard (H), Vulnerability (V), and Coping capacity (C) which may be expressed as:

$$R = \frac{H \times V}{C}$$

where the disaster risk (R) is directly proportional to the consequences of hazard and vulnerability ($H \times V$) when the coping capacity is constant. On the contrary it is to be noted that disaster risk (R) is proportionally inversed to the coping capacity when the outcome of hazard and vulnerability ($H \times V$) is constant.

18.3   Disaster Risk Reduction

Disaster risk reduction may be defined as the preparation of reducing potential risks via systematic efforts for managing and analyzing the underlying factors of disasters and protective measures against risks. Disaster risk management is the organized progression by means of proper administrative directives, organizations, assistances, and capacities to execute policies and improve decision-making as well as handling capacities so as to negating the influences of probability of upcoming disaster. It avoids the hostile effects of hazards through preventive and precautionary measures, mitigation and preparedness as well (UNISDR 2009).

Disaster risk reduction concerns the activities that more focused on a strategic level of management, whereas management of disaster risk is the tactical and operational implementation of the reduction of hazard. Disaster risk reduction deals with the focused activities on strategic management whereas the disaster management is the operational implementation of risk reduction. Therefore, the appropriate conceptual framework for preparedness through early warning system followed by prevention and rapid recovery must be the goal at all level of disaster management cycle.

This risk management cycle has basically two parts: risk reduction phase at pre-disaster level and damage recovery phase at post-disaster level (Fig. 18.1).

18.4   Natural Disasters and Outbreak of Epidemic Diseases

The whole world is facing various types of hazardous problems resulting from disasters since long. Reduction of disaster risk and adaptation with changed climatic conditions has become the core development issues nowadays. The trend in the loss of life due to exposure toward hazards (Fig. 18.2) clearly reveals the death rates (per 10,000) of human population due to different types of natural disasters occurred around India and the whole world during last 27 years from 1990 to 2017.

From the above diagram it is also clear for making a comparative study of Indian scenario with the world during the specified time period and for making policies and strategies regarding disaster risk reduction and very crucially the conceptual program for risk reduction. Adaptation to climate change will be a vital challenge to

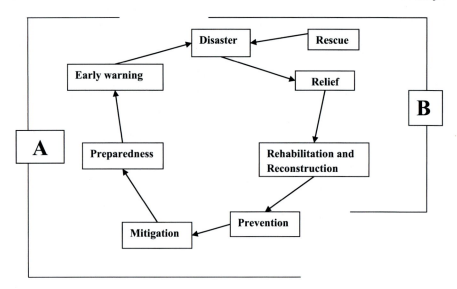

Fig. 18.1 Traditional disaster management cycle

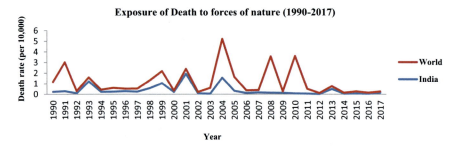

Fig. 18.2 Death exposure to forces of nature (1990–2017) of India and World. *Source* https://our worldindata.org/natural-disasters#all-charts-preview

the community exposed to the hazards in near future. Climate change adaption is the human's response to the changing climatic pattern that becomes an important milieu for human security. The risk reduction and climate change must be closely linked to human development as well as national development addressing the local need.

The climatic information must be studied rigorously to analyze the complexity as well as uncertainty in order to adaptation and disaster risk reduction. This climate change adaptation reduces the vulnerability to climate change and securing human livelihood. Alleviation of hazard risk and risk management significantly consider the climate change adaptation. To develop the consciousness of hazard risk and its moderation, the concept of climate change and its adaptation is a valid topic in this context.

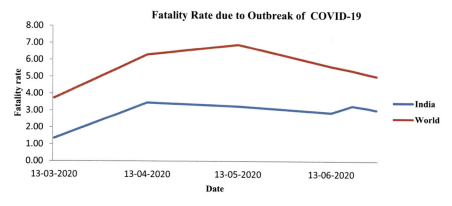

Fig. 18.3 Fatality rate due to outbreak of COVID-19 in India and World

The worldwide outbreak of novel Corona virus (COVID-19) viral pneumonia created global economic and healthcare crisis leading to the death of millions of human beings. The World Health Organization (WHO) has declared it as a global pandemic in March 2020. After World War II, probably the whole world is facing the most alarming epidemic situation aroused due to COVID-19.

The present scenario reveals that the fatality rate in India is 3.04 at the end of June 2020 whereas the rate is 5.04 in world (Fig. 18.3) (Source: https://www.who. int/emergencies/diseases/novel-coronavirus-2019/situation-reports). Even it is also clear that the overall trend of fatality rate in India is much less as compared to global fatality rate during last few months indicating a better scenario as compared to global situation.

The Government of India has taken several fruitful and significant preventive and precautionary steps to control the proliferation and to reduce the risk of this disaster by implementation of lockdown, social awareness and social distancing, self-isolation, home quarantine, and identification of clusters of cases, etc.

18.5   Geospatial Solution for Disaster Risk Management

Being acute event juxtaposition to physical conditions and social agony, disaster becomes more and more complex regardless of their origin, extent, and consequential phenomena although disaster management denotes a set of processes designed to implement before, during, and after disasters to prevent or mitigate their effects.

Therefore, the role of geospatial technology is vital for effective management to control the hazard risk that works both in active and passive mode through an efficient Information System in a collaborative manner with a good interoperability. It includes input of data or information with a systematic integration for capturing, storing, displaying, updating, manipulating, and analyzing spatial data for visualization and analysis followed by implementation program. Even there is a scope to develop an additional platform for follow up interaction mechanism with active participation of local people and the officials involved in the disaster reduction process.

Several researchers are working on identifying the hot-spot areas, monitoring and analysis of critical damage situation with time series analysis, risk location with developing information through damage inventory and assessment of life and infrastructure, predicting for future disaster locations using Remote Sensing data from different satellites with different sensors (Table 18.1).

From Table 18.1, we can find that globally accepted sensors, satellites, and technologies are used for the geospatial solution to reduce the risk. Even in case of health and disease spread analysis the role of geospatial technology at global or local scale through data capturing system using UAV and other devices are also vital. Similarly these technologies are also useful for model development, analysis, monitoring, planning, and forecasting for future prediction of the disasters with time series analysis.

Table 18.1 Sensor and satellites used for disaster risk management

Disaster type	Sensors/satellites	Applications
Earthquakes	PALSAR, IKONOS 2,InSAR, SPOT, IRS	Hazard mapping, damage assessment, rescue planning routes, rehabilitation, etc.
Cyclones	KALPANA 1, radar, INSAT-3A, QuikScat, Meteosat	Risk modeling, vulnerability analysis, impact assessment, spatial planning, etc.
Floods	AMSR-E, KALPANA I, Tropical Rainfall Monitoring Mission	Flood detection, early warning, flood mapping, damage assessment, spatial planning, etc.
Landslides	PALSAR, IKONOS 2, InSAR, SPOT, IRS	Rainfall monitoring, risk modeling, DEM models, hazard mapping, damage assessment, slope stability, etc.
Volcano	MODIS, Hyperion AVHRR	Risk modeling, DEM models, hazard mapping, damage assessment, spatial planning, emission monitoring, etc.
Drought	FEWS NET, AVHRR, MODIS, SPOT	Weather forecasting, vegetation monitoring, drought mitigation, land and water resource management, risk modeling, vulnerability analysis, damage assessment, etc.
Fire	MODIS, SERVIR, AFIS, Sentinel Asia	Risk modeling, damage assessment, Fire detection and prediction, fuel load monitoring, etc.

18.6 Disaster Risk Reduction Framework

Several policies have been developed, new institutes have come up, new groups have been created, existing institutes and organizations have been modernized and entrusted with additional responsibilities to work on planning for preparedness, response, and recovery process following Disaster Management Cycle to combat the devastating situations and disaster resilience. In spite of all efforts the failure mainly comes from the decision-making process at organization level due to lack of coordination that affects culture of complacency, poor administrative structures, lack of adaptive leadership with uncertainties and ambiguities. The failure also comes from physical disruption as well as from hierarchical barrier.

A study on Hiroshima Landslide of 2014 reveals that efforts should make proactively with community participation for effective solution to catastrophe reduction in general and death reduction in particular (Nibedita et al. 2014). Therefore, an appropriate methodology and networking system is very much important for moderating and managing the hazard risk. The concept of sustainable development is being introduced for better understanding of the network system to reduce or moderate the risk. Some significant and widely used frameworks for disaster risk reduction following UNDRR model for reduction of disaster risk, PEOPLES framework suggested by team researchers of University of Buffalo and associates for disaster risk reduction, etc.

The seven facets of community resilience are defined within *PEOPLES* Reliance Framework (UNISDR 2004) as Population and Demographics (P), Environmental/Ecosystem (E), Organized Governmental Services (O), Physical Infrastructure (P), Lifestyle and Community Competence (L), Economic Development (E) and Social-Cultural Capital (S). This framework is studied and developed by the team of researchers of the University of Buffalo in 2010. In recent time, this *PEOPLES* frame is the widely used and globally accepted framework for community reliance. This *PEOPLES* Framework requires the combination of qualitative and quantitative data sources at various temporal and spatial scales (Table 18.2).

The information gathered through remotely sensed devices and to analyze them in digital and or GIS platform will play a significant role in assessing the resilience of all integrated systems and establish an effective resilience model. Dimension Indicators include Population and Demographics (P) with composite distribution of socio-economic-cultural status, etc.; Environmental/Ecosystem (E) with air quality, soil health, biomass, and biodiversity status, etc.; Organized Governmental Services (O) with administrative-legal and security services, health services, etc.; Physical Infrastructure (P) with necessary amenities, lifelines, etc.; Lifestyle and Community Competence (L) with quality and well-being of life, etc.

This *PEOPLES* framework will help us to reduce the disaster risk in the changing scenario. The dimensions of the framework very significantly consider local, regional, state, multi-state, national, multi-national, continental, and global

Table 18.2 *PEOPLES* framework

Dimension	Indicators
Population and Demographics (**P**)	Composition, distribution, socio-economic status, etc.
Environmental/Ecosystem (**E**)	Air quality, soil, biomass, biodiversity, etc.
Organized Governmental Services (**O**)	Legal and security services, health services, etc.
Physical Infrastructure (**P**)	Facilities, lifelines, etc.
Lifestyle and Community Competence (**L**)	Quality of life, etc.
Economic Development (**E**)	Financial, production, employment distribution, etc.
Social-Cultural Capital (**S**)	Education services, child and elderly care services, etc.

geographic scale. This technological skill will run in the cycle of disaster management using different layers like risk assessment, problem area identification, establishment of linkage, strategic formulation, testing and implementation, gap identification at different level of management process with a good interoperability for effective disaster management.

18.7 Conclusion

This chapter aims to provide a broad apprehension of the reduction of disaster risk and management for mitigation in the changing scenario. The conceptual understanding of the terminologies viz. hazards, vulnerability, disaster, and resilience are discussed. This study focuses on the link between climate adaptation, risk reduction, and human security and its geospatial solution.

Finally, it may be concluded that the disaster risk can be moderated in a holistic manner by integrating scientists, environmentalists, industry houses, administrators along with the participation of local people. Different conceptual models like *PEOPLES* Framework as discussed may be considered at different levels of implementation program. Overall this will also help the community, the government, planner, and other decision-making departments to enhance disaster resilience using modern tools and techniques and to come up with an enhanced model structure for disaster risk reduction in this changing scenario.

References

Gunderson L (2000) Ecological resilience–in theory and application. Annu Rev Ecol Syst 31:425–439.

Nibedita R, Shiroshita H (2014) Disasters, deaths and the Sendai Framework's target one–a case of systems failure in Hiroshima Landslide 2014, Japan. Disaster Prevention and Management in October 2019

Renschler CS, Frazier AE et al (2010) Developing the 'peoples' resilience framework for defining and measuring disaster resilience at the community scale. In: Proceedings of the 9th U.S. national and 10th Canadian conference on earthquake engineering. Paper no 1827

Tierney K (2009) Disaster response: research findings and their implications for resilience measures. CARRI Research Report 6. Community and Regional Resilience Initiative (CARRI), Oak Ridge, TN. www.resilientUS.org

UNDP (United National Development Programme) (1992) An overview of disaster management. UNDP-DMTP, Geneva, 125 p

UNISDR (United Nations International Strategy for Disaster Reduction) (2004) Living with risk: a global review of disaster reduction initiatives. ISDR Secretariat, Geneva. Access 12 September 2011, http://www.unisdr.org/we/inform/publications/657

UNISDR (United Nations International Strategy for Disaster Reduction) (2009) UNISDR terminology on disaster risk reduction. UNISDR, Geneva, 30 p. Access 8 September 2011. http://www.preventionweb.net/files/7817_UNISDRTerminologyEnglish.pdf

Chapter 19
Exploring the Impacts of River Morphology Change Associated Natural Disasters on Teesta Riparian Environment of Bangladesh

Rebeka Sultana and Shitangsu Kumar Paul

Abstract Natural disaster destroys socio-economic conditions and it impacts on riparian environment of Bangladesh. The objective of the study is to examine the effects of natural disasters such as flood, river bank erosion, sedimentation, and river channel shifting. The study also seeks to explore both the positive and negative impacts of river morphology change allied natural disasters on environment. Teesta River within Bangladesh has been chosen as study area. Using simple random sampling technique, data has been collected through household questionnaire survey. Data analysis and interpretation has been completed with SPSS and Excel software. The study result unveils 33.1% responses on the negative impacts of flood is it erodes river banks. Moreover, the positive impact of sedimentation on environment shows 34.3% replies on silt increases soil fertility. Hence, the study explored significant positive and negative impacts on environment due to river morphology change-related natural disasters. The present research advocates consciousness of community inhabiting in the Teesta riverine environment to reduce the negative impacts of natural disasters owing to river morphology change. Besides, government should take initiatives toward minimization of river morphology change accompanying natural disaster impacts on riparian environment using modern technology.

Keywords River morphology change · Impact · Natural disaster · Teesta riparian environment · Bangladesh

19.1 Introduction

Bangladesh is considered as the largest delta of the world occupied by the three mighty river system of the Ganges–Brahmaputra-Meghna (GBM). About 85% run off of the country is flowed through the GBM river system (Allison 1998a). These rivers made the country vulnerable to river morphology change-related natural disaster

R. Sultana (✉) · S. K. Paul
University of Rajshahi, Rajshahi-6205, Bangladesh
e-mail: rebeka.sultana@ru.ac.bd

© The Author(s), under exclusive license to Springer Nature Singapore Pte Ltd. 2022
N. C. Jana and R. B. Singh (eds.), *Climate, Environment and Disaster in Developing Countries*, Advances in Geographical and Environmental Sciences, https://doi.org/10.1007/978-981-16-6966-8_19

which impacts on society, culture, economy, and environment. Fluvial morphological disaster has huge impact on economy and environment and Bangladesh along with other south and south east Asian countries are affected by recurrent floods during the last decades (Khan et al. 2014). Bangladeshi rivers are subject to frequent flood accompanied by sedimentation, bank erosion, and channel shifting mechanism. In socio-environmental perspectives, recurrent channel migration is a growing concern regarding the large river systems of Bangladesh (Dewan et al. 2017). The Teesta is a dynamic river for its repeated morphological adjustments. It is a multithreaded trans-boundary river which entered into Bangladesh through Nilphamari district and falls as a most active affluent of the Brahmaputra river at Fulchhari ghat. The Teesta plays significant role in the livelihood of the community of its confluence zone both in northern Bangladesh and the northern part of West Bengal of India. Therefore, the river is geo-strategically important. But the frequent river morphology change in the downstream location of the Teesta in Bangladesh is predominant to natural disasters which impacts on the riparian environment. River bank erosion, channel migration, and sedimentation usually occur in the monsoon that threatens the environment of Teesta river basin. Moreover, flood intensifies the river morphology change incidents. Flood is the foremost cause of channel changes in the downstream river flow (Chakraborty and Datta 2013). One of the most influential factors regarding channel change is sediment accumulation. River morphology change has been directed by sediment transportation process during flood (Gharbi et al. 2016). Besides, upstream flow of sediment also changes the river regime. The upstream flow of Teesta supplies a considerable amount of sediment in the downstream and determines the river as a braided stream (Ghosh 2014). Similarly, the river developed a large fan in the plains of India and Bangladesh. The sedimentation process of Teesta river formed a megafan with 145 km breadth and extent of 166 km (Chakraborty and Ghosh 2010). Among the natural disasters, river bank erosion, channel shifting, and sedimentation originate from fluvial geomorphic processes and flood initiates from both climatic and geomorphic evolution of a river system. Riverine environment alteration occurs due to hydrological and sedimentological variations as well as erosional activities of alluvial rivers in the floodplains. River bank erosion, channel change and sediment deposition are the typical feature of alluvial river (Uddin et al. 2011).

Riparian environment largely depends upon the geomorphological characteristics of a river valley. Fluvial geomorphic process controls the equilibrium of riparian ecology (Steiger et al. 2005). Likewise, water regime and erosional activities and sediment concentration in river bed impacts on vegetation and riparian habitat. The hydrology and morphology of a river contributes to riparian vegetation growth (Gurnell et al. 2012). Channel behavior has been directed by erosional activities in the downstream river valleys. Erosion in any river bank modifies channel morphology and channel configuration (Florsheim et al. 2008). The leading factors behind river morphology change associated natural disasters impact on environment are hydrological variation in space and time. The spatio-temporal characteristics of flow regime of river control riparian environment (Nilsson and Svedmark 2002). On the contrary, bank erosion and sediment deposition in both river bed and bank fluctuates seasonally and also year to year with the variation on water discharge. Erosion–deposition adjustment varies with time (Chakraborty and Hait 2014). The environment of riparian zone changes through hydrological changes in dry and wet season and also changes in sedimentation process. The aspects of channel cutting and sedimentation are interlinked with each other and these geomorphological adjustment controls the riparian habitat. Riparian environment alteration occurs with the changes in sediment accumulation, incision in channel, and changes in channel width (Loheide and Booth 2011).

Little attention has been paid in fluvial disaster impact on riparian environment though channel morphology change impact on vegetation, soil and ecosystem has been explored (Loheide and Booth 2011). Social impacts of channel change also have been investigated (Alam et al. 2007). However, invaluable works have been carried out regarding the impact of riparian area on development of fluvial sediments (Keesstra et al. 2012), modification of riparian environment by land use change (Nagasaka and Nakamura 1999), and condition of sediment transportation in the floodplains during floods (Toda et al. 2005). The state of flood, river channel shifting, and sediment transportation in highland environment through numerical simulation expanded the views regarding climate change effects on this phenomenon (Lane et al. 2007). Geomorphologists and river engineers, along with scholars of geography and environment identified river morphology change previously. Notable works have been done on river bank erosion and channel changes through GIS and Remote Sensing Techniques (Allison 1998b; Aher et al. 2012; Thakur et al. 2012; Hossain et al. 2013; Mallick 2016). Review of the former literatures extended our understanding on river morphology change and natural disasters. But there is no relevant literature regarding river morphology change linked natural disasters impact on environment. Since, the viewpoint of river morphology change associated natural disasters impact on environment; the study is quite different from other literatures. It will enhance our understanding on river morphology change; natural disasters interrelated with river and their impacts on environment.

19.2 Objectives of the Study

Flood, river bank erosion, and channel shifting directly influence the riparian environment but sedimentation indirectly effects on the natural phenomenon. This study preliminarily investigates the positive and negative impacts of flood, river bank erosion, and sedimentation. The existing study also tries to explore the influence of channel shifting in the riverine environment with a particular emphasis on the downstream location of river valley such as the Teesta river in Bangladesh. Hence, the specific objective of the present study is to focus on the effects of river morphology change accompanying natural extreme events on the surroundings of Teesta River.

19.3 Methodology

The study area is a part of Teesta floodplain. The Teesta river drains out the eastern part of Rangpur division that covers the districts of Nilphamri, Lalmonirhat, Kurigrm, Rangpur, and Gaibandha. The study purposively selected seven unions (lower most administrative unit) from these five districts which are situated in the confluence Zone of Teesta River and severely affected by river morphology change. The existing study is the outcome of a rigorous field survey. The survey method includes questionnaire survey from determined sample size of 381. Apart from that, observation and FGDs have been carried out. Data collection has been followed by selecting the households through simple random sampling techniques. The collected primary data has been analyzed with the help of SPSS and Excel softwares.

19.4 Teesta River Morphology Change and Natural Disaster Impacts

The Teesta is an unpredictable river and it has a tendency of channel shifting. The channel shifting process of Teesta river has been followed through east–west directions (Chakraborty and Dutta 2013). It is a most influential right bank tributary of the Brahmaputra river. Though the catchment area of Teesta is larger in India than Bangladesh, it is considered as the heart of northern Bangladesh. In Bangladesh the catchment area of Teesta is 2004 km^2 and in India the river covers 10,155 km^2 catchment (Khan and Islam 2015). Typically, Teesta is an alluvial river which has a long history of channel changes accompanied by socio-economic and environmental consequences. Alluvial rivers are exposed to morphological changes and these changes have significant effects on socio-economic condition, environment, and infrastructures (Hossain et al. 2013; Surian and Rinaldi 2003). The interrelationship among river morphology change and natural disasters has been illustrated in Fig. 19.1. Fluvial morphology change-related natural disaster impacts on riparian environment have been depicted in Fig. 19.2.

The history of channel morphology change and floods of Teesta River is well documented. The Teesta is vulnerable to flooding and has historical evidences of channel change (Rahman et al. 2011). The flow of Teesta river changed during the 1787 flood and the Brahmaputra river moved westward (Hutton and Haque 2003). The Teesta basin is the breeding ground of river morphology change associated multi-hazard such as flood, bank erosion, sedimentation, and channel shifting and that is the foremost cause of disaster in the riparian system. Human interference, water discharge, sediment supply as well as channel pattern control the morphology of fluvial systems (Mugade and Sapkale 2015). The Teesta is a braided stream and the river widens its valley through lateral erosion. The Teesta in the downstream locations of Bangladesh is considered as multithreaded stream channel with huge bed load deposition and wide river valley (Ghosh 2014).

Fig. 19.1 Teesta River morphology change associated natural disasters

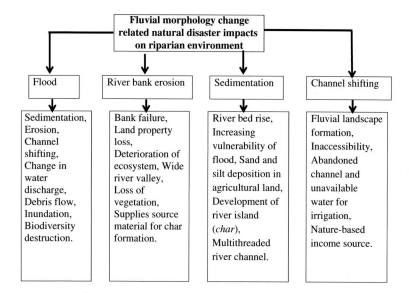

Fig. 19.2 Fluvial morphology change-related natural disaster impacts on riparian environment. *Source* Developed by the author

19.5   Results and Discussion

The study areas are exposed to recurrent fluvial disasters. The present study was based on household questionnaire survey in the natural disaster affected community of Teesta riparian environment. Therefore, the respondents of the study are asked about the impacts of riverine disasters on environment. The study result indicates that both positive and negative impacts are exposed to change in natural settings owing to river morphology change of Teesta River.

19.5.1   Impact of Flood on Riparian Environment

River morphology change controls anthropogenic environment but it also impacts on the nature. The present study attempts to examine the positive and negative effects of flood on riparian habitat. The positive impacts of flood are demarcated in Fig. 19.3. The figure represents about 47.0% respondents responded regarding the fertility of agricultural land increases after flood incidences. Floodplain soil becomes fertile with the fine grained sediments (Nilsson and Svedmark 2002). Flood supplies nutrients on soil and thereby modifies channel morphology (Toda et al. 2005). Thus, the study explores the fact that flood is the prominent factor which controls the balance of soil nutrients. During FGDs the farmers mentioned flood hampers agricultural production but it exaggerates crop production through siltation in the agricultural land after flood water recession. Moreover, the enhanced fertility of floodplain soil grounds for the growth of different plant species in the riverine ecosystem. About 30.0% respondents replied regarding flood nourishes plants and animals. During floods, the branch of trees floats in the river current and gathers in the bank. Therefore, the people of the study area collect fuel wood after flood. Among the respondents 23.0% opined about fuel wood gathered in the bank due to flood occurrences. The community collects the fuel wood from river bank for cooking.

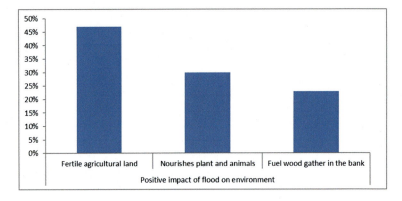

Fig. 19.3 Positive impact of flood on environment

Table 19.1 Negative impact of flood on environment

Union	Negative impact of flood on environment									
	Inundation		Erodes river bank		Agricultural production loss		Destroys biodiversity		Total	
	f	%	*f*	%	*f*	%	*f*	%	*f*	%
Purba chatnai	20	24.7	34	42.0	13	16.0	14	17.3	81	100
Saulmari	42	26.5	64	40.5	29	18.4	23	14.6	158	100
Bidyananda	29	24.4	41	34.5	31	26.0	18	15.1	119	100
Khuniagachh	62	25.4	72	29.5	70	28.7	40	16.4	244	100
Kolkonda	60	26.7	64	28.4	64	28.4	37	16.5	225	100
Sindurna	26	25.2	30	29.1	27	26.2	20	19.4	103	100
Belka	52	23.6	76	34.5	61	27.7	31	14.0	220	100
All	291	25.3	381	33.1	295	25.6	183	15.9	1150	100

Source Field survey, 2017(Percentages are based on responses)

The negative impact of flood is acute in the environment of Teesta basin. The study finds extreme river bank erosion which shows highest percentage (33.1%) regarding the negative impacts of flood in the study area (Table 19.1). Among the respondents 25.6% answered about flood destroys agricultural production.

Inundation in agricultural land causes crop production loss. The respondents mentioned that prolonged flood hampers in agricultural production through sand accumulation, debris deposition and inundation in the cropped land. Debris flow in farmland during flood disrupts farming practices (Prokop and Sarkar 2012). Around 25.3% respondents answered regarding inundation in flooding period mortifies standing crops and it causes homestead submergence too. Observations from the field unveil that the birds, animals, and trees have been disrupted in flood water. Accordingly, destruction of bio-diversity has been claimed by 15.9% of respondents of the study area. Hence, the study result indicates that, though flood has noticeable positive impact but its negative impacts also mostly influence on environment.

19.5.2 Impact of Erosion on Environment

Erosion is a destructive factor effecting on environment. The people of the study area were asked about the positive impacts of river bank erosion. But they were extremely disagreed (96.9% of respondents) about the positive impacts of erosion. Only 3.1% respondents replied about the positive impacts of river bank erosion is, it supplies source material for char formation. Table 19.2 represents the multiple responses of negative impact of erosion on environment. The multiple response analysis regarding negative impact of erosion on environment denotes 28.2% replies about damage of vegetation. The river banks were occupied by trees before bank erosion. Field

Table 19.2 Negative impact of erosion on environment

Union	Negative impact of erosion on environment							
	Damage of vegetation		Loss of land property		Wide river valley		Total	
	f	%	f	%	f	%	f	%
Purba chatnai	19	22.9	34	40.9	30	36.1	83	100
Saulmari	38	23.8	64	40.0	58	36.2	160	100
Bidyananda	25	25.0	41	41.0	34	34.0	100	100
Khuniagachh	61	30.5	71	35.5	68	34.0	200	100
Kolkonda	55	30.4	64	35.4	62	34.2	181	100
Sindurna	28	32.5	30	34.9	28	32.6	86	100
Belka	59	29.8	74	37.4	65	32.8	198	100
All	285	28.2	378	37.6	345	34.2	1008	100

Source Field survey, 2017 (Percentages are based on responses)

observation during river bank erosion points toward the damage of vegetation into the river gorge. Damage of natural vegetation, social forests, and orchards have large-scale monetary value in both local and national economy. In terms of environment, the respondents answered that damage of vegetation due to river bank erosion obstructs the compactness of soil and instigates further bank material collapse. Loss of land is the main issue of the study area and 37.6% respondents opined about land property loss owing to river bank erosion. FGD result shows that increasing velocity of current during flood and non- compact bank soil are the base of recurrent river bank erosion in the study area. Frequent bank collapse executes river valley widening in the Teesta riparian environment. Among the respondents, 34.2% responded about river valley widening occurs on bank erosion incidents.

19.5.3 Sedimentation Effects on Environment

As a braided river Teesta produces *char*[1] through sedimentation process. Sedimentation is the ultimate cause of *Char* land development in river bed (Rahman 2010). Different types of *char* such as stable and unstable *char*, small and big *char*, old and new *char,* etc., are found in the Teesta basin due to its progression of sediment. Sediment accumulation in river bed and bank has been observed by the respondents of the study area. The respondents distinguish the positive and negative impacts of sedimentation on riverine environment. The agrarians of Teesta basin mentioned that agricultural production hampers for sand deposition in agricultural field but if flood brings silt then it is the blessing of river morphology change for soil productivity. After silt accumulation in the floodplains, crop production increases due to nutrient

---

[1] *Char* is a local term used in riverine Bangladesh to designate river islands.

Fig. 19.4 Negative impact
of sedimentation on
environment

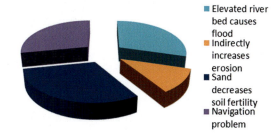

concentration in the soils. Among the replies highest percentage (34.3%) has been noticed in silt increases soil fertility. Similarly, sedimentation in river bed and bank brings the hope of gaining land for dwelling places of the river victims. If the land owners do not get back their land property, after accretion, they can easily reside in the *khas*[2] land. Hence, 25.4% respondents replied regarding the positive impact of sedimentation is it increases dwelling place for accretion of sediment. Moreover, sedimentation in river bed enhances the opportunity of domestic animal rearing in the study area.

Therefore, about 22.9% respondents mentioned that the positive impact of sedimentation on environment is it forms cattle grazing fields and brings the prospect of animal husbandry. Various livelihood activities such as paddy drying, cattle grazing, and collection of animal fodder have been noticed in the *char* lands of the study area during field survey. Besides, *char* development maintains the balance of river morphology after bank erosion. Thus, 17.4% respondents mentioned the positive impact of sedimentation is it forms *char* lands.

Optimistic views of the respondents regarding sedimentation have been discussed in the previous section but it also impacts negatively on environment. Figure 19.4 shows the negative impact of sedimentation on environment. Highest percentage regarding negative impact of sedimentation on environment has been depicted 30.8% in sand decreases soil fertility which governs the agricultural sector and makes the farmers production loss. Crop production hampers owing to sand deposition and debris accumulation in farm land. Debris deposition in the agricultural land causes agricultural land unproductive (Prokop and Sarkar 2012). The farmers of the study area become hopeless when agricultural land covers with sand after downturn of flood water. The respondents replied that if thick sand covers the agricultural land it needs 4–5 years to transform the land productive again. As a result, the lands become inactive and farmers fall into economic loss. Moreover, sedimentation causes the increase of river bed height. About 29.4% respondents replied that elevated river bed causes flood. During FGDs the respondents argued about repeated sediment

---

[2] *khas* land means acquisition of land by the Government of Bangladesh which is accreted from river bed after 30 years of bank erosion.

deposition lowered down the capacity of Teesta River to hold the flood water and it causes over bank spilling and floodplain inundation. Likewise, uneven distribution and concentration of sediment owing to discharge variability in different seasons impacts on channel pattern. Thus, the present study identifies multithreaded river channel in Teesta River. The channel type also causes navigation problem in the dry season. The third highest (28.2%) negative impact of sedimentation on environment has been found in navigation problem. Moreover, the multithreaded river is the main reason for remoteness of the study area. Sedimentation indirectly increases erosion has been replied by 11.6% of respondents. The people of the study area argued that when sedimentation occurs in one portion of the river, it indicates that the river is eroding in another side for balance and it's a game of nature.

19.5.4 Diversified Landscape Development Through Channel Shifting Impact

River morphology change grounds for channel shifting and this kind of hazard is closely associated with modification of environmental settings. Diversified landscape development is the impact of channel shifting mechanism which increases the scenic beauty of riparian environment. The present study explores the impact of channel shifting on environment through multiple responses from household questionnaire survey. The study result indicates presence of braided stream channel morphological features such as zigzag channel pattern (26.3%), active river channel (26.3%), abandoned channel, and back swamp (26.3%) in all the study unions (Table 19.3). One study union (Belka) has not been observed oxbow lake. Therefore, oxbow lake

Table 19.3 Impact of channel shifting on environment

Union	Impact of channel shifting on environment									
	Zigzag channel		Active channel		Abandoned channel and back swamp		Oxbow lake		Total	
	f	%	f	%	f	%	f	%	f	%
Purba chatnai	34	25.0	34	25.0	34	25.0	34	25.0	136	100
Saulmari	64	25.0	64	25.0	64	25.0	64	25.0	256	100
Bidyananda	41	25.0	41	25.0	41	25.0	41	25.0	164	100
Khuniagachh	72	25.0	72	25.0	72	25.0	72	25.0	288	100
Kolkonda	64	25.0	64	25.0	64	25.0	64	25.0	256	100
Sindurna	30	25.0	30	25.0	30	25.0	30	25.0	120	100
Belka	76	25.0	76	25.0	76	25.0	0	0	228	100
All	381	26.3	381	26.3	381	26.3	305	21.0	1448	100

Source Field survey, 2017 (Percentages are based on responses)

      Zigzag channel   Active channel   Back swamp

Fig. 19.5 Impact of channel shifting on environment

represents 21.0% of replies from the respondents. Thus, the study finds out that channel shifting impact on environment produces variety of landforms in the study area. Hence, the study found that active river channel, oxbow lake, and back swamp plays vital role in the socio-economic condition of the respondents. Because this type of landscape provides nature-based income opportunity such as fishing, water-borne vegetable selling, agriculture, etc., to the community of the study area. On the other hand, zigzag channel pattern increases inaccessibility and appearance of abandoned channel hampers agricultural production for scarcity of water.

However, channel shifting breaks the monotonousness of everyday life through varied morphological landscape formation in Teesta river basin. The impact of channel shifting on environment has been illustrated in Fig. 19.5

19.6 Concluding Remarks

River morphology change is a remarkable feature of natural environment and the dynamism of a river is controlled through its geomorphology, discharge, velocity of water, and also with human intervention. Flood intensifies the change in river morphology through erosion, sedimentation, and channel shifting. The study result synthesizes the positive and negative impacts of flood and sedimentation, negative impacts of erosion, and impacts of channel shifting through variety of landform development in the riparian zone. The present research explored the dimensions of river morphology change and its impact on natural environment. The study will be very helpful to understand the impact of flood, erosion, sedimentation, and channel shifting influence on natural settings. It will enhance our understanding on riparian environment of downstream locating countries. The study advocates conducting an Environmental Impact Assessment (EIA) regarding river morphology change in Teesta riparian environment. Finding of the present study suggests community awareness to diminish the negative impacts of disasters in Teesta floodplain. Furthermore, the present study emphasizes on tree plantation in the bank for sediment trapping and routing system development toward river bank stabilization. The study also recommends strong government initiative to reduce natural disaster impacts using modern technology.

References

Aher SP, Bairagi SI, Deshmukh PP, Gaikwad RD (2012) River change detection and bank erosion identification using topographical and remote sensing data. Int J Appl Inf Syst 2:1–7

Alam JB, Uddin M, Ahmed UJ, Cacovean H, Rahman HM, Banik BK, Yesmin N (2007) Study of morphological change of river old Brahmaputra and its social impacts by remote sensing. Geographia Technica 2:1–11

Allison MA (1998a) Geologic framework and environmental status of the Ganges-Brahmaputra Delta. J Coastal Res 14(3):826–836

Allison MA (1998b) Historical changes in the Ganges-Brahmaputra delta front. J Coastal Res 14(4):1269–1275

Chakraborty S, Datta K (2013) Causes and consequences of channel changes–a spatio-temporal analysis using remote sensing and GIS—Jaldhaka-Diana River System (Lower Course), Jalpaiguri (Duars), West Bengal, India. J Geogr Nat Disasters 3(107):2167–587

Chakraborty S, Dutta K (2013) Causes and consequences of Fluvial Hazards-A Hydro-Geomorphic analysis in Duars Region, India. Indian Streams Res J 2(12):1–10

Chakraborty T, Ghosh P (2010) The geomorphology and sedimentology of the Tista megafan, Darjeeling Himalaya: implications for megafan building processes. Geomorphology 115(3–4):252–266

Chakraborty S, Hait MM (2014) Impact of fluvio-morphic controls on channel configuration- A fluvio-morphic analysis of Ranikhola river (Tista drainage system) Sikkim, India. Soc Sci Nat 5(3):577–589

Dewan A, Corner R, Saleem A, Rahman MM, Haider MR, Rahman MM, Sarker MH (2017) Assessing channel changes of the Ganges-Padma River system in Bangladesh using Landsat and hydrological data. Geomorphology 276:257–279

Florsheim JL, Mount JF, Chin A (2008) Bank erosion as a desirable attribute of rivers. Bioscience 58(6):519–529

Gharbi M, Soualmia A, Dartus D, Masbernat L (2016) Floods effects on rivers morphological changes application to the Medjerda River in Tunisia. J Hydrol Hydromech 64(1):56–66

Ghosh K (2014) Planform pattern of the lower Teesta River after the Gazaldoba Barrage. Indian J Geogr Environ 13:127–137

Gurnell AM, Bertoldi W, Corenblit D (2012) Changing river channels: the roles of hydrological processes, plants and pioneer fluvial landforms in humid temperate, mixed load, gravel bed rivers. Earth Sci Rev 111(1–2):129–141

Hossain MA, Gan TY, Baki ABM (2013) Assessing morphological changes of the Ganges River using satellite images. Quatern Int 304:142–155

Hutton D, Haque CE (2003) Patterns of coping and adaptation among erosion-induced displacees in Bangladesh: implications for hazard analysis and mitigation. Nat Hazards 29(3):405–421

Keesstra SD, Kondrlova E, Czajka A, Seeger M, Maroulis J (2012) Assessing riparian zone impacts on water and sediment movement: a new approach. Neth J Geosci 91(1–2):245–255

Khan MS, Islam ARMT (2015) Anthropogenic impact on morphology of Teesta River in Northern Bangladesh: an exploratory study. J Geosci Geomat 3(3):50–55

Khan MMA, Shaari NAB, Bahar AMA, Baten MA, Nazaruddin DAB (2014) Flood impact assessment in Kota Bharu, Malaysia: a statistical analysis. World Appl Sci J 32(4):626–634

Lane SN, Tayefi V, Reid SC, Yu D, Hardy RJ (2007) Interactions between sediment delivery, channel change, climate change and flood risk in a temperate upland environment. Earth Surf Process Landf J Br Geomorphol Res Gr 32(3):429–446

Loheide SP, Booth EG (2011) Effects of changing channel morphology on vegetation, groundwater, and soil moisture regimes in groundwater-dependent ecosystems. Geomorphology 126(3–4):364–376

Mallick S (2016) Identification of fluvio-geomorphological changes and bank line shifting of river Bhagirathi-Hugli using remote sensing technique in and around of Mayapur Nabadwip area, West Bengal. Int J Sci Res (IJSR) 5(3):1130–1134

Mugade UR, Sapkale JB (2015) Influence of aggradation and degradation on river channels: a review. Int J Eng Tech Res 3(6):209–212

Nagasaka A, Nakamura F (1999) The influences of land-use changes on hydrology and riparian environment in a northern Japanese landscape. Landscape Ecol 14(6):543–556

Nilsson C, Svedmark M (2002) Basic principles and ecological consequences of changing water regimes: riparian plant communities. Environ Manage 30(4):468–480

Prokop P, Sarkar S (2012) Natural and human impact on land use change of the Sikkimese-Bhutanese Himalayan piedmont, India. Quaes Geogr 31(3):63–75

Rahman MR (2010) Impact of riverbank erosion hazard in the Jamuna floodplain areas in Bangladesh. J Sci Found 8(1–2):55–65

Rahman MM, Arya DS, Goel NK, Dhamy AP (2011) Design flow and stage computations in the Teesta River, Bangladesh, using frequency analysis and MIKE 11 modeling. J Hydrol Eng 16(2):176–186

Steiger J, Tabacchi E, Dufour S, Corenblit D, Peiry JL (2005) Hydrogeomorphic processes affecting riparian habitat within alluvial channel–floodplain river systems: a review for the temperate zone. River Res Appl 21(7):719–737

Surian N, Rinaldi M (2003) Morphological response to river engineering and management in alluvial channels in Italy. Geomorphology 50(4):307–326

Thakur PK, Laha C, Aggarwal SP (2012) River bank erosion hazard study of river Ganga, upstream of Farakka barrage using remote sensing and GIS. Nat Hazards 61(3):967–987

Toda Y, Ikeda S, Kumagai K, Asano T (2005) Effects of flood flow on flood plain soil and riparian vegetation in a gravel river. J Hydraul Eng 131(11):950–960

Uddin K, Shrestha B, Alam MS (2011) Assessment of morphological changes and vulnerability of river bank erosion alongside the river Jamuna using remote sensing. J Earth Sci Eng 1:29–34

Chapter 20
An Assessment on Effects of Coastal Erosion on Coastal Environment: A Case Study in Coastal Belt Between Kalu River Mouth and Bolgoda River Mouth, Sri Lanka

Kirishanthan Punniyarajah

Abstract The Southwest coast of Sri Lanka has become more prone to extreme coastal erosion, posing numerous environmental and social-economic threats. Therefore, the main focus of this research was on assessing the effects of coastal erosion on the coastal environment in the coastal belt between the mouths of Kalu River and Bolgoda River in Sri Lanka. For the purpose of data collection, Thermal Infrared Sensor images (2003–2018) were collected through the United States Geological Survey. In addition, questionnaire survey and key informant interviews were also conducted. The data collected via questionnaire survey and interviews were analysed statistically and thematically respectively. The results of the study revealed that destroying coastal landforms, saltwater intrusion, loss of green belt and wildlife and coastal flooding were recognized as the key environmental issues due to coastal erosion. Among them, coastal landforms are more prone to erosion in the study area, especially at Kalido beach, Kalutara. The current research is clearly highlighted that accelerating coastal erosion severely affects the balance and values of the coastal environment. Thus, coastal conservation and protection bodies, plans, and programmes should immediately take necessary action to mitigate the environmental issues in the coastal belt.

Keywords Coastal erosion · Coastal environment · Loss of landforms · Kalido beach · Sri Lanka

20.1 Introduction

Sri Lanka has a coastline of 1600 km and of this, nearly a third (approximately 500 km) is subjected to varying degrees of coastal erosion (Wickremaratne 1985). Being an island, coastal zones are very significant to Sri Lanka, because it highly

K. Punniyarajah (✉)
Department of Geography, University of Colombo, Colombo, Sri Lanka
e-mail: krishanthan@geo.cmb.ac.lk

© The Author(s), under exclusive license to Springer Nature Singapore Pte Ltd. 2022   375
N. C. Jana and R. B. Singh (eds.), *Climate, Environment and Disaster
in Developing Countries*, Advances in Geographical and Environmental
Sciences, https://doi.org/10.1007/978-981-16-6966-8_20

influences the country's economy, society, culture, security, transportation, biodiversity, and disaster prevention. Rapid population growth, booming settlement, urbanisation, industrialisation, infrastructure, and tourism developments are taking place in several coastal regions around the Island. For instance, the coastal area of Sri Lanka represents about 24% of the island's land area and is home to 25% of Sri Lanka's population with a range of activities (Costal Conservation Department of Sri Lanka 2006). In addition, the coastal environment is very cruel for the balance of the ecosystem in the environment in terms of their ecosystem services. Coastal ecosystems such as landforms; sand dunes, dunes, rocks, mangroves, and species play a vital role in coastal protection, resource management, as well as socio- economic benefits of coastal communities.

Wijayawardane et al. (2013) revealed that coastal erosion has become an extreme risk in many coastal zones, particularly along the Southwest coastline of Sri Lanka. Recent studies show that the highest intensity of erosion in Sri Lanka has been recorded from the Southwestern coastal zone, which also induced Kalutara district. However, in Sri Lanka there are very few studies that have emphased the impact of the coastal erosion on the coastal environment. Hence, this study is trying to fill the gap through the analysis of the effects of the coastal erosion on coastal environment.

20.2   Literature Review

Coastal erosion is a global problem with available evidence showing the escalating environmental threat caused by the phenomenon making it a cause for global concern (Cai et al. 2009). Devoy (2000) said that if the sea level rises in tandem with the occurrence of greater and more frequent storms, coastal flooding and erosion problems will become exacerbated in vulnerable coastal areas. Luijendijk et al. (2018) discovered that 24% of the world's sandy beaches are eroding at rates exceeding 0.5 m/yr, whereas 28% are accreting and 48% are steady. The many sandy shorelines in marine protected regions are eroding, raising cause for major concern. In addition to that, the team added that the continent with the largest accretion rate (1.27 m/yr) is Asia, likely due to the artificial development of the Chinese coast and large land reclamations in, for example, Singapore, Hong Kong, Bahrain, and UAE. The world's shorelines have accreted on appropriately 0.33 m/yr over the past three decades.

As Sri Lanka is being an island in the Indian Ocean, it has a long coastal belt around it and it contributes 43% of the nation's Gross Domestic Product (GDP). In Sri Lanka, nearly, 25% of its population living in coastal areas, 62% of industrial units, and more than 70% of tourist infrastructure are located in the coastal areas (Ministry of Mahaweli Development and Environment 2016). According to the Regional Technical Assistance for Coastal and Marine Resources Management and Poverty Reduction in South Asia: Situation Analysis Report (2002) coastal erosion continues to remain a severe problem in Sri Lanka despite the attention paid to it by the Coastal Conservation Department over the years. On average it assessed that there is a net erosion rate of 0.2–0.35 m per annum.

Coastal landforms, influenced by short-term perturbations: storms, generally return to their pre-disturbance morphology, implying a simple, morphodynamic equilibrium. Several coasts experience continual adjustment toward a dynamic equilibrium, typically adopting different states in reaction to shifting wave energy and sediment supply (Woodroffe 2003). In addition, Woodroffe (2003) added that coasts are highly prone to extreme events, such as storms, which impose substantial costs on coastal societies. Annually, approximately one hundred twenty million people are exposed to tropical cyclone hazards that killed 250,000 individuals from 1980 to 2000. Through the twentieth century, the rising mean sea level contributed to expanded coastal inundation, erosion, and ecosystem losses, however with significant local and regional variation due to different factors. Late twentieth-century impacts of rising temperature incorporate loss of ocean ice, defrosting of permafrost and associated coastal retreat, and more frequent coral bleaching and mortality (IPCC 2007). Coastal flooding is among the many negative results of coastal erosion. Coastal erosion in particularly low-lying communities reduces the effectiveness of the beach to act as because of the initial line of defence against storm surges and freak tidal waves.

Wolf (2009) describes coastal flooding as being popular caused by a combination of high water levels, which can be caused by tides and storm surges, along with waves, which may result in overtopping of coastal defences and inundation of low-lying areas, potentially causing damage to life and property. Therefore, coastal erosion, its negative impacts on the coastal environment should be examined clearly in order to protect and conserve the coastal environment as well as resources in the climate change era. Hence, the main focus of this study was to assess the effects of coastal erosion on coastal environment in coastal belt between Kalu River mouth and Bolgoda River mouth, Sri Lanka. The findings of the study will be useful to the relevant stakeholders such as the Coastal Conservation Department, Ministry of Disaster Management, and local authorities for the coastal protection and management, planning, and development in near future.

20.3 Study Area

The research area is located in Kalutara district falls in the Southwest coastline of Sri Lanka. It is about 92 km of shoreline (Wickramaarachchi, 2000) and covered two coastal district secretariat divisions (DSD) namely: Kalutara and Panadura. Figure 20.1 indicates the location of the study area.

The coasts of the study location are predominantly sandy and are, therefore, dynamic (Master Plan Coast Erosion Management 1986), and this coast too, like most coasts of Sri Lanka, have been subject to erosion since time immemorial. The Kalu River, Panadura River, and Bentota River all bring in beach sediment to this coast. The KaluRiver is by far the most important source of beach sediment for this area. The coastal plain in Southwest Sri Lanka is broadly divided into two parts as the upper coastal plain and the lower coastal plain. The study area has been

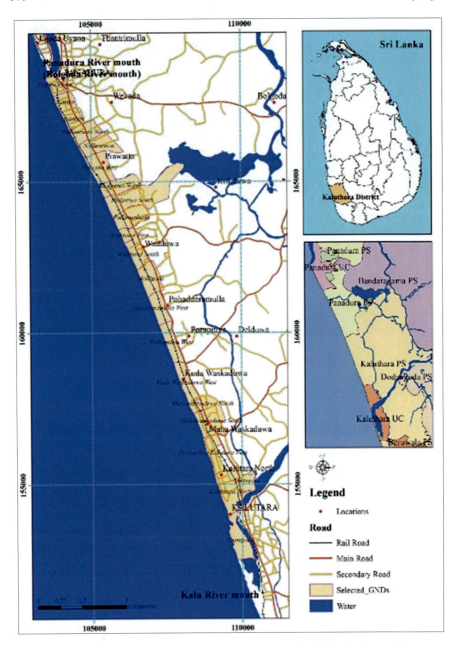

Fig. 20.1 Location of the study area (*Source* Prepared by author in 2018 using Survey Department data, 2014)

divided into six geomorphological or landform units, namely: the modern beaches, raised beaches, residual hills, silted up lagoons/ estuaries, levees, and fluvial terraces (Wickramagamage 2011).

The most critical period of the year for the Southwest coast of Sri Lanka is the period of the Southwest monsoon, the waves coming from the Southwest reach the coastline at an oblique angle generating currents that flow in north direction. These currents are responsible for intense erosion along the coastline. According to the latest population census the total population of the study area was 40,596 in 2011. Further, it is a highly populated and urbanised coastal area. Industry, services, and agriculture are major economic activities in the study area.

20.4   Methodology

In terms of nature and data used for the current study, it can be considered as a mixed method combination of both quantitative and qualitative approaches. The mixed methods provided an away from of the examination of a research problem than either approach alone. Both primary and secondary data were collected for the study. Primary data were collected through questionnaire survey, key informant interviews, field investigation, and observation, while secondary data such as topographic sheets, aerial photographs, Landsat images, digital data, and population census data were obtained from relevant departments and institutions.

20.4.1   Data Collection

Primary Data

Questionnaire Survey

Primary data used for the study included a questionnaire survey to determine the people's perceptions on environmental impacts of coastal erosion. The structured questionnaire survey was administered at household level within the coastal *Grama Niladhari* Divisions (GNDs). For the questionnaire survey, two hundred (200) households were selected using stratified sampling from all the GNDs of the study area. Table 20.1 indicates the sampling process for the questionnaire survey.

Both males and females with ages ranging 18–65 years participated in the questionnaire survey. Participants were asked about their perception and knowledge about coastal changes that have been occurring in the area near them, the potential causes, the perceived effects of such changes in environment, their livelihoods as well as the actions that they have been taken to control those impacts. Data from the questionnaire survey were analysed using both simple statistical analysis and descriptive methods. The findings of the analysis are represented through percentage, pie, line and bar charts, and also tables.

Table 20.1 Sampling for questionnaire survey

Name of the GND	Total housing units	No. of questionnaires
Deshathra Kalutara West	487	10
Kalutara North	632	13
Kudawaskaduwa West	437	9
Mahawaskaduwa North	353	7
Mahawaskaduwa South	644	13
Pohoddaramulla West	241	5
Pothupitiya West	372	7
Thotupala	366	7
Molligoda	512	10
Nalluruwa	750	15
Nalluruwa North	481	10
Palliyamankada	710	14
Pattiya	356	7
Pinwatta West	538	11
Sagara Place	131	3
Thalpitiya North	614	12
Thalpitiya South	879	17
Uyankale	453	9
Wadduwa South	744	15
Wadduwa West	354	7
Total	10,054	200

Key Informant Interview

The key informant interview was another primary tool that was used to collect data for this study. It knows that qualitative in-depth interview with people who know well about the research problem. The purpose of key informant interviews was to collect information from a wide range of people: community leaders, residents, professionals, or experts who have a better understanding, knowledge and attitudes about this particular research problem. Totally twenty (20) key informant interviews were carried out using structured questions. It included engineer (1), community leaders (2), environment officer (1), Grama Niladhari officers (village officer) (2), development officer (1), a monk (1), fishermen (3), elderly people above sixty years old (5), teachers (2), and hotel managers (2). The key informant interviews helped to understand the current problem of the coastal erosion and its effects on the environment as well as the socio-economic of the coastal community. Data gathered through key informant interviews were analysed using the thematic analysis method, which was supported to analyse the qualitative data properly. In addition, field observation was also conducted to investigate the nature, causes, and environmental impacts of coastal erosion in the study area.

Secondary Data

The secondary data were one of the key sources of the study, since the coastal erosion analysis data were mostly gathered through secondary sources such as topographic maps, digitized images, and satellite images. Digital images were obtained from the Survey Department, Sri Lanka, and Landsat data were downloaded from the United States Geological Survey website. Other sources such as: journals, statistical reports, and research reports were also used for the study.

20.4.2 Data Analysis

Data were analysed in both quantitative and descriptive manner. To identify the environmental changes due to coastal erosion, both 2010 and 2018 satellite images from Google Earth helped in visually identifying the various land uses along the coastline in the study location in the ArcGIS platform. These identified land use types were then verified in the field through a GPS field survey and observation. After that to determine the density of green space on the earth's surface, observe colours of visible and near infrared sunlight reflectance by the plants such as grassland, scrub land, barren land, forest land, and wetland were used. Finally, the land use matrix was prepared using GIS to identify the land cover changes over the last ten year period. This land use information along the coastline supported the identification of the influences of the coastal changes, especially land loss and formation as well as loss of vegetation covers.

20.5 Results

The findings of the study revealed that the coastal belt between Kalu River mouth and Bolgoda River mouth includes, beaches, sand dunes, sand bars, lagoons, estuaries, mangroves, salt marshes, coastal marshy wetlands as well as different water bodies, which ecologically, environmentally, economically, and socially play a vital role. However, currently all these ecosystems are under threat due to the accelerating coastal erosion and coastal environmental degradation. The analysis of the questionnaire survey depicted that about 79% (158) of the households commented that they had observed certain environmental degradation along the coastline in the study area,

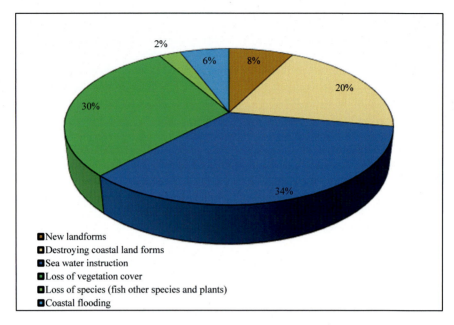

Fig. 20.2 The key environmental effects of coastal erosion in the study area (*Source* Field survey, 2018)

while only about 21% (42 households) had not observed any change in the coastal environment.

Figure 20.2 demonstrates the key effects of coastal erosion on the coastal environment of the study area. In reference to the public's viewpoint, six major environmental issues were recognised, namely: destroying coastal landforms, saltwater intrusion, loss of green belt, loss of species, formulating new landforms, and coastal flooding as the primary effects of coastal erosion. The elaborative explanations of the key environmental issues are as follows.

Sea Water Intrusion

The majority (34%) of the households considered seawater intrusion is the primary environmental problem, among the others, which has arisen due to increasing coastal erosion. Key informant interviews also revealed that the water level of the sea has increased as the results of climate change and the sea level rising have mixed with freshwater in the area through Kalu River. Furthermore, they added that it may lead to drinking water scarcity in Kalutara area in the near future. Some places in the study area are already facing drinking water issues. In addition to that, they added that "seawater intrusion is a major risk along the study area particularly, Kalutara lagoon boundaries because of the current deepening of lagoon estuary and coastal erosion". Moreover, Kalutara district's irrigation engineer said that "they have made arrangements to place sandbags to minimize saltwater intrusion after a discussion with the officials of the Coastal Conservation Department to prevent saltwater intrusion into land areas". However, the local people of the study area said sandbags won't work well.

Loss of Green Cover

Next to the seawater intrusion, the loss of the green belt in the coastal area is another environmental problem in many coastal areas of the study location. About 30% of households reported that the decline of the vegetation cover has been escalating as the result of coastal erosion. Meanwhile, there are many human activities such as fire-wood collection, cutting mangroves for economic benefits, and construction of build-ings, and hotels are also contributing to the loss of vegetation cover. The remaining coconut trees near the coastline are on the border of devastation because of erosion phenomena as well as development activities. The following Figs. 20.3a–f portrait the vegetation cover loss at Kalutara area.

Beach Erosion

The coastal landforms such as sandbars, sandy beaches, and estuaries are prone to erosion as a result of accelerating coastal erosion in the study area. Nearly 20% of households have observed the loss of coastal landforms, more specifically beaches near Kalu River mouth area. When considering the losses of beaches, Kalido beach at Kalutara is deemed as more at erosion risk. Kalido beach, is a strip of beach, which runs between Kalu River and the Indian Ocean. It is one and a half kilometre long and 200–250 m wide strip of land. The area is a huge tourist attraction, because of its unique location and scenic beauty. Some parts of the beach have waned due to erosion and other environmental threats. Figures 20.4a, b, 20.5a–d, 20.6a, b exhibit how strip and Kalido beach have been eroded respectively during the recent years.

According to public opinion, erosion along the study area is a natural process for a long time ago. The environmental officer of the study area said that "we can see the changes in coastal area over the past years, particularly, in Kalutara strip". Further, residents of the study area reported that over one metre of sand reef between the Kalutara lagoon and the ocean has been washed away to the ocean and it would cause a serious threat to the environment soon. The people of the study location strongly believe that this situation can bring some negative consequences to Kalutara Navy Camp, Kalutara main bridges, and coastal railway lines if precautionary measures are not taken.

The land use and land cover changes matrix determined that about 0.0203 and 0.0036 km^2 of previous open areas were converted as ocean and river respectively in 2018. Similarly, 0.0865 and 0.0022 square kilometres of earlier sand areas were converted into the ocean and river respectively in 2018. Thus, the above two scenarios of land use and land cover changes clearly revealed that coastline/land loss has been occurred due to coastal erosion in the study area. Figure 20.7 presents the beach areas that have been eroded during the last ten years in the study area.

Coastal Natural Hazards

A few natural hazards such as Tsunami, storms, or cyclones with high winds and flooding (river) also were distinguished as the environmental impacts, triggered by

Fig. 20.3 **a** and **b** Coastal vegetation covers in the study area (*Source* Field survey, 2018). **c** and **d** Natural environmental degradation in the study area (*Source* Field survey, 2018). **e** and **f** Vegetation cover loss in the study area (*Source* Field survey, 2018)

coastal erosion in the study area. Figure 21.8 illustrates the key coastal hazards that occurs due to erosion. Low-lying areas are the most direct and obvious consequences of sea level rising. Sea level rising may also increase the danger of coastal erosion and floods. Approximately, 63% of the households mentioned that flooding is the primary hazard within most of the coastal area of the study.

Loss of Biodiversity

From in key informative informers'point of view, coastal erosion is also a threat to wildlife, especially sea turtles and mangroves. Moreover they elaborated that

a   b

Fig. 20.4 a and **b** Coastal erosion at the study area (*Source* Field survey, 2018)

"the population of turtles of the area is gradually decreasing in recent years due to coastal erosion and other coastal environmental degradations". Further, below are some listed effects on biodiversity due to coastal erosion and degradation: net loss of wetland area, adverse impacts of sea level rise and storm surges on mangroves, coral reefs, and seagrass beds could affect marine organisms' important breeding grounds, disturbed coastal dynamics and habitats that would cause changed species composition and distribution, communities, and ecosystem services, increased spread of marine invasive species, the loss of freshwater and brackish habitats, changes in salinity concentration of lagoons and estuaries might collapse the balance ecosystem services and the species they contain, as well as the abundance of fish and crustaceans vital for the food fishery and changes in current patterns due to climate change, may affect the commercial fishery sector was observed and listed out during the field survey.

In addition to the above described environmental problems, formulating new land-forms due to depositions, and coastal flooding were also identified as environmental issues in the research area.

Coastal Management in the Study Area

Thought the questionnaire survey, approximately 66% of the people reported that relevant authorities have been already taken several mitigation measures such as groin, sandbags, tree planting, etc., to control the coastal erosion in the study area. In contracts with the above, nearly 44% of the people mentioned that none of the preventive measures have been taken yet. Figure 20.9 shows the proposed coastal erosion management measures by the local coastal communities.

From public's opinion, most of the respondents (37%) pointed out that main-taining vegetation cover such as mangroves, sea grasses, and other trees would help to reduce the impact of coastal erosion, while nearly 21% of the people mentioned that reduction or limitation of the tourism activities also might support to control the erosion. Further, 16%, 15%, and 7% of people stated that construction of sea walls, moving buildings back from coastal areas and dune planting would help to mitigate

Fig. 20.5 **a**, **b**, **c** and **b** Temporal loss of sand strip near to the Kalu river mouth (*Source* Google Earth pro)

Fig. 20.6 Temporal loss of Kalido beach in Kalutara (*Source* Google Earth pro)

the acceleration of coastal erosion respectively. Minority (2%) people mentioned that beach nourishment may protect the coastal from erosion.

20.6   Conclusion

In conclusion, this study clearly enlightened that the coastal environment of the study area is under severe threat due to coastal erosion and coastal environmental degradation. There were six (06) primary environmental issues due to coastal changes that are recognized namely: destroying coastal landforms, saltwater instruction, loss of green belt, formulating new landforms, loss of species, and coastal flooding. With the reference to the land use and land cover changes matrix, a considerable amount of earlier sand areas became as the ocean and river respectively in the last ten year period.

20.7   Recommendations

Participatory resource management is included consulting, planning together with the local community, and involving them locally in the monitoring of resource utilisation. The government authorities set-up an information desk to maintain transparency between communities. In order to educate the community members, awareness sessions or education sessions must conduct with the coastal communities. The awareness programs focus on the recent climate change impact in the coastal zone,

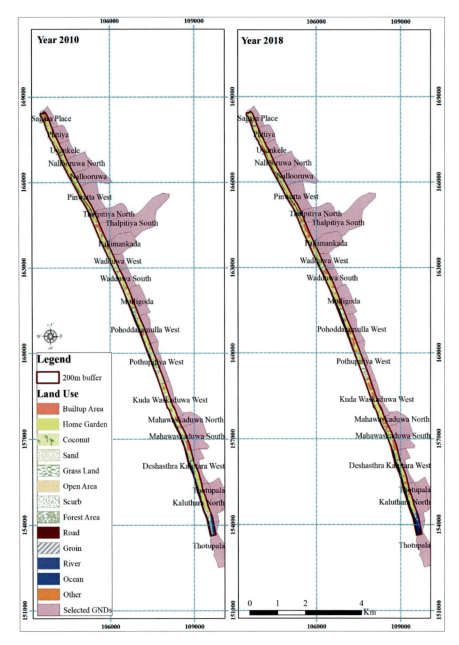

Fig. 20.7 Land use and land cover changes in the study area over the period of 2010–2018 (*source* Prepared by author, 2018)

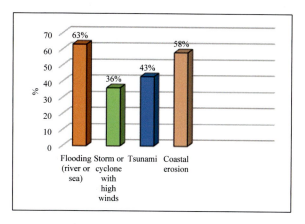

Fig. 20.8 Key natural hazards in the study area (*Source* Field survey, 2018)

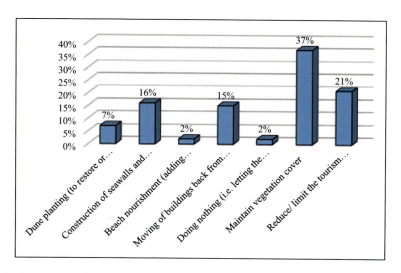

Fig. 20.9 Measures of coastal erosion management in the study area (*Source* Field survey, 2018)

the magnitude and trend of the damages, the possible adaptation mechanisms, and the importance of increasing ownership/ participation of the local community in the management system. Moreover, provide effectively reasonable and understandable information regarding coastal hazards such as storms, tsunamis, and other sea level related-hazards and protection measures, particularly to individuals, who are at high risk, to urge and empower individuals or communities to take plans and actions to mitigate the potential dangers or impacts and build resilience. The information should fuse pertinent customary and indigenous knowledge and also culture legacy and be customized for various target groups, taking into consideration local socio-cultural factors and values.

Furthermore, strengthen networks among coastal specialists, managers, and planners across various sectors and between regions, and make or fortify methods for utilising accessible ability when organisations and other significant stakeholders develop local risk mitigation plans. Advance and improve discourse and participation among mainstream researchers and specialists working on natural coastal risk reduction and mitigation, and also motive partnerships among relevant stakeholders, remembering those working on the socio-economic dimensions of disaster risk management.

References

Asian Development Bank and The World Conservation Union Sri lanka (2002) Regional technical assistance for coastal and marine resources management and poverty reduction in compendium report of high priority areas–Sri Lanka Component. Asian Development Bank and IUCN-The World Conservation Union

Cai F, Su X, Liu J, Li B, Lei G (2009) Coastal erosion in China under the condition of global climate change and measures for its prevention. Progress Nat Sci 19:415–426

Coast Conservation Department (1986) Sri Lanka mater plan: coast erosion management, coast conservation department. Sri Lanka, Colombo

Costal Conservation Department of Sri Lanka (2006) Coastal Zone Management Plan (revised). Coast Conservation Department, Sri Lanka, Colombo.

Devoy RJN (2000) Implications of sea level rise for Ireland. In: Proceedings of the SURVAS expert workshop. ZMK University of Hamburg, pp 33–46

Intergovernmental Panel on Climate Change (2007) Climate change: impacts, adaptation and vulnerability: contribution of working group II to the fourth assessment report of the IPCC. Cambridge University Press, Cambridge

Ministry of Mahaweli Development and Environment (2010) Sri Lanka: national climate change adaptation strategy 2011–2016. Climate change secretariat, Ministry of Mahaweli Development and Environment, Sri Lanka

Luijendijk A, Hagenaars G, Ranasinghe R, Baart F, Donchyts G, Aarninkhof S (2018) The State of the World's Beaches. Scientific Reports, 8 (1).

Wickremaratne H (1985) Environmental problems of the coastal zone in Sri Lanka. Econ Rev 8–16

Wickramarachchi B (2000) Risk profiling for sea level rise in South-West coast, Sri Lanka.

Wickramagamage P (2011) Evolution of the Kalu Ganga-Bolgoda Ganga flood plain system, Sri Lanka. J Geol Soc Sri Lanka 14:41–53

Wijayawardane ISK, Ansaf KMM, Ratnasooriya AHR, Samarawickrama SP (2013) Coastal erosion: investigations in the Southwest coast of investigations on sediment transport. In: international conference on sustainable built environment, pp 13–14

Wolf J (2009) Coastal flooding: impacts of coupled wave-surge-tide models. Proudman Oceanographic Laboratory, Liverpool, U.K.

Woodroffe CD (2003) Coasts: form, process and evolution. Cambridge University Press, Cambridge

Chapter 21
Vulnerability and Exposures to Landslides in the Chittagong Hill Region, Bangladesh: A Case Study of Rangamati Town for Building Resilience

Md. Iqbal Sarwar and **Muhammad Muhibbullah**

Abstract Landslide is a major significant natural disaster in the hilly regions of Bangladesh. Over the last three decades, the Chittagong Hilly Region (CHR) has been vulnerable to this disaster at an increasing rate in terms of frequency, magnitude, and damage. On June 13, 2017, a devastating landslide has hit the Rangamati hilly region. Rapid urbanization, improper land use, hill cutting, increased population density, deforestation, and unsustainable *Jhum* cultivation were triggered by the landslides vulnerability in CHR. This study attempts to identify the causes and consequences of landslides in the lives and livelihoods of local people, i.e. inhabitants of CHR, after assessing the household coping mechanism. Conducting questionnaire survey with 250 respondents, and secondary information, the present study was conducted in eleven (11) landslide prone areas in Rangamati town. These respondents were selected randomly based on the landslides susceptible locations. Several observations, opinion surveys, and FGDs have also been conducted in the Rangamati area for identifying the vulnerability, causes, and impacts of 2017 landslide. The findings of the study suggest that the emergency preparedness strengthened response, rescue mechanism and risk mitigation are important for sustainable mountain development, and reducing the economic and physical losses due to future landslides events in CHTs area.

Keywords Vulnerability · Landslide · Disaster management · Resilience · Chittagong hill tracts

Md. I. Sarwar (✉) · M. Muhibbullah
Department of Geography and Environmental Studies, University of Chittagong, Chittagong-4331, Bangladesh
e-mail: iqbalsrwr@cu.ac.bd

21.1 Introduction

Bangladesh is frequently affected by a number of natural disasters (cyclones, floods, drought, seasonal storms, thunderstorm, tornadoes, etc.) as well as a few human induced disasters, (ferry and road accidents, building collapses, fire in residential and commercial buildings, etc.). Due to its geographical location, Chittagong Region suffers from numerous natural disasters like landslide, water logging, cyclone, flash flood, etc. But at present, landslide is the most burning issues in respect of Rangamati hilly region. The commercial capital, Chittagong and Chittagong hilly region of Bangladesh are vulnerable to landslides similar to other hilly regions of the world. Heavy rainfall during monsoon, hill slopes altered by human, soil erosion, adverse hydrological conditions and unplanned land use practice all these factors trigger landslides in Chittagong hilly region. During the last 20–25 years, the intensity and frequency of landslide disasters has sharply increased, which has resulted in loss of property and deaths of more than 300 people. Mountains/hilly regions are typically exposed to multiple hazards (Kohler and Maselli 2009). A landslide is a type of "mass wasting". Mass wasting is down slope movement of soil and/or under the influence of gravity. The term landslide is generally used to describe the downward movement of soil, pillar of strength, inorganic and organic materials, etc., under the effects of gravity and also the terrain that results from such movement (Highland and Bobrowsky 2008). Many physical and anthropogenic factors contributed to landslide, these include prolonged rainfall in short period, wave action, geological condition, gravitational force, slopes, land-cover changes, and hill cutting. The hills of Chittagong and Chittagong hill region are situated in the folded belt of Bengal Fore-deep and most of the hills of Rangamati are in North–south alignment. Geologically these hills were formed during the tertiary period. The hills are mainly composed of loose and un-stratified sedimentary rocks and primarily consist of sand particles. The soils are weak and also have steep slopes, which increase the risk of vulnerability to landslide. According to Durham International Landslide Centre, till now more than 500,000 people have been killed due to landslide and most of them are mountainous less developed countries. According to EM-DAT (EMDAT 2014) shows that between 2000 and 2014 more than 6000 people have been killed in South Asia due to Landslide. Chittagong city has already been known as one of the most vulnerable cities to landslide (Ahmed and Rubel 2013). The city dwellers of Chittagong have experienced a number of devastating landslides. Different studies show that more than 500,000 impoverished people are living in informal settlements on the risky foothills of Chittagong city (Khan and Chang 2008). Since 1997, landslides have caused the death of nearly 235 people and injured more than 500 people in various hilly areas within Chittagong city and adjacent small urban centers (Technical Committee Report 2008). Chittagong hills are susceptible to landslides by different man-made stresses such as, hill cutting for construction, housing, filling-up wetlands, establishment of infrastructures, deforestation, etc. Along with these, there are several other reasons responsible for creating vulnerability to landslide. Since 1990, Chittagong region suffered about 20 times landslides. During the last thirty (30) years the death toll is approximately 600+ that occurred massive economic and property damages. The heavy rainfall of June 2017 triggered 140 landslides that destroyed the roads and settlements in the Chittagong Hilly regions (Especially in the Rangamati, Bandarban, and Khagrachari), Bangladesh. Landslides on the hilly steep slopes are quite common but massive landslides damage the roads, infrastructures, and housing with numerous deaths and injuries. Besides, the landslides that caused heavy loss and destruction include thousands of acres of garden trees, agricultural production, fruits, and homesteads. However, measures show that Rangamati town area suffered the least affected. In order to reduce the landslide vulnerability in hilly region of Bangladesh, this study tries to explore the causes and impacts of landslides as well as tries to put forward some recommendations after assessing the household coping mechanism for sustainable land management and building resilience in the Rangamati hilly region.

21.2 Materials and Methods

21.2.1 Study Area

Rangamati subdivision was turned into a district in 1983. The present study was conducted in Rangamati Town. The study area (Rangamati Town) is located at the South-Eastern side of Bangladesh and it lies between $22°27'$ and $23°44'$ North latitudes and $91°56'$ and $92°33'$ East longitudes (Fig. 21.1). The total area of the Rangamati town is 546.49 km^2. and the total population is 595979 according to 2011 census of which 52% are tribal and 48% are non-tribal people. Among them, the number of males is 313027 and the females is 282,952 (Bangladesh Bureau of Statistics (BBS) 2011). The Rangamati town was selected by researchers because it has had a historic massive landslide in June, 2017.

21.2.2 Methodology

The study was carried out by the combination of both primary data and secondary data sources. The primary information has been collected by conducting field survey in Rangamati town. The questionnaire survey consisted of 250 questionnaires conducted through face to face interview from different types of stakeholders during the primary data collection.

Several opinion surveys (different professionals including District Commissioner, Mayor, president of press club, civil surgeon, etc.) and three FGDs have also been conducted in the Rangamati town for identifying the vulnerability, causes, and impacts of 2017 landslide. The collected quantitative data were checked and then entered into the spreadsheet in MS Excel. Using statistical software SPSS, frequency

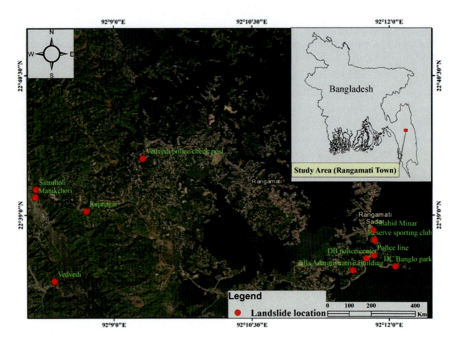

Fig. 21.1 Study Area and Landslides Location Map of Rangamati Town

and percentage were analyzed from the collected data according to the objectives. Finally, all the secondary sources of data were collected from different journals, newspapers, books, research papers, and other published and unpublished documents.

21.3   Results and Discussion

21.3.1   *Socio-economic Condition of the Respondents*

To realize the vulnerable condition of the people of Rangamati, the Socio-economic condition of the respondents is necessary. The socio-economic condition was done by household questionnaire survey. The respondents were both male (56%) and female (44%). Respondents ages from 18 to more than 60 years though maximum (42%) respondents were middle aged (31–44 years old). Maximum (45%) of the interviewee had primary level (Class one to class five) of education. Among rest of the respondents, 18% had secondary and 11% had high school level of education and 26% were illiterate. The respondents are from different occupations (Farmer-29%, Business-18%, Labor-16%, Driver-12%, Housewife-25%) and their monthly income also differs (ranging from taka 5000 to more than taka 20,000). Some of them are local and some are migrated and maximum (nearly 55%) live in tin-shed house. So, the respondents were of different backgrounds and that's why they could provide diversified information of different perspectives.

21.3.2 Nature of Landslide

Landslide is a common phenomenon in the hilly areas of Bangladesh which is accelerated by physical and man-made interventions. The first record of noticeable landslide was from Kaptai road area in 1968, but there was no casualty. However, the landslide of 2017 is severely affected incidents in the history of landslide in Bangladesh and received great concern in south-eastern part of Bangladesh as Chittagong and Chittagong Hill Tracts region. The physiographic characteristics of Bangladesh make the country more susceptible to landslide. According to geological time scale, hilly area of Bangladesh developed in tertiary age and mainly composed of unstratified sedimentary rocks such as sandstone, siltstone, shale, and conglomerate. The mountain areas are underlained by tertiary and quaternary deposits that have been folded, faulted, and elevated, then deeply cut up by stream channel (Brammer 2012). Unsustainable land-use practice and conversion in the hills including indiscriminate removal of trees and hill cutting are two major factors in Bangladesh that accelerated the landslide susceptibility in the hilly areas.

According to field survey, the respondents mentioned the nature of landslide vulnerability in the Rangamati town (Table 21.1). About 61% respondents reported that very severe nature; in contrast 31 and 08% respondents said that severe and moderate nature of landslide vulnerability respectively. Vulnerability narrates the characteristics and conditions of a community, system, or asset that make it susceptible to the harmful effects of a hazard. Vulnerability is mainly four types: physical, economic, social, and environmental vulnerability. Physical vulnerability is set on by aspects such as population density, remoteness of a settlement, the area, design, and materials used for critical foundation and for housing. According to the local administrations (Chittagong Hill District, Rangamati Municipality), about 15,000 families including 12,450 households in Rangamati district alone have bear losses in 2017 landslide. According to UNDP, it is found that a total of 38,177 households were affected, and 40,387 households were damaged.

Table 21.1 Nature of landslide at Rangamati

SL	Nature	Percentage (%)
1	Very severe	61
2	Severe	31
3	Moderate	08

Source Field Survey, 2018

Table 21.2 Causes of
Rangamati landslide

SL	Causes	Percentage (%)
1	Rainfall	37
2	Hill morphology	26
3	Hill cutting	21
4	Deforestation	10
5	Unplanned road construction	06

Source Field Survey, 2018

21.3.3 Causes of Rangamati Landslide

In a landslide or rock falls, movements of the materials depend on the slope. Several geological, morphological, and human-induced changes cause these slope instabilities. According to field survey, the major causes of Rangamati landslides are shown in Table 21.2 where heavy rainfall (37%), hill morphology (26%), hill cutting (21%), deforestation (10%), and unplanned road construction (06%), etc., are mentioned by the respondents. Among other natural factors such as soil properties, slope, geological features, and earthquake, rainfall is considered to be the major triggering factor for landslides. The Rangamati landslide happened due to too much rain in a short time. The Chittagong Hill Regions saw over 590 mm of rainfall in the 10th to 12th June 2017.

The hills of Bangladesh are composed of loose and un-stratified sedimentary rocks and primarily consist of sand particles. Both stratified and un-stratified sands are found. The latter is more susceptible to landslides. Landslides are found in hills having more than 35° slope. On the other hand, indiscriminate hill cutting particularly close to the town areas for housing and settlements and in remote areas for lateral expansion of agricultural land is one of the major causes of recent landslide in CHT. Similarly, cutting of forest in the hilly areas is another major reason for landslide in Chittagong region. Deforested areas are more prone to landslide than a forested area. Vegetation protects the soils and makes slope stable which reduces the risk of landslides. The Rangamati town was developed on hilltops. Thousands of people live on slopes of hills on both sides of roads in Rangamati town and surrounding areas. Development activities such as construction of roads without enough impact assessment are also a factor in landslides occurring in the CHT.

21.3.4 Impact of Rangamati Landslide

The landslide on June 13, 2017 happened in several regions of Rangamati. The impacts of Rangamati landslide exceed all previous occurrences in Bangladesh. According to field survey and secondary sources of data, the impacts and damage level of these landslides are discussed below.

Destruction of houses and Roads: Over 6,000 shelters and other buildings (36 schools damaged, 11 adolescent clubs severely damaged) have been reported destroyed. Destruction at 50 locations along 20 km part of the Rangamati-Chittagong road. Twenty-five spots where the slides caused 50 percent damage that has made the road totally unusable.

Power lines damaged/electricity shortages: The power supply got snapped in the district as several poles of main transmission line were destroyed in the landslide. Most areas in the Rangamati district were experiencing a power blackout for 36 h. Due to power outages, people were reportedly living without supply of water in most part of the town. Mobile network also appeared to have been seriously disrupted.

Deaths and Injuries: According to field survey, human life was destroyed due to landslide in Rangamati district which is very alarming. The landslide disaster contended 164 people in the region of Bandarban, Khagrachari, Chittagong, and Cox's Bazar, including 120 people in Rangamati hill district alone, while pull out of 227 people were injured.

Shortages of key commodities: The stock of fuel ran out in the hill district due to landslide and shortage of food and medicine. There has been widespread damage to household food stocks and non-availability of cooking facilities. Essential food materials also reduce as access to markets is challenging. Generally, women and girls are most severely affected, particularly pregnant and aged women. Crops and plantation, vegetable and fruit gardens have been inundated, resulting in economic loss and livelihood uncertainty. Automobile fuel is not available. Lack of electricity triggers a serious water crisis.

Education: Landslide in Rangamati has its immediate impacts on education. Government and Nongovernment thirteen (13) primary level schools in Rangamati town, twenty-one (21) schools in *Jurachari,* and 2 schools in *Langudu upazila*, were damaged by landslide disaster. Two informal schools supported by Save the Children in Rangamati district were damaged. About 40 schools reported that facilities and/or educational materials are also damaged. Eight schools are being used as temporary shelters.

21.3.5   Coping Mechanism for Building Resilience and Sustainable Management Options in CHT

It is not possible to stop landslide in the hilly regions of Bangladesh, but it is possible to reduce the risk level. Relief, resettlement, rehabilitation, and rebuilt may be post-disaster responses but not the sustainable solution. For sustainable landslide disaster management, it is necessary to develop a comprehensive and effective hill management system. According to field survey and expert opinion a few of the priority suggestions are as given below:

1. Awareness generation: Educate the public about signs that a landslide is imminent so that personal safety measures may be taken. Bring education to school students at community level so that they are aware of the significance and that they can spread knowledge between their group members.

2. Inclusive approach in Disaster Risk Reduction (DRR): Existing policies, rules, and regulations are not adequate to effective control of sustainable landslide disaster management. It is needed to ensure that the dweller Chittagong Hill Region (CHR) is seen as comprehensive in terms of involvement, decision-making, and recognition of diversity, tailored approach, and removal of barriers in the context of landslide in Bangladesh. Recognize the capabilities of the community and identify the community resources to prevent any incident.

3. Retention walls: In a large hilly region, it is not possible to protect all landslide vulnerable sites. Although, retention wall and engineering measures can be undertaken to protect landslides areas in CHR, otherwise it would never be possible to reduce the occurrence of landslides in future.

4. Landslide vulnerability zoning: Landslide vulnerability assessment and zoning is a prerequisite for sustainable management. At present, there is no landslide vulnerability zoning map for CHT. On the basis of geomorphologic biological and socio-economic analysis the zoning should be administered.

5. Land-use planning: Comprehensive land use planning of the vulnerable areas, a hill database, zoning, and geophysical analysis of the Rangamati area is crucial to minimize landslides and their consequences in the hilly region.

6. Proper coordination and communication: Extensive and well-coordinated steps will be needed for sustainable landslide disaster management. Proper communication amidst the Bangladesh Meteorological Department (BMD), Community Based Organization (CBO), the Civil defense wings and local government authorities is needed to receive regular data relevant to the area as soon as the monsoon season sets.

7. Feasibility Study: Appropriate feasibilities studies, along with assessments of risk, uncertainty, possible consequences, constructability, environmental impacts, and cost–benefit analysis by independent authorities are needed for landslide mitigation measure.

8. Stop hill cutting: Hill cutting is one of the main causes of landslides hazard in CHT area. Landslide-related hazards can be reduced or even probability of land sliding can be reduced if only hill cutting could have been prohibited. In cases of the places, where hill cutting has already taken place, sustainable structural measures such as retaining wall can be explored as mitigation options.

21.4   Conclusion

Landslide is a very common and visible incident in the Chittagong hill region. The substructure and soil condition of these tertiary hills are not stable, for which impetus these territories are highly vulnerable to landslide. Landslides in Rangamati can be reduced by avoiding construction of steep slopes, re-vegetation, maintaining proper watershed management and keep an eye on geotechnical control. These steps will be fruitful by restricting, embargo, or imposing conditions on landslides disaster activities. Long-term landslide mitigation plan may be prepared and should be carried put through phase by phase. In a nutshell, to enhance the city governance, policy and institutional aspects are to be reviewed considering the issues mentioned above. Appropriate facilitating roles of the concerned Government agencies can create an enabling situation for an effective public–private community partnership toward achieving sustainable mountain development in Bangladesh.

References

Ahmed B, Rubel YA (2013) Understanding the issues involved in urban landslide vulnerability in Chittagong metropolitan area, Bangladesh

Bangladesh Bureau of Statistics (BBS) (2011) Population and Housing Census, Socio-economic and Demographic Report, National Series, Ministry of Planning, Dhaka, Bangladesh

Brammer H (2012) The physical geography of Bangladesh. The University Press Limited, Dhaka

EMDAT (2014) The international disaster database, Durham, UK

Highland LM, Bobrowsky P (2008) The landslide handbook- a guide to understanding landslides, vol 1325. Virginia: U.S. pp 129p

Khan YA, Chang C (2008) Landslide hazard mapping of Chittagong City Area, Bangladesh. Indian Soc Eng Geol 35(1–4):303–311p

Kohler T, Maselli D (2009) Mountains and climate change—from understanding to action. Produced with an international team of contributors. Bern, Switzerland

Technical Committee Report (2008) Identification of landslide causes and recommendation for risk reduction. Chittagong Divisional Office, Chittagong, Bangladesh

Chapter 22
Hydro-Meteorological Analysis of 2015 Rarh Bengal Flood in the Lower Gangetic Plain of India: Exceptional, Fast and Furious

Soumen Chatterjee◉ and Narayan Chandra Jana◉

Abstract Rarh Bengal, the part of Lower Gangetic Plain (LGP) is one of the most affected flood-prone regions in the world as well as in India with immense social impact, including human life loss and also major infrastructural damage. The region is constituted of five major tributary systems of the Bhagirathi-Hooghly River in the western part viz. the Kopai-Mayurakshi-Dwarka, the Ajay, the Damodar, the Dwarakeswar-Silabati-Rupanarayan, and the Kangsabati-Keleghai-Haldi System. The majority of rivers in this region are short and have small catchments, but these all often cause significantly high flooding in terms of unit peak discharges. If we unfold the flood history since 1960, it can be observed that the Rarh Bengal has experienced only 11 flood-free years. The 2015 Rarh Bengal Flood was exceptional not only in terms of hydro-meteorological aspects but also in terms of duration, damage, and death toll. In this context, an attempt has been made to analyse the hydrological variability of Rarh Bengal's rivers during the 2015 flood with detail analysis of synoptic conditions. This study has also explained the extension of the 2015 flood (inundated area) by using remote sensing data (MODIS and Landsat 8 OLI) with its other devastating consequences in Rarh Bengal.

Keywords Rarh Bengal · LGP · Hydro-meteorological aspects · Inundated area · Death toll · MODIS · Landsat 8 OLI

22.1 Introduction

Flood occurs when the rivers overflow due to intense rainfall and cause inundation of the lower catchment areas that not only endanger lives but also increase human tragedy as well as economic losses (EUFD 2007; WBDMD 2020; Natural Disaster: Flood 2020). It is one of the most frequently worldwide occurring disastrous hydro-meteorological hazards (Jonkman 2005; Kvočka et al. 2016) which accounts for 43.4% (1998–2017) and ranks first among all extreme natural events (UNISDR

S. Chatterjee (✉) · N. C. Jana
Department of Geography, The University of Burdwan, Golapbag 713104, West Bengal, India

© The Author(s), under exclusive license to Springer Nature Singapore Pte Ltd. 2022   401
N. C. Jana and R. B. Singh (eds.), *Climate, Environment and Disaster in Developing Countries*, Advances in Geographical and Environmental Sciences, https://doi.org/10.1007/978-981-16-6966-8_22

CRED 2018). Flooding also reckons 70% of the natural disasters that occurred in Asia between 1950 and 2008 (Guha-Sapir et al. 2013). This catastrophe hazard is also very common in the monsoon-dominated Asian countries like India where 75% of the total rainfall occurs during four monsoonal months (June to September) due to successive weather disturbances like low pressure and depression (Dhar and Nandargi 2003). If we analyse the historical statistics of natural hydrological hazards of India (1960–2020), it can be observed that the land has experienced 331 floods with 76,691 human lives lost (EM-DAT 2020) because Indian rivers irrespective of their drainage area increase their discharge during monsoon season and experience large floods almost every year (Kale 2003, 2012; Mirza et al. 2001). A more detailed study by Kale (Kale 2014) also explores that Anthropocene floods have higher flood levels than the late Holocene floods. Being the part of Lower Gangetic Plains (LGP) of India, Rarh Bengal (Fig. 22.1) has witnessed several flooding due to a large amount of monsoonal rainfall with immense social impact, including loss of human life and major damage to property and infrastructure. In the 2015 summer monsoon parts of the Rarh Bengal experienced a catastrophic flood, in which more than 10 million people suffered due to inundation of the lower catchment area of Rarh Bengal Rivers.

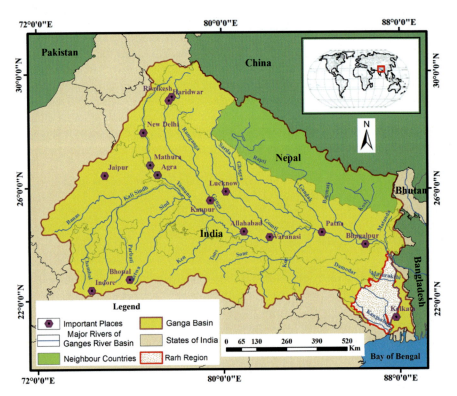

Fig. 22.1 Geographical location of the Rarh Bengal along with the major tributaries of the entire Ganga basin

In this work, an attempt has been made so that we can explain the reason behind the 2015 Rarh Bengal Flood by analysing the hydro-meteorological variability throughout the monsoon season, and also the cause of this exceptional, fast & furious flood by scrutinizing different types of data (meteorological data, hydrological data & food damage data) and examining various satellite images (flood & non-flood time) in 2015. So, this article gives detailed information on the hydrological and meteorological drivers that triggered the 2015 monsoon flooding and inundated vast areas of the Rarh Bengal.

22.2 Study Area

The Rarh Bengal (lat. ext.: 24°51′00″ N to 21°45′36″ N; long. ext.: 86°24′00″ E to 88°30′00″ E) is the part of Lower Gangetic Plains (LGP) of the Ganga–Brahmaputra-Meghna (GBM) delta, extending from the Chhotanagpur plateau in the west to the Bhagirathi-Hooghly River in the east and covering 35,400 km^2 (39.89% of the total geographical area of West Bengal). The east–west extension of the study area is nearly 195 km, whereas the north–south is nearly 360 km wide.

The Rarh region is constituted of five major tributary systems of the Bhagirathi-Hooghly River (a distributary of the Ganges system) in the western part (Fig. 22.2) viz. the Kopai-Mayurakshi-Dwarka System, the Ajay, the Damodar, the Dwarakeswar-Silabati-Rupanarayan System and the Kangsabati-Keleghai-Haldi System. On account of limited geographical area, only four rivers (Mayurakshi, Ajay, Damodar and Kangsabati) have basin area more than 5,000 km^2. The lower catchment areas of all these tributary systems are flood-prone (Bandyopadhyay et al. 2014; Bose 1970; Sen 1970). Some other small tributaries in the northern part of Rarh Bengal (Bansloi and Pagla), in the Damodar-Bhagirathi interfluve region (Khari, Banka and Saraswati) and in the southern part of Rarh Region (Keleghai and Kapaleswari) also contribute water to the Bhagirathi-Hooghly River and cause inundation due to drainage congestion at the time of monsoon rains frequently (Chatterjee and Majumdar 1972; Rudra 2018).

Geophysically, the entire Rarh Bengal has been classified into two:

(1) The Holocene deltaic silt as riverine flood deposit can be found in the eastern part along the Bhagirathi- Hooghly River (Rudra 2018), and

(2) The western portion is composed of Pleistocene laterite and older alluvium (Bandyopadhyay et al. 2014; Rudra 2018; Chakraborty 1970; Sengupta 1970, 1972).

Surface topography also helps to reflect the geophysical as well as the physiographical environment of the area truly. The undulating terrain (Altitude range: 20–190 m and average slope range: 0.4°–1.5°) and the lower riverine floodplain (Altitude range: 5–20 m and average slope range: 0.1°–0.5°) can be found in the western and eastern portion of the Rarh Bengal respectively. The cross-section (ab) which has been taken from SRTM DEM (Fig. 22.3) helps to establish the same

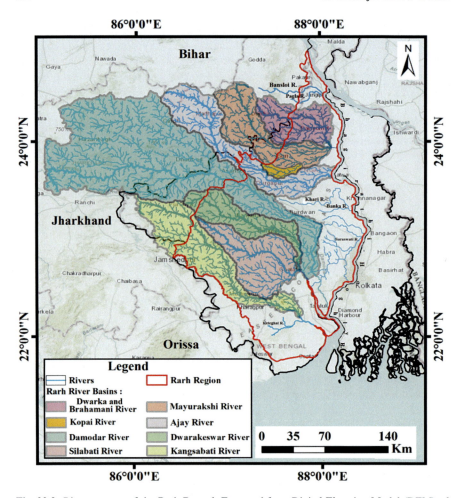

Fig. 22.2 River systems of the Rarh Bengal: Extracted from Digital Elevation Model (DEM) of NASA's Shuttle Radar Topography Mission (SRTM). The base map indicates national and international boundaries with major locations (*Source* Esri, HERE, Garmin, Intermap, increment P Corp., GEBCO, USGS, FAO, NPS, NRCAN, GeoBase, IGN, Kadaster NL, Ordnance Survey, Esri Japan, METI, Esri China: Hong Kong, © OpenStreetMap contributors and the GIS User Community)

fact. Monsoon time copious rainfall from low-pressure systems and cyclones also create high magnitude flood discharge and cause drainage congestion in the lower catchment area (lower riverine floodplain) of the Rarh Bengal Rivers.

From the administrative viewpoint, the Rarh Bengal is constituted of four entire districts (Purba Bardhaman, Hooghly, Howrah & Bankura); the lion's share of four districts (Birbhum, Murshidabad, Purba Medinipur & Paschim Medinipur) and the minor portion of two districts (Paschim Bardhaman and Purulia). Among them, Murshidabad, Purba Bardhaman, Hooghly, Howrah, Purba Medinipur and Paschim Medinipur district of the Rarh Bengal are highly flood-prone (WBDMD 2020; Natural Disaster: Flood 2020).

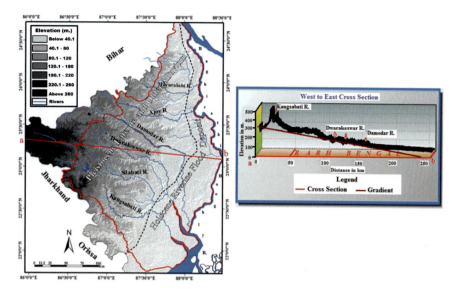

Fig. 22.3 Altitudinal differences of the Rarh Bengal and surroundings from SRTM DEM. The cross-profile has been taken from West to East along the line 'ab' for showing the topographic variation and regional gradient of the Rarh Bengal, India

22.3 Anthropocene Flood History of the Rarh Bengal

An extensive area (16,545 km^2, 46.7%) of Rarh Bengal is highly susceptible to floods. That value is much higher than Indian standards, i.e., 12% (NDMA-GoI 2007) as well as West Bengal's standards, i.e., 42.3% (IWD-GoWB 2013, 2015, 2017). If we unfold almost the last 60 year's flood history since the 1960s, it can be observed that the Rarh Bengal as a part of West Bengal has witnessed only 10 flood-free years (1985, 1989, 1992, 1994, 1997, 2001, 2005, 2006, 2013 and 2014). It is very hard to find out even a single year when West Bengal has been flooded but the Rarh Bengal has not experienced any flood. Four devastating floods in the history of Rarh Bengal occurred in 1978, 1984, 1991 and 2000 (Bandyopadhyay et al. 2014; IWD-GoWB 2015; Chapman and Rudra 2004, 2007). The highest affected area in all these cases was more than 22.5% of the state's geographical area (Fig. 22.4). The most noticeable thing is that there are fourteen such occasions while floods have covered 500 to 2000 km^2 of the state's area (IWD-GoWB 2017). The 2015 flood in West Bengal was such a flood in which 1395 km^2 area of the Rarh Region was affected badly, and the minor inundated area was nearly 1220 km^2. So, it is clear from the given statistics that the 2015 flood was exclusively confined within the Rarh Bengal.

Most of the Rarh Bengal Rivers (RBRs) cause high flooding at the time of summer monsoon in terms of unit peak discharges (0.91 m^3/s/km^2 for Damodar River at Rondiha, 0.53 m^3/s/km^2 for Ajay River at Nutanhat, 1.21 m^3/s/km^2 for Dwarakeswar River at Patakola and 2.72 m^3/s/km^2 for Mayurakshi River at Tilpara Barrage) even

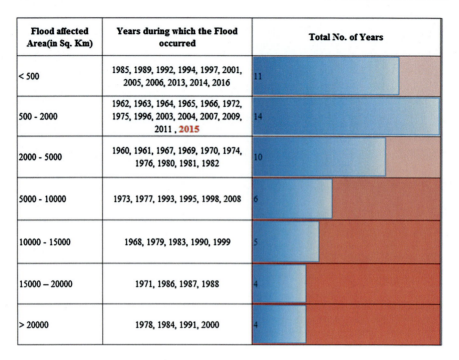

Flood affected Area(in Sq. Km)	Years during which the Flood occurred	Total No. of Years
< 500	1985, 1989, 1992, 1994, 1997, 2001, 2005, 2006, 2013, 2014, 2016	11
500 - 2000	1962, 1963, 1964, 1965, 1966, 1972, 1975, 1996, 2003, 2004, 2007, 2009, 2011 , 2015	14
2000 - 5000	1960, 1961, 1967, 1969, 1970, 1974, 1976, 1980, 1981, 1982	10
5000 - 10000	1973, 1977, 1993, 1995, 1998, 2008	6
10000 - 15000	1968, 1979, 1983, 1990, 1999	5
15000 – 20000	1971, 1986, 1987, 1988	4
> 20000	1978, 1984, 1991, 2000	4

Fig. 22.4 Diagrammatic representation of the frequency of similar intensity flood in terms of inundated area of West Bengal (*Source* IWD-GoWB 2017)

which are significantly higher than World Large Rivers (WLRs), Indian Peninsular Rivers (IPRs) and North Bihar Rivers (NBRs). Some studies on floods of WLRs include the works by Blum (2007), O'Connor and Costa (2004); Rodier and Roche (2003); Vörösmarty et al. (2000) and Wohl (2007). The works of Sinha with other co-authors in 1994, 1998 and 2005 (Sinha and Friend 1994; Sinha and Jain 1998; Sinha et al. 2005) help to gather the required flood-related information of NBRs. The effect of floods on the catchment area of IPRs has been widely studied by Kale (1999) and Kale et al. (1997). On the other hand, the unit peak discharge of four RBRs viz. Damodar, Dwarakeswar, Ajay and Mayurakshi have been calculated based on ample of the literature survey (Bhattacharyya 2013, 2011) and collected data from CWC of Govt. of India and IWD of Govt. of West Bengal. The basin area and the area above the discharge site of these four rivers have been extracted from SRTM DEM for the calculation of Unit Peak Discharge and making the comparison with other rivers (Fig. 22.5).

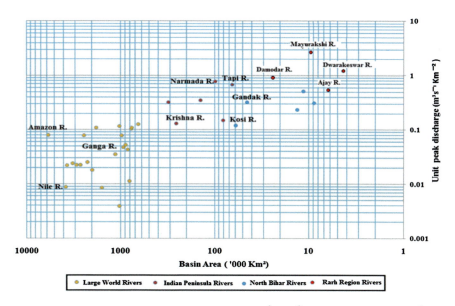

Fig. 22.5 Relationship between Unit Peak Discharge (m^3/s/km^2) and basin area ('000 km^2) for making comparison among World Large Rivers (WLR), Indian Peninsula Rivers (IPR) and North Bihar Rivers (NBR) with Rarh Bengal Rivers (RBR)

22.4 Materials and Methods

A comprehensive database is required for analysing the hydrological and meteorological variability of the 2015 Rarh Bengal Flood. Various free-access satellite products like Landsat 8 OLI from USGS for identification of flood-inundated areas by comparing flood-time and normal-time images; MODIS Flood Water Map from NASA for real-time floodwater extent mapping (Nigro et al. 2014); SRTM DEM from USGS for getting the topographical view of Rarh Bengal and TRMM (TRMM: Tropical Rainfall Measuring Mission 2011, 2016) gridded datasets for observation of rainfall events have been used to conduct the entire study (Table 22.1).

Many other datasets have been utilized to trace the 2015 Rarh Bengal Flood. The data of Southern Oscillation Indices (SOI) have been used to link the Rarh Bengal Floods with global phenomena like El-Niño and La-Niña events (https://www. ncdc.noaa.gov/teleconnections/enso/indicators/soi/). Other meteorological data like cyclone and depression track related informations (http://www.rsmcnewdelhi. imd.gov.in/index.php?lang=en) and rainfall informations (http://hydro.imd.gov. in/hydrometweb/(S(4nrdce55fiv1wpbmsqbvsovt))/DistrictRaifall.aspx) have been collected from IMD websites and CWC office. The hydrological data like monsoontime gauge level of different hydrological sites and inflow-outflow data of various dams and barrages on Rarh Bengal Rivers have been gathered from CWC-GoI (2015) and IWD-GoWB (2015) reports. To access the vulnerability of 2015 flood on mankind, the block-wise population density per unit area has been calculated

Table 22.1 Elevation and other satellite images used in the study for the analysis of 2015 Rarh Bengal Flood

Data (Type)	Spatial resolution	Imaging date		Entity ID	Data source
SRTM DEM (.tif)	89.51 m	01 February, 2005		SRTM3N21E087V1, SRTM3N21E088V1, SRTM3N22E086V1, SRTM3N22E087V1, SRTM3N22E088V1, SRTM3N23E085V1, SRTM3N23E086V1, SRTM3N23E087V1, SRTM3N23E088V1, SRTM3N24E085V1, SRTM3N24E086V1, SRTM3N24E087V1, SRTM3N24E088V1	https://earthexpl orer.usgs.gov
TRMM (.cdf)	27.75 km	Real Time	3 Hourly (during three spells daily)	TRMM_3B42_Daily	https://disc.gsfc. nasa.gov
		Real Time	Monthly (June–September)	TRMM_3B43	
MODIS Flood Water Map (.shp)	250 m	14 day composite (01 August–14 August, 2015)		MFW_2015226_080E030N_A14 × 3D3OT_V	https://floodmap. modaps.eosdis. nasa.gov
Landsat 8 OLI (.tif)	30 m	06 August, 2015(Flood Time)		LC08_L1TP_139043_20150806_20170406_01_T1, LC08_L1TP_139044_20150806_20170406_01_T1, LC08_L1TP_139045_20150806_20170406_01_T1	https://earthexpl orer.usgs.gov
		07 September, 2015(Non-flood Time)		LC08_L1TP_139043_20150907_20170404_01_T1, LC08_L1TP_139044_20150907_20170404_01_T1, LC08_L1TP_139045_20150907_20170404_01_T1	

based on ORG & CCI (Cci 2011) data. Besides the 2015 flood damage data from the Irrigation and Waterways Department of Govt. of West Bengal and also the West Bengal Disaster Management & Civil Defence Department help to give a complete shape to that study.

22.5 Hydro-Meteorological Variability of the 2015 Rarh Bengal Flood

22.5.1 Meteorological Set-Up

The Indian summer monsoon rainfall is significantly correlated with the Southern Oscillation Index (SOI). The large positive (negative) SOI value indicates strengthening (weakening) the Walker Circulation that causes large excess (deficient) of rainfall over Rarh Bengal, a part of Lower Gangetic Plain in India (Bhalme and Jadhav 1984; NOAA-NCEI 2019). But the most interesting thing is that despite being an El Niño year (2015), successive weather disturbances, more specifically heavy downpour in July caused terrible floodings like 1986, 1987, 1990 and 1991 Rarh Bengal Floods (Fig. 22.6).

If we closely examine all those floods, it will give information that most of the floods in Rarh Bengal are closely associated with the seasonal rainfall caused by deep depression and cyclonic storms. The 2015 Rarh Bengal flood was also the result of intense rainfall due to depressions and cyclones during summer monsoon season. In 2015 the monsoon was found vigorously active during three spells (Fig. 22.7).

The first two spells caused sparse rainfall throughout the Rarh Bengal due to depressions (Fig. 22.8). The onset of the south-west monsoon occurred over the Bay of Bengal and adjoining Bangladesh coast on 19 June 2015 and the monsoon trough arrived at the Lower Gangetic Plain of India on 25 June. Then it began to move northwestward through Bardhaman and arrived over Bihar (Gaya) on 29 June (CWC-GoI 2015). On the other hand, the second spell that continued from 16 to 20 July, caused heavy rainfall over the Rarh Region. An upper air cyclonic circulation laid over North Bay of Bengal and along with the coastal areas of Gangetic West Bengal on 16 July. That helped to create a low-pressure centre over the North Bay of Bengal. Then it began to move north and north-westward (after 18 July) and finally arrived over the northern part of Madhya Pradesh, as well as adjoining southern part of Uttar Pradesh on 20 July. The lower catchment area of all Rarh Bengal Rivers received more than 120 mm rainfall in that second spell (Fig. 22.8). But the intense cyclonic storm, Komen (26 July–02 August 2015) caused large downpour that helped to cross the threshold and flooded the lower catchment area of Rarh Bengal Rivers (Fig. 22.8). This cyclonic storm (CS) developed from a low-pressure area which lay over northeast Bay of Bengal (BoB) and adjoining Bangladesh & Gangetic West Bengal on 25th July evening and concentrated into a depression over the same area in the morning of 26th July (IMD 2015a, b, c). It made landfall near Bangladesh coast (lat. 91° 24′

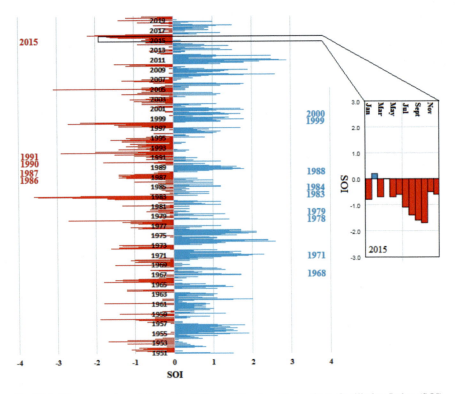

Fig. 22.6 The connection of Major Rarh Bengal Floods with Southern Oscillation Index (SOI). The negative SOI values indicate El Nino episodes, whether the positive SOI values are of La Nina episodes (*Source* Modified after NOAA-NCEI 2019)

E) at noon on 30 July (WMO/ESCAP 2015) following a semi-circular track. After landfall, Komen began to move initially north-northwest wards, then westwards and west-southwest wards across Bangladesh, Rarh Bengal (Gangetic West Bengal) and Jharkhand respectively. The upper and lower catchment areas of Rarh Bengal Rivers received an excessive amount of rainfall (avg. 221.5 mm and 293.6 mm respectively) during that last spell. It weakened gradually into a well-marked low-pressure area over Jharkhand and adjoining north Odisha and north Chhattisgarh on 02 August.

On the other hand, monthly rainfall analysis (June to September) helps to reveal some facts that: (a) the lower catchment areas of Rarh Rivers (monsoonal avg. 279.1 mm) have received little bit higher rainfall than the upper catchment areas (monsoonal avg. 261.1 mm) throughout the monsoon season of 2015 (Fig. 22.9a, b) heavy rainfall in July that caused flooding in Rarh Bengal was abnormally higher than other monsoonal months, and (c) NASA's TRMM gridded monthly datasets (Fig. 22.9b) also help to establish the monthly monsoonal rainfall variability (June to August). The TRMM image of July (2015) shows that almost no place in the lower catchment areas of Rarh Rivers has received less than 500 mm rainfall and the CS Komen have contributed at least 50% of the total rainfall in July.

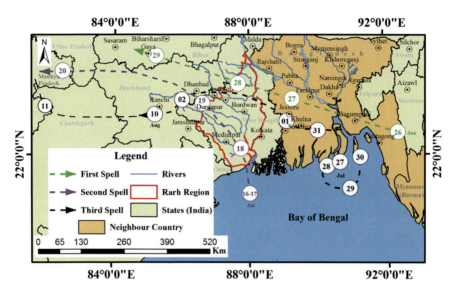

Fig. 22.7 The track of the depressions and cyclones (three spells) which caused intense rainfall during 2015 monsoon (*Source* CWC-GoI 2015; IMD 2015a, b, 2016b; WMO/ESCAP 2015)

22.5.2 Hydrological Variability

The Indian rivers are dramatically controlled by seasonal rainfall because most of the rivers carry a very high proportion of their annual discharge (80–95%) in just four to five months of the monsoon season (Kale 2002). The Rarh Bengal Rivers (RBR) are not exceptional in that case. A large volume of water was released from various dams and barrages (avg. 1248 cumec for Durgapur Barrage on Damodar River, 232 cumec for Massanjore Dam on Mayurakshi River, 543 cumec for Tilpara Barrage on Mayurakshi River and 379 cumec for Mukutmanipur Dam on Kangsabati River) during Komen. If we closely examine the flood-time discharge statistics, it can be observed that all dams and barrages have exceeded their Danger Level (DL) because of the large volume of inflow (peak inflow rate in cumec: 2640 for Durgapur Barrage on 3 Aug., 701 for Massanjore Dam on 1 Aug. 1296 for Tilpara Barrage on 3 Aug. and 1133 for Mukutmanipur Dam on 29 Jul.) in a single day due to excessive rainfall in and around the Rarh Bengal and began to release substantial volume of water (max. peak discharge in cumec: 3318 for Durgapur Barrage on 4 Aug., 708 for Massanjore Dam on 3 Aug. 1479 for Tilpara Barrage on 31 Jul. and 854 for Mukutmanipur Dam on 30 Jul. in 2015) during the last spell or just after the last spell (Fig. 22.10a). Besides, the comparison between maximum inflow and maximum outflow data of several dams and barrages (Fig. 22.10b) also shows a positive relationship (R = 0.72) which helps to establish that the higher amount of inflow due to large amount of rainfall in the upper catchment areas causes the huge volume of discharge. That makes flood situation more miserable in the lower catchment areas of the Rarh Bengal Rivers.

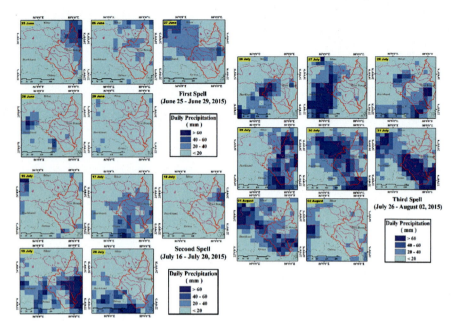

Fig. 22.8 Daily rainfall distribution from 3 hourly (temporal resolution) and 0.25 degree (spatial resolution) gridded data of NASA's Tropical Rainfall Measuring Mission (TRMM): First Spell (June 25- June 29, 2015), Second Spell (July 16- July 20, 2015) and Third Spell (July 26- August 02, 2015). Major river basins of Rarh Bengal have been shown as: **a** Dwarka-Brahamani Basins, **b** Mayurakshi-Kopai Basins, **c** Ajay Basin, **d** Damodar-Barakar Basin, **e** Dwarakeswar Basin, **f** Silabati Basin and **g** Kangsabati Basin

On the other hand, each Gauge Station (except Barkisuraiya) attained their Warning Level (WL) and was closer to Danger Level (DL) during the 2015 Rarh Bengal Flood. The rising water level of the Ajay crossed the DL at Gheropara (40 m on 3 Aug., DL: 39.41 m) submerging vast areas of the Rarh Bengal (Fig. 22.11). Besides, the water level of some other Gauge Stations which are located at the active riverine floodplain zone in the Rarh Bengal (27.20 m where the DL: 27.98 m at Narayanpur of Mayurakshi River, 12.50 m where the DL: 12.80 m at Harinkhola of Lower Damodar River, 24.52 m where the DL: 25.72 m at Mohanpur of Kangsabati River) also help to understand the flood situation during the 2015 summer monsoon. Besides most of the Gauge Stations of Irrigation and Waterways Department under Govt. of West Bengal (e.g., Ranagram on Dwarka River, Bazarshow on Babla River, Katwa on Bhagirathi and Ajay River, Champadanga and Amta on Lower Damodar River, Bandar on Rupnarayan River, etc.) in the lower catchment area of Rarh Rivers attained Extreme Danger Level (EDL) and also made the flood situation worse to worst.

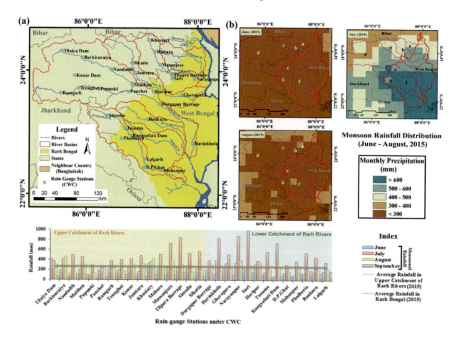

Fig. 22.9 **a** Monthly rainfall distribution of various Rain Gauge (RG) Stations under CWC throughout the catchment area of the Rarh Rivers with monsoonal average (June- September) rainfall lines. The stations which are confined only within the Rarh Bengal Plain have been considered under 'Lower Catchment of Rarh Rivers', but those RG stations which are located outside the Rarh Region have been contemplated under 'Upper Catchment of Rarh Rivers.' **b** Monthly monsoonal rainfall distribution map (June- August) from NASA's TRMM gridded datasets. Major river basins of Rarh Bengal have been denoted as: a. Dwarka-Brahamani Basins, b. Mayurakshi-Kopai Basins, **c** Ajay Basin, **d** Damodar-Barakar Basin, **e** Dwarakeswar Basin, **f** Silabati Basin and **g** Kangsabati Basin

22.6   Results and Discussion

22.6.1   The 2015 Rarh Bengal Flood: Exceptional

The entire East and North-East India have received 74.23% (1,343.4 mm) of the total rainfall (i.e., 7% less in terms of long-period average) in just four months during the 2015 summer monsoon (IMD: India Meteorological Department 2016a). Monthly monsoonal rainfall distribution throughout the entire region also depicts almost the same scenario (e.g., 1% deficit in June, 16% deficit in July, only 14% surplus in August and 25% deficit in September). Although the Rarh Region belongs from the East and North-East India, the monsoonal rainfall statistics gives a completely different scenario. The detailed analysis of rainfall data helps to understand the exceptional behaviour of the south-west monsoon because the Rarh Bengal has experienced abnormally high rainfall during the two successive weather disturbances in July. The

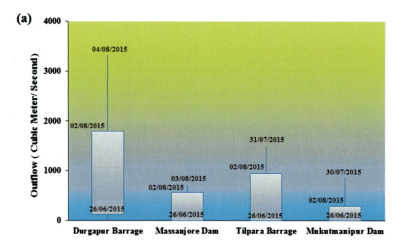

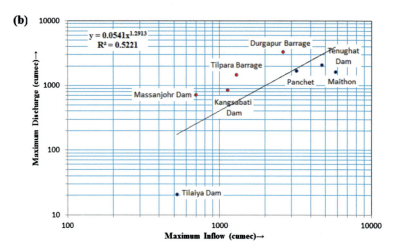

Fig. 22.10 **a** Graphical representation of the outflow in m^3/s from Durgapur Barrage, Massanjore Dam, Tilpara Barrage and Mukutmanipur Dam: 26 June 2015 (opening day of the first spell); 2 August 2015 (last day of the third spell) and the maximum outflow dates (*Source* IWD-GoWB 2015). **b** Relationship between maximum inflow and maximum outflow data of various dams and barrages throughout the entire catchment area of Rarh Rivers (*Source* Red dots and black dots indicate the data source from IWD-GoWB 2015 and CWC-GoI 2015 respectively)

spatial variation of rainfall in different districts of the Rarh Bengal also helps to illustrate the actual situation of monsoon (Table 22.2). If we want to correlate the July rainfall with the 2015 Rarh Bengal Flood, it will give a clear message that more departure of rainfall from the long-period average in case of any particular district has a greater chance of flooding than other districts. That's why Howrah, Hooghly, East Medinipur and Birbhum have faced terrible flooding in terms of vulnerability and risk during 2015 monsoon season.

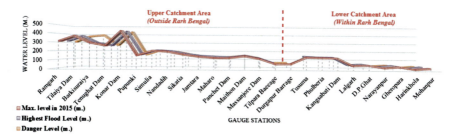

Fig. 22.11 Comparison among Danger Level (DL), Highest Flood Level (HFL) and Maximum Gauge Level (MGL) in 2015 flood at different gauge stations of Rarh Bengal Rivers (*Source* CWC-GoI 2015)

Table 22.2 District wise monthly monsoonal rainfall variation from June to September in 2015 within the Rarh Bengal under Lower Gangetic Plains (LGP). The actual district rainfall value of a particular month represents the arithmetic average of rainfall, measured from all IMD stations under that particular district and the departure in percentage shows the departures of rainfall from the long period rainfall averages for the district (*Source* IMD 2016a)

Name of the districts	Distribution of Rainfall							
	June		July		August		September	
	Actual (mm)	Departures (%)	Actual (mm)	Departures (%)	Actual (mm)	Departures (%)	Actual (mm)	Departures (%)
Bankura	152.2	−29	467.5	54	230.6	−21	96.5	−60
Burdwan	338.1	71	587.3	100	285.8	00	111.8	−55
Hooghly	299.5	23	671.6	112	188.0	−29	215.3	−11
Birbhum	321.5	45	698.5	123	296.8	−01	144.2	−47
East Medinipur	270.6	07	757.3	166	286.0	−16	162.4	−53
West Medinipur	138.1	−43	526.8	60	238.2	−25	177.1	−36
Murshidabad	341.5	44	640.4	95	274.6	07	193.2	−25
Howrah	227.9	−02	854.4	149	180.8	−45	202.2	−34

22.6.2 The 2015 Rarh Bengal Flood: Fast

For monitoring flood situation, Central Water Commission (CWC) under Govt. of India and Irrigation & Waterways Department (IWD) under Govt. of West Bengal have set up several hydro-meteorological observation stations. Out of which 18 CWC and 27 IWD active Gauge Stations (GSs) are located within the Rarh Bengal that helps to monitor, collect and compile the river stage (flood stage) data throughout the year (during monsoon). CWC generally forecast flood level based on changing water level concerning Warning Level (WL), Danger Level (DL) and Highest Flood Level (HFL). On the other hand, IWD does the same with respect to Primary Danger Level (PDL), Danger Level (DL) and Extreme Danger Level (EDL). When the river water level is at or above WL (PDL) but below DL, it is called as Above Normal Flood; when the river water level is at or above DL and below HFL (EDL), it is termed as Severe Flood and when the river water level attain HFL (EDL), it is called Extreme Flood. The data collected from CWC shows that the GSs located nearer to the mouth of all Rarh Rivers (Narayanpur on Mayurakshi River, Gheropara on Ajay River, Harinkhola on Lower Damodar River and Mohanpur on Kangsabati River) had crossed the WL mark. Even Gheropara GS on Ajay River crossed the DL mark also. It signifies that the GSs nearer to the mouth of all the Rarh Bengal Rivers have experienced Above Normal Flood (except the Ajay River near Gheropara that faced Severe Flood) during monsoon in 2015. To get more prominent and actual scenario, we have taken help of all the available GSs data from IWD (Table 22.3) that are located in the active riverine flood plain zone of the Rarh Bengal because limited CWC GSs on that area are not enough to illustrate the true flood situation of the entire Rarh Bengal.

Table 22.3 Flood level at different Gauge Stations under IWD on various Rarh Bengal Rivers above Extreme Danger Level (EDL) due to effect of Komen in 2015 (*Source* Modified after IWD-GoWB 2015)

Districts	River	Gauge station	EDL (m)	Duration of flow above EDL (July 27-August 07)			Peak Level attained (m.)
				From	To	Stability (Hrs.)	
Murshidabad	Dwarka	Ranagram	17.86	July 27	August 07	273	18.67
	Kuye	Angarpur	20.05	August 03	August 03	14	20.46
	Babla	Bazarshow	15.63	July 27	August 07	275	16.56
Burdwan	Bhagirathi	Katwa	14.32	August 04	August 05	24	14.85
	Ajay	Katwa	15.09	August 03	August 04	37	15.50
Hooghly	Mundeswari	Muchighata	6.76	August 03	August 07	108	8.15
	Lower Damodar	Champadanga	13.50	August 04	August 06	36	14.11
Howrah	Lower Damodar	Amta	6.24	July 30	August 07	191	6.75
West Medinipur	Rupnarayan	Bandar	7.46	August 02	August 07	132	8.29
	Old Cossye	Kalmijole	9.90	August 03	August 04	27	10.24
East Medinipur	New Cossye	Panskura	9.90	July 31	August 03	72	9.97

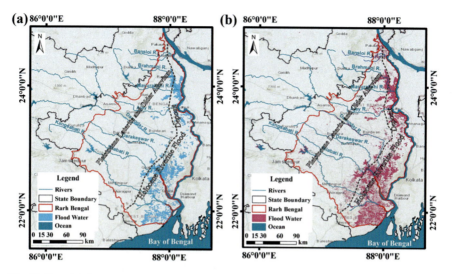

Fig. 22.12 **a** 14-day composite (August 1- August 14, 2015) of NASA's Terra and Aqua satellites images, extracted by Moderate Resolution Imaging Spectroradiometer (MODIS) for identifying the inundated area during 2015 flood (*Source* NASA GSFC NRT Global Flood Mapping). **b** Image classification of the Rarh Bengal for identification of floodwater during 2015 monsoon season from Operational Land Imager (OLI) of Landsat 8. The base maps for both images indicate national and international *boundaries* with major locations (*Source* Esri, HERE, Garmin, Intermap, increment P Corp., GEBCO, USGS, FAO, NPS, NRCAN, GeoBase, IGN, Kadaster NL, Ordnance Survey, Esri Japan, METI, Esri China: Hong Kong, © OpenStreet Map contributors and the GIS User Community)

Although three successive weather events helped to cause flooding in the Rarh Bengal, only the last spell, CS Komen contributed most for making the flood so devastating due to excessive rainfall. The water level at all the IWD GSs located nearer to the mouth on various rivers crossed the EDL mark and stabled for a longer period that made flood condition more miserable. For mapping the flooded area that caused inundation, two types of satellite products have been used viz. MODIS Flood Water Map from NASA (Fig. 22.12a) and Landsat 8 OLI images from USGS (Fig. 22.12b).

22.6.3 The 2015 Rarh Bengal Flood: Furious

An attempt has been made for mapping flood extent by using satellite images. For that purpose, two types of satellite images with different spatial resolution have been utilized to get the utmost true scenario of the 2015 Rarh Bengal Flood. The MODIS Near Real Time (NRT) flood products provide the extent of flood water base on NASA's Terra and Aqua satellite images with approximately 250 m spatial resolution (Nigro et al. 2014). But the main problem is that the satellite is enabled to detect flood feature smaller than 250 m. Besides, MODIS satellite product has also another limitation, i.e., we are unable to determine surface water extent when the area is cloudy. To overcome that particular problem, 14 days (3D3OT) composites product has been taken into consideration because it simply adds all previous 14 days' 3-day output product. For validation of these reliable satellite products, another type of satellite images (Landsat 8 OLI) has been used with spatial resolution of 30 m. The analysis of those Landsat images gives more detailed information regarding flood inundated areas of the Rarh Bengal during 2015 monsoon. For making the study more reliable, we have fused the output of both satellite images and prepared the final inundation map of the 2015 Rarh Bengal Flood (Fig. 22.13).

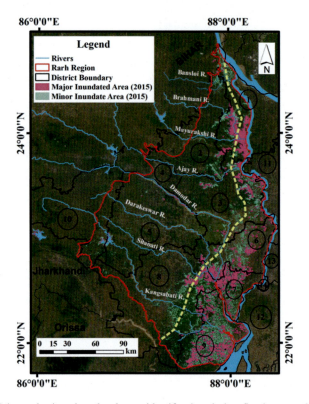

Fig. 22.13 Major and minor inundated area identification during flood season 2015 based on MODIS Terra and Aqua satellites images and Landsat 8 OLI satellite images. The north–south extended dotted line divides the Rarh Bengal into two, namely Holocene Riverine Flood Deposit (in the east) and Pleistocene Laterite & Older Alluvium (in the west). Name of the districts within and outside the Rarh Bengal have been denoted numerically (Districts within the Rarh Bengal: 1. Western portion of Murshidabad, 2. Part of Birbhum, 3. Purba Bardhaman, 4. Eastern Part of Paschim Bardhaman, 5. Bankura, 6. Hooghly, 7. Howrah, 8. Portion of Paschim Medinipur and 9. Part of Purba Medinipur/ Districts outside the Rarh Bengal: 10. Purulia, 11. Nadia, 12. South 24 Parganas, 13. North 24 Parganas and 14. Kolkata)

Table 22.4 Most affected C.D. Blocks of the Rarh Bengal during flood season 2015

Districts	Affected C.D. Blocks
Murshidabad	Kandi, Bharatpur- I & II, Khargram, Nabagram and Southern portion of Burwan block
Birbhum	Nanoor and Labhpur
Burdwan	Purbasthali- I & II, Manteswar, Katwa- I & II. Besides, the part of Kalna- I & II, Memari and Jamalpur (N.B., all these blocks presently in Purba Bardhaman district)
Hooghly	Khanakul- I & II, Jangipara, Haripal, Singur, Arambag, Tarakeswar, Pursurah and the part of Dhanikhali block
Howrah	Amta- I & II, Shyampur I & II, Uluberia- I & II, Panchal, Udaynarayanpur, Sankrail, Bagnan- I & II, Jagatballavpur, Domjur and Howrah Municipal Corporation Area
East Medinipur	All the blocks effected in 2015 flood. Mostly inundated blocks are Panskura- I & II, Tamluk, Sahid Matangini, Nandigram- I & II, Egra- II, Khejuri- I, and Kolaghat
West Medinipur	Daspur- I & II, Ghatal, Debra, Kharagpur, South-Eastern part of Keshpur and part of Pingla & Sabang

This helps to classify the flood inundated zones into two viz. Major Inundated Area based on NASA GSFC NRT Global Flood Mapping and Minor Inundated Area from Landsat 8 OLI images by comparing water features extracted from flood-time and normal-time Landsat 8 OLI images (for getting the total flood water extent areas) and omitting the Major Inundated Area from the total flood water extent areas. The 2015 Rarh Bengal Flood inundated nearly 2615 km^2 area, out of which 1395 km^2 was Major Inundated Area and 1220 km^2 was Minor Inundated Area. The list of the most effected C.D Blocks of the Rarh Bengal (Table 22.4) during flood season 2015 has been made based on the fusion map.

The most interesting thing which comes out from the study is that the blocks of those districts affected most are densely populated and located at the active riverine floodplain zone of the Rarh Bengal along the Bhagirathi-Hooghly River (Fig. 22.14).

The intense rainfall due to CS Komen increased the water velocity on the downstream of all Rarh Bengal Rivers and inundated the active riverine floodplain zone in the lower catchment area by causing embankment damage (317.64 km in Rarh Bengal out of 395.20 km effected embankment in West Bengal, 80.37%) due to overtopping of the flood water (Fig. 22.15).

As most of the floodplain is occupied by the common people for their habitat and livelihood purpose (agricultural), it increases the risk of damage caused by flooding (Sinha 2008). The 2015 Rarh Bengal flood also affected lives and livelihood of the common people. This unprecedented flood snatched 338 people lives and displaced nearly 10.84 million people. A large number of houses (nearly 8.30 lack) had been destroyed due to this flood as reported in the Flood Damage Report of the Principal Secretary of West Bengal Disaster Management, Government of West Bengal and published in CWC-GoI (2015).

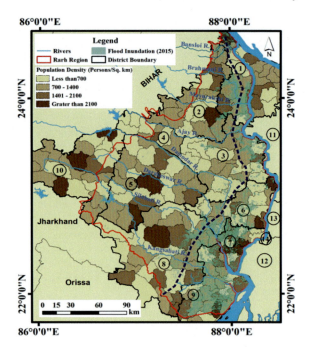

Fig. 22.14 Map showing affected C. D. Blocks of Rarh Bengal with population density due to inundation during flood season 2015 (*Source* Block wise population density calculated from 2011 Census). Name of the districts within and outside the Rarh Bengal have been denoted numerically (Districts within the Rarh Bengal: 1. Western portion of Murshidabad, 2. Part of Birbhum, 3. Purba Bardhaman, 4. Eastern Part of Paschim Bardhaman, 5. Bankura, 6. Hooghly, 7. Howrah, 8. Portion of Paschim Medinipur and 9. Part of Purba Medinipur/ Districts outside the Rarh Bengal: 10. Purulia, 11. Nadia, 12. South 24 Parganas, 13. North 24 Parganas and 14. Kolkata)

22.7 Conclusion

The 2015 Rarh Bengal Flood constitutes a landmark in terms of hydro-meteorological characteristics and also the extent of flood damage caused by inundation. The study confirms that the excessive rainfall at the end of July due to CS Komen has caused such type of severe flood. The river gauging records located at the active riverine floodplain zone of the Rarh Bengal indicate that the flood peak has occurred from the last week of July to the beginning of August. A large volume of water has also been released from various dams and barrages due to heavy downpour in the upper catchment area of the Rarh River Basins. That has made the flood situation more complex in the Rarh Bengal. Besides, the satellite remote sensing techniques have also been used to get the flood inundated area over the Rarh Bengal. This study does not recommend any particular management plan, but it urges that some suitable remedial measures have to be taken to combat with such type of unavoidable hazards to get off from the death toll and large amount of damages for the sake of more than 38 million population of the Rarh Region, one of the most flood prone region of the

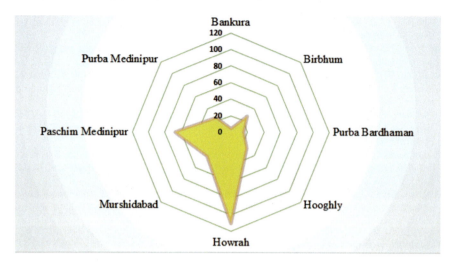

Fig. 22.15 Damaged embankment length (in kilometre) in various districts of the Rarh Bengal during the 2015 monsoonal flood (*Source* IWD-GoWB 2015)

world. We have to take zero casualty approach (not possible in reality, but helps to decrease the damages caused by floods) to manage that type of extreme flood events. Sufficient number of Flood Relief Centre (FRC) have to be built up so that evacuated people can be moved to the safest place based on flood forecast by CWC and IWD.

Acknowledgements This research work was presented at XIV International Geographical Union (IGU)—India, an International Conference on "Agriculture, Food, Water, Biodiversity and Health in Changing Climate" which was organised by The University of Burdwan, West Bengal from 06 March to 08 March 2020. We are indebted to Dr. Subhajit Sinha, Session Chairperson from Department of Geology, University of Calcutta and Co-Chairperson for their valuable suggestions. The authors acknowledge Central Water Commission (CWC) under Govt. Of India and Irrigation & Waterways Department under Govt. of West Bengal for cooperating with us at the time of data collection. Finally, we also express gratitude to the anonymous reviewers and the editor for their thoughtful and thorough reviews that improved the clarity of the manuscript.

References

Bandyopadhyay S, Kar N, Das S, Sen J (2014) River systems and water resources of West Bengal: a review. In Vaidyanadhan R (ed) Rejuvenation of surface water resources of india: potential, problems and prospects. Geological Society of India Special Publication, 3, pp 63–84
Bhalme HN, Jadhav SK (1984) The Southern Oscillation and its relation to the monsoon rainfall. Int J Climatol 4(5):509–520. https://doi.org/10.1002/joc.3370040506
Bhattacharyya K (2011) The Lower Damodar River: Understanding the human role in changing fluvial environment. In: Nusser M (ed) Advances in Asian human-environmental research, Springer, Dordrecht, Heidelberg, London, & New York

Bhattacharyya A (2013) Evaluation of hydro-geonomic characteristics of flood in the Mayurakshi river Basin India. Doctor of Philosophy Thesis, Visva-Bharati University

Blum MD (2007) Large river systems and climate change. In: Gupta A (ed) Large Rivers: geomorphology and management, 1st edn. Wiley, Chichester, pp 627–659

Bose NK (1970) Rivers of West Bengal and their control. In Chatterjee AB, Gupta A, Mukhopadhyay PK (eds) West Bengal, 1st edn. The Geographical Institute, Presidency College, Calcutta, pp 59–65

Chakraborty S (1970) Some consideration on the evolution of physiography of Bengal. In: Chatterjee AB, Gupta A, Mukhopadhyay PK (eds) West Bengal, 1st edn. The Geographical Institute, Presidency College, Calcutta, pp 16–29

Chatterjee KD, Majumdar NG (1972) Drainage problems of the Bhagirathi Basin. In: Bagchi KG (ed) The Bhagirathi-Hooghly Basin: Proceedings of the interdisciplinary symposium, 1st edn. Kanjilal SN, Calcutta, pp 78–88

Chapman GP, Rudra K (2004) Bengal's Millennium Flood: The Question of Appropriate Development. In: Singh S, Sharma HS, De SK (eds) Geomorphology & Environment, 1st edn. abc publications, Kolkata, pp 366–385

Chapman GP, Rudra K (2007) Water as foe, water as friend. J South Asian Dev 2(1):19–49. https://doi.org/10.1177/097317410600200102

CWC-GoI: Central Water Commission, Govt. of India (2015) The 2015 Flood Report. Damodar Division, Asansol

Dhar ON, Nandargi S (2003) Hydrometeorological aspects of floods in India. Nat Hazards 28:1–33. https://doi.org/10.1023/A:1021199714487

EM-DAT (2020) The emergency events database—Universite catholique de Louvain (UCL)—CRED. https://www.emdat.be/emdat_db

EUFD: EU Floods Directive (2007) Directive 2007/60/EC of the European Parliament and of the Council of 23 October 2007: On the Assessment and Management of Flood Risks. Official J Europ Union L 288:27–34. http://data.europa.eu/eli/dir/2007/60/oj

Guha-Sapir D, D'Aoust O, Vos F, Hoyois P (2013) The frequency and impact of natural disasters. In: Guha-Sapir D, Santos I (eds) The economic impacts of natural disasters, 1st edn. Oxford University Press, New York, pp 7–27

IWD-GoWB: Irrigation and Waterways Directorate, Govt. of West Bengal (2013) Annual Flood Report for the Year 2013. Directorate of Advance Planning, Project Evaluation &Monitoring Cell, Kolkata. https://wbiwd.gov.in/uploads/anual_flood_report/ANNUAL_FLOOD_REPORT_2015.pdf

IWD-GoWB: Irrigation and Waterways Directorate, Govt. of West Bengal (2015) Annual Flood Report for the Year 2015. Directorate of Advance Planning, Project Evaluation & Monitoring Cell, Kolkata. https://wbiwd.gov.in/uploads/anual_flood_report/ANNUAL_FLOOD_REPORT_2015.pdf

IMD: India Meteorological Department (2015a) Cyclonic Storm, KOMEN over the Bay of Bengal (26 July-02 August, 2015): A Report. Cyclone Warning Division, IMD, New Delhi. http://www.rsmcnewdelhi.imd.gov.in/images/pdf/publications/preliminary-report/KOM.pdf

IMD: India Meteorological Department (2015b) Special Tropical Weather Outlook. DEMS-RSMC Tropical Cyclones, New Delhi. http://www.rsmcnewdelhi.imd.gov.in/images/pdf/archive/bulletins/2015/RKO

IMD: India Meteorological Department (2016a) Rainfall Statistics of India—2015. Hydromet Division, IMD, New Delhi. http://hydro.imd.gov.in/hydrometweb/(S(qwluyvfwmga5piyqyzrlif55))/PRODUCTS/Publications/Rainfall%20Statistics%20of%20India%20-%202015/Rainfall%20Statistics%20of%20India%20-%202015.pdf

IMD: India Meteorological Department (2016b) Report on Cyclonic Disturbances over North Indian Ocean During 2015. RSMC-Tropical Cyclones, New Delhi. http://www.rsmcnewdelhi.imd.gov.in/images/pdf/publications/annual-rsmc-report/RSMC-2015.pdf

IWD-GoWB: Irrigation and Waterways Directorate, Govt. of West Bengal (2017) Annual Flood Report for the Year 2017. Directorate of Advance Planning, Project Evaluation & Monitoring Cell, Kolkata. https://wbiwd.gov.in/uploads/ANNUAL_FLOOD_REPORT_2017.pdf

Jonkman SN (2005) Global Perspectives on loss of human life caused by floods. Nat Hazards 34:151–175. https://doi.org/10.1007/s11069-004-8891-3

Kale VS (1999) Long-period fluctuations in monsoon floods in the Deccan Peninsula, India. J Geol Soc India 53:5–15

Kale VS, Hire P, Baker VR (1997) flood hydrology and geomorphology of monsoon-dominated rivers: the Indian Peninsula. Water Int 22(4):259–265. https://doi.org/10.1080/02508069708686717

Kale VS (2003) geomorphic effects of monsoon floods on Indian rivers. Nat Hazards 28:65–84. https://doi.org/10.1023/A:1021121815395

Kale VS (2012) On the link between extreme floods and excess monsoon epochs in South Asia. Clim Dynam 39(5):1107–1122. https://doi.org/10.1007/s00382-011-1251-6

Kale VS (2014) Is flooding in South Asia getting worse and more frequent? Singap J Trop Geogr 35(2):161–178. https://doi.org/10.1111/sjtg.12060

Kvočka D, Falconer RA, Bray M (2016) Flood hazard assessment for extreme flood events. Nat Hazards 84:1569–1599. https://doi.org/10.1007/s11069-016-2501-z

Kale VS (2002) Fluvial geomorphology of Indian rivers: an overview. ProgPhysGeog 26(3):400–433. https://doi.org/10.1191/0309133302pp343ra

Mirza MMQ, Warrick RA, Erickson NJ, Kenny GJ (2001) Are floods getting worse in the Ganges, Brahmaputra and Meghna Basins? Environ Hazards 3(2):37–48. https://doi.org/10.3763/ehaz.2001.0305

NDMA-GoI: National Disaster Management Authority, Govt. of India (2007) National disaster management guidelines: preparation of State disaster management plans. Ministry of Home Affairs, New Delhi. https://nidm.gov.in/PDF/guidelines/sdmp.pdf

Nigro J, Slayback D, Policelli F, Brakenridge R (2014) NASA/ DFO MODIS Near Real-Time (NRT) Global flood mapping product evaluation of flood and permanent water detection. Science Systems and Applications, NASA Goddard Space Flight Center and Dartmouth Flood Observatory. https://floodmap.modaps.eosdis.nasa.gov/documents/NASAGlobalNRTEvaluationSummary_v4.pdf

NOAA-NCEI: National Oceanic and Atmospheric Administration, National Centers for Environmental Information (2019) Southern Oscillation Index (SOI). https://www.ncdc.noaa.gov/teleconnections/enso/indicators/soi

O'Connor JE, Costa JE (2004) The world's largest floods, past and present: their causes and magnitudes. U.S. Geological Survey Circular 1254. https://pubs.usgs.gov/circ/2004/circ1254/pdf/circ1254.pdf

ORG & CCI: Office of Registrar General & Census Commissioner of India (2011) C.D. Block Wise Primary Census Abstract Data (PCA)—WEST BENGAL. Ministry of Home Affairs, Government of India. http://censusindia.gov.in/pca/cdb_pca_census/Houselisting-housing-WB.html

Rodier JA, Roche M (2003) World catalogue of maximum observed floods. IAHS Publ. 284. IAHS Press, Wallingford, UK (37)

Rudra K (2018) Rivers of the Ganga-Brahmaputra-Meghna Delta: a fluvial account of Bengal. Springer, Switzerland

Sen S (1970) The importance of drainage in agriculture of West Bengal. In: Chatterjee AB, Gupta A, Mukhopadhyay PK (eds) West Bengal, 1st edn. The Geographical Institute, Presidency College, Calcutta, pp 75–84

Sengupta S (1970) Geology of the southwestern Bengal. In: Chatterjee AB, Gupta A, Mukhopadhyay PK (eds) West Bengal, 1st edn. The Geographical Institute, Presidency College, Calcutta, pp 1–6

Sengupta S (1972) Geological framework of the Bhagirathi-Hooghly Basin. In: Bagchi KG (ed) The Bhagirathi-Hooghly Basin: proceedings of the interdisciplinary symposium, 1st edn. Kanjilal SN, Calcutta, pp 3–8

Sinha R, Friend PF (1994) River systems and their sediment flux, Indo-Gangetic plains, Northern Bihar. India. Sedimentology 41(4):825–845. https://doi.org/10.1111/j.1365-3091.1994. tb01426.x

Sinha R, Jain V (1998) Flood Hazards of North Bihar Rivers, Indo-Gangetic Plains. In: Kale VS (ed) Flood Studies in India, memoir geological society of India, Bangalore 41, pp 27–52

Sinha R, Jain V, Babu GP, Ghosh S (2005) Geomorphic characterization and diversity of the fluvial systems of the Gangetic Plains. Geomorphology 70(3–4):207–225. https://doi.org/10.1016/j.geo morph.2005.02.006

Sinha R (2008) Flood hazard: a GIS based approach. Geography and You 8:6–11

TRMM: Tropical Rainfall Measuring Mission (2011) TRMM (TMPA/3B43) Rainfall Estimate L3 1 month 0.25 degree x 0.25 degree V7. Goddard Earth Sciences Data and Information Services Center (GES DISC). https://doi.org/10.5067/TRMM/TMPA/MONTH/7

TRMM: Tropical Rainfall Measuring Mission (2016) TRMM (TMPA) Precipitation L3 1 day 0.25 degree x 0.25 degree V7. Goddard Earth Sciences Data and Information Services Center (GES DISC). https://doi.org/10.5067/TRMM/TMPA/DAY/7

UNISDR & CRED: United Nations Office for Disaster Risk Reduction & Centre for Research on the Epidemiology of Disasters (2018) Economic Losses Poverty Disast, 1998–2017. https://www. preventionweb.net/files/61119_credeconomiclosses.pdf

Vörösmarty CJ, Fekete BM, Meybeck M, Lammers RB (2000) Geomorphometric attributes of the global system of rivers at 30-minute spatial resolution. J Hydrol 237:17–39. https://doi.org/10. 1016/S0022-1694(00)00282-1

WBDMD: West Bengal Disaster Management & Civil Defence Department (2020) Natural Disaster: Flood. http://wbdmd.gov.in/Pages/Flood2.aspx

Wohl EE (2007) hydrology and discharge. In: Gupta A (ed) Large Rivers: Geomorphol Manag, 1st edn. Wiley, Chichester, pp 29–44

WMO/ESCAP: World Meteorological Organization/Economic and Social Commission for Asia and the Pacific (2015) Annual Cyclone Review—2015. WMO/ESCAP Panel on Tropical Cyclones. http://www.rsmcnewdelhi.imd.gov.in/images/pdf/publications/annual-cyclone-review/annual-%20review-%202015.pdf

Chapter 23
Application of Remotely Sensed Data for Estimation of Indices to Assess Spatiotemporal Aspects of Droughts in Bankura District of West Bengal, India

Asraful Alam⊙, **Rajat Kumar Paul**⊙, and **Lakshminarayan Satpati**⊙

Abstract Drought is a complex environmental issue, defined as the condition of severe water shortage in any spatiotemporal condition. This study primarily aims to deliberate upon the use of Geographic Information System and Remote Sensing techniques for analysis of meteorological data to find out drought risk assessment in Bankura district of West Bengal. The study incorporates multispectral band ratio to estimate vegetation density and vegetation health for evaluation of spatiotemporal aspects of drought conditions. Land Surface Temperature (LST) and Normalized Difference Vegetation Index (NDVI) have been worked out to measure three other indices, namely: Vegetation Condition Index (VCI), Temperature Condition Index (TCI), and Vegetation Health Index (VHI). Moreover, an attempt has been made to assess spatiotemporal issues of drought risks associated with agriculture through analysis of temporal images for NDVI and Standardized Precipitation Index (SPI) of the area. Three drought years i.e., 2000, 2010, and 2018 have been selected for assessment of drought in the district. Overall, the results showed that the district has experienced moderate to severe drought situations in the recent past.

Keywords Drought · Bankura district · NDVI · VHI · TCI · SPI

A. Alam (✉)
Department of Geography, Serampore Girls' College, University of Calcutta, 13, T.C. Goswami Street, Serampore, Hooghly, West Bengal 712201, India

R. K. Paul
Department of Geography, University of Calcutta, Kolkata, India

L. Satpati
Department of Geography and Director, UGC-HRDC, University of Calcutta, Kolkata, India

© The Author(s), under exclusive license to Springer Nature Singapore Pte Ltd. 2022   425
N. C. Jana and R. B. Singh (eds.), *Climate, Environment and Disaster in Developing Countries*, Advances in Geographical and Environmental Sciences, https://doi.org/10.1007/978-981-16-6966-8_23

23.1 Introduction

Occurrence of drought is characterized by its slow progress during a full hydrological cycle and its persistent consequences are usually exposed after its completion (Vogt et al. 2011). Drought is one of the different types of extreme climatic disasters which make the environment, agriculture, and socioeconomic conditions suffer severe damages (Wilhite 2000). Compared to any other form of natural disasters, droughts affect more areas and have longer durations, with the main characteristic feature being more frequent in occurrence (WMO 1997). Droughts are basically classified into four types: meteorological, hydrological, agricultural, and socioeconomic (Wilhite and Glantz 1985). These are having slow on-set and often not quite distinguishable when these started or ended, making the phenomena difficult to study (Ali and Nezar 2007). Drought occurs when rainfall in a region is less than statistical multi-year average for that region over an extensive time period (Mala et al. 2014); and due to shortage of rainfall, groundwater discharge is more than recharge, and it creates huge problem for the following periods (Nag and Ghosh 2013). Remote sensing and GIS can play an important role in detecting, assessing, and managing droughts as these offer up-to-date information on spatial and temporal scales (Hayes et al. 2012). Hydrological drought is a condition of deficiency in water availability of surface and subsurface water reservoirs. Socioeconomic drought is the final phase of drought, which is caused by prolonged shortage in agricultural production and food, thus affecting overall economy of a region or country (Linsley et al. 1975). Meteorological drought has been defined as a 'period of more than some particular number of days with precipitation less than some specified small amount' (Great Britain Meteorological Office 1951). Agricultural drought denotes that there is not enough moisture to support average crop production. Socioeconomic drought, which expresses features of the socioeconomic effects of the drought, can also incorporate features of meteorological, agricultural, and hydrological droughts (Kifer et al. 1938). Agricultural drought may become detrimental in association with meteorological drought, as in case of India (Murthy et al. 2015; Agutu et al. 2017). Subrahmanyam (1967) has identified six types of droughts: meteorological, climatological, atmospheric, agricultural, hydrologic, and water management. Many other scholars have included economic or socioeconomic factors as an essential factor in the determination of occurrence of droughts (Hoyt et al. 1942; Guerrero Salazar 1975). In this background, the south-western part of West Bengal is often likely to receive less rain in monsoon season, and thus its experiences both agricultural and meteorological droughts. The districts of Bankura and Purulia may likely experience a 1 °C rise in average temperature during 2025–2099 (Ghosh 2016, 2018). In Bankura district particularly, drought has become more frequent during the recent past. The problems are further compounded with the growing population, lack of water resources, and introduction of water-intensive commercial crops in the district. Climate change especially manifested by drought will be one of the most challenging issues for this part of West Bengal in future.

23.2 The Objectives

The average annual rainfall of Bankura and Purulia districts is about 1400 mm, but 90% of it occurs during the months of June to September (Ghosh 2019). Hot westerly winds prevail in the district from March to June (Ghosh 2018). This study primarily aims to deliberate upon the use of RS and GIS techniques for analysis of meteorological data for drought risk assessment of Bankura district of West Bengal. This is to find out how multispectral band ratio can be useful to estimate vegetation density and vegetation health for evaluation of spatiotemporal aspects of drought conditions.

23.3 The Study Area

Physiographically West Bengal is divided into three parts, such as the Northern Hill region, the Western Plateau region, and the Gangetic-Delta plain land. The Western Plateau is one of the most important regions for the planners in terms of water management. Administratively at present Bankura district belongs to Medinipur division. The district is the connecter of plateau and plain lands of West Bengal. The eastern and north-eastern parts are mainly plain, and undulating parts are found in the western side of it, which is actually the extended part of the Chhotonagpur plateau (Das 2013). Average rainfall of the district is about 140 cm (occurring predominately from the months of July to September), but groundwater recharge is very poor due to high-intensity rainfall in a short span and presence of low permeable undulating lateritic land surface (Fig. 23.1). Except in the monsoon period, the problem of water scarcity is very much prominent in this region, and it has severe impacts on agricultural production although more than 70% of the district's income comes from agriculture sector. Demand for water is almost 63 L per day per capita for drinking and cooking purposes, but the deficit of water is mostly historical (Maiti 2015). One of the main problems of this district is fluoride contamination in groundwater, as may be found in pockets (Chakrabartiand Bhattacharya 2013). Due to improper management of water supply, the Bankura is designated as a drought-prone district of the state of West Bengal.

23.4 Datasets Used

To fulfil the above objectives, remote sensing as well as station based climatic data have been used. Multi-temporal satellite images of Bankura district have been downloaded from the USGS website. To justify the trends of drought situation in the concerned district, three temporal images (years: 2000, 2010, and 2018) have been used, and all the images have been taken to be cloud and noise-free to serve the

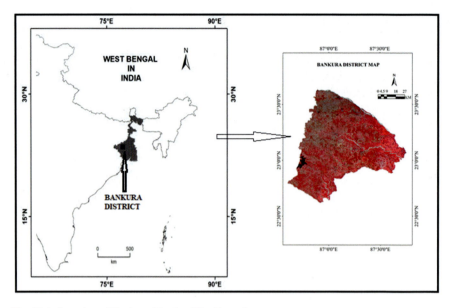

Fig. 23.1 Location of Bankura District, West Bengal

Table 23.1 Brief description of the used satellite data

District	Path	Row	Date of acquisition with satellite sensor		
			2000 (TM)	2010 (ETM+)	2018 (OLI)
Bankura	139	044	21 March	09 March	24 April

purpose (Giri et al. 2003). In case of Landsat TM/ETM + satellite imagery, Bands 3, 4, and 6, and for Landsat OLI image Bands 4, 5, and Band 10 have been used. Detailed information about the satellite images used is given below (Tables 23.1, 23.2 and 23.3).

23.5   Methodologies

23.5.1   *Standardized Precipitation Index (SPI)*

Stationed rainfall data has been used for Standardized Precipitation Index (SPI). SPI (McKee et al. 1993) is considered to be a very useful index to measure the moisture surplus and deficit of a region. World Meteorological Organization (WMO) recommends all Metrological Organizations (MO) to use SPI to measure the drought condition, apart from the individual MO's own indices. The most important benefit of using is that it is multi-temporal in nature and different temporal SPIs can be

Table 23.2 Spectral Bands of Landsat ETM+/TM Image for the years 2006 and 2011

Channel	Spectral range (μm)	Spatial resolution (m)	Electromagnetic region
Band 1	0.45–0.52	30	Visible Blue
Band 2	0.53–0.63	30	Visible Green
Band 3	0.63–0.69	30	Visible Red
Band 4	0.78–0.90	30	Near Infrared
Band 5	1.55–1.75	30	Short Wave Infrared (1)
Band 6	10.4–12.5	120(60)[a]	Thermal Infrared
Band 7	2.09–2.35	30	Short Wave Infrared (2)
Band 8	0.52–0.90	15	Panchromatic

[a]The resolution of Thermal Infrared is 120 m and 60 m in Landsat TM and Landsat ETM+, respectively, while it can be re-sampled in 30-m resolution

Source www.usgs.gov retrieved on 18 February 2020

Table 23.3 Spectral Bands of Landsat 8 OLI

Channel	Spectral range (μm)	Spatial resolution (m)	Electromagnetic region
Band 1	0.43–0.45	30	Coastal Aerosol
Band 2	0.45–0.52	30	Visible Blue
Band 3	0.53–0.59	30	Visible Green
Band 4	0.64–0.67	30	Visible Red
Band 5	0.85–0.88	30	Near Infrared
Band 6	1.57–1.65	30	Short Wave Infrared (1)
Band 7	2.11–2.29	30	Short Wave Infrared (2)
Band 8	0.50–0.68	15	Panchromatic
Band 9	1.36–1.38	30	Cirrus
Band 10	10.60–11.19	100	Thermal Infrared (1)
Band 11	11.50–12.51	100	Thermal Infrared (2)

Source www.usgs.gov retrieved on 18 February 2020

used to demarcate different kinds of drought scenarios, like: meteorological drought, agricultural drought, and hydrological drought. Here SPI of 6 months' time step has been used to demarcate the drought of the study area. Computation of the SPI requires fitting of a probability distribution to the historical precipitation records for the timescale(s) of interest in order to define the relationship of the probability

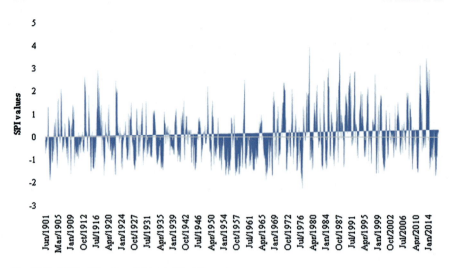

Fig. 23.2 SPI6 of Bankura District, West Bengal

to the occurred precipitation. The fitted probability distribution is then normalized to a standard normal distribution using the inverse normal (Gaussian) function. In a standard normal distribution, the mean and variance of SPI for the location and desired time period are 0 and 1, respectively (Fig. 23.2). Therefore, for any observed precipitation data, the SPI value is the deviation from the entire standard normal distribution (McKee et al. 1993; Edwards and McKee 1997; Heim 2002; Mishra and Singh 2010). SPI can be estimated using the following empirical formula:

$$\mathrm{SPI} = S\frac{t - (c2t + c1)t + c0}{((d3t + d2)t + d1)t + 1.0}$$

$$t = \sqrt{\ln\frac{1}{H(X)^2}}$$

$$G(x) = \frac{1}{\beta^\gamma T(\gamma)_0}\int_0^x x^{\gamma-1}e^{-x/\beta}dx, \ x > 0$$

23.5.2 Estimation of Drought Using Remote Sensing Technique

After determination of the most important drought year of the district, geospatial technology has been used to determine the effect of drought on the agriculture and vegetation. Multispectral and thermal data from Landsat are used to measure Normalized Difference Vegetation Index (NDVI) and estimate Land Surface Temperature

(LST) (Rizqi et al. 2015), and all data have been converted to sensor spectral-radiance through radiometric calibration (Chander et al. 2009). NDVI and LST time series have prospect to explain the various dynamics of dry conditions (Wang et al.2014). Lastly, the two sets of data have been used to calculate Vegetation Condition Index (VCI) and Temperature Condition Index (TCI), respectively. Collective VCI and TCI data have been used to calculate Vegetation Health Index (VHI) as the index of vegetative drought which incorporates overall vegetation health, and its adherence to designate agricultural drought level at any time of the year (Rizqi et al. 2015). Calculation of VHI is presented in Fig. 23.3. The VCI has been calculated from NDVI to monitor vegetation condition (Kogan 1995) and the VCI data has been calculated by using the following equation:

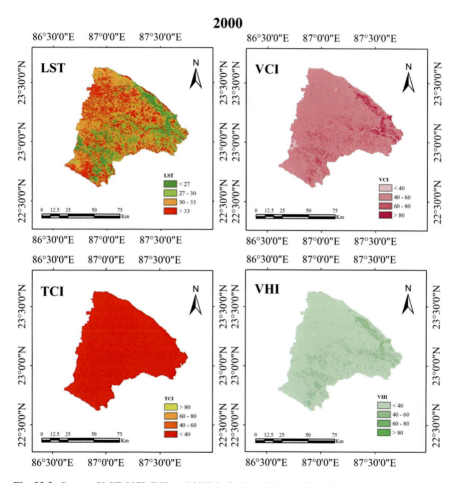

Fig. 23.3 Status of LST, VCI, TCI, and VHI in Bankura District, West Bengal for the year 2000

$$VCI = \frac{(NDVI\alpha - NDCI_{min})}{(NDVI_{max} - NDVI_{min})} \times 100$$

Here, $NDVI\alpha$ represents NDVI value of the current month, while NDVI*min* and NDVI*max* denote the minimum and maximum NDVI values, respectively, for the period of observation. VCI is recommended as a drought tool; but only using of VCI is not enough to explain drought quite perfectly, and for that one needs to develop TCI to capture different responses of vegetation to in-situ temperature as supplementary information. Thus, to fulfil the whole process application of thermal channels for drought monitoring (Kogan 1995) (Table 23.1) has been done. TCI has been calculated with the help of the following formula:

$$TCI = \frac{(BT_{max} - BTa)}{(BT_{max} - BT_{min})}$$

Here, LST α is the LST value of the current or present month, LST$_{min}$ represents the minimum LST values, and LST$_{max}$ denotes the maximum LST values considered from multi-year time series data. Lastly, VHI has been calculated to evaluate both category of vegetation stress and temperature to weigh up drought severity. VHI is estimated by the following equation:

$$VHI = \alpha VCI + (1 - \alpha)TCI$$

Here, VHI is related to VCI and TCI by α, and α equals to 0.5 (Rizqi et al. 2015).

23.6 Results and Discussions

The seasonal and inter-annual vegetation growth can be monitored by NDVI (Paul et al. 2021), which ranges between -1 and $+1$. Negative NDVI denotes severe drought proneness; and if NDVI becomes positive, it denotes wet condition (Jenson 2016). From NDVI, Vegetation Condition Index (VCI) is derived. NDVI demonstrates that the indices are consequential from remotely sensed data under visible and near-infrared bands of the electromagnetic spectrum (Ghosh et al. 2018). The vegetation responded independently to the different wavelengths of electromagnetic radiation. The leaf pigments (chlorophyll) absorb the red (and the blue) segment of the visible spectrum, while the near-infrared part is reflected. It is worth noting that the phenological phase and the environmental conditions (heat and moisture stress) are primary factors in shaping the rate of reflection. In autumn, the leaves reflect less NIR (and more visible) radiation than in spring (Molnar 2016). On the basis of the classified three temporal LST maps, it is observed that LST values are increasing rapidly for the region. In the year 2010, less than 27 °C temperature had been recorded in almost 14% of the total area of the district, which was not found in the year 2018 (Table 23.4). On the other hand, temperature zone of more than 33 °C was found almost 20% of total area in the year 2010 (Fig. 23.4), whereas it gained

Table 23.4 Status of LST, VCI, TCI, VHI for the period 2000–2018 in Bankura District, West Bengal

Indices		Year		
		2000	2010	2018
LST	<27	963.8604	593.775	0.009
	27–30	1261.798	1731.033	888.7743
	30–33	3474.371	3870.196	2687.856
	33	1259.323	764.3484	3382.692
VCI	<40	141.6078	179.4231	148.5711
	40–60	5668.7022	5955.045	2097.945
	60–80	1016.5995	759.78	4302.323
	>80	132.4431	65.1042	410.4927
TCI	<40	6959.312	0.0252	595.9422
	40–60	0.0243	0.0495	1781.312
	60–80	0.009	5143.354	3456.283
	>80	0.0072	1815.924	1125.795
VHI	<40	6329.592	0.0153	196.8813
	40–60	629.7426	922.8951	2135.408
	60–80	0.0171	5878.064	3942.046
	>80	0.0009	158.3784	684.9972

quickly in the year 2018 which was almost 45% of the total area (Fig. 23.5). The LST value had been increasing from western and north-western sides of the district, because those sides were facing the problems of deforestation and improper land surface management. More than 85% of the area were facing high LST (>30 °C), which was above the state average temperature for the month of March and the 1st week of April. Comparing the year between 2010 and 2018, it has been observed that the vegetation health was going to worse condition especially in the north-eastern part of the district. Due to severe temperature in 2000, the VHI of the district was almost equally distributed in that time spread (Fig. 23.3). But after that more than 70% of the district was having healthy vegetation for the year 2010, but subsequently the vegetation health index of more than 60 was less than 45% (in the year 2018) of the whole district. The reason behind it is the potentially low ground and soil water recharge, and unscientific forest-cutting. From the calculation of LST and VHI, it has been observed that eastern and northern sides of the district were facing severe droughts compared to the rest of the district.

LST has been found to have a negative correlation with NDVI (vegetation cover). This means wherever there was high NDVI, the surface temperature was low and vice versa. In the year 2010 high LST and low VHI were recorded in the western part of the district, particularly in the CD Blocks of Chhatna, Bankura I, Bankura II, and Indpur out of 22 blocks of the district. Intensity of the LST gradually increased, and high LST and low NDVI were found in the western, north-western, and northern

2010

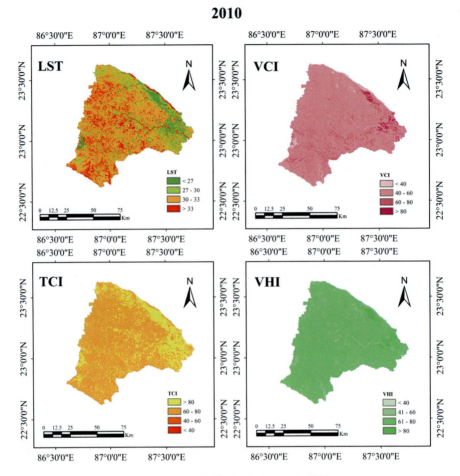

Fig. 23.4 Status of LST, VCI, TCI, and VHI in the Study area in 2010

parts of the study area, i.e., in the blocks of Saltora, Chhatna, Bankura I, Bankura II, Gangajalghati (Fig. 23.5). NDVI values were found to be good in the eastern and southern parts of the district. Thus, the impact of drought directly goes on to water and vegetation. Due to insufficient water recharge soil gets dried up and becomes unusable for agricultural production, and failure of crop production adversely affects the people.

23.7 Conclusions

In this work, we have tried to evaluate the capability of remotely sensed LST, VCI, TCI and derived VHI to represent status of drought in the district of Bankura, West Bengal, with the aim of using LST and VCI within the context of a near-real-time

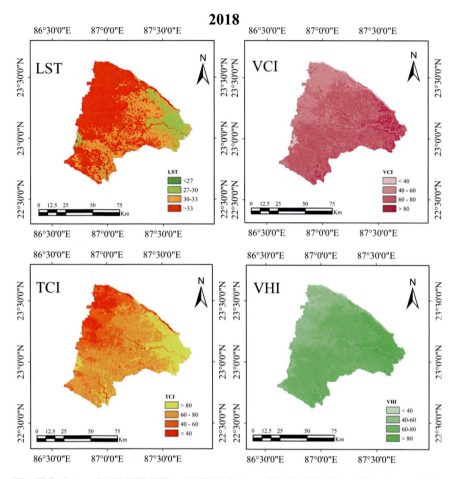

Fig. 23.5 Status of LST, VCI, TCI, and VHI in Bankura District, West Bengal for the year 2018

drought monitoring system. Monsoon countries are largely dependent on rainfall for their agriculture, but a slight deviation of it from the normal usually has a great impact on their agricultural production. The results of this study show that the district had been experiencing moderate to severe drought situation during the analysis period. Improper water management and deforestation make the situation more serious. It is predicted that if the present situation continues then drought will engulf the eastern side of the district too, in the near future. New policy may be taken up based on the ground reality through participatory water management with the local people in confidence that will definitely reduce deforestation and augment usable water through construction and renovation of more number of surface structures to hold surface run-off.

References

Azim Uddin AFM, J. K. Effects of Riverbank Erosion on Livelihood. Dhaka-1215, Bangladesh: Unnayan Onneshan-The Innovators.

Ali E-N, Nezar H (2007). Drought assessment using GIS and remote sensing in amman-Zarqa Basin, Jordan. Jordan J Civil Eng 1(2)

Chakrabarti S, Bhattacharya HN (2013) Inferring the hydro-geochemistry of fluoride contamination in Bankura district, West Bengal: a case study. J Geol Soc India 82(4):379–391

Chander G, B, M. L., & L, H. D. (2009) Summary of current radiometric calibration coefficients for Landsat MSS, TM, ETM+, and EO-1 ALI sensors. Remote Sens Environ 113:893–903

Das B (2013) Geo-spatial assessment of agricultural drought by remote sensing & GIS tecnnique (A case study of Bankura, West Bengal). Department of Remote Sensing & GIS Vidyasagar University Midnapure-721102

Ghosh A, Alam A, Ghosh, S (2018) Remote sensing image-based analysis of the relationship between land surface temperature and vegetation index: a statistical correlation in Indpur and Hirbandh block of Bankura district, West Bengal, India. Hill Geographer, NEHU, Shilling, vol XXXIV no 2, 63–72pp

Giri C, Defourny P, Shrestha S (2003) Land cover characterization and mapping of continental Southeast Asia using multi-resolution satellite sensor data. Int J Remote Sens 24(21):4181–4196

Ghosh KG (2018) Analysis of rainfall trends and its spatial patterns during the last century over the Gangetic West Bengal, Eastern India. J Geovisualization Spat Anal 2(2):15. https://doi.org/10.1007/s41651-018-0022-x

Ghosh KG (2016) Long range climatic variability over Birbhum district, west bengal and their impact on rainfedAman crop in the context of climate change: adoption and mitigation. In Rural Health, women empowerment and agriculture: issues and challenges, Chap 21, ed. P.K. Chattopadhyay and D.S. Kushwaha, 1st ed., 277–298. India: New Delhi Publishers

Ghosh KG (2019) Spatial and temporal appraisal of drought jeopardy over the Gangetic West Bengal, eastern India. Geoenvironmental Disasters 6.https://doi.org/10.1186/s40677-018-0117-1

Great Britain Meteorological Office (1951) The meteorological glossary. Chemical Publishing Co., New York

Guerrero Salazar P, Yevjevich V (1975) Analysis of drought characteristics by the theory of runs. Hydrology Paper No. 80, Department of Civil Engineering, Colorado State Univ., Fort Collins, CO

Hayes MJ, Svoboda MD, Wardlow BD, Anderson MC, Kogan F (2012) Drought monitoring: historical and current perspectives. Drought Mitigation Center Faculty Publications 94. http://digitalcommons.unl.edu/droughtfacpub/94

Hoyt WG (1942) Droughts. In: Meinzer OE (ed) Hydrology. Dover Publications, New York, p 579

Kifer RS, Steward HL (1938) Farming hazards in the drought area, Monograph XVI, Works Progress Administration, Washington, DC

Kogan FN (1995) Application of vegetation index and brightness temperature for drought detection. Adv Space Res 15(11):91100

Linsley RK, Kohler MA, Paulhus JLH (1975) Hydrology for Engineers, 2nd edn., McGrawHill, Kogukusha, Tokyo

Maiti MM (2015) Role of roof–top water harvesting to manage drought in Bankura district of West Bengal. Int Res J Basic Appl Sci 1(2015):18–23

McKee TB, Doesken NJ, Kleist J (1993) The relationship of drought frequency and duration to time scales. In: Proceedings of the 8th conference on applied climatology. Boston, MA: American Meteorological Society

Molnar G (2016) Analysis of Land Surface Temperature And NDVI Distribution For Budapest Using Landsat 7 ETM+ Data. UniversitatisSzegediensis, Tomus, Acta ClimatologicaEtChorologica

Murthy CS, Chakraborty A, Seshasai MVR, Roy PS (2011) Spatio-temporal analysis of the droughts of kharif 2009 and 2002. Curr Sci 100:1786–1788

Nag SK, Ghosh P (2013) Delineation of groundwater potential zone in Chhatna Block, Bankura District, West Bengal, India using remote sensing and GIS techniques. Enviro Earth Sci 70(5):2115–2127

N, V., & P, P. (2017) gis based agricultural drought assessment for the State of Tamilnadu, India using vegetation condition index (VCI). Int J Civil Eng Technol (IJCIET) 8(5):1185

Paul RK, Baidya A, Alam A, Satpati L (2021) An assessment of cyclone-induced vulnerability and change in land use and land cover (LULC) of G-plot in Patharpratima CD block of south 24 Parganas district, West Bengal. Indian J Geogr Environ Manag 17–18:1–13

Rizqi IS, Trisasongkoa BH, Shiddiqa D, Imana LO, Kusdaryantoa S, Manijoa D et al (2015). Identification of agricultural drought extent based on vegetation health indices of Landsat data: case of Subang and Karawang, Indocase of Subang and Karawang, Indonesia. (T. 2.-I. Environmental, Ed.) Procedia Environmental Sciences, Elsevier

Subrahmanyam VP (1967) Incidence and spread of continental drought, WMO/IHD Report No. 2, Geneva

Vogt JV, Safriel U, Von Maltitz G, Sokona Y, Zougmore R, Bastin G, Hill J (2011) Monitoring and assessment of land degradation and desertification: towards new conceptual and integrated approaches. Land Degrad Dev 22(2):150–165

Wang H, Lin H, Liu D (2014) Remotely sensed drought index and its responses to meteorological drought in Southwest China. Remote Sensing Letters 5(5):413–422

Wilhite DA, Glantz MH (1985) Understanding: the drought phenomenon: the role of definitions. Water Int 10(3):111–120

Wilhite DA (2000) Drought as a natural hazard: concepts and definitions. In: Wilhite D (ed) drought: a global assessment, vol 1. Routledge, London & New York, pp 3–18

World Meteorological Organization (WMO) (1997) Climate, drought and desertification. Geneva, WMO No. 869, 12 p

Chapter 24
Temporal Variability of Discharge and Suspended Sediment Transport in the Subarnarekha River Basin, Eastern India: A Geomorphic Perspective

Sunanda Banerjee⑩, Arup Kumar Roy⑩, and Asraful Alam⑩

Abstract In the world, over the last 50 years, many rivers had experienced decreasing trend in runoff and sediment load. In Subarnarekha, after analysing the data of discharge, runoff and sediment load of hydrological stations, it is clear that Subarnarekha also has declining trend. In the last few decades, precipitation rate has also declined day by day. In 2010–2011 precipitation rate was very low in Subarnarekha basin. Runoff, water discharge and sediment load are proportionately related to precipitation rate. Anthropogenic factors are also responsible for declining rate in Subarnarekha. Previously water discharge was closely related to the monsoonal rainfall but in few decades discharge and sediment load gradually decrease due to the human intervention like barrage, dam construction and other activities. In Ghatshila station, before dam construction the peak of highest suspended sediment concentration occurs first, then the rising limb of hydrograph occurs but after dam construction, peak suspended concentration and peak of flood hydrograph coincide with each other. Satellite imageries show the channel bar changes in Adityapur, Jamshedpur and Ghatshila.

Keywords Discharge · Hydro-sedimentary · Sedimentation · Geomorphological templets · Anthropogenic factors

24.1 Introduction

Channel geomorphology mainly depends on water discharge and sediment supply (Wang et al. 2011). In anthropocene epoch, human interventions are also determining factor. During the last few decades, several studies have been made in worldwide rivers like Amazon (Marengo 2005), Danube (Ovcharuk 2019), Yellow (Wei et al.

S. Banerjee (✉) · A. K. Roy
Department of Geography, Kazi Nazrul University, Paschim Barddhaman, Asansol, West Bengal 713340, India

A. Alam
Department of Geography, Rampurhat College, Rampurhat, Birbhum, West Bengal 731224, India

© The Author(s), under exclusive license to Springer Nature Singapore Pte Ltd. 2022   439
N. C. Jana and R. B. Singh (eds.), *Climate, Environment and Disaster in Developing Countries*, Advances in Geographical and Environmental Sciences, https://doi.org/10.1007/978-981-16-6966-8_24

2016), Indian rivers like Narmada (Gupta and Chakrapani 2005), Mahanadi (Bastia and Equeenuddin 2016). From their study, it is seen that there is significant decrease in water discharge and sediment load. In the world, over the last 50 years, many rivers also experienced decreasing trend in runoff and sediment load (Walling and Fang 2003). So, it is clear that in past decades climate change effects rainfall and play dominating role in water discharge and sediment transport. In Subarnarekha, water discharge and sediment load are decreasing (Das 2019) but determining factors and its geomorphic implications are still unknown. So, the present study aims to analyse the water discharge, runoff and sediment load data and examine its determining factors i.e., climate change, anthropogenic factors.

An important issue in hydrology is sedimentation in water bodies like rivers, streams, dams, etc. According to Central Water Commission Report, in the last few decades, many large dams and barrages have been constructed. Among these dams and barrages, Chandil dam and Giludih barrage are most important. Distance between Chandil dam to Jamshedpur is only 31 km. Giludih barrage has been constructed 13 km. away from Ghatshila. So, it may be of great impact on runoff and sediment load in two hydrological stations.

24.2   The Subarnarekha Basin: Geomorphological and Hydrological Setting

The Subarnarekha river located in the Peninsular India is very important interstate east-flowing river. It originates near Nagri village (23°18' N, 85°11'E) in Ranchi plateau at an elevation of ~610 m. After flowing 480 m long distance, it drains into Bay of Bengal. The entire basin of Subarnarekha has covered approximately 19,000 km^2 area and spread over three states viz. Jharkhand (Ranchi, Purbi Singbhum and Saraikela Kharsawan districts), West Bengal (Purulia and West Midnapore districts) and Odisha (Mayurbhanj and Balasore districts). It is naturally bordered the north by Damodar basin; South koel, Baitarani, Burhabalang and Jamira in the west, Kasai and Kaliaghai in the east, Bay of Bengal in the south. The most important tributaries are Raru, Kanchi, Karkari, Kharkai and Dulung. Most of the tributaries flow in the right bank of the river (Das 2019). So right bank consists most of the total basin area and constructs bottle-neck asymmetric river basin shape (Fig. 24.1).

The Subarnarekha basin accounts for 0.6% of the geographical area of India (Giri and Singh 2018). The Geology of the basin area and its surroundings can be subdivided into three units i.e., The Archean-Precambrian formation, Tertiary deposits, and Quaternary sediment deposits (Guha and Patel 2017). The upper catchment of the basin area (Purulia, Ranchi, Singbhum and Mayurbhanj district) consists of Granitic-gneissic complex of Archean-Precambrian formation. Middle catchment area is formed by Dharwar, where oldest sedimentary and metamorphic rocks are formed (Mukhapadhyay 1980). Lower catchment shows recent alluvium deposition.

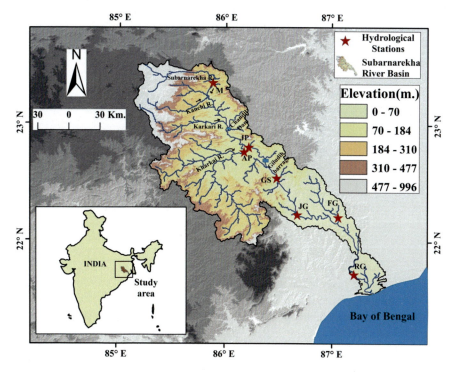

Fig. 24.1 Location of Subarnarekha river basin and seven hydrological stations viz. Muri (M), Adityapur (AP), Jamshedpur (JP), Ghatshila (GS), Jamsholaghat (JG), Fekoghat (FG) and Rajghat (RG)

River terraces has been developed along the lower section of the river. Due to gully erosion, the edge of terraces further formed badland topography (Mukhapadhyay 1980).

As Subarnarekha river basin is a part of Chhotanagpur plateau, it has peculiar type of geomorphic history. The Subarnarekha basin is characterised by polycyclic landscape i.e. ridge-valley topography, rejuvenated river valley, terraces and floodplain deposits. So, it has a complex geomorphic history (Mukhapadhyay 1980). Geomorphic map (Fig. 24.2) of Subarnarekha basin shows five major geomorphic units, namely (i) Pediment pediplain complex, which occupied 80% of the study area; (ii) Highly dissected plateau in the south-eastern portion of study area; (iii) Moderately dissected hills and valleys distributed in the western and south-eastern portion; (iv) Alluvial plain in the lower catchment of Subarnarekha. In alluvial plain river terraces have been formed, which is a imprint of very recent upliftment (Mukhapadhyay 1980); (v) Flood plain, before Subarnarekha drains in Bay of Bengal. Topography of this area depends upon this type of complex geomorpic settings. So that, undulating topography has been seen in this area. Physiographically, the Subarnarekha basin area has two major divisions viz. plateau region, varies between 996 and 300 m. elevation and plain region varies between 300 and 0 m. elevation (Fig. 24.1).

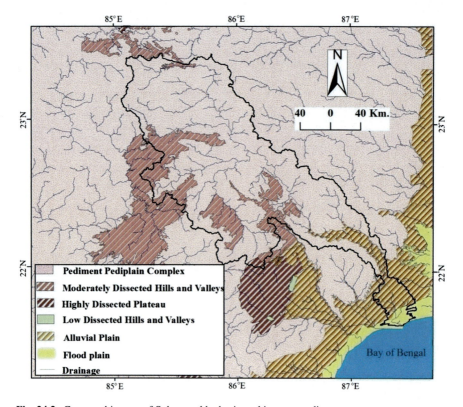

Fig. 24.2 Geomorphic map of Subarnarekha basin and its surroundings

Generally, climate of a region depends on its geographical location and physiography. The Subarnarekha River drained in the south-eastern side of Chhotanagpur plateau in eastern India. This region experiences tropical monsoon type of climate. So entire Subarnarekha basin is dominated by northeast and southwest monsoon (Mukhapadhyay 1980). During monsoonal period the rainfall becomes very high. The range of mean annual rainfall of the entire basin and its surroundings is 2967 mm (Fig. 24.3). It has transitional pattern of weather because it is located between Chhotanagpur plateau in the west and Bay of Bengal in the east. So, the lower catchment of the basin is strongly affected by tropical cyclone and maritime affect. Tropic of Cancer (23°30′) passes through the northern margin of the Subarnarekha basin (Mukhapdhyay 1980). So, northern region of this basin has experienced special type of climatic phenomena like alteration of wet and dry seasons. The average annual maximum temperature of this basin for 36 years (1969–2004) is 31.46 °C. The maximum temperature was recorded in the month of April and May and the minimum temperature has been recorded in the month of December and January.

Subarnarekha is a monsoon-dominated river. The average annual rainfall varies from 3508 to 541 mm. Middle catchment experienced highest rainfall. Towards upper and lower catchment rainfall decreases (Fig. 24.3). Historical records indicate the

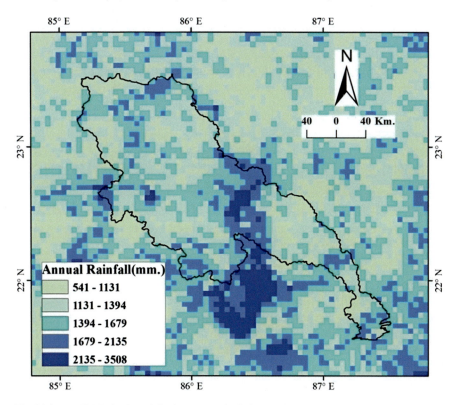

Fig. 24.3 TRMM derived precipitation pattern for Subarnarekha Basin and its surroundings (http://www.geog.ucsb.edu.wbodo.TRMM/)

high magnitude flood affecting some parts of middle and lower catchment, in the following years: 1973 in Jamshedpur, 1997 in Ghatshila, 1988 in Fekoghat, 1976, 1987 and 2008 Rajghat (CWC).

24.3 Data and Methods

Central Water Commission (CWC), under Ministry of Water Resources, Govt. of India and Indian Space Research Organization (ISRO), under department of Space, collaboratively developed web-enabled Water Resources Information System of India (INDIA-WRIS). For multipurpose planning and management based on SRTM DEM, India is divided into 25 basins and 101 sub-basins (Subarnarekha Bain Report, 2014). Subarnarekha is only non-classified river in Eastern India. So, its data are available in WRIS.

Daily discharge data are available in WRIS for five stations, namely Muri (M), Jamshedpur (JP), Adityapur (AP), Ghatshila (GS), Jamsholaghat (JG), Fekoghat

(FG). But sediment concentration data are available for only three stations namely Adityapur (AP), Jamshedpur (JP) and Ghatshila (GS). As it is an ephemeral river, only monsoonal sediment concentration data has been analysed. Though only three stations' sediment concentration data is not sufficient to get proper result, it gives a general view of this river. Peak discharge (Qmax) data are available in Water Year Book for seven stations, namely Muri (M), Jamshedpur (JP), Adityapur (AP), Ghatshila (GS), Jamsholaghat (JG), Fekoghat (FG) and Rajghat (RG). Available peak discharge has been analysed for extreme flow characteristics. Landsat Sattelite imageries have been used to understand geomorphic changes.

24.4 Results

24.4.1 Peak Discharge Variability

Extreme flow or peak flow is observed by highest discharge in every year (Bandyopadhyay et al. 2016; Roy and Sinha 2016). As Subarnarekha is monsoon dominated river, its peak discharge occurred in monsoon period (June to October). In Subarnarekha, large records of peak discharge are available for seven hydrological stations, namely Muri (M), Jamshedpur (JP), Adityapur (AP), Ghatshila (GS), Jamsholaghat (JG), Fekoghat (FG) and Rajghat (RG) (Table 24.1).

Figure 24.4 shows the trend of peak discharge in six hydrological stations from upper to lower reach (Muri, Adityapur, Jamshedpur, Ghatshila, and Fekoghat). In the last four decades, highest peak discharge (Qmax) in Muri station is 481 cumecs (2017–2018) and lowest peak discharge is 44.80 cumecs (2010–2011). In Adityapur highest peak discharge occured in the year of 1997–1998 with 6700 cumecs discharge and lowest discharge was seen in the year of 2010–2011 with around 139.67 cumecs. In Jamshedpur, highest peak discharge was seen in the year of 2013–2014 with 8769.78 cumecs water discharge and lowest with 240 cumecs in the year of 2010–2011.

24.4.2 Temporal Variability of Suspended Sediment Transport

Annual sediment load data have been analysed for three stations (Adityapur, Jamshedpur and Ghatshila) from 1973 to 2018. Change in sediment load over this time period shows interesting trend (Fig. 24.5). The plotted graph (Fig. 24.5) shows the decreasing trend of Sediment load. As it is a monsoon-dominated river, 99% runoff and sediment load occured from June to October. So it has seasonal charcteristics (Das 2019). In Adityapur, lowest annual sediment load has been seen in the year of 2010–2011 (0.012×10^6 metric tons) highest sediment load has been seen in the year of 1977–1978 (4.84×10^6 metric tons). In Jamshedpur, the highest sediment

Table 24.1 General information of hydrological station of Subarnarekha basin

Name of hydrological station	River/stream	State	Latitude (N), Longitude (E), Elevation	Reaches	Catchment area (10^3 km^2)	Channel type
Muri (M)	Subarnarekha	Jharkhand	23°21'46", 85°10'28", 231	Upper Reach	1.33	Bed rock channel
Jamshedpur (JP)	Subarnarekha	Jharkhand	22°47', 86°12', 111	Middle Reach	12.649	Bed rock channel
Adityapur (AP)	Kharkai	Jharkhand	22°47'29", 86°10'06", 123	Middle Reach	6.309	Bed rock channel
Ghatshila (GS)	Subarnarekha	Jharkhand	22°34'49", 86°10'06", 72	Middle Reach	14.176	Bed rock channel
Jamsholaghat (JG)	Subarnarekha	Odisha	22°13'08", 86°43', 42	Lower Reach	16	Bed rock channel, Transition between plateau and plain
Fekoghat (FG)	Dulung	West Bengal	22°18'28", 86°55'11", 40	Lower Reach	0.7	Flat alluvial channel
Rajghat (RG)	Subarnarekha	Odisha	21°46'04", 87°09'51", 3	Lower Reach	18.26	Meandering alluvial channel with floodplain

load has been seen in the year of 1984–1985 (13.6 10^6 metric tons) and lowest load in the year 2010–2011 (0.1 10^6 metric tons). In Ghatshila the highest load was seen in the year of 2007–2008 (11.2 10^6 metric tons) and lowest in the year of 2010–2011 (0.1 10^6 metric tons).

In large river like Amazon, sediment transport is not proportionate to runoff. Because most of the sediment supply are from Andes mountains (Chakrapani 2005). But in case of Subarnarekha river, there is a relationship between runoff and annual sediment load (Fig. 24.6). Figure 24.7 shows when runoff increases annual sediment load also increases and when runoff decreases anual sediment also decreases. Runoff is low in the year of 2010–2011, in Adityapur 573 million cubic metre, in Jamshedpur 1808 million cubic metre and in Ghatshila 838 million cubic metre. So, sediment load was proportionately low in this year.

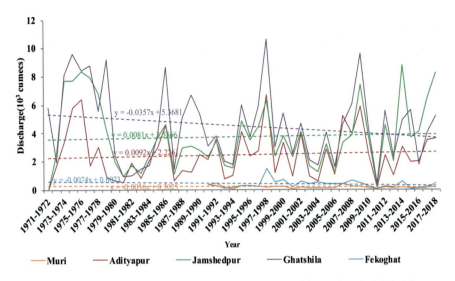

Fig. 24.4 Graph showing annual peak discharge and their trends of five station viz. Muri, Adityapur, Jamshedpur, Ghatshila and Fekoghat

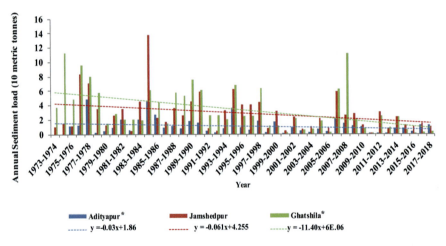

Fig. 24.5 Temporal variations of total annual sediment load in three stations viz. Adityapur, Jamshedpur and Ghatshila, * significant at 95% level and P value less than 0.05%

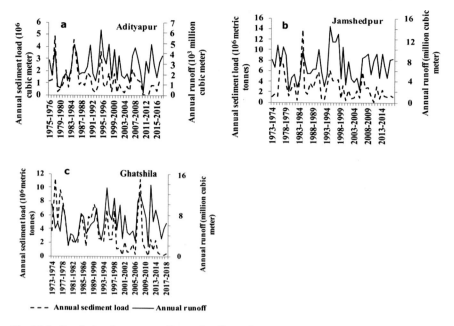

Fig. 24.6 Graph showing pattern of annual sediment load and annual runoff

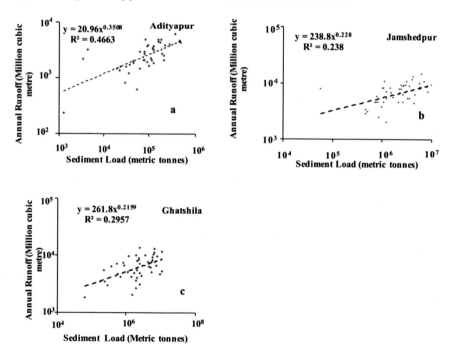

Fig. 24.7 Graph showing the relationship between Annual sediment load and Annual runoff of three stations

24.5 Discussion

24.5.1 Hydrological Variability and Its Geomorphic Implications

The results of this investigation into the downstream hydrologic and geomorphic effects of dams consist of some additional issues like regional variability of the rivers which are affected by dams, drivers of variability regionally in geomorphology affected by dams. Hydrologic conditions modified the interactions between geomorphic signatures and geomorphic landscapes. Regional variations in rivers, dams and responses are the main regulator of the geomorphic changes. Annual discharge as well as sedimentation variability effects on downstream.

Adityapur, Jamshedpur and Ghatshila are situated in plateau region, where elevation is higher than lower catchments hydrological stations like Jamsholaghat, Fekoghat, Rajghat. It is seen from last four decades peak discharge analysis, the site like Jamshedpur, Ghatshila, observed high magnitude flood in past and the lower catchments site like Jamsholaghat, Rajghat experienced high frequency but low magnitude flood (Banerji and Mukhopadhyay 2018). So, in Rajghat, there is several times cross danger level, but no extreme flood records. The plotted trend given in Fig. 24.4, shows the slight decreasing trend in peak discharge. The most important feature is maximum years in the last two decades experienced lower peak discharge than mean annual peak discharge. In 1997–1998, highest peak discharge is seen with 10,582 cumecs discharge and in 2010–2011 lowest peak discharge is seen. The peak discharge of Ghatshila in 1997 (10582 cumecs) was 2.5 times greater than mean annual peak discharge (Q_m) and 49 times greater than average water discharge (Q_a) (Table 24.2). In Fekoghat, highest peak discharge is seen in the year of 1997–1998 with 1488 cumecs water discharge and lowest peak discharge is seen in the year of 2010–2011 with 40 cumecs discharge. 1997 is extreme flood year. There are no extreme magnitude flood records after 1997 (Figs. 24.8, 24.9, 24.10 and 24.11).

Subarnarekha River a monsoon-dominated river wheat adequate discharge disrupted with dam construction also have some temporal changes in the course of the river. Through the satellite image, it has been easy to interpret the changes of geomorphological template. Landsat 1-5MSS C1 Level 2, Landsat TM 4/5, Landsat 8 OLI/TIRS C1/Level 1 satellite image has been used to identify the temporal geomorphic change. Figure 24.12 shows the temporal changes in channel bar from January, 1976 to January, 2020. It is seen that in Adityapur, Jamshedpur and Ghatshila the bar area is gradually decreasing.

Table 24.2 Flow characteristics of the Subarnarekha River and its major tributaries derived and recorded in the present study

River/site	Qmax data availability	Annual Peak discharge (m³ s⁻¹) Qmax	Qm data availability	Mean annual peak discharge (m³ s⁻¹) Qm	Qa data availability	Average water discharge (m³ s⁻¹) Qa
Subarnarekha–at Muri	1990–2018 (28 years)	481	1990–2018 (28 years)	183.21	1990–2013 (23 years)	20.42
Subarnarekha–at Jamshedpur	1971–2018 (47 years)	8769.7	1971–2018 (47 years)	3810.89	1972–2013 (41 years)	219.86
Kharkhai–at Adityapur	1972–2018 (46 years)	6700	1972–2018 (46 years)	1339.06	1972–2013 (41 years)	87.83
Subarnarekha–at Ghatsila	1972–2018 (46 years)	10,582	1972–2018 (46 years)	4510.31	1972–2013 (41 years)	214.35
Subarnarekha–at Jamsholaghat	2014–2018 (4 years)	3750	2014–2018 (4 years)	2858	Daily discharge data not Available	–
Dulung–at Fekoghat	1990–2018 (28 years)	1488	1990–2018 (28 years)	360.19	1990–2013 (23 years)	15.60
Subarnarekha–at Rajghat	2014–2018 (4 years)	4143.9	2014–2018 (4 years)	3412.1	Daily discharge data not Available	–

24.5.2   Controls on Discharge and Sediment Load

24.5.2.1   Temporal Variability and Role of Climate

Rainfall is the most influencing factor on water discharge, runoff and sediment load in river basin area. As Subarnarekha is located in the tropical region, 99% of monsoon occur in June to October (153 days). Subarnarekha has occupied large area with plateau and plain so, rainfall is not uniformly occur in this Basin. Rainfall data are available in Meteorological Handbook, published by Central Water Commission. Figure 24.8 shows the total annual rainfall variability in all the hydrological stations from upstream to downstream in Subarnarekha river basin. In long term scale, change in precipitation rate has occurred due to climate change (Wang 2018). From 1951 to 2004, rainfall rates decrease in India over 16 river basins (Kumar and Jain 2011). This plot (Fig. 24.8) shows the temporal change in precipitation in some hydrological station of Subarnarekha basin. There is a negative trend in precipitation rate in all stations. This plot (Fig. 24.8) also shows the slightly decreasing rainfall rate change all over the basin. Figure 24.9 shows the positive relation between rainfall, total annual runoff and total sediment load in Adityapur, Jamshedpur and Ghatshila.

There was a significant link between the annual rainfall and discharge, sediment load. In 1997–1998, Adityapur received high rainfall (1999 mm). So, in this year discharge and sediment load are also high 6700 cumecs and 1,823,387 metric tons. In 2010–2011, from last few decades rainfall rate gradually decreased. In 2010–2011

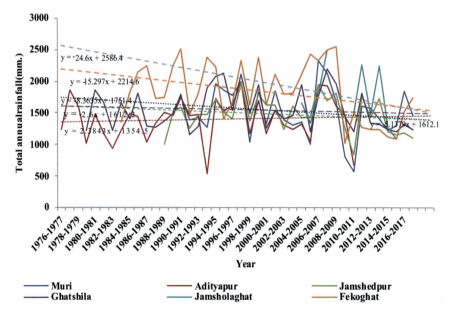

Fig. 24.8 Graph showing temporal variations of total annual rainfall of six station of Subarnarekha river basin

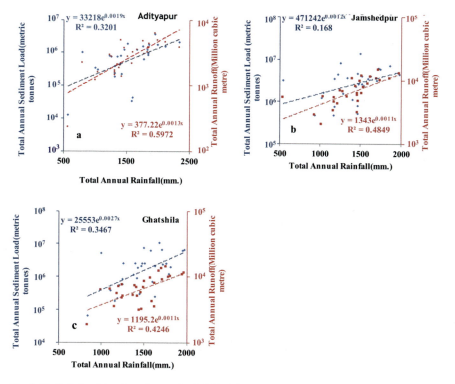

Fig. 24.9 Relationship of total annual rainfall with total annual sediment load and total annual rainfall in three station

Adityapur received lowest rainfall (573 mm) so that in this year discharge and sediment load become lowest 139.7 cumecs and 234 metric tonnes. From 1988 to 2018 Ghatshila received highest rainfall (1968 mm) in 1996–1997. In this year devastating flood has been occurred with 10,582 cumecs discharge. In 2011, Ghatshila received lowest rainfall with 838.1 mm, so that discharge and sediment load are also lowest 410cumecs and 1856 metric tons.

24.5.2.2   Construction of Dam

Generally, dams have been constructed to control flood, irrigation purpose, generate hydropower, drinking water supply. But it has great impact on river geomorphology (Skalak et al. 2013). Dam construction directly affects runoff, because it increases the residence time of water (Chakrapani 2005). It also influences downstream sediment loads (Skalak et al. 2013). In Subarnarekha many dams had been constructed in last four decades (Table 24.3). Among them, Chandil dam and Giludih barrage have been constructed near Jamshedpur and Ghatshila. Distance between Chandil dam to

Table 24.3 General information of major dams and barrages in Subarnarekha river and its tributaries

Dam/Barrage	River	Completion year
Getalsud Dam	Subarnarekha	1971
Kakudajodi Dam	Kakudajodi	1976
Nesa Dam	Nesa	1978
Kharkai Dam	Kharkai	1984
Jambhira Dam	Jambhira	1986
Sunei Dam	Sunei	1990
Naku Dam	Bijay	2010
Icha Dam	Kharkai	Under Construction
Raisa Dam	Kanchi	Under Construction
Chandil Dam	Subarnarekha	Under Construction (started on 1982–83 and almost completed on 2001)
Giludih Barrage	Subarnarekha	Under Construction (started on 1982–83 and almost completed on 2001)

Jamshedpur is only 31 km. Giludih barrage has been constructed 13 km. away from Ghatshila. So, it has great impact on runoff and sediment load in two hydrological stations.

Construction of Chandil dam and Giludih barrage was started in 1982–1993, but due to land acquisition issue and lack of funding, construction work has been stopped many times. Till now this dam was under construction but 99.5% of its work was done in 2001. So it effects on total water discharge and sediment concentration. Figure 24.10 shows pre-dam and post-dam scenario of total water discharge in Jamshedpur. From 1973 to 2001 average total water discharge was 84,719 cumecs and from 2002 to 2013 average total water discharge is 68250 cumecs. Peak of total water discharge occur in 1994 with 163,664 cumecs before completion of dam construction.

Ghatshila is affected by Chandil dam and Giludih barrage. It has dramatic effect on discharge and sediment load. In 1976, the peak of highest suspended sediment concentration occurs first, and then the rising limb of hydrograph occurs. It is a normal case (Fig. 24.11a). After dam construction peak of highest suspended sediment concentration and peak discharge coincide with each other (Fig. 24.11b).

24.6   Conclusion

From this analytical study, it is seen that a decreasing trend is seen in water discharge, runoff and sediment load in all hydrological station of Subarnarekha river basin. Landsat imageries also prove that. Though it is not enough to analyse whole basin, because except rainfall variability and anthropogenic factors, sediment load also

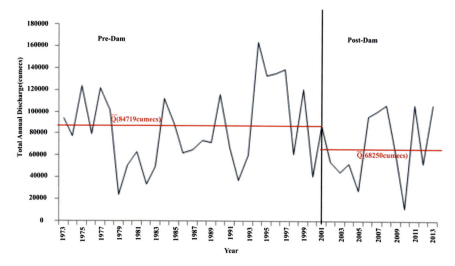

Fig. 24.10 Change of total annual discharge in pre-dam and post-dam period in Jamshedpur

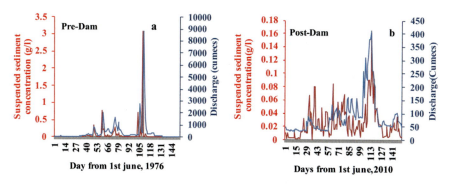

Fig. 24.11 Pre and post dam construction scenario in Ghatshila station

depends on its geology, structure, relief, weathering rate, etc. (Das 2019). So, further research is important to analyse rest of variables which control the water discharge, runoff and sediment load in this basin. But in this result, it can be said that in last few decades precipitation rate has slightly decreased, so there is climatic signals in the basin. There is also good relationship between annual runoff and annual sediment load. In all hydrological station sediment increases and decreases with precipitation rate. Peak discharge trend in the last four decades has also decreased. In peninsular India, there is also rise in sea level (Das 2019). So, climate plays a great role in this river basin.

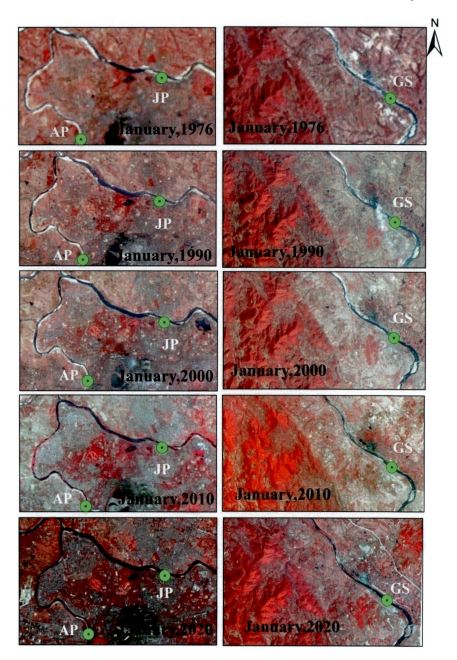

Fig. 24.12 Channel bar changes in three station (viz. Adityapur, Jamshedpur and Ghatshila) from 1976 to 2020

Acknowledgements The authors are deeply indebted to Dr. Sujay Bandyopadhyay, Department of Geography, Kazi Nazrul University for his constant encouragement for better performance in the present work. The authors are truly thankful to their family and almighty God for showers of blessings.

References

Bandyopadhyay S, Ghosh PK, Jana NC, Sinha S (2016) Probability of flooding and vulnerability assessment in the Ajay River, Eastern India: implications for mitigation. Environ Earth Sci 75:578

Banerji D, Mukhopadhyay S (2018) Spatial and temporal variations of the hydrological characteristics of the Subarnarekha River, Eastern India. Earth Sci India 11 (IV):183–200

Bastia F, Equeenuddin SK (2016) Spatio-temporal variation of water flow and sediment discharge in the Mahanadi River India. Glob Planet Change 144

Chakrapani GJ (2005) Factors controlling variations in river sediment loads. Curr Sci 88:570–575

Das S (2019) Four decades of water and sediment discharge records in Subarnarekha and Burhabalang basins: an approach towards trend analysis and abrupt change detection. Sustain Water Resour Manag 5:1665–1676. https://doi.org/10.1007/s40899-019-00326-1

Guha S, Patel PP (2017) Evidence of topographic disequilibrium in the Subarnarekha River Basin, India: A digital elevation model based analysis. J Earth Syst Sci 126. https://doi.org/10.1007/s12040-017-0884-1

Gupta H, Chakrapani G (2005) Temporal and spatial variations in water flow and sediment load in Narmada River Basin, India: Natural and man-made factors. Environ Geol 48(4):579–589. https://doi.org/10.1007/s00254-005-1314-2

Kumar V, Jain SK (2011) Trends in rainfall amount and number of rainy days in river basins of India (1951–2004). Hydrol Res 42(4):290–306

Marengo JA (2005) Characteristics and spatio temporal variability of the Amazon River basin water budget. Clim Dyn 24:11–22. https://doi.org/10.1007/s00382-004-0461-6

Mukhapadhyay SC (1980) Geomorphology of the Subarnarekha Basin: the Chota Nagpur Plateau, eastern India

Ovcharuk VA (2019) Temporal variation of water discharges in the lower course of the Danube River across the area from Reni to Izmail under the influence of natural and anthropogenic factors. Energetika 144–160

Roy NG, Sinha R (2016) Linking hydrology and sediment dynamics of large alluvial rivers to landscape diversity in the Ganga dispersal system, India. Earth Surf Process Landforms

Singh AK, Giri S (2018) Subarnarekha River: the Gold Streak of India. In: Singh DS (ed.) The Indian Rivers. Springer Hydrogeology. https://doi.org/10.1007/978-981-10-2984-4_22

Skalak KJ, Benthem AJ, Schenk ER, Hupp C, Galloway JM, Nustad RA, Wiche GJ (2013) Large Dams and Alluvial Rivers in the Anthropocene: the impacts of the Garrison and Oahe Dams on the Upper Missoiri River. Anthropocene 2: 51–64

Walling DE, Fang D (2003) Recent trends in the suspended sediment loads of the world's rivers. Elseviar 39 (1–2), https://doi.org/10.1016/S0921-8181(03)00020-1

Wang R (2018) Runoff and sediment load change and its influencing factors in the Xiaonanchuan River Basin China. Earth Environmental Sci. 170:3–4

Wang S, Yunxia Y, Yingkui L (2011) Spatial and temporal variations of suspended sediment deposition in the alluvial reach of the upper Yellow River from 1952 to 2007. CATENA 92:30–37

Wei Y, Zhao G, Jiao J (2016) Spatial–temporal variation and periodic change in streamflow and suspended sediment discharge along the mainstream of the Yellow River during 1950-2013. Catena 140:105–115

Chapter 25
Assessment of Jiadhal River Basin Using Sedimentary Petrology and Geospatial Approach

Akangsha Borgohain[ID]**, Kusumbor Bordoloi**[ID]**, Dhrubajyoti Sahariah**[ID]**, Santonu Goswami**[ID]**, Anup Saikia, and Ashok Kumar Bora**

Abstract The river Jiadhal is one of the significant right bank tributaries of the Brahmaputra. It starts from the Arunachal Himalaya, crosses a few km through Arunachal Pradesh and enters the Brahmaputra valley (Assam part) near Jiadhal-mukh of Dhemaji district, Assam. The river Jiadhal recurrently inundates vast areas of its basin during the monsoon and put a significant impact on the people and their livelihood. This river carries a huge amount of sediment load and debris triggered by continuous and heavy rainfall within the basin mainly in the upper catchment region causing severe loss of fertile agricultural land and infrastructure of the basin. This study aims to analyze the morphological characteristics of the river Jiadhal using sedimentary petrology, Remote Sensing techniques and Geographic Information System (GIS). The study gives a better prospect toward understanding the morphometric characteristics and sediment dynamics of the basin. It will help the concerned authorities for better planning and mitigation of the issues related to this river.

Keywords River basin assessment · Jiadhal River · Sedimentary petrology · Remote sensing · GIS

A. Borgohain · K. Bordoloi (✉) · D. Sahariah · A. Saikia · A. K. Bora
Department of Geography, Gauhati University, Guwahati, Assam, India
e-mail: kusumbor@gauhati.ac.in

D. Sahariah
e-mail: dhrubajyoti@gauhati.ac.in

A. Saikia
e-mail: asaikia@gauhati.ac.in

A. K. Bora
e-mail: ashokkumarbora@gauhati.ac.in

S. Goswami
Earth and Climate Science Area, National Remote Sensing Centre, Indian Space Research Organization, Hyderabad, India
e-mail: goswami_s@nrsc.gov.in

25.1 Introduction

Tropical rivers have characteristics of high sediment yield (Syvitski et al. 2014; Sinha et al. 2014). The growing disturbances due to land-use, grazing, agriculture and construction activities in the catchment have resulted in increased sediment yield (Zimbres et al. 2018; Sushanth and Bhardwaj 2019). Although scientific studies on sediment started long back around the 1930s the nature and dimension of the study have changed in recent times (Guzmán et al. 2013; Wilkinson et al. 2015). Increased sophistication, methodological improvements have led to the growing interest of geomorphologists, geologists and remote sensing scientists to look into the subject (Fagundes et al. 2020; Bertin and Friedrich, 2016). However, this increased works on sediment are compartmental and only countries like the USA, China, Italy and the UK contribute to the bulk of the research (www.bibliometrix.org). The works are also not equally distributed spatially among the major rivers of the world. While some of the rivers have multiple studies, the others have scanty attention. Discharge and sediment yield are two significant parameters in river basin studies. Basin level sediment process identification, land-use and sediment yield relationship, remote sensing applications in sediment studies, the enquiry of soil and water quality parameters, climate uncertainties and related sediment transport, patterns and the dynamics of sediments, etc. are studied worldwide (Nichols et al. 2011; Zhou et al. 2018; Walia et al. 2020). Sediment-related hazards (Brakenridge et al. 2017; Bordoloi et al. 2020) have been identified as one of the major contributors to changing land-use, livelihood, especially in tropical regions. Sediment yield in any basin is dependent on geology, vegetation types, topographic character and climate. Although topographic character and geology remain the same in the short term, vegetation and climate parameters show variability (Easterling et al. 2000). Over time these parameters contribute to the sediment dynamics of the basin. Increasing concerns among the human community that the rate of soil erosion and sediment yield is on the rise due to increasing anthropogenic intervention in the basin (Saikia et al. 2013). For effective river basin planning, therefore, identification of sediment sources, nature of yield and its impact on land-use are needed to be ascertained (Hancock and Revill 2013; Wilkinson et al. 2015).

The river Brahmaputra carries excessive sediments compared to many major rivers of the world. Several studies have been undertaken to describe the fluvial pattern of the river (Goswami 1985; Sarma 2005; Saikia et al. 2019). However, scanty of work is devoted to measuring the basin sediment output. The river Jiadhal, which is a significant right-bank tributary of the Brahmaputra, is known for its growing sediment yield, both in the form of suspended and overbank forms. The accessive sedimentation in the basin also has contributed to vast agricultural fields becoming unsuitable, leading to the loss of livelihood of the dependent community. The present research is an attempt to gather information related to (a) the morphometric characteristics of the basin and (b) assessment of the basic qualities of sediment and fluvial processes and their relationship with sediment yield. Although fieldwork is an integral part of geomorphic studies, remote sensing data are also widely used to supplement the information gathered through fieldwork and laboratory analysis.

25.2 Study Area

Jiadhal River measures ~1385 km² basin area. The river originates from the West Siang district (Arunachal Pradesh), flows through the Dhemaji district (Assam) and finally meets the KherkatiaSuti, a distributary of the Brahmaputra River. It is confined between the Subansiri and Moridhol River sub-watershed in the west and east direction, respectively, while the southern side is bounded by the KherkatiaSuti (Borgohain and Sahariah 2017). As per the report of the Water Resource Department, Govt. of Assam, the average annual rainfall varies between 2,965 and 4,386 mm, with a mean of about 3,150 mm. The annual temperature varies from 8 to 39 °C (IMD, Guwahati Airport). The characteristics of the climatic parameters of the region resemble the subtropical monsoon climate (Fig. 25.1).

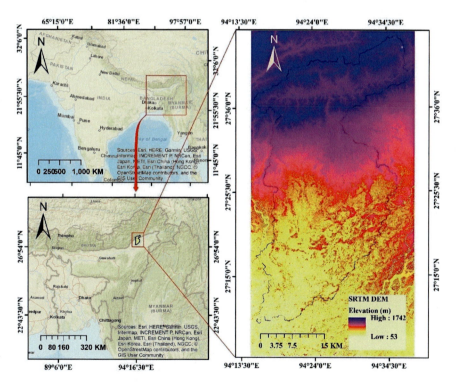

Fig. 25.1 Study area map: the Jiadhal River basin

Table 25.1 Datasets used in this study

Sl. No	Data	Details	Source
1	SRTMDEM	30 m	https://earthexpl orer.usgs.gov/
2	Toposheets	1:50,000	Survey of India
3	Sediment Samples	23 nos. of samples	Collected from the field
4	Landsat 8 OLI	30 m (08–12–2020)	https://earthexpl orer.usgs.gov/

25.3 Database and Methodology

25.3.1 Data Used

For this study, both the geospatial approach and sedimentary petrology based approach are taken into consideration. Morphometric characteristics of the basin are analyzed using the Digital Elevation Model (DEM). Here, the Shuttle Radar Topography Mission (SRTM) DEM of 30 m spatial resolution is used within the GIS environment. To cross-validate, the accuracy of SRTM DEM in morphometric analysis, Topographic sheets (1:50,000 scale) are considered published by the Survey of India (SOI). Apart from DEM, Landsat 8 OLI data are considered to generate Soil Adjusted Vegetation Index (SAVI), Normalised Difference Vegetation Index (NDVI) and Bare Soil Index (BSI) for the study area. SRTM DEM and Landsat 8 OLI datasets are downloaded from https://earthexplorer.usgs.gov/. On the contrary, sediment samples are collected from the field for sedimentary analysis (Table 25.1).

25.3.2 Methodology

To attain the required objectives, the complete study is concentrated into two broad sets, i.e., (i) Morphometric analysis and (ii) Sedimentary Analysis.

25.3.2.1 Morphometric Analysis

Morphometric analysis is an integral part of a drainage basin study which is done with the help of mathematically derived parameters from a river basin. The main three aspects of river basin morphometry, i.e., Linear aspect, Areal aspect and Relief aspects are considered here. Table 25.2 represents the considered parameters under all three aspects.

Table 25.2 Parameters of morphometric analysis

Sl. no	Parameter	Symbol	Formula	Description	References
Linear aspects					
1	Stream order	N_u	N_u	Hierarchical rank	Strahler (1964)
2	Bifurcation ratio	R_b	$Rb = Nu / Nu + 1$	N_u = Number of stream segment present in the given order	Horton (1945)
				$N_u + 1$ = Number of segments of the next higher order	
3	Stream length	L_u	L_u	$L\mu$ = Length of the stream (km)	Horton (1945)
4	Mean stream length	L_{sm}	$L_{sm} = L_u / N_u$	Lsm = Mean Stream length	Strahler (1964)
				Lu = Total stream length of the order	
				Nu = Total number of stream segments of order	
5	Stream length ratio	R_L	$R_L = L_u / L_u - 1$	R_L = Stream length ration	Horton (1945)
				Where, L_u = Total stream length of order (U),	
				$L_u - 1$ = She total stream length of its next lower order	
				N_u = Number of stream segment present in the given order and $N_u + 1$ = Number of segments of the next higher order	
Areal aspect					
6	Drainage Density	D_d	$D_d = Lu/A$	Dd = $L\mu$/A	Horton (1932)
				Where Dd = Drainage density (km/km^2)	
				$L\mu$ = Total stream length of all orders and	
				A = Area of the basin (km^2)	

(continued)

Table 25.2 (continued)

Sl. no	Parameter	Symbol	Formula	Description	References
7	Stream Frequency	F_S	$F_S = N/A$	F_S = Stream frequency	Horton (1945)
				Where L = Total number of stream,	
				A = Area of the basin	
8	Drainage Texture	D_T	$D_T = F*D$	D_T = Drainage texture	Strahler (1964)
				Where, F is the drainage frequency	
				D is the Drainage density	
9	Drainage Texture Ratio	T	$(T) = N_1/P$	Where N_1 = The number of first-order streams	Horton (1945); Smith (1950)
				P = The perimeter of the basin(km)	
10	Elongation Ratio	R_e	$R_e = \sqrt{(Au/\pi)}/Lb$	Where A = Area of the basin(km^2)	Schumn (1956)
				π = 3.14	
				Lb = Basin length(km)	
11	Circulatory Ratio	R_c	$R_c = 4\pi A/P^2$	Where, A = Basin area (km^2)	Miller (1953); Sarma et al. (2013)
				π = 3.14	
				and	
				P = Perimeter of the basin (km)	
12	Form Factor Ratio	R_F	$R_f = A/(Lb)^2$	Where, A = Area of the basin	Singh and Singh (1997)
				Lb = basin length	
13	Length of Overland Flow	L_g	$L_g = 1/2D_d$	Lg = 1/2D km	Horton (1945)
				Where D = Drainage density (km/km^2)	

(continued)

Table 25.2 (continued)

Sl. no	Parameter	Symbol	Formula	Description	References
Relief aspects					
14	Basin relief	B	B = h-i,	Where, h = the Maximum elevation of the basin (m) above the mean sea level	Rudraiah et al. (2008)
				i = is the minimum elevation of the basin in (m) above the mean sea level	
15	Relief Ratio	R_h	R_h = Bh/Lb	Where, Bh is the basin relief in (m)	Schumn (1956)
				Lb is the length of the basin in (m)	
16	Ruggedness Number	R_n	Rn = Bh X Dd	Bh = Basin Relief in (m)	Schumn (1956)
				Dd = Drainage density (km/km^2)	

25.3.2.2 Sedimentary Analysis

To identify the sediment characteristics in the Jiadhal basin, samples of overbank sediments are collected from 23 different locations. All the samples are collected randomly from the sand/silt deposited areas and also non-deposited areas. On average, there are 4 nos. of sediment samples that are collected for each location and later mixed into one sample. To understand the textural class and grain size of the samples, methodology is opted as per American Society for Testing Materials (ASTM 1964) and Stoke's law (Stokes 1851). The samples are air-dried and reduced to about 100 gm by the process of coning and quartering. The air-dried samples are now subjected to sieving (Krumbein and Pettijohn 1938) in ½ π sieve interval ASTM sieves for sieving in the Ro-Tap sieve shaker for 10 min for optimum separation. A representative-weighted sample that has been poured through different sizes of sieves and material retained in each sieve is weighed. The percentage of total mass retained and the average particle size in each sieve is then calculated and used for further interpretation of statistical size parameters (Folk and Ward, 1957; Cadigan 1961; Passega 1964; Moiola and Weiser 1968; Passega and Byramjee 1969).

Apart from this, the remote sensing-based approach has been applied to understand the vegetation cover in the study area using Normalised Difference Vegetation Index (NDVI = [(NIR − Red) / (NIR + Red)]) (Rouse et al. 1973; Tucker 1979), Soil Adjusted Vegetation Index (SAVI = [(NIR − Red)/(NIR + Red + L)] × (I + L)) (Huete 1988), Bare Soil Index (BSI = [((SWIR + RED)−(NIR + BLUE)) / ((SWIR + RED) + (NIR + BLUE))]) (Jamalabad and Abkar 2004).

25.4 Results and Discussion

25.4.1 Results of Morphometric Analysis

The linear aspects, as well as the topographical characteristics of the stream segments of the basin network, are analyzed which includes: Stream Order (Nu), which denotes the determination of the hierarchical position of a stream within a drainage basin. Here in the Jiadhal river basin system, a total number of 1499nos. of streams are found. The highest 7th order stream is calculated in this basin (Borgohain and Sahariah 2017). According to the estimation of stream orders, 1154 nos. of streams were identified under the 1st order, 261 nos. of 2nd order streams, the 3rdorder consists of 63 streams, the 4th order has 14 streams, the 5th order consists of 4 streams, in the 6th order, only 2 streams and the main river channel has been ranked as 7th order. Here in this river basin, the streams show the dendritic pattern of the stream network. The first order streams almost dominate the whole basin in the upper part that falls under Arunachal Pradesh and they remain active even in the winter season. However, the 5th and 6th order streams remain ephemeral in the winter season, and these streams become more active during monsoon seasons and add a huge volume of water and sediments to the basin which aggravates the flood and siltation in the lower part of the basin.

The Bifurcation Ratio (R_b) which is an important indicator of the channel pattern is analyzed for the basin. A higher value of bifurcation ratio (more than 5) is a hint of structural influence over the basin (Strahler 1964). On the other hand, the lower Rb value specifies less influence of structural controls. The most common bifurcation ratio is ranging between 3 and 5 in such a basin where the geological structure does not distort the drainage pattern (Strahler 1964) and it shows that the drainage system characterizes with homogeneous lithology. The bifurcation ratio for the Jiadhal River basin is 3.76, i.e., the basin is structurally controlled (Das 2016; Borgohain and Sahariah 2017).

The ratio of the mean stream length of a given order to the mean stream length of the next lower order has an important relationship with the surface flow and discharge and it is defined by the Stream Length ratio (R_L) of the basin (Horton 1945). The R_L value of the 1st order stream is 1.96, the 2nd order stream is 1.84, the 3rd order stream is 2.21, the 4th order stream is 1.54, the 5th order stream is 0.27, the 6th order stream is 9.21, and 7th order stream is 18.49, respectively. In elongated basins, the values will be high in higher-order, and in circular basins, the length ratio will be low. The length ratio in the basin ranges from 0.27 to 18.49. In the Jiadhal basin due to an elongated shape and uneven distribution of streams the length ratio shows a decreasing trend in the 2nd order, increase in the 3rd order, again decrease in the 4th order and thereafter increase in the 6th order. Although stream length ratio increases as the order of the stream increases, Jiadhal basin experiences differently because there are significant variations in slope and topography of the basin.

The total stream length per unit area indicates the Drainage Density (D_d) of a river basin. Horton (1945) introduced the drainage density as an important indicator of the linear scale of landform elements in the stream eroded topography. The calculated values for the Jiadhal river basin varied from 0.035 to 8.38. The mean value of Dd in the Jiadhal river basin is 1.05 km/km^2 which specifies that the region has extremely permeable subsoil, dense vegetation cover characterized by moderate relief.

Stream frequency (F$_S$) is an important measure for drainage basin analysis which gives an idea about geologic structure, topographic condition, vegetation, hydrology and terrain characteristics. Horton (1945) defined stream frequency as the measure of the number of streams per unit area. From the analysis, it is revealed a low stream frequency of 1.08 streams/ km^2 in the Jiadhal basin. It shows a positive relationship with the drainage density (Das 2016).

Drainage Texture (D$_T$) is another important parameter to study basin characteristics. The ultra-fine drainage texture is usually found in high drainage density region, medium texture is found in medium drainage density areas, whereas low drainage density regions are characterized by coarse drainage texture (Strahler 1964; Das 2016). The result shows that the present basin has a low drainage texture of 1.13.

Drainage Texture Ratio (T) is another important parameter that has been highlighted by (Smith 1950; Rao et al. 2010). In the present study, it is recorded as 3.49 which denotes a moderate infiltration and relief of the area.

Elongation ratio is defined by Schumn (1956), as the ratio of the diameter of a circle of the same area as the drainage basin and the maximum length of the basin. It is an important index for analyzing the shape of the basin. According to Strahler (1964), a higher value of the elongation ratio indicates very low relief. In the present basin, the value of the elongation ratio is 0.27, which is a lower value. Thus, it indicates high relief and elongated shape of the basin. Similarly, the Circularity Ratio (R$_C$) is 0.4, which indicates the elongated shape of the basin.

The Form Factor Ratio (R$_F$) is defined as the ratio of the basin area to the square of the basin length. The factor indicates the flow intensity of a basin of a defined area (Horton 1945). The form factor value should be always less than 0.7854. The smaller the values of the form factor, the more elongated will be the basin. Basin with high form factors experiences larger peak flows of shorter duration, whereas elongated watersheds with low form factors experience lower peak flows of longer duration. Form factor ratio for the Jiadhal river basin is 0.23, indicating an elongated basin with lower peak flows of longer duration.

The length of the overland flow (L$_g$) is the length of water over the ground surface before it gets concentrated into a definite stream channel (Horton 1945). The smaller value of overland flow is an indication of quicker surface flow that enters the streams. Less amount of rainfall with a relatively homogeneous surface area may contribute a significant amount of discharge to the streams when the length of overland flow is small. The length of overland flow for the Jiadhal basin is found to be 0.49 km indicating a very low value that favors frequent flood events in the basin.

The vertical properties of a basin are indicated by Relief aspects. The word relief in a true sense stands for the relative vertical inequality of land surface (Strahler 1952). In the Jiadhal basin, the highest point (h) is 1526 m and the lowest point (i) is 91 m from the mean sea level. Therefore a high basin relief (B) is measured, i.e., 1435 m indicates high sediment transport and runoff.

The Relief Ratio (R_h) is an indicator of erosional processes (Schumn 1956) and the overall steepness of the basin. The high value of the relief ratio (~20) indicates a high erosivity of the basin (Das 2016).

Ruggedness Number (R_n) is the product of the basin relief and drainage density. Generally, the ruggedness number becomes higher with the higher value of relative relief (Rr) and drainage density (Dd) values. Extremely high values of ruggedness number generally occur when both the values are large, *i.e.*, when slopes are steep as well as long in nature (Strahler 1956). The ruggedness number of the basin is 1.50. The high value of ruggedness number indicates higher erosional intensity in the basin.

25.4.2 Results of SedimentaryAnalysis

The sedimentary analysis illustrates the high concentration of sand percentage found in the samples collected from Mingmang, Laluki, Dirpai, Dihiri, Bordoi, Pehiyoti, Bogipung and Ajuha village (Fig. 25.2a). On the other hand, Chawaldhua, Hapekhaiti, Garubandha, Dirpai and Balisori village showed less concentration of sand in our study. Figure 25.2b depicts the percentage concentration of Silt. The

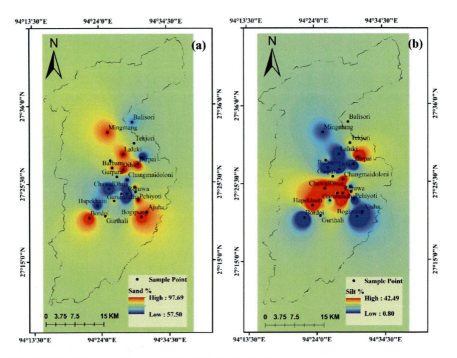

Fig. 25.2 Distribution of Sand and Silt within Jiadhal basin

samples were collected from the village Bordoi, Ajuha, Bogipung, Pehiyoti, Chama-rajan, Barbam, Dihiri, Kosutoli, Dirpai, Laluki and Mingmang having the maximum concentration of Silt. On the contrary samples from Hapekhaiti, Gohainchapori, Nepalikhuti, Rotuwa, Chawaldhuwa, Changmaidoloni and Tinighoria village showed less concentration of silt in our study.

Frequency distribution curves of the overbank sediments of the Jiadhal River are shown in Fig. 25.3, which indicates the nature of sediments by the representation of the weight percentage of different fraction of sediments. In Fig. 25.3a, It is observed that in Tekjuri, Lalukijan, Dihiri village shows bimodal classification with no primary mode and have the secondary modes 2, 3, and 4ϕ, respectively, which indicates that there are variations from the medium, fine to very fine sand in the samples and Barbam village shows unimodal classification with one primary mode 4ϕ, which indicates that there are no variations in the size of the grains and the grains are very fine sand. In Fig. 25.3b, Changmaidoloni, Chamarajan, Nepalikhuti and Rotuwa all are bimodal with secondary modes varying from 1.5 to 4.5ϕ, which shows variations of grain size from the medium, fine to very fine sand in each sample. Again in Fig. 25.3c, Gorubandha, Pehiyoti, Hapekhaiti all the samples show bimodal classification having secondary modes from 1.5, 2.3, and 4ϕ with variations of grain size from the medium to the very fine sand, and Gohainchapori shows unimodal characteristics with one primary mode, i.e., 4ϕ which shows maximum dominant of very fine sand particles. In Fig. 25.3d, Ajuha and Garpara show unimodal classification with the same primary mode, i.e., 4ϕ which shows the dominant classification of grain size are very fine sand whereas Kosutoli and Balisori show bimodal classification 1.5, 2 2.5, and 4ϕ, where there are multiple distributions of grain size ranging from medium to very fine sand. In Fig. 25.3e, Bogipung shows bimodal classification ranging from 1.5, 2.9 and 4ϕ, which indicates the grain size was varying from medium, fine and very fine sand, whereas Mingmang, Dirpai, Gurthali gives unimodal classification with one primary mode ranging from 3, 3.5 and 4ϕ, respectively, which indicates that Mingmang has the major dominant of fine sand, and Dirpai and Gurthalidominate very fine sand. In Fig. 25.3f, Bordoi and Chawaldhua show unimodal classification with one primary mode 4ϕ, which indicates that the major dominant of grain is of fine sand in these two areas, whereas Tinigharia shows bimodal classification with secondary mode ranging 1.5, 3, and 3.5ϕ that means in Tinigharia the grain size were of medium sand and fine sand. Out of the total 23 nos. of over-bank samples, a dominant group comprising 70% of the total show bimodal character in its distributions with no clear peak and having modal class ranging from 1.5 to 4ϕ (Coarse to very fine sand), while a few (30%) shows slightly unimodal characters in the distribution of the grain size, with a primary mode ranging 3–4ϕ (fine sand to coarse silt).

In Fig. 25.4, NDVI of the study area showed the value ranges between a minimum of −0.90 and a maximum of 0.85. Similarly, SAVI value ranges between a minimum of −0.21 and a maximum of 0.72. On the other hand, Fig. 25.5 represents the BSI value ranges between a minimum of −0.60 and a maximum of 0.31 in the study area.

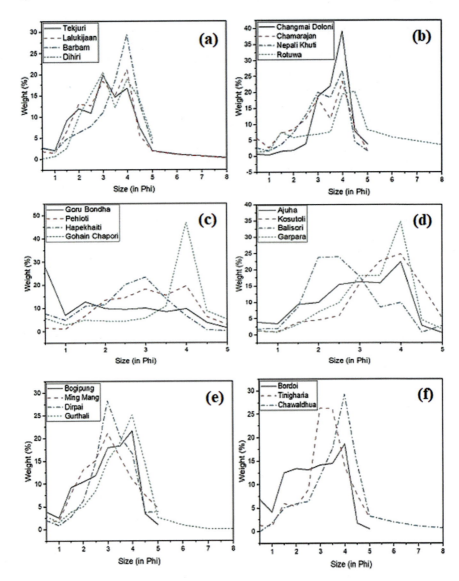

Fig. 25.3 The frequency distribution curve of overbank sedimentation in the Jiadhal River basin

To find the correlation among the different raster layers including the sand percentage layer and the silt percentage layer, the Band Collection Statistics of ArcGIS Spatial Analyst tool is used. The results of this spatial analysis show that Sand and Silt distribution are positively correlated with the BSI of the region and vice versa. On the other hand, NDVI and SAVI show a highly positive correlation in this analysis (Tables 25.3 and 25.4).

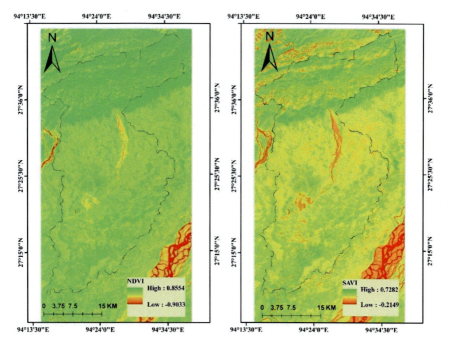

Fig. 25.4 Normalized Difference Vegetation Index (NDVI) and Soil Adjusted Vegetation Index (SAVI) of the study area

25.5   Conclusion

In this study, a special attempt has been made to assess the morphometric charac-teristics and the sedimentary behavior of the Jiadhal River using both geospatial and sedimentary petrology based approach. The morphometric analysis of the basin covers all the three main aspects, i.e., linear, areal and relief aspects. Linear aspects suggest that the basin comprises of the highest 7th order of stream controlled by the dendritic pattern of the stream network. Bifurcation ratio and stream length ratio indicates the structural control as well as variation in the slope and topography in the region. Under the areal aspects, it is seen that the region has highly permeable subsoil, dense vegetation cover having moderate relief, course drainage texture and elongated in shape. Similarly, the relief aspects indicate the higher erosional intensity within the study area. The results of the sedimentary investigation in the Jiadhal basin clearly show a correlation between the nature of sediment and vegetation. The sand and silt dominated deposition have changed the physiographic and surface hydrology of the region. Excessive silt has also contributed to the changing agricultural practices of the region. Although sediment yield is a natural process in any floodplain river, its high yield often becomes a hazard for the dwelling community. While making a

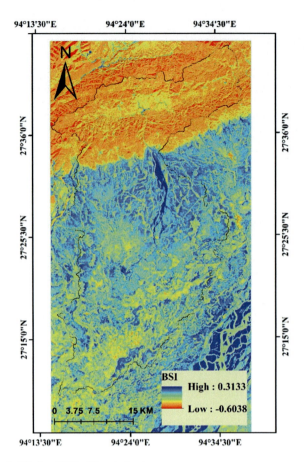

Fig. 25.5 Bare Soil Index (BSI) of the study area

Table 25.3 Statistics of individual raster layers

Layer	Min	Max	Mean	STD
1. SAND	57.5041	97.6990	89.0303	1.4319
2. SILT	0.8001	42.4959	10.5146	1.3963
3. NDVI	−0.9034	0.8555	0.4704	0.2292
4. BSI	−0.6038	0.3134	−0.2110	0.1608
5. SAVI	−0.2149	0.7283	0.2742	0.1422

developmental plan and hydrologic interventions, sediment should become one of the major considerations specially basin like Jiadhal. Remote sensing methods nicely supplement the field data. Further, these findings need to be considered for livelihood planning of the basin.

Table 25.4 Correlation matrix

Layer	SAND	SILT	NDVI	BSI	SAVI
1. SAND	1.0000	−0.9519	−0.0384	0.0376	−0.0315
2. SILT	−0.9519	1.0000	−0.0182	0.0352	−0.0041
3. NDVI	−0.0384	−0.0182	1.0000	−0.5793	0.9193
4. BSI	0.0376	0.0352	−0.5793	1.0000	−0.4291
5. SAVI	−0.0315	−0.0041	0.9193	−0.4291	1.0000

Acknowledgements Authors acknowledge National Remote Sensing Centre, Indian Space Research Organization, Hyderabad, India for generous funding of the project under ISRO-RESPOND Programme (Grant No. DS-2B-13012(2)/45/2018). Thanks to the United State Geological Survey for providing free Landsat 8 OLI and SRTM data for this study. We acknowledge Dr Jayanta Jivan Laskar, Associate Professor, Department of Geology, Gauhati University for his support in this research.

References

ASTM (1964) Standard method for grain size analysis of soils. In: Procedures for testing soils. American society for testing materials, Philadelphia, Pennsylvania

Bertin S, Friedrich H (2016) Field application of close-range digital photogrammetry (CRDP) for grain-scale fluvial morphology studies. Earth Surf Proc Land 41(10):1358–1369. https://doi.org/10.1002/esp.3906

Bordoloi K, Nikam BR, Srivastav SK, Sahariah D (2020) Assessment of riverbank erosion and erosion probability using geospatial approach: a case study of the Subansiri River, Assam, India. Appl Geomat https://doi.org/10.1007/s12518-019-00296-1

Borgohain A, Sahariah D (2017) A study on morphometric characteristics of Jiadhal River basin using Remote Sensing and GIS. Int J Res Manag Soc Sci 5(3), 13–19.

Brakenridge GR, Syvitski JPM, Niebuhr E, Overeem I, Higgins SA, Kettner AJ, Prades L (2017) Design with nature: causation and avoidance of catastrophic flooding, Myanmar. Earth Sci Rev 165:81–109. https://doi.org/10.1016/j.earscirev.2016.12.009

Cadigan RA (1961) Geologic interpretation of grain-size distribution measurements of Colorado Plateau sedimentary rocks. J Geol 69(2):121–144

Das LM (2016). Morphometric analysis of Jiya Dhol river basin. Int J Sci Res (IJSR) 5(3), 1482–1486. https://doi.org/10.21275/v5i3.nov162194

de Fagundes de HO, de Paiva RC, Fan FM, Buarque DC, Fassoni-Andrade AC (2020) Sediment modeling of a large-scale basin supported by remote sensing and in-situ observations. CATENA 190 https://doi.org/10.1016/j.catena.2020.104535

Easterling DR, Meehl GA, Parmesan C, Changnon SA, Karl TR, Mearns LO (2000). Climate extremes: observations, modeling, and impacts. Science. https://doi.org/10.1126/science.289.5487.2068

Folk RL, Ward WC (1957) Brazos River bar [Texas]; a study in the significance of grain size parameters. J Sediment Res 27(1):3–26

Goswami DC (1985) Brahmaputra River, Assam, India: physiography, basin denudation, and channel aggradation. Water Resour Res 21(7):959–978. https://doi.org/10.1029/WR021i007p00959

Guzmán G, Quinton JN, Nearing MA, Mabit L, Gómez JA (2013) Sediment tracers in water erosion studies: current approaches and challenges. J Soils Sediments. https://doi.org/10.1007/s11368-013-0659-5

Hancock GJ, Revill AT (2013) Erosion source discrimination in a rural Australian catchment using compound-specific isotope analysis (CSIA). Hydrol Process. https://doi.org/10.1002/hyp.9466

Horton RE (1932) Drainage-basin characteristics. EOS Trans Am Geophys Union 13(1):350–361

Horton RE (1945) Erosional development of streams and their drainage basins; hydrophysical approach to quantitative morphology. Geol Soc Am Bull 56(3):275–370

Huete AR (1988) A soil-adjusted vegetation index (SAVI). *Remote Sens Environ* 25(3), 295–309. https://doi.org/10.1016/0034-4257(88)90106-X

Jamalabad MS, Abkar AA (2004) Forest canopy density monitoring using satellite images. In Geo-Imagery Bridging Continents XXth ISPRS Congress. Istanbul, Turkey

Krumbein WC, Pettijohn FJ (1938) Manual of sedimentary petrography

Miller VC (1953) Quantitative geomorphic study of drainage basin characteristics in the Clinch Mountain area, Virginia and Tennessee. In Technical Report (Columbia University. Department of Geology); No. 3

Moiola RJ, Weiser D (1968) Textural parameters; an evaluation. J Sediment Res 38(1):45–53

Nichols JD, Koneff MD, Heglund PJ, Knutson MG, Seamans ME, Lyons JE, Williams BK (2011) Climate change, uncertainty, and natural resource management. J Wildl Manag. https://doi.org/10.1002/jwmg.33

Passega R (1964) Grain size representation by CM patterns as a geologic tool. J Sediment Res 34(4):830–847

Passega R, Byramjee R (1969) Grain-size image of clastic deposits. Sedimentology 13(3–4):233–252

Rao NK, Latha SP, Kumar AP, Krishna HM (2010) Morphometric analysis of Gostani river basin in Andhra Pradesh State, India using spatial information technology. Int J Geomat Geosci 1(2):179

Rouse JW Jr, Haas RH, Schell JA, Deering DW (1973) Monitoring the vernal advancement and retrogradation (green wave effect) of natural vegetation

Rudraiah M, Govindaiah S, Vittala SS (2008) Delineation of potential groundwater zones in the Kagna river basin of Gulburga district, Karnataka, India using remote sensing and GIS techniques. Mausam 59(4):497–502

Saikia A, Hazarika R, Sahariah D (2013) Land-use/land-cover change and fragmentation in the Nameri Tiger Reserve India. Geografisk Tidsskrift-Danish J Geog 113(1):1–10

Saikia L, Mahanta C, Mukherjee A, Borah SB (2019) Erosion-deposition and land use/land cover of the Brahmaputra river in Assam, India. J of Earth Syst Sci 128(8) https://doi.org/10.1007/s12040-019-1233-3

Sarma PK, Sarmah K, Chetri PK, Sarkar A (2013) Geospatial study on morphometric characterization of Umtrew River basin of Meghalaya, India. Int J Water Resourc Environ Eng 5(8):489–498

Sarma JN (2005) Fluvial process and morphology of the Brahmaputra River in Assam, India. Geomorphology *70*(3–4 SPEC. ISS.), 226–256. https://doi.org/10.1016/j.geomorph.2005.02.007

Schumn SA (1956) Evolution of drainage basins and slopes in bund land of Peth Amboy, New Jersey. Bull Geol Soc Am 67:597–646

Singh S, Singh MC (1997) Morphometric analysis of Kanhar river basin. Natl Geograph J India 43(1):31–43

Sinha R, Kale VS, Chakraborty T (2014). Tropical rivers of south and southeast Asia: landscape evolution, morpho-dynamics and hazards. Geomorphology https://doi.org/10.1016/j.geomorph.2014.08.020

Smith KG (1950) Standards for grading texture of erosional topography. Am J Sci 248(9):655–668

Stokes GG (1851) On the effect of internal friction of fluids on the motion of pendulums. Trans Cambridge Philos Soc

Strahler AN (1952) Dynamic basis of geomorphology. Geol Soc Am Bull 63(9):923–938

Strahler AN (1956) Quantitative slope analysis. Geol Soc Am Bull 67(5):571–596

Strahler AN (1964) Quantitative geomorphology of drainage basins and channel networks. Handbook of Applied Hydrology. McGraw-Hill, New York, pp 4–39

Sushanth, K., & Bhardwaj, A. (2019). Assessment of landuse change impact on runoff and sediment yield of Patiala-Ki-Rao watershed in Shivalik foot-hills of northwest India. Environ Monitor Assess 191(12). https://doi.org/10.1007/s10661-019-7932-z

Syvitski JPM, Cohen S, Kettner AJ, Brakenridge GR (2014) How important and different are tropical rivers?–An overview. Geomorphology 227:5–17. https://doi.org/10.1016/j.geomorph.2014.02.029

Tucker CJ (1979) Red and photographic infrared linear combinations for monitoring vegetation. Remote Sens Environ 8(2), 127–150. https://doi.org/10.1016/0034-4257(79)90013-0

Walia S, Singh S, Babbar R (2020) Runoff induced soil erosion and its impact on the quality of water for Upper-Patiala-Ki-Rao catchment lying on Shivalik hills. J Geol Soc India 95(4):385–392. https://doi.org/10.1007/s12594-020-1447-7

Wilkinson SN, Olley JM, Furuichi T, Burton J, Kinsey-Henderson AE (2015) Sediment source tracing with stratified sampling and weightings based on spatial gradients in soil erosion. J Soils Sediments. https://doi.org/10.1007/s11368-015-1134-2

Zhou Y, Huang HQ, Ran L, Shi C, Su T (2018) Hydrological controls on the evolution of the Yellow River Delta: an evaluation of the relationship since the Xiaolangdi Reservoir became fully operational. Hydrol Process 32(24):3633–3649. https://doi.org/10.1002/hyp.13274

Zimbres B, Machado RB, Peres CA (2018) Anthropogenic drivers of headwater and riparian forest loss and degradation in a highly fragmented southern Amazonian landscape. Land Use Policy 72:354–363. https://doi.org/10.1016/j.landusepol.2017.12.062

Chapter 26
An Assessment of RUSLE Model and Erosion Vulnerability in the Slopes of Dwarka–Brahmani Lateritic Interfluve, Eastern India

Sandipan Ghosh

Abstract The present study focuses on the hillslope erosion by running water which is observed in the lateritic badlands of Dwarka–Brahmnai Interfluve (eastern part of Rajmahal Basalt Traps). Using Revised Universal Soil Loss (RUSLE) model and field measured data (developing dams and sedimentation pits), the annual erosion rates (i.e., 8.12–24.01 kg m^{-2} y^{-1}) of eighteen sample slopes are estimated and validated using different quantitative and statistical techniques. The linear regression of experimental results shows a positive correlation ($r = 0.72$) and high increment between observed (X) and predicted (Y$_C$) erosion rate (Y$_C$ = 5.90 + 0.659 X; R^2 = 0.521). Based on eighteen dam sites and sedimentation data (2016–17), the average erosion rate (A$_P$) is 16.63 kg m^{-2} y^{-1} which is beyond the soil tolerance value of laterite (i.e., 1.0 kg m^{-2} y^{-1}), showing high level of erosion vulnerability (i.e., loss of land, soil bareness, ferruginous crusting, deterioration of crop, and other biomass). Alongside the dynamics and susceptibility of water erosion is assessed here to emphasis the role and function of gullies in the evolution of lateritic badlands. In this region 52.51% of gullies are developed by overland flow erosion while 27.96% belongs to mass movement.

Keywords Soil Erosion · Gully · Model efficiency coefficient · Model effectiveness coefficient · RUSLE · Laterite

26.1 Introduction

During a long period of time a drainage system or its components can be considered as an open but dynamic system which is progressively losing potential energy and mass, but over shorter spans of time self-regulation is important, and components of system may be graded or in dynamic equilibrium (Schumm and Lichty 1965). Models can be considered as essential simplifications of earth's reality, and it is developed as system approach in geomorphology (Morgan 2005). The root of erosion models is

S. Ghosh (✉)
Department of Geography, Chandrapur college, West Bengal Purba Bardhaman 713145, India

© The Author(s), under exclusive license to Springer Nature Singapore Pte Ltd. 2022
N. C. Jana and R. B. Singh (eds.), *Climate, Environment and Disaster in Developing Countries*, Advances in Geographical and Environmental Sciences, https://doi.org/10.1007/978-981-16-6966-8_26

found in the scientific studies of 1950s–1960s (Strahler 1952; Chorley 1962, 1964; Schumm and Lichty 1965) and with the passage of time it is developed as a research theme for the erosion-prediction technology and soil conservation planning (Baas 2017). Researchers seek models that describe how the system functions in order to enlighten understanding of the system and how it responds to change (Morgan 2005). In the soil erosion study, the development of a model is associated with the erosion-prediction technology which is a powerful tool used for more than half a century in policy development, erosion inventories, conservation planning, and engineering design (Toy et al. 2013). It is not possible to take measurements at every point in the landscape, and it also takes time to build up a sufficient database and long-term measurements. In order to overcome these deficiencies, models can be used to predict erosion under a wide-range of conditions. In this regard, one of the important methods of soil erosion science, used to evaluate the effectiveness of model, is to compare the predictions given by the selected model to measured data from soil loss collected on plots taken under natural rainfall conditions (Boardman and Favis-Mortlock 1998; Nearing 2000; Torri et al. 2013).

Modeling and prediction of water erosion has long history of subsequent researches and experiments with first studies published in renowned journals more than seven decades ago using North American data sets (Bennett 1939). Since 1980s many studies have documented the magnitude of soil erosion problems using various models in different parts of the world (especially India), expressed as billions of tons of eroded soil or billions dollars of erosion and sedimentation damage each year (Narayana and Babu 1983; Kothyari 1996; Lal 1990; Singh et al. 1992; Wasson 2003; Pimentel 2006; Reddy and Galab 2006; Borrelli et al. 2013; Kumar and Pani 2013; Pimentel and Burgess 2013; Sharda et al. 2013; Sharda and Dogra 2013; Froehlich 2018; Sharma 2018; Poesen 2018; Pennock 2019). Often it is observed that gullies are highly visible and severe form of soil erosion, with seep-sided, incised and drainage lines greater than 30 cm deep, and it signifies instability in the landscape, regarded as threshold phenomena under certain physical conditions, relating to flow erosivity and surface resistance (Horton 1945; Patton and Schumm 1975; Begin and Schumm 1984; Banadyopadhyay 1988; Vandaele et al. 1996; Singh and Dubey 2002; Posen et al. 2003; Valentin et al. 2005; Morgan and Mngomezulu 2003; Sharma 2009; Jha and Kapat 2009; Samni et al. 2009; Dong et al. 2013; Torri and Poesen 2014; Shit et al. 2015; Ghosh and Guchhait 2017). It was found that ravines along the banks of the Yamuna, Chambal, Mahi, Tapti, and Krishna Rivers revealed soil losses exceeding $18\,t\,ha^{-1}\,y^{-1}$ and the erosion rates on the alluvial Indo-Gangetic Plains of Punjab, Haryana, Uttar Pradesh, Bihar, and West Bengal were 5 to $10\,t\,ha^{-1}\,y^{-1}$ (Singh et al. 1992). In India soil erosion is taking place at the rate of $16.35\,t\,ha^{-1}\,y^{-1}$ and about twenty nine percent of the total eroded soil is lost permanently to the sea (Narayana and Babu 1983; Singh et al. 1992). Ten percent of it is deposited in the reservoirs, and the remaining 61% is dislocated from one place to the other (Narayana and Babu 1983; Singh et al. 1992). About 69.5% area of India has soil loss tolerance limit of $<10\,t\,ha^{-1}\,y^{-1}$, while about 13.3% area has a soil loss tolerance limit of only up to $2.5\,t\,ha^{-1}\,y^{-1}$ (Sharda and Dogra 2013).

The essential purpose of quantitative assessment is that erosion control targeted toward the areas with the highest rates can markedly reduce erosion averages. Before taking any erosion protection measures, the estimation of annual erosion rate at plot to basin scale is the fundamental step toward achieving soil conservation and sustainable development (Toy et al. 2013). Erosion protection measures should start from micro scale to get long-term soil productivity and long-term sustainable agriculture in the developing counties, like India where erosion protection technologies are limited by economic and other cultural conditions. In addition, it is necessary to state that the laterite terrain of West Bengal (known as *Rarh* Bengal, i.e., the land of red soil) is severely dissected by the dense network of rills and gullies (Ghosh and Guchhait 2017), developing badland topography and there are very few databases of accurate annual erosion rates and empirical model applications. In this regard, the present study can give light on the aspect of soil erosion modeling using minimal data inputs and measured plots at basin scale to estimate annual erosion rate in the slopes of lateritic badlands. Two major objectives of the study are set forth as (1) estimating annual soil erosion rate and (2) evaluating the suitability and effectiveness of erosion model. Alongside, the vulnerability of erosion is also investigated here to emphasize the role of gullies in the evolution of lateritic badlands.

26.2   Brief Description of Study Area

The selected study area of Dwarka–Brahmani interfluve (about 176 km^2, encompassed by 24°08′N to 24°14′ N and 87°38′ E to 87°44′ E), covers Shikaripara block (Dumka, Jharkhand), Rampurhat I and Nalhati I blocks (Birbhum, West Bengal) (Fig. 26.1). The region belongs to southern fringe of RBT (50–448 m from msl) which were chemically weathered due to lateritization processes from Palaeogene to Late Pleistocene. Field study reveals successive occurrences of fresh quartz-normative tholeiite Rajmahal basalt, weathered coarse saprolite, kaolinite pallid zone, mottle zone, and pisoliticferricrete in the litho-sections (Ghosh and Guchhait 2015). Each

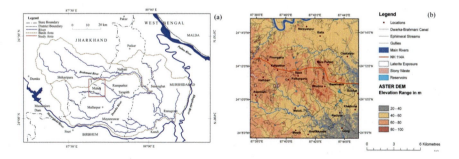

Fig. 26.1 a Location of study area in Dwarka–Brahmani Interfluve and **b** spatial arrangement of streams and gullies in different elevation zones of study area (Ghosh and Guchhait 2020)

laterite section reflects both primary *in-situ* type palaeogenesis of high-level plateau laterites and secondary ex-situ evolution of piedmont slope laterites which are prone of to water erosion, forming patches of badlands in the *Rarh*Plain. Theses badlands of Himalayan Foreland Basin, Chambal valley, and Chotanagpur Plateau Fringe are believed to have developed due to neo-tectonic activities, strengthening of south-west monsoon and intensive fluvial erosion since Late Pleistocene–Late Holocene (Sinha et al. 2012; Ghosh and Guchhait 2015).

The climate of this region has been identified as sub-humid and sub-tropical monsoon type, having strong influence to tropical cyclones (June to October) and thunderstorms (April to June). The region receives mean annual rainfall of 1300 to 1437 mm. The peak monsoon and cyclonic rainfall intensity of 21.51 mm h^{-1} (minimum) to 25.51 mm h^{-1} (maximum) is the most powerful climate factor to develop this lateritic badlands (Ghosh and Guchhait 2020). In and around the study area the soil series of Bhatina, Raspur, and Jhinjharpur (Sarkar et al. 2017) has been developed in the present geo-climatic setting, having high degree of ferruginization. Generally, the thin solum is loamy-skeletal and hypothermic in nature developing on the barren lateritic wastelands with sparse bushy vegetation and grass. The loose secondary laterite (16 to 34 cm) is developed as cementation (low cohesion and weak structure) of derived materials over mottle and kaolinte horizon, and it is much prone to overland flow erosion, tunnel erosion, and bank failure. The natural vegetation of the study area belongs to the tropical moist and dry deciduous type with few evergreen types. The areal coverage of major land use–land cover (for the year 2016–17) is estimated as (1) natural green vegetation and grassland cover an area of 58.83 km^2, (2) the pediment barren surface of laterite covers (including *morrum* quarries) an area of 16.78 km^2, (3) stone quarries, roads, built-up area, and permanent fallow land (including basalt exposure and non-arable land) cover an area of 69.56 km^2, (4) arable land covers an area of only 15.25 km^2, and (5) rivers (seasonal flow) and water bodies cover an area of only 15.58 km^2.

26.3   Methodology

Soil erosion is the gradual process of movement and transport of the upper later of soil by different agents (mainly water), causing its deterioration in the long term. In this context soil erosion models are considered as efficient tools for estimating and predicting erosion rate, sediment budget, and the design and effectiveness of erosion control practices in the humid tropical environments (Avwunudiogba and Hudson 2014). Many mathematical models (Table 26.1), categorized as empirical or index-based, conceptual, physically based, or process-oriented, are variable to estimate soil erosion (mainly rill and inter-rill erosion) at different spatial and temporal scales (Wischmeier and Smith 1978; Morgan et al. 1998; Flanagan et al. 2001; Morgan 2001; Merriti et al. 2003; Renard et al. 2011; Avwunudiogna and Hudson 2014; James et al. 2017; Morgan and Duzant 2008; Alewell et al. Pennock 2019).

Table 26.1 A concise summary of soil erosion models

Model type	Form	Derivation method	Strengths	Example
Regression-derived	A single or a few equations having a form that best fits the data	Derived by fitting an equation to an empirical database representing field conditions	Generally simply and easy to use; input values can be simple and easy to obtain	Multivariate logistic regression model (MLRM)
Index-based	Uses indices, usually in a multiplicative form, to represent how climate, soil, topography and land use affect erosion	Values for indices determined from large empirical database representing field conditions	Simple and easy to use: input values can be simple and easy to obtain; very powerful in relation to simplicity and input values	Universal soil loss equation (USLE)
Process-based	Uses sophisticated equations to represent how variables controlling erosion change between storm values and a steady-state equation to compute the erosion for each storm	Derived from theory and empirical database for erosion processes and land use sub-factors and theory, validated against field database	Powerful; can represent a wide range of conditions, capture main effects of erosion processes	European soil erosion model (EUROSEM)

Lumped Parameter Models (LPMs) or index-based models use averaging techniques to lump the influence of non-uniform spatial processes of a given area, such as a basin-averaged precipitation for run off computation (Toy et al. 2013; Avwunudiogba and Hudson 2014). It is found that LPMs linked to Geographical Information System (GIS) are very useful and user-friendly method for erosion risk assessment and conservation planning than other sophisticated distributed parameter models. The Revised Universal Soil Loss Equation (RUSLE) is a popular empirical equation for predicting long-term average soil erosion from agricultural field under specific cropping and management practice (Renard et al. 1994). There are few hindrances or problems to implement distributed parameter or process-based models (like WEPP, EURSOEM, etc.) in the study area. Three key problems or hindrances are stated as follows:

(1) the amount of money, man-power, and time devoted to collection of the data in justifying the application for the watershed of gully;

(2)   the lack of institutional framework, personnel, and financial commitment to undertake the long term research necessary for implementations of process-based models; and

(3)   the difficulty with instrument-based data requirements and complex mathematical equations in process-based modeling.

RUSLE type modeling is widely used throughout the world because this model has high degree of flexibility and data accessibility, parsimonious parameterization, extensive scientific literature, and comparability of results allowing the model to variablegeo-environmental settings (Kinnell 2014; Alewell et al. 2019). In this study RUSLE model is applied with various stages of geomorphic research methodology, from data collection and sample sites selection to model execution and result analysis (Fig. 26.2).

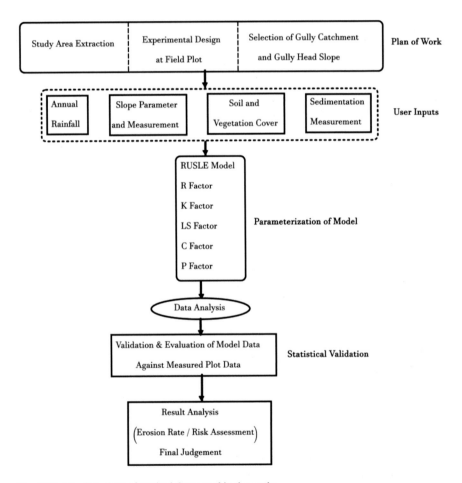

Fig. 26.2 The flowchart of methodology used in the study

26.3.1   Secondary Database

In this study the main sources of secondary data are mentioned as follow: (1) National Bureau of Soil Service and Land Use Planning (NBSS and LUP) soil report of Birbhum, (2) unpublished geology report and maps of Geological Survey of India (GSI) from official website (https://www.gsi.gov.in), (3) Land Use–Land Cover database of Census of India, (4) district gazetteer of Birbhum, (5) official websites of IMD (Indian Meterological Department) Pune (https://www.imdpune.gov.in/) and Kolkata (http://imdkolkata.gov.in/), and (6) official website (https://wbiwd.gov.in/) of Irrigation and Waterways Dept. of Govt. of West Bengal (IWD). To understand the minute topographic anomalies, drainage network, and reginal association of landforms the SOI (Survey of India) topographical sheets (72 P/12/NE, R.F. 1:25,000 and 72 P/12, R.F. 1:50,000) are very useful tool. Alongside, other spatial information is gathered from the Google Earth Engine using multi-temporal database. Landsat TM (Thematic Mapper, 30 m resolution) image and ASTER (*Advanced Spaceborne Thermal Emission and Reflection* Radiometer, 30 m resolution) DEM data are used to develop geomorphic and land use–land cover maps in the GIS platform (ArcGIS 9.2, Erdas Image 9.1, and MapInfo Professional 11.5). To assess annual soil erosion rate, it is essential to gather daily and monthly rainfall data from one or more permanent stations. For soil erosion modeling the rainfall database (period 2016–17) was collected from three IWD rain-gauge stations (Fig. 26.3)—(1) Nalhati (24°17'25"N, 87°49'44"E), (2) Rampurhat (24°10'13"N, 87°46'50"E), and (3) Mollarpur (24°04'35"N, 87°42'36"E). Using Thiessen polygon method for point-base rainfall calculation the rainfall of study area which is situated around areal distance of 18 to 25 km from three stations. The calculated mean annual rainfall for this region is 1510 mm in 2016 (considering maximum intensity of rainfall is 25.21 mm h^{-1}), and the per day rainfall amount is 17.48 mm, considering total rainfall and rainy days in a year.

26.3.2   Spatial Scale

A well-defined basin of gully is selected here as a fundamental and smallest unit of geomorphic study. The basin or catchment or watershed, bounded by its drainage divide and subject to surface and sub-surface drainage under gravity to interior lake or to basin of higher order, forms the logical areal unit of hydro-geomorphic study (Chorley 1969). The various components of a gully are identified as (1) the head, (2) gully sides or bank gullies, (3) gully bed, (4) lateral gully branch, (5) gully drainage area, and (6) gully outlet. The methodology has focused on the hillslope erosion by water (i.e., sheet erosion, rill, and inter-rill erosion) which is observed in the upper catchment drainage area of a gully. The selection of upstream drainage area is based on the fact that the gully head receives water and sediments through rills from this drainage area along the direction of slope and the location of gully head reflects the

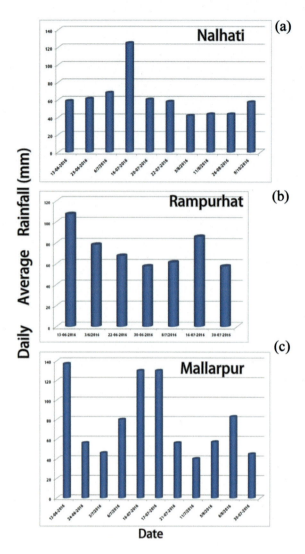

Fig. 26.3 The rainfall graphs of three stations showing distribution of daily maximum rainfall events (above 40 mm day^{-1}) in 2016

position of concentrated overland flow convergence, received eroded sediments from upslope during rainfall. Selecting the slope segments above gully heads, the model based erosion assessment is applied and verified with measured database of erosion rate. In the study three catchments of gully are selected (Fig. 26.4).

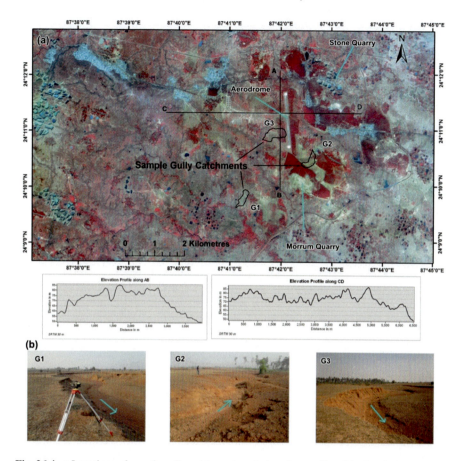

Fig. 26.4 a Locations of sample gully catchments and elevation profile of the laterite terrain, and **b** field photographs of gully catchments 1 (G1), catchment 2 (G2), and catchment 3 (G3) (*Note* blue arrow showing direction of ephemeral flow)

26.3.3 *Experimental Design and Erosion Measurement at Field*

The RUSLE model demands slope as spatial scale (e.g., erosion plot) of erosion study. Therefore, the plan present study is to select the vulnerable slopes in a small catchment of gully. It is found that gully head slope is the most active erosion part of the basin where sheet erosion and flow erosion are very dominant (at convex part) to form permanent gullies at the base of slope (i.e., concave part). The total slope length is the overland flow part between the gully head and water divide. In this lateritic terrain three high erosion risk catchment of gully are firstly selected and it has well defined basin area (about 1,09,250 to 2,16,050 m^2) and dense network of gullies (7.57 to 8.33 km^{-2}) (Fig. 26.5). At fundamental phase of research, eighteen gully heads

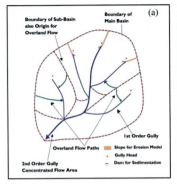 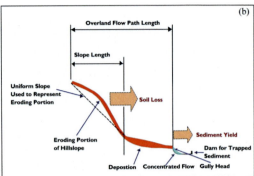

Fig. 26.5 Experimental design for erosion model **a** selection of slope, gully head, and dam site in a gully catchment, and **b** process of erosion in hillslope and location of dam for trapping sediments

of three catchments were identified and then eighteen gully head slope elements (considering 2 m width of slope strip to incorporate soil–land use parameters) were selected, denoting S1 to S18 respectively. The steepness of hillslope was measured using Leica Sprinter 150 m digital leveling instruments (accuracy— ± 0.7 mm of the 250 m distance) and other parameters of model were estimated in the recurrent field survey (2016–2017) and guided values of RUSLE model (Renard et al. 1997; Morgan 2005).

Maintaining a certain distance (1.5 to 2 m) from active gully head, eighteen check dams (used as sedimentation pits) were developed (denoting Dam 1 to Dam 18) at the base (i.e., gully floor) of representative slope elements to trap eroded sediments coming from upslope in a year (2016–2017).The dams were developed in January, 2016 with the help of local manpower and resources. The dams were height range of 40–55 cm and width range of 92–190 cm. Mostly during monsoon period (June to October) of 2016 the eroded material of these slope elements or upstream drainage areas of gullies were trapped behind the dams. Then after one year of observation the sedimentation was measured in January, 2017 and the mass volume was measured as multiplying the area of sedimentation behind dam and mean depth of sedimentation at eighteen dam sites. The observed rate of erosion (unit as kg m^{-2} y^{-1}) was measured by dividing the total mass weight of eroded materials by strip area of slope element or erosion plot for one year (2016–17).

26.3.4 RUSLE Model Description

The precise description of RUSLE is found in the writing of Renard et al. (1997), predicting soil erosion by water for conservation planning in the geo-climatic condition of Untied States. Smith (1999), Svorin (2003), Babu et al. (2004), Jain and Das (2012), Sinha and Joshi (2012), Bayramov et al. (2013), Kinnell (2014), Karydas et al. (2014), Devatha et al. (2015), Mondal et al. (2017), and Bhattacharya et al. (2020) have successfully applied RUSLE to assess erosion rate in different environmental settings and they have found the suitability and effectiveness of RUSLE in comparison to other models, e.g., Soil Loss Estimation Model for Southern Africa (SLEMSA), Morgan Morgan Finney Model (MMF), Water Erosion Prediction Project (WEPP), and European Soil Erosion Model (EUROSEM). The applied version of RUSLE is mentioned as follows (Renard et al. 1994, 1997) (Table 26.2):

$$A_p = RKLSCP \qquad (26.1)$$

where

- A_P is the computed soil loss per unit area (t ha^{-1}y^{-1} or kg m^{-2} y^{-1}); it can transform into SI unit
- R, the rainfall and runoff factor, is the number of rainfall erosion index units, i.e., EI_{30}
- K, the soil erodibility factor, is the soil loss rate per erosion index unit for a specified soil as measured on a unit-plot, which defined as a 72.6 ft length of uniform 9% slope continuously in clean-tilled fallow

Table 26.2 Operating functions in RUSLE

Description	Operating functions	Parameter definitions	Source
Rainfall Erosivity Index (R)	$R = (R_1 + R_2) / 2$ $R_1 = P (0.119 + 0.0873$ $\log_{10}I_m). \log_{10} I_{30}$ $R_2 = 79 + 0.363 P$	P is the mean annual rainfall, I_m is the average rainfall intensity (i.e., 25.21 mm hr^{-1}), I_{30} is the maximum 30 minute rainfall intensity (i.e., 75 mm hr^{-1}, recommended by Wischmeier and Smith 1978),	Renard et al. (1997); Sarkar et al. (2005); Jha and Paudel (2010); Ganasri and Ramesh (2016)
Soil Erodibility Index (K)	$K = 1.2917 [2.1 \times 10^{-4}$ $(12—OM) M^{1.14} +$ $3.25 (s–2) + 2.5 (p–3)]$ $/ 100$ $M = \%$ silt $(100–\%$ clay)	OM is the percentage of organic matter in soil, M is the particle size parameter, s is the soil structure code and p is permeability code recommended by (Wischmeier and Smith 1978)	Sarkar et al. (2005); Bayramov et al. (2013)
Slope-Length Index (LS)	$LS =$ $(L/22.13)^{0.5}.(0.065 +$ $0.045 \theta + 0.0065 \theta^2)$	L is the slope length (m) and θ is slope steepness in percent	Sarkar et al. (2007); Rahaman et al. (2015)

- L, the slope-length factor, is the ratio of soil loss from the field slope length to that from a 72.6 ft length under identical conditions
- S, the slope-steepness factor, is the ratio of soil loss from the field slope gradient to that from a 9% slope under otherwise identical conditions
- C, the cover and management factor, is the ratio of soil loss from an area with specified cover and management to that from an identical area in tilled continuous fallow
- P, the support practice factor, is the ratio of soil loss with a support practice like contouring, strip cropping, or terracing to that with straight-row farming up and down the slope.

26.3.5 Model Validation and Evaluation Techniques

An important part of validation is to determine how well the model fits measured data. The model efficiency coefficient (MEC), firstly proposed by Nash and Sutcliffe (1970), is now increasingly used an alternative to the correlation coefficient to express the performance of model (Morgan 2011). Generally, a MEC value of greater than 0.5 is considered that the model performs satisfactorily in the region, and one should not expect values to exceed 0.7 (Quinton and Morgan 1998; Morgan 2005, 2011).

$$MEC = 1 - \sum(X_{obs} - X_{pred})^2 / \sum(X_{obs} - X'_{pred})^2 \qquad (26.2)$$

where X_{obs} is the observed value, X_{pred} is the value predicted by the model, and X'_{obs} is the mean of a set of observed values.

A 'model effectiveness coefficient' (E_C) was defined by Nearing (2000) for studies undertaken, and it provides a quantitative criterion for taking into account natural variability and uncertainty in measured erosion plot data when those data are used to evaluate erosion models (Nearing 2000). Null hypothesis is that RUSLE prediction (P_s) is equal to the measured value (M). The relative difference (R_{diff}) between predicated and measured values are calculated and then a particular set of conditions that 95% of the values for differences in erosion (fall within a certain range) is calculated.

$$R_{diff} = (P_s - M)/(P_s + M) \qquad (26.3)$$

Relative difference values (*Y*-axis) are plotted against measured values (*X*-axis) to get a trend in the scatters.

$$R_{diff} = m \log_{10}(M) + b \qquad (26.4)$$

It is defined E_c as the fraction of simulation model predictions for which a model is effective in predicting the measured erosion, using the acceptance criteria. Using the

95% occurrence intervals from the replicated erosion data would result in a value, $E_{c(a\,=\,0.05)}$. The value of $E_{c(a\,=\,0.05)}$ signifies that the percentage of the difference between measured and predicted soil loss fell within the expected range of difference for two measured data points within the same population (Nearing 2000).

For the statistical judgment and significant inter-relationship of observed and predicted values, Chi-square test, linear regression, Pearson's product moment correlation, t-test of correlation, and regression slope are applied (Table 26.3). In addition,

Table 26.3 Statistical parameters used for RUSLE model evaluation

Statistical parameter	Description	Statements of null (H_0)/alternate (H_1) hypothesis
Chi-square Test (x^2 Goodness of fit)	$x^2 = \Sigma\,(O_i - E_i)/E_i,$ n–1 degree of freedom O_i = observed erosion rate, E_i = predicted erosion rate of RUSLE or RMMF	H_0 (O_i–E_i = 0)—no difference between observed and predicted erosion rate H_1 (O_i–$E_i \neq 0$)—Significant difference between observed and predicted erosion rate
Linear Regression	$Yc = a + b\,X$ Y_c = predicted erosion rate of RUSLE or RMMF X = observed erosion rate a = intercept b = slope R^2 = coefficient of determination	–
Pearson's product moment correlation (r)	$r = Cov(X,Y)/\sigma_X\sigma_Y$ Cov (X,Y) = Covariance of X and Y σ_X = standard deviation of X σ_Y = standard deviation of Y Range = $+\,1 < r < -\,1$	–
t-test of b-value	$t_b = b/SE_b$ SE_b = standard error of b $SE_b = \sigma_X/\,\sigma_Y\,\sqrt{(1-r^n)/(n\text{-}2)}$ n = number of sample degree of freedom (n–2)	H_0—Regression slope (based on observed and predicted erosion rate) is significant, having close resemblance of X–Y relationship H_1—Slope is insignificant
t-test of r-value	$t_r = r\,\sqrt{(n\text{–}2)/(1\text{–}r)}$ degree of freedom (n–2)	H_0—There is a zero correlation H_1– There is a significant correlation, i.e., no zero
Confidence Interval	$C_i = X_m \pm (Z.\sigma_X/\sqrt{n})$ X_m = mean of observed erosion rates Z = the Z value (1.96) for desired confidence level α ($\alpha_{0.05}$—95% confidence level) (obtained from normal curve)	–

correlation coefficient value (r) also confirms the degree of resemblance in the X–Y relationship. Then the estimated values of b and r are tested using t-test statistics at 0.05 significance level.

26.4 Results and Discussion

26.4.1 Measurement of Annual Erosion Rate at Field

The field survey using digital instruments (Leica Sprinter 150 m, Leica Disto S610 and Garmin GPS Monatna 680) and GIS mapping have produced following important information on the lateritic terrain:

- Channel magnitude of sample badlands (Bhtaina) is 22 (M), having 12 S links (source links), 10 TS links (tributary source links), 6 B links (bifurcating links), 10 T links (tributary links), 2 TB links (tributary bifurcating links), and 3 CT links (cis–trans links);
- Linear and relief properties of gully basin: gully density—6.93 km km^{-2}, length of overland flow -72.16 m, constant of channel maintenance -144.32 m^2, hypsometric integral -0.43, sinuosity index -1.39, and circularity ratio -0.63;
- Basin relief varies from 19 to 11 m;
- Maximum dissection of terrain is 17.5 m;
- Steepness of slope elements varies from $03°15'$ to $12°15'$;
- Gully head upstream slope length varies from 23.2 m to 116.7 m.

The field experiment has presented that in eighteen dam sites (Fig. 26.6) the estimated weight of trapped sediments (i.e., mostly ferruginous nodules, gravels, and coarse sands) varies from 566 to 3581 kg, and the eroded sediments are controlled by the relative activeness of hillslope erosion, slope steepness and overland flow length. The observed annual erosion rate (O) of three sample catchments was finally measured as (Table 26.4).

(a)   10.50 to 24.27 kg m^{-2} y^{-1} (gully catchment 1),
(b)   8.12 to 20.82 kg m^{-2} y^{-1} (gully catchment 2), and
(c)   11.87 to 20.82 kg m^{-2} y^{-1} (gully catchment 3).

The erosion analysis of lateritic terrain suggests that water erosion occurs on two types of soil texture— (1) sandy loam and (2) sandy clay loam(Ghosh et al. 2021). In the lateritic badlands the average measured rate of soil erosion is near about 16.27 kg m^{-2} y^{-1} which is much greater (16 times greater) than the soil loss tolerance T-value of the *Rarh*region (i.e., 1.0 kg m^{-2} y^{-1}). Therefore, it can be said that the lateritic badlands of study area have high erosion risk (rendering organic rich top-soil development and increasing Fe-crusting, badlands area, and degradation of biomass) and the region needs immediate protective measures to check erosion and land degradation at basin scale (Ghosh et al. 2021). After getting the measured erosion

Fig. 26.6 a Location of dam sites in the gully catchment 1, **b** developing dam to trap sediment using laterite boulders and cements in catchment 2, **c** two dams at the base of gully heads in catchment 1, and **d** measuring the topographic parameter in the dam site of catchment 3

data the analysis is carrying forward to fulfill the key purpose of study which is to compare the predicted data of erosion model (RUSLE) with the observed data at field scale.

26.4.2   *Predicted Erosion Rate by RUSLE Modeling*

The input parameters of RUSLE are mean annual rainfall (P), average rainfall intensity (I_m), soil erodibility (K) based on soil organic matter content and percentage of sand, silt and clay particles, crop cover, and management factor (C) and protective erosion control factor (P) (Table 26.5). It is observed that the land use and land cover of the catchments do not change too much throughout the year, and the region has

Table 26.4 A brief summary of slope and dam's parameters and observed rate of hillslope erosion (for the year 2016) at eighteen dam sites of three gully catchments

Gully catchments	Check dam	Slope length (m)	Slope in degree	Sin S	Annual mean runoff (mm)	Dam width (cm)	Dam height (cm)	Mean sedimentation depth at dam (m)	Observed annual erosion rate, O_E (kg m^{-2} y^{-1})
Gully 1	1	22.1	10° 09′	0.176	845.09	102	42	0.13	14.10
	2	25.4	11° 06′	0.192	965.74	110	50	0.20	19.94
	3	45.4	8° 30′	0.147	973.08	87	41	0.25	24.27
	4	65	6° 11′	0.107	905.25	114	48	0.15	14.45
	5	74.5	5° 50′	0.101	847.55	98	35	0.11	10.50
	6	50.8	8° 05′	0.14	813.21	94	30	0.23	24.01
Gully 2	7	44.2	4° 30′	0.069	821.78	98	37	0.22	08.12
	8	106.8	5° 30′	0.088	854.99	110	41	0.27	15.81
	9	84	7° 20′	0.125	756.07	108	46	0.15	15.23
	10	86.7	3° 45′	0.057	905.77	122	48	0.11	10.15
	11	35.2	8° 30′	0.142	845.58	107	45	0.19	20.82
	12	65	8° 45′	0.139	829.96	125	51	0.23	14.71
Gully 3	13	75.2	5° 20′	0.092	774.35	98	32	0.19	15.05
	14	55.2	7° 30′	0.13	808.23	105	45	0.23	16.12
	15	62	7° 00′	0.121	830.38	102	48	0.27	19.24
	16	90.5	6° 40′	0.116	721.63	112	54	0.14	11.87
	17	58.1	8° 10′	0.142	770.31	94	40	0.17	20.62
	18	55	7° 30′	0.133	808.48	90	38	0.20	17.97

Table 26.5 Input parameters and model outputs in RUSLE modeling

Gully catchments	Check Dam	LS	K	R	P	C	Predicted annual erosion rate, A_P (kg $m^{-2} y^{-1}$)
Gully 1	1	1.33	0.28	654	0.1	0.61–0.91	15.89
	2	1.34					16.74
	3	1.53					18.70
	4	1.16					13.22
	5	1.19					13.07
	6	1.50					20.87
Gully 2	7	0.67	0.23	654	0.1	0.65–0.83	07.86
	8	1.3					15.37
	9	1.68					19.71
	10	0.72					8.44
	11	1.33					15.72
	12	1.38					16.19
Gully 3	13	1.07	0.28	654	0.1	0.68–0.82	16.06
	14	1.39					20.87
	15	1.33					19.97
	16	1.42					17.68
	17	1.63					24.47
	18	1.35					18.34

minimum human disturbance (Ghosh et al. 2021). The most important phenomenon is that the study area is not protected under any erosive control measures, except few patches of Accacia plantation (P Factor—0.1). Based on the above estimation of inputs, multiplied R, K, LS, C, and P factors are taken to get potential or predicated values of annual soil erosion rate (A_P) in the study area (Table 26.5). The average range of predicated values varies from 7.80 to 11.85 kg m^{-2} y^{-1} in the eighteen sample sites. In the experimental sites mean A_P of three gully catchments is estimated as.

(1) Catchment 1: 13.07 to 20.87 kg m^{-2} y^{-1} (mean -16.415 kg m^{-2} y^{-1}),
(2) Catchment 2: 07.86 to 19.71 kg m^{-2} y^{-1} (mean -13.881 kg m^{-2} y^{-1}), and
(3) Catchment 3: 16.06 to 24.47 kg m^{-2} y^{-1} (mean -19.565 kg m^{-2} y^{-1}).

It is derived from database that A_P of hillslope yields maximum erosion value due to high LS-factor (>1.50) and high bareness of lateritic surface (C factor <0.3) acts as catalysts for hillslope erosion. It is found that if the slope is recognized as short length and high steepness, it has high potential for erosion (at sample sites 1, 2, 6, 11, and 14). Based on eighteen sample sites of study area, the mean A_P is 16.63 kg m^{-2} y^{-1} which is beyond the soil tolerance T-value limit (1.0 kg m^{-2} y^{-1}), showing high risk of erosion at present (Lenka et al. 2014).

26.4.3 Analyzing Model Validation and Evaluation

The absolute error between observed and predicted data is measured, showing positive anomaly (over estimation of erosion in response to observed rate) and negative anomaly (under estimation of erosion in response to observed rate) (Table 26.6). It is learned that 55.55% of predicated sample (i.e., ten dam sites) provide under estimation of erosion phenomena and 44.45% of data gives over estimation of erosion phenomena in RUSLE modeling.

At 0.05 level of significance and 17 (n–1) degree of freedom, the Chi-square (x^2) test statistic sets forth the null hypothesis (H_o, O—A_P or $S_P = 0$) which states that there is no difference between certain characteristics of a population, i.e., difference between predicted and observed value is zero and good correlation. The alternate hypothesis (H_1, O—A_P or $S_P \neq 0$) reflects significant difference between predicted and observed value. The value of x^2 statistic is assigned as 27.59 at 0.05 significance level with 17 degree of freedom. The x^2 statistic value of RUSLE is estimated respectively as 10.43 which are much lower than the tabulated x^2 value at 0.05 level. Therefore, it is concluded that H_o is accepted and H_1 is rejected. So, there is no significant difference between observed and predicted values in the study.

Now, applying model efficiency coefficient (MEC) (Nash and Sutcliffe 1970; Morgan and Duzant 2008) into the relation between observed and predicted data, the MEC value for RUSLE is calculated as 0.48. The MEC > 0.50—0.70 signifies good and satisfactory performance of model in reference to observed erosion results

Table 26.6 Summarizing the results of testing statistics

Statistical parameter	Tabulated testing statistical value	Calculated value RUSLE	Remarks on hypothesis
Chi-square (x^2)	x^2 statistic is assigned as 27.59 at 0.05 significance level with 17 degree of freedom	10.43	H_0 is accepted and H_1 is rejected (RUSLE model are accepted and predicted values resemblance with measured values)
t-test statistic of r value	t statistic is assigned as 2.120 at 0.05 significance level with 16 degree of freedom	5.44	H_0 is rejected and H_1 is accepted (RUSLE model are accepted and there is god correlation)
t-test statistic of b value	t statistic is assigned as 2.120 at 0.05 significance level with 16 degree of freedom	2.99	H_0 is rejected and H_1 is accepted (regression value of RUSLE give desired result than RMMF)
Confidence interval, $\alpha_{0.05}$	14.15 to 18.29 kg m^{-2} y^{-1} at 0.05 significance level	61% of sample fallen within interval	RUSLE model can provide more satisfactory results than RMMF

(Morgan 2011). The result shows that MEC value of RUSLE is much closer to 0.50. So, it can be decided that RUSLE model can be applied in this geo-climatic setting in place of RMMF model.

To get the trend of inter-relation between observed and predicted database, now the scatter plot and linear regression trend line ($Yc = a + bX$) are prepared, taking observed data as X and predicted data of RUSLE as Y (Fig. 26.7). It is finally estimated that the predicted values of RUSLE is statistically inter-related with the

Fig. 26.7 **a** Line graph showing the deviation of RUSLE predicted data (A_P) from observed erosion data (O_E) at eighteen sample sites, **b** linear regression showing the positive incremental relationship between O_E and A_P, and **c** semi-log relationship between O_E and difference o results to show over-predicted and under-predicted zone

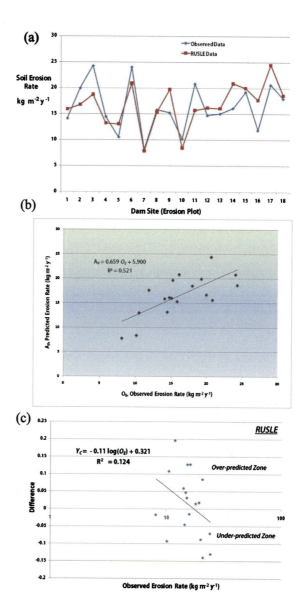

observed values ($A_P = 5.90 + 0.659\,O_E$), having good coefficient of determination (R^2) of 0.521 (i.e., inter-relation explained 52.10% in population) and notable slope (*b*) value of trend line, i.e., 0.659 (Fig. 26.7). The *b*-value of regression line (i.e., slope) reflects the quantitative judgment (indicating a change on response variable caused by a unit change of respective explanatory variable) of *Y* dependence on *X*. The *t*-test statistic of *b* value is 2.120 at 0.05 significance level with 16 (n—2) degree of freedom (H_o: b = 0, *Y* does not depend on *X*; H_1–*Y* depends on *X*). The estimated value of *t*-test statistic is 2.99 (RUSLE). This analysis reflects that test statistic of RUSLE *b*-value is greater than the tabulated *t*-value, and it means high dependence of predicted values on the observed values (i.e., RUSLE predicted values resemblance with observed erosion rates).

The Pearson's product moment correlation (*r*) of this analysis is estimated as 0.72 which reflect good correlation between predicted and observed values. Here, again the *t*-test statistic of *r* value is 2.120 at 0.05 significance level with 16 (n−2) degree of freedom (H_o: b = 0, *Y* does not correlate with *X*; H_1—there is a good correlation between *Y* and *X*). The estimated value of *t*-test statistics is 5.44 (RUSLE) which is much greater than the tabulated *t*-value at 0.05 significance level. Here, it can be concluded that *r* value or correlation between observed and predicted value is statistically significant in this study.

At last, the effectiveness coefficient (E_C) of erosion model is applied on the basis of linear regression database, 0.05 confidence interval of observed erosion rate (O_E) and Z-value of 1.96. The calculated R_{diff} value (relative difference) varies from + 0.196 to −0.139 in RUSLE respectively. It is found from the regression analysis (R_{diff} = m $\log_{10}O_E$ + b) that 55.55% of RUSLE results is placed in over-predicted zone, whereas 77.7% results of RMMF is located in over-predicted zone (Fig. 26.7). The confidence interval of observed erosion rate is 14.15 to 18.39 kg m^{-2} y^{-1}. If the large number of predicted values is fallen within this confidence interval, then E_C yields high value, signifying the good performance of the model. In general, E_C is the ratio between number of samples fallen within confidence interval and total number of samples. E_C of RUSLE modeling is 0.61 which reflects that 61% of predicted results are fallen with observed confidence interval. In other words, it can be concluded that at 0.05 significance of confidence interval RUSLE model can provide satisfactory results and the RUSLE is finally validated in this laterite terrain.

26.4.4 Erosion Vulnerability of Laterite Terrain

26.4.4.1 Classification of Gully

Water erosion is started from rain splash erosion and progressive accumulation of runoff aggravates rill and inter-rill erosion, gully erosion, puddle erosion, pedestal erosion, pinnacle erosion, piping and slumping, or bank failure (Bocco 1991). From the recurrent field survey, it is clearly observed that the overland flow gets concentrated in thread like channels forming sub-parallel rills in the laterites. Most of the rills are formed at a critical distance downslope from the crest, because at that point the kinetic energy of overland flow overcomes the soil resistance and the flow is channeled. As the water continue to concentrate and acquire additional energy for scouring, these rills become deeper and broader, and eventually some of them coalescence to form deep gullies. The prime cause of gully initiation is the too much water at a time and if the velocity of runoff or tractive force of runoff exceeds a critical threshold value, the gully will form. The size (depth), drainage area, and average runoff discharge rate are estimated in the 118 sample gully heads and three types of gully system are identified (Table 26.7). About 48.25% of gullies are associated with medium gullies which have basin area of 2400–7200 m^2, 5–10 m width and 2–5 m depth and 13.03% of gullies are recognized as large gully which have basin area of >7200 m^2, >10 m width, and >5 m depth.

In addition, it is essential to mention that the visual impression of badland landscape reflects the recognition of these sites as geomorphosites, having deep geotouristic value (Kale 2014; Zglobicki et al. 2018). In general, Zglobicki et al. (2018) have defined badlands as intensely dissected landforms mainly by fluvial erosion with least vegetation cover at the time of formation. The gullied landscape is a miniature form of badlands where a set of values (i.e., scientific, cultural, or aesthetic) can be ascribed as geomorphosites of West Bengal, focusing on the cognitive tourism in nature sites. These badlands of laterite can perform as geotourist sites to promote the conservation of geo-diversity and an understanding of earth's surface processes (for learning and research). The badlands of study area, having relief of 10 to 20 m, are characterized by tropical weathering profiles on Rajmahal Basalt Traps (Early–Late Cretaceous) and intense gullying of Holocene Epoch on the Neogene–Early Pleistocene Laterites, having storehouse of kaolin (i.e., china clay), ichnofossils, and petrified woods (Fig. 26.8).

Table 26.7 A scheme of gully classification based on hydro-geomorphic parameters

Gully Class	Depth (m)	Width (m)	Total Basin Area (m^2)	Basin Average Slope in degree	Fluvial Network Density (m m^{-2})	Annual Runoff Yield (mm)	Runoff Curve Number	Percentage of Gullies
Small Gully	<1–2	<3–5	<2400	3° 20′	0.37	<100–350	75	38.72%
Medium Gully	2–5	5–10	2400–7200	5° 45′	0.78	350–600	83	48.25%
Large Gully	>5	>10	>7200	4° 10′	1.12	>600	87	13.03%

Note Total sample of gully heads—118; Statistical technique of clustering–Dendogram Analysis

Fig. 26.8 Geotouristic
significance of lateritic
badlands–field photographs
showing dissected landscape
of gully catchment 1 (**a**),
catchment 2 (**b**), and
catchment 3 (**c**) in the
Dwarka–Brahmani Interfluve

26.4.4.2   Glimpses of Active Water Erosion

In addition to rill and gully, few other forms of active water erosion are observed in the
different parts of gully catchments—(1) *Puddle Erosion* (as the drops beat the bare
surface during violent storms, shatters the clods and soil crumbs and breaks down
the soil structure into puddle condition), (2) *Pedestal Erosion* (when as easily erode
soil is protected from rain splash erosion by a stone or tree root, isolated pedestals

capped by the resistant materials are left), (3) *Pinnacle Erosion* (this erosion occurs in gullies as the result of deep vertical rills widening until pinnacles are left like islands in the bed of gully and more resistant soil layer of ferruginous crust often caps the pinnacle), (4) *Soil Piping* (if there is an outlet so that the infiltrated water can flow laterally through the less permeable layer of sub-surface soil and the fine particles of the more porous laterite are washed out), and (5) *Slumping* (once the gully head has stared, erosion can continue by slumping, as mass movement, in the time of rainfall) (Fig. 26.9).

26.4.4.3   Erosion Dynamics of Rill and Gully

There are a number inter-related processes operated in the gully erosion and an understanding of water erosion should take into account the strong relationship between hydrologic and erosion processes (Bocco 1991). The processes of rill initiation and development generally involves four stages (Bocco 1991): (1) unconcentrated overland flow, (2) overland flow with concentrated paths, (3) microchannels without headcut, and (4) microchannels with headcut (Fig. 26.10). Rill erosion is a function of the forces applied to the soil by rill flow in relation to the forces within the soil that resist detachment (Morgan 2005; Toy et al. 2013). Kirkby and Bracken (2009) examines the condition of gully formation—a sharp step to initiate a headcut, a sufficiently low effective bedload fraction to evacuate eroded material, and the potential to maintain steep sidewalls, usually dominated by mass movement processes (Fig. 26.10). Gully initiation and development, in contrast to rills, generally involve multiple episodes of channel erosion—(1) downward scour, (2) headward cutting, (3) rapid enlargement, and (4) stabilization (Ghosh and Guchhait 2020).

Topography is an important factor controlling gully erosion rates. Study of self-enhancing feedback mechanisms generally is encouraged by a threshold approach, whereas study of self-regulating feedback mechanisms generally is encouraged by an equilibrium approach (Bull 1980). The present situation of badlands is quantitative described by the empirical equation, $S_{cr} = 0.0207 A^{-0.2784}$, where the critical slope of gully drainage area is defined as S_{cr} (i.e., minimum amount slope required to initiate gully) and gully-head upstream drainage area is described by A. The b value, greater than 0.2, usually signifies the dominancy of overland flow erosion in the gully initiation. The exponent value of $S–A$ model suggest that the vegetation of gully basins exerts a strong control on water movement and flow concentration, obliging the overland flow to be laminar also as high Reynolds number ($b \sim 0.25$). It is confirmed from the topographic threshold model that the secondary laterites of *Rarh*, having mean critical slope range of 0.0222–0.0407 m m^{-1}, are very susceptible to rill and gully erosion, particularly in the upstream deforested catchment of barren surface.

The gully head and sidewall collapse of gully (Fig. 26.11) is a composite and cyclical process resulting from downslope creep, tension crack development, crack saturation by overland flow, head or wall collapse followed by debris erosion which facilities the next failure (Poesen et al. 2003; Valentin et al. 2005; Ghosh et al. 2021).

Fig. 26.9 Field photographs of soil erosion vulnerability showing **a** dissection of terrain and expansion of gullies, **b** initiation of rills to erode gully bank, **c** bank collapse due to mass movement, **d** glimpse of sheet erosion, **e** measuring depth of erosion in landscape, **f** initiation of secondary gully heads at main gully, **g** active part of gully channel with devoid of vegetation, and **h** gully deposition at the outlet of basin

Fig. 26.10 Development of
1st order ephemeral gully by
overland flow (Ghosh and
Guchhait 2020)

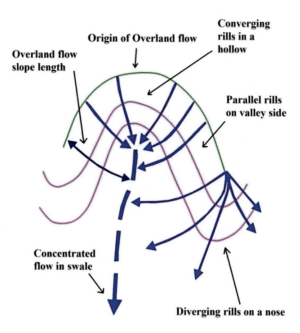

At the gully head the formation of grooves and tension cracks in the laterite curst layer develop several tunnels or pipes as seepage lines into the mottled zone, but the due to presence of impermeable kaolinte clay layer the pipes are restricted only in the upper two zones. During rainstorm the tunnel erosion expands the pipes and the overhanging mass of laterite is destabilized over the pallid zone (Ghosh et al. 2021). The roof of gully head has been collapsed and slumped in the gully floor, enhancing the upward migration of head. In this study area 52.51% of gullies are affected by overland flow erosion while 27.96% belongs to landslide erosion (Ghosh and Guchhait 2020). Only 15.25% of are affected by tunnel erosion or seepage erosion (Table 26.8). The glimpses of rills and gullies along with bare erodible surface are the sign of active and intensive water erosion. In the study area three broad vulnerable sites of erosion are identified and these should get proper attention (Fig. 26.12):

(1) Above gully head many sub-parallel rills are converged downstream from the water divide. These sites of flow convergence need immediate vegetative cover to diverge the flow.

(2) High slope steepness (>5°) and long stretch of semi-convex runoff slope (>70 m) (i.e., increasing overland flow path) are much prone to deep incision in the torrential rain.

(3) Bank failure (>70°) due to mass wasting, pipe flow, rill erosion, and undercutting by gully channel are one of key problems in gully expansion.

Fig. 26.11 **a** Different components of a gully, **b** erodible dark ferruginous layer of laterites at gully bank, **c** observed cyclic and composite processes of gully head retreat, **d** evidence of gully head retreat with expansion of length and width, and **e** developing dam to trap transported sediments from upslope of gully head

Table 26.8 Distribution of gully heads in respect of dominant erosion process (Ghosh and Guchhait 2020)

Dominant gully erosion process	Percentage of gully heads	Slope range	Area range
1. Overland Flow Erosion	52.51%	1.2 to 5.2°	2129.1 to 10,513.9 m²
2. Seepage Erosion	15.25%	2.2 to 4.6°	685.5 to 3843.7 m²
3. Landslide Erosion	27.96%	5.2 to 9.5°	457.1 to 5702.5 m²
4. Diffusive Erosion	4.28%	4.4 to 5.3°	483.2 to 879.9 m²

Note Sample size of gully—118

Fig. 26.12 Field photographs of vulnerable sites showing: **a** active channel part and vulnerable banks of main gully with high erosion depth, **b** down wasting of water divides and expansion gully width, and **c** vegetation growth on deposited materials rendering growth of gully head

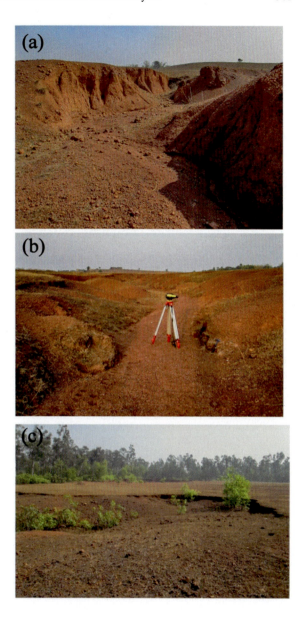

26.5   Conclusion

The model-based erosion study has fulfilled the objectives and recognized the laterite region as high potential erosion risk at basin scale using measured data and RUSLE model (16.27 to 18.63 kg m^{-2} y^{-1}). It is obtained from result database that RUSLE model is statistically useful in the region with minimal input data and it can couple with geographic information system to prepare an erosion map. It is also found that 52.51% of gullies are affected by overland flow erosion (Slope steepness, S $-1.2°$ to 5.2° and Drainage area, A -2129.1 to 10,513.9 m^2) while 27.96% belongs to landslide erosion or mass movement (S $-5.2°$ to 9.5° and A -457.1 to 5702.5 m^2). Here, the predicted annual erosion rate (A$_P$) of hillslope yields maximum erosion value due to high LS-factor (>1.50). It is finally understood and verified that if the slope is recognized as short length and high steepness, it has high potential for water erosion (at dam sites 1, 2, 6, 11, and 14). Based on eighteen dam sites and sedimentation data, the average A$_P$ is 16.63 kg m^{-2} y^{-1} which is beyond the soil tolerance value (i.e.,1.0 kg m^{-2} y^{-1}), showing high level of erosion vulnerability (i.e., loss of land, crop, and other biomass). The S–A model of topographic threshold reflects that the gullies which have larger drainage area (high potential yield of runoff volume), has required minimum angle of slope steepness to trigger a gully headcut through high concentration of overland flow (providing kinetic energy to erosion over resistance of laterite).

As a part of erosion control practices the most challenging task is to grow new plants in the infertile and heavily eroded surface of laterites where progressive expansion of rills and gullies, surface crusting, water crisis in lean period (November to April) and bareness are the key issues. No part of upper gully catchment (above gully headcut) should be not be left barren to decrease the amount of overland flow and flow convergence. The convex slope of barren lateritic land shows active erosion phases; so this land should be protected through grass plantation using flow barriers. The transported sediments can be trapped by the check dams which had been seized a massive amount of eroded materials in this experimental design, ranging from 566 to 3581 kg during 2016–17. Based on total analysis it can be suggested here that in each gully catchment should be treated as unique or distinctive erosion system and the focus of erosion control should centered on five aspects—(1) reduction of discharge rate, (2) reduction of channel grade, (3) control of headcut and sidewall erosion, (4) downstream sediment control by smalls scale dams, and (5) vegetative measures. Apart from erosion management, there is also a need of research to recognize the laterite badlands as geomorphositeorgeotourism site, because it is a unique landform to which a value can be attributed.

Acknowledgements The author is very much grateful to Dr. Sanat Kumar Guchhait (Professor, Department of Geography, The University of Burdwan, West Bengal, India) for research guidance. The author is indebted to R.P.C. Morgan (Retired Professor, National Soil Resource Institute, Cranfield University) and J. Poesen (Professor, Division of Geography & Tourism, KU Leuven) for their valuable articles and suggestions in the field-based data collection, analysis, and review of results. The author is very much grateful Rahman Ashiq Illahi, Suvendu Roy, SubhankarBera, and Subhamay Ghosh for their rigorous supports in the field survey and data collection. In addition, the author is indebted to the local inhabitants of Maluti (Dumka, Jharkand) for field assistance and manpower support.

References

Alewell C, Borrelli P, Meusburger K, Panagos P (2019) Using the USLE: chances, challenges and implications of soil erosion modelling. Int Soil Water Conserv Res 7:203–225

Avwunudiogba A, Hudson PF (2014) A review of soil erosion models with special reference to the needs of humid tropical mountainous environments. J Sustain Develop 3(4):299–310

Babu R, Bhyani BL, Kumar N (2004) Assessment of Erodibility status and refined iso-erodent map of India. India J Soil Conserv 32:171–172

Bandyopadhyay S (1988) Drainage evolution in a badland terrain at Gangani in Medinipur district. West Bengal. Geogr Rev India 50(3):10–20

Bayramov E, Buchroithner MF, McGurty E (2013) Differences of MMF and USLE models for soil loss prediction along BTC and SCP pipelines. J Pipeline Syst Eng Pract 4(1):81–96

Begin ZB, Schumm SA (1984) Gradational thresholds and landform singularity; significance for quaternary morphology. Geol Soc Amer Bullet 56(3):267–274

Bennett HH (1939) A permanent loss to New England soil erosion resulting from the hurricane. Geogr Rev 29(2):196–220

Bhattacharya RK, Chatterjee NC, Das K (2020) Estimation of erosion susceptibility and sediment yield in ephemeral channel using RUSLE and SDR model: tropical plateau fringe region, India. In: Shit PK, Pourghasemi HR, Bhunia GS (eds) Gully Erosion Studies from India and Surrounding Regions. Springer, Switzerland, pp 163–185

Boardman J, Favis-Mortlock D (1998) Modelling soil erosion by water: some conclusions. In: Boardman J, Favis-Mortlock D (eds) Modelling Soil Erosion by Water. Springer, Berlin, pp 515–520

Bocco G (1991) Gully erosion: processes and models. Prog Phys Geogr 15(4):392–406

Borrelli P et al (2013) An assessment of the global impact of 21st century land use change on soil erosion. Nat Commun 8:1–13

Chorley RJ (1962) Geomorphology and general systems theory. Geological Survey Professional Survey 500-B, Washington

Chorley RJ (1964) Geography and analogue theory. Ann Assoc Am Geogr 54(1):127–137

Chorley RJ (1969) The drainage basin as the fundamental geomorphic unit. In: Chorley RJ (ed) Water Earth and Man. Methuen, London, pp 77–100

Devatha CP, Despande V, Renukaprasad MS (2015) Estimation of soil loss using USLE model for Kulhan watershed, Chhattisgarh–a case study. Aquatic Procedia 4:1429–1436

Dong Y, Xiong D, Su Z, Li J, Yang D, Zhai J, Lu X, Liu G, Shi L (2013) Critical topographic threshold of gully erosion in Yuanmou dry—hot valley in southwestern China. Phys Geogr 34(1):50–59

Froechlich DC (2018) Estimating reservoir sedimentation at large dams in India. E3S Web of Conference 40:1–8

Ganasri BP, Ramesh H (2016) Assessment of soil erosion by RUSLE model using remote sensing and GIS—a case study of Nethravathi Basin. Geosci Front 7:953–961

Ghosh S, Gucchait SK (2017) Estimation of geomorphic threshold to permanent gullies of lateritic terrain in Birbhum, West Bengal. India. Curr Sci 113(3):478–485

Ghosh S, Guchhait SK (2015) Characterization and evolution of primary and secondary laterites in northwestern Bengal Basin, West Bengal. India. J Palaeogeography 4(2):203–230

Ghosh S, Guchhait SK (2020) Geomorphic threshold and SCS-CN based runoff and sediment yield modelling in the gullies of Dwarka-Brahmani interfluve, West Bengal, India. In: Shit PK, Pourghasemi HR, Bhunia GS (eds) Gully Erosion Studies from India and Surrounding Regions. Springer, Switzerland, pp 45–68

Ghosh S, Guchhait SK, Illahi RA, Bera, Roy S (2021) Geomorphic character and dynamics of gully morphology, erosion and management in laterite Terrain: few observations from Dwarka—Brahmani Interfluve, Eastern India. Geology, Ecology and Landscapes. https://doi.org/10.1080/24749508.2020.1812148

Horton RE (1945) Erosional development of streams and their drainage basins; hydrophysical approach to quantitative morphology. Geol Soc Am Bull 56:275–370

Jain MK, Das D (2012) Estimation of sediment yield and areas of soil erosion and deposition for watershed prioritization using GIS and remote sensing. Water Resour Manag 24:2091–2112

James C, Ascough II, Dennis CF, Tatarko J, Nearing MA, Kipka H (2017) Soil erosion modeling and conservation planning. Agronomy Monograph 59, Madison

Jha MK, Paudel RC (2010) Erosion predictions by empirical models in a mountainous watershed in Nepal. J Spat Hydrol 10(1):89–102

Jha VC, Kapat S (2009) Rill and gully erosion risk of lateritic terrain in south-western Birbhum district, West Bengal. India. Sociedade & Natureza 21(2):141–158

Kale VS (2014) Geomorphosites and geoheritage sites in India. In: Kale VS (ed) Landscapes and landforms of India. Springer, Heidelberg, pp 247–267

Karydas CG, PanagosP GI (2014) Classification of water erosion models according to their geospatial characteristics. Int J Digital Earth 7(3):229–250

Kinnell PIA (2014) Applying the QREI30 index within the USLE modeling environment. Hydrol Process 28(3):591–598

Kirkby KJ, Bracken LJ (2009) Gully processes and gully dynamics. Earth Surf Proc Land 34(14):1841–1851

Kothyari UC (1996) Erosion and sedimentation problems in India. IAHS Publication No. 236:531–540

Kumar H, Pani P (2013) Effects of soil erosion on agricultural productivity in semi-arid regions: the case of lower Chambal valley. J Rural Dev 32(2):165–184

Lal R (1990) Soil erosion and land degradation: the global risk. In: Lal R, Stewart BA (eds) Soil Degradation, vol II. Soil Advances in Soil Science. Springer, New York, pp 129–172

Lenka NK, Mondal D, Sudhishri S (2014) Permissible soil loss limits for different physiographic regions of West Bengal. Curr Sci 107(4):665–670

Merriti WS, Letcher RA, Jakeman AJ (2003) A review of erosion and sediment transport models. Environ Model Softw 18:761–799

Mondal A, Khare D, Kundu S, Mukherjee S, Mukhopadhyay A, Mondal S (2017) Uncertainty of soil erosion modelling using open source high resolution and aggregated DEMs. Geosci Front 8:425–436

Morgan RPC (2001) A simple approach to soil loss prediction: a revised Morgan-Morgan-Finney model. CATENA 44:305–322

Morgan RPC, Duzant JH (2008) Modified MMF (revised Morgan-Morgan-Finney) model for evaluating effects of crops and vegetation cover on soil erosion. Earth Surf Proc Land 32:90–106

Morgan RPC, Mngomezulu D (2003) Threshold conditions for initiation of valley-side gullies in the Middle Veld of Swaziland. CATENA 50:401–414

Morgan RPC (2005) Soil erosion and conservation. Blackwell Publishing, Oxford

Morgan RPC, Quinton JN, Smith RE, Givers G, Poesen J, Auerswald K (1998) The European Soil Erosion Model (EUROSEM): a process-based approach for predicting soil loss from fields and small catchments. Earth Surf Proc Land 28(3):591–598

Narayana DVV, Babu R (1983) Estimation of soil erosion in India. J Irrig Drain Eng 109(4):419–434

Nash JE, Sutcliffe JV (1970) River flow forecasting through conceptual model. J Hydrol 10:282–290

Nearing MA (2000) Evaluating soil erosion models using measured plot data: accounting for variability in the data. Earth Surf Proc Land 25(9):1035–1043

Patton PC, Schumm SA (1975) Gully erosion, north-western Colorado: a threshold phenomenon. Geology 3:88–90

Pennock D (2019) Soil erosion: the greatest challenge for sustainable soil management. FAO, Rome

Pimentel D (2006) Soil erosion: a food and environmental threat. Environ Dev Sustain 8:119–137

Pimentel D, Burgess M (2013) Soil erosion threatens food production. Agriculture 3:443–463

Poesen J (2018) Soil erosion in the Anthropocene: research needs. Earth Surf Proc Land 43(1):64–84

Poesen J, Nachtergaele J, Verstraeten G, Valentin C (2003) Gully erosion and environmental change: importance and research needs. CATENA 50:91–133

Quinton JN, Morgan RPC (1998) EUROSEM: an evaluation with single event data from the C5 watershed, Oklahoma, USA. In: Boardman J, Favis-Mortlock . (eds), Modelling soil erosion by water, Springer, Berlin, 65–74

Rahaman SA, Aruchamy S, Jegankumar R, Ajeez SA (2015) Estimation of annual average soil loss, based on RUSLE model in Kallar watershed, Bhavari Basin, Tamil Nadu, India. ISPRS Ann Photogramm Remote Sens Spat Inform Sci 2:207–214

Reddy VR, Galab S (2006) Looking beyond the debt trap. Econ Pol Wkly 41(19):1838–1841

Renard KG, Foster GA, McCool DK, Yoder DC (1997) Predicting soil erosion by water: a guide to conservation planning with the revised universal soil loss equation. USDA Agriculture handbook No. 703

Renard KG, Laflen JM, Foster GR, McCool DK (1994) The revised universal soil loss equation. In: Lal R (ed) Soil Erosion: Research Methods. CRC Press, Boca Raton, pp 105–126

Renard KG, Yoder DC, LightleDT, Dabney SM (2011) universal soil loss equation and revised universal soil loss equation. In: Morgan RPC, Nearing MA (eds), Handbook of erosion modelling, Wiley, Chichester, 9–32.

Samni AN, Ahmadi H, Jafari M, Boggs G, Ghoddousi J, Malekian A (2009) Geomorphic threshold conditions for gully erosion in southwestern Iran (Boushehe–Samal watershed). J Asian Earth Sci 35(2):180–189

Sarkar D, Nayak DC, Dutta D, Dhyani BL (2005) Soil erosion of West Bengal. National Bureau of Soil Survey and Land use Planning, NBSS Publication 117, Nagpur

Sarkar D, Nayak DC, Dutta D, Gajbhiye KS (2007) Optimizing Land Use of Birbhum District (West Bengal) Soil Resource Assessment. NBSS & LUP, NBSS Publ.130, Nagpur

Schumm SA, Lichty RW (1965) Time, space and causality in geomorphology. Am J Sci 263:110–119

Sharda VN, Dogra P (2013) Assessment of productivity of monetary losses due to water erosion in rainfed crops across different states of India for prioritization and conservation planning. Agric Res 21(4):382–392

Sharda VN, Mandal D, Ojasvi PR (2013) Identification of soil erosion risk areas for conservation planning in different states of India. J Environ Biol 34:219–226

Sharma HS (2009) Progress of researches in ravines and gullies geomorphology in India. In: Sharma HS, Kale VS (eds), Geomorphology in India (Prof. Savindra Singh Felicitation Volume), PrayagPustak Bhavan, Allahabad, 441–458

Sharma RK (2018) Soil loss setbacks to Indian agriculture. Acta Scientific Agric 2(6):95–97

Shit PK, Nandi AS, Bhunia GS (2015) Soil erosion risk mapping using RUSLE model on Jhargram sub-division at West Bengal in India. Model Earth Syst Environ 1:28

Singh G, Babu R, Narain P, Bhushan LS, Abrol IP (1992) Soil erosion rates in India. J Soil Water Conserv 47(1):97–99

Singh S, Dubey A (2002) gully erosion and management: methods and application. New Academic Publishers, New Delhi

Sinha D, Joshi VU (2012) Application of Universal Soil Loss Equation (USLE) to recently reclaimed badlands along the Adula and Mahalung Rivers, Pravara basin, Maharashtra. J Geolog Soc India 80:341–350

Sinha R, Jain V, Tandon SK, Chakraborty T (2012) Large river systems of India. Proc. Indian Natn Sci Acad 78(3):1–17

Smith HJ (1999) Application of empirical soil loss models in southern Africa: a review. South African J Plant Soil 16(3):158–163

Sovrin J (2003) A test of three soil erosion models incorporated into a geographical information system. Hydrol Process 17:967–977

Strahler AN (1952) Dynamic basis of geomorphology. Geol Soc Am Bull 63:923–938

Torri D, Poesen J (2014) A review of topographic threshold conditions for gully head development in different environments. Earth Sci Rev 130:73–85

Torri D, Santi E, Marignani M, Rossi M, Borselli L, Maccherini S (2013) The recurring cycles of biancana badlands: Erosion, vegetation and human impact. Catena 106:22–30

Toy TJ, Foster GR, Renard KG (2013) Soil Erosion: Processes, Prediction. Measurement Control. Wiley, New York

Valentin C, Poesen J, Li Y (2005) Gully erosion: impacts, factors and control. CATENA 63:132–153

Vandaele K, Poesen J, Govers G, Wesemael B (1996) Geomorphic threshold conditions for ephemeral gully incision. Geomorphology 16(2):161–173

Wasson RJ (2003) A sediment budget for the Ganga—Brahmaputra catchment. Curr Sci 84(8):1041–1047

Wischmeier WH, Smith DD (1978) Predicting Rainfall Erosion Losses—A Guide to Conservation Planning. USDA, Agricultural Handbook No. 537:1–5

Zglobicki W, Poesen J, Daniels M, Monte MD, Guerra AJT, Joshi V, Paterson G, Shellberg J, Sole-Benet A, Su Z (2018) Geotouristic value of badlands. In: Nadal-Romero E, Martinez-Murillo J, Kuhn N (eds) Badland Dynamics in the Context of Global Change. Elsevier, Amsterdam, pp 277–313

Chapter 27
Urban Flooding Scenario and Human Response in Guwahati, India

Sutapa Bhattacharjee and **Bimal Kumar Kar**

Abstract Rapid growth of Guwahati, the largest urban center in India's north-east, in terms of area, population, and functionality, is contributing to the complexity of the urban environment. Although due to the influence of monsoon the amount of rainfall in the city almost remains same, it has undergone distributional change, with decrease in rainy days and increase in high intensity rain events. Variety of factors typical to the dynamic environment of Guwahati characterized by surrounding hills, Brahmaputra flowing through it, the ever-expanding concrete surface, and high-rise buildings amidst somewhat rugged terrain across the city; induces severe urban flooding problems. The gravity of the situation can be marked by the fact that, moderate to heavy rainfall for about 2–3 h often results in high intensity urban flooding during the monsoon season. It has become a recurring problem which even transcends to be devastative in certain localities within the city and the life and living of the city dwellers become deplorable. Therefore, this study attempts to analyze the flooding pattern in Guwahati with respect to its intensity, identify the major causes and consequences associated with it, and understand the human response to deal with the resulting situation; primarily on the basis of field observation and investigations.

Keywords Rainfall behavior · Topography · Urban flooding · Flood intensity · Human response

27.1 Introduction

Urbanization could presently be assumed as one of the most dynamic phenomena globally. The intensive structural and functional characteristics of urban spaces embark a sharp contrast with their surroundings, developing them into a unique

S. Bhattacharjee (✉)
Department of Civil Engineering, Indian Institute of Technology Guwahati, Guwahati, India

B. K. Kar
Department of Geography, Gauhati University, Guwahati, India
e-mail: bimalkar@gauhati.ac.in

system with typical environmental characteristics (Oke 1987; Batty 2005, 2008; Imhoff et al. 2004; Grimm et al. 2008). The interaction of these highly heterogeneous and complex urban systems with the surroundings is extremely potent and rigorous, which leads to significant modifications in the environment (Voogt and Oke 1997; Hidalgo et al. 2008; Mirzaei and Haghighat 2010). Along with the surface conditions, the climatic properties overlapping the urban entity also develop considerable alterations (Voogt and Oke 1997; Hidalgo et al. 2008; Mirzaei and Haghighat 2010). Rainfall is one of the most significant climatic elements in the case of urban environment, as it has a direct impact on the water budget of the cities. The replenishment of the surface/sub-surface water demands as well as its abundance leading to surface accumulation is significantly related to the rainfall received. Besides the amount of rainfall and its intensity, the local physiography and urban structures exert a definite influence on the nature and magnitude of surficial flow of rain water in the city and its surroundings (Pappas et al. 2007; Tingsanchali 2012; Nkwunonwo et al. 2016; McGrane 2016; Bhattacharjee et al. 2021).

Urbanization, as has been mentioned has a burgeoning effect and engulfs more and more areas from their natural surroundings rapidly converting them into impermeable land surfaces. It induces decrease in lag time and rain water detention capability of the built-up landscapes followed by increment in surface run off generation (Shi et al. 2007; Kim and Yoo 2009; Dewi 2007). With the increasing incidences of heavy and unevenly distributed rainfall in the recent times, occurrence of floods has almost become a common phenomenon in the city environment (McGrane 2016; Shi et al. 2007; Kim and Yoo 2009; Dewi 2007). Urban floods can be described as an abnormal flooding or water-logging condition within the built-up zone for a certain duration caused either by intensive downpour for a short period or incessant rainfall for a considerable period of time; overwhelming the capacity of the drainage system of an urban area, with varying magnitude across the globe (Dewi 2007; Chatterjee 2010; Bhattacharjee et al. 2021). It usually accompanies hazardous consequences like submergence of dwelling units and city streets, loss of life and property, disruptions in mobility, health and hygiene problems, chaotic city functions, huge wastage of time and money, physical and psychological disturbances for the city dwellers (Smith and Ward 1998; M¨uller et al. 2011).

This study aims to understand the role of physiography, drainage system, and other urban characteristics in the occurrence of urban floods and to assess the vulnerability of human lives and properties with respect to varying intensity of flooding in different localities of Guwahati city. It also aims to understand the behavior and response of the city dwellers pertaining to this recurring and intensifying problem in major parts of the city.

27.2 Study Area

Guwahati, is a major urban center in India due to its location as well as strategic, economic, and administrative importance (Bhattacharyya et al. 1972; Sharma 2014).

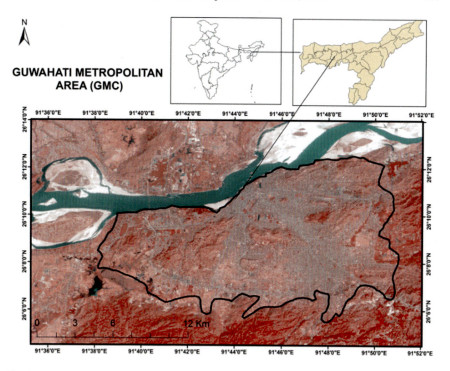

Fig. 27.1 Location of the study area

It is the only city of this dimension and capability in the entire north-eastern region of India lying between 25°5' N and 26°12' N and 91°34' E to 91°51' E (Fig. 27.1). It covers an area of about 216.79 km^2 with over a million people living within its administrative boundary at present. It has an average altitude of 55 m above the mean sea level. The city is situated on the southern bank of the Brahmaputra making the river its northern boundary; and toward south it is bounded by the Meghalaya plateau. Khanajan and Bondajan, tributaries of Brahmaputra flank the city on the west and east, respectively, along with some scattered hill ranges and wetlands. Thus, Guwahati can be designated as a riverside settlement roughly bearing a crescent-shape sprawling in a curvilinear fashion; and dotted with intermittent denudational hills and depositional plains along with low-lying areas and marshes, conferring it an undulating surface characteristic (Bhattacharyya et al. 1972; Sharma 2014). It is a typical sub-tropical city lying in the most active monsoon belt of India, and receives the majority share of its rainfall during the monsoon months. The average annual temperature ranges from 24 to 30 °C and the annual rainfall received lies within 160 to 200 cm (Sharma 2001; Barman and Goswami 2009).

The spatial expansion and rapid urban development of Guwahati is caused by huge population explosion over a short period of time due to highly heterogeneous functional characteristics of the city. According to the Census records, population increased from 43,615 in 1951 to 9,68,549 in 2011, with its average population density as high as 4468 persons per sq. km. The city center, old city, business centers as well as the newly developed residential areas show highest population density. This is indicative of the growing compactness and complexities in the landuse of the city with time (Bhattacharyya et al. 1972; Sharma 2014). Like any other growing city the lifestyle of the people is in a phase of transition being influenced by the process of development. This led to evident change in the city's local environmental condition compared to its surroundings as well as spatial variation within it, initiating other allied problems. It was observed that urban flooding has become a recurring problem within the city with some areas being flooded almost throughout the monsoon season. A major part of the city core is flooded with a very short duration of high-intensity downpour, causing serious inconvenience to the city dwellers; and is experienced a number of times annually (Sharma 2001; Barman and Goswami 2009).

27.3 Material and Methods

Guwahati bears a highly potent as well as volatile character of urban flooding of rising intensity. To understand the nature and impact of the problem specific to this city, complicated by its structure; both physical and social factors were systematically evaluated. Besides some secondary data relating to topography, rainfall, drainage, and population; information associated with pattern and magnitude of urban floods, the consequences and human response etc., were collected through field study (Fig. 27.2). The data so collected from the different secondary and primary sources have been applicably processed, analyzed, and presented by using appropriate statistical and cartographic techniques (Fig. 27.2). A total of 24 localities were identified throughout the city which are flooded every year, according to various secondary sources (scientific reports, documents provided by different authorities, newspapers, etc.). The residents were interviewed to extract a detailed account of the scenario, including their experiences (Survey period-April—September 2014). Since, there are no instrumental measurement techniques in place, to gauge the urban flood water depth and its intensity, the responses from around 120 participants (5 from each locality) who have resided in the area for a long time (more than 10 years) were considered authentic; and a database was generated regarding the flooding pattern in the individual localities. Accordingly these locations were categorized into different flood intensity zones by assigning suitable weightage to the different flood indicators and their prominence in each locality (Fig. 27.2). The causes, direct and indirect impact of the different magnitude of water logging on the dwellers and their attitude toward it were then assessed for the individual localities, based on the responses (Fig. 27.2).

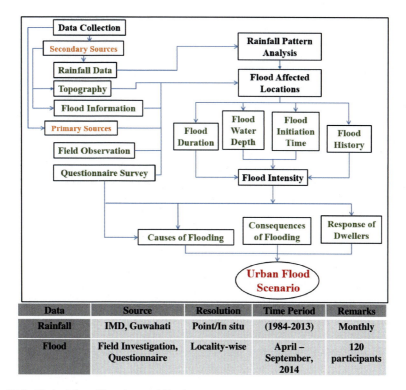

Fig. 27.2 Methodology Flowchart and Database

27.4   Results, Analysis and Discussion

27.4.1   Rainfall Pattern in Guwahati

The average annual rainfall received by Guwahati (recorded by IMD Guwahati station at Borjhar, Guwahati) follows the general fluctuating trend similar to that of the entire region (Bhattacharjee 2015; Bhattacharjee and Kar 2015). However, a gradual decline has been observed in the overall annual rainfall received by the city over a period of 30 years (1984–2013) (Fig. 27.3). The average annual rainfall for the said period is 1718 mm, and only a few years have received above average rainfall since the year 2000. A similar trend is observed in the monthly rainfall received by the city, as during the recent years the total monthly rainfall received is mostly below the average limit (Bhattacharjee 2015; Bhattacharjee and Kar 2015). Rainfall is heavy during the monsoon months (June, July, August, September) during which about 70 percent of the total annual rainfall occurs. The total monsoon rainfall ranges from about 1200–1600 mm, followed by the Pre-Monsoon season (March, April, May) ranging between 250 and 600 mm. The Post-Monsoon (October, November) season also receives considerable amount of rainfall, with an average of about 100–250 mm.

Fig. 27.3 Trend of Annual
Rainfall, 1984–2013

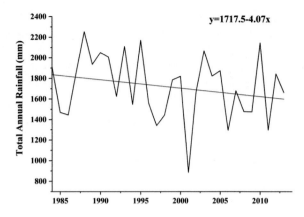

Fig. 27.4 Trend of Average
Monthly Rainfall,
1984–2013

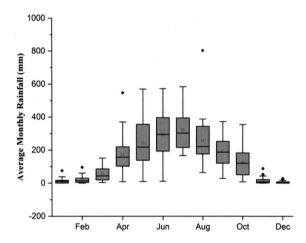

The winter months (December, January, February) get the least amount of rainfall and is the driest season of the region as a whole (Fig. 27.4). The average rainfall ranges from 30 to 100 mm and most part of it is concentrated in the month of February during winter. The monsoon months also have least variation from normal in the amount of rainfall received with the value of CV (co-efficient of variation) being 22.86, followed by pre-monsoon (38.20), post-monsoon (61.30), and winter (77.31).

27.4.2 Flood Situation of the City

Urban flooding in Guwahati has become a colossal and continuous problem resulting from a chain of events which have been taking place since the last few decades with rapid development of the city. To assess this extremely serious condition; constantly increasing in spatial extent, frequency, and intensity with every passing year; a detailed observation was carried, especially in the areas experiencing recurring water logging. Accordingly, 24 flood affected locations were selected based on secondary information with varying flood dimensions (Fig. 27.5), and field investigation was conducted to understand the pattern, causes, and consequences of flooding along with the plight of the local inhabitants. It was observed that all the surveyed localities lie within the most urbanized parts of the city, with low relief and slope (Fig. 27.6). This results in the accumulation of the rain water received and also rushed in from the surrounding high altitude areas, due to low impermeability of the surface. Various important indicators such as, the duration and depth of flooding, flood initiation duration, and flood history, etc. (Fig. 27.7); which specify the severity of the problem were extracted from the responses of the inhabitants.

Flood Duration: In the most severely affected areas, the water remains stagnant for a period of minimum 2 days to maximum 7–8 days. The flood duration extends from at least 5–6 h to about 2–3 days in the moderately affected parts. However, the relatively less affected areas remain flooded for about 6–7 h and not less than 1 h at a row (Fig. 27.7).

Flood Water Depth: The flood water level fluctuates all throughout the city depending on the local topography, amount of rainfall, drainage condition, and density of population. However, the average water level in the highly affected regions

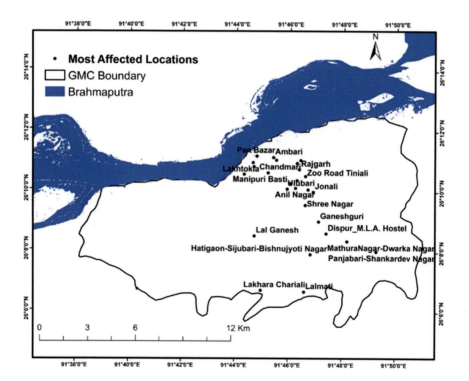

Fig. 27.5 Flood affected locations in Guwahati

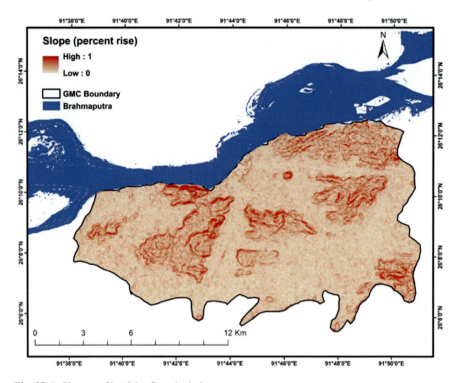

Fig. 27.6 Slope profile of the Guwahati city

vary from a minimum of 2 ft to a maximum of 4.5 ft. In some cases, it even rises to 6 ft. On an average, the water level rises up to 2 ft in the moderately affected and 1 ft in the least affected zones (Fig. 27.7).

Flood Initiation Time: The most affected parts of the city begin to flood within a minimum 20–30 min of extremely heavy shower (30 mm/hour) during the monsoon period when the drainage infrastructure and sub-surface water table is almost saturated. On the other hand, rainfall for 45–60 min and 2–3 h of the same intensity (30 mm/hour) has the potential to flood the medium and least affected areas, respectively. The ability of 60 mm/day rainfall to begin flooding is found to be about 2 h (Fig. 27.7). This fairly low time period since the beginning of heavy showers to initiate and sustain flooding conditions, poses the biggest threat pertaining to this problem and its management.

Flood History: In most of the surveyed locations it has been learned that the problem of flooding has increased remarkably during the last 7–8 years tremendously. However, in some localities lying in the moderately affected parts of the city, the problem persists since a long period (above 20 years) due to other influencing factors such as relief, lack of proper drainage facility, and unawareness of the people. In some of the highly affected regions, the problem has begun only about 6–7 years prior to the survey period and has aggravated drastically; while in most other locations the

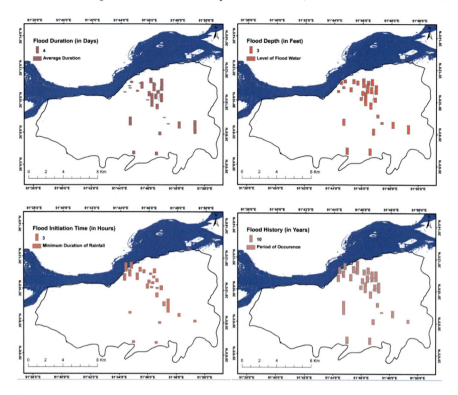

Fig. 27.7 Flood indicators

problem prevails for more than 10 years and a consistent rise in its severity is observed (Fig. 27.7).

Therefore Multi Criteria Decision Analysis (MCDA) was performed considering different weights for the various flood indicators according to their significance at a certain location and finally all the localities were further categorized into different flooding intensity zones (Figs. 27.7 and 27.8). On the basis of combined weightage; it was found that, out of the 24 locations, 8 are categorized under High Intensity, 8 under Medium Intensity, and 7 under Low Intensity categories (Fig. 27.8). Thus, it can be concluded that in the High Intensity flood affected localities, the flood water generally rises above 3–4 feet on an average with a heavy shower of few hours or constant low intensity rainfall continuing for a few days. The gravity of the situation intensifies with the duration of the water-logging condition expanding over a few hours in the Low Intensity areas to even more than a week in the worst cases. In few areas such as Shankardev Nagar of Panjabari, it was also observed that the flood problem has aggravated in short period of time due to unprecedented and unplanned development resulting into conversion of more and more areas into impenetrable built-up surfaces. Among the worst affected areas, the most prominent are Anil

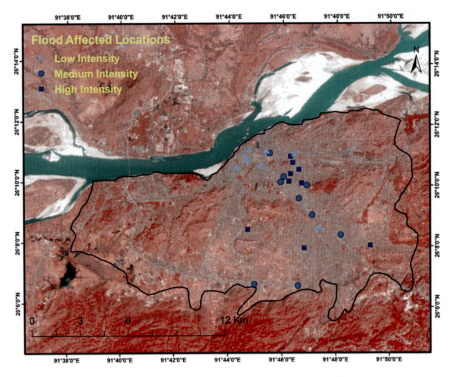

Fig. 27.8 Flood affected locations of different intensity and the associated causes

Nagar, Nabin Nagar, Pub-Sarania, Zoo Tiniali, Lalganesh, Panjabari (Figs. 27.5 and 27.8).

27.4.3 Causes of Urban Floods

The primary causes responsible for flooding in Guwahati are both physical and anthropogenic. Some of the most important and direct causes are undulating physiography, unplanned drainage network, large-scale encroachment of the low-lying areas and water bodies, rapid concretization, heavy siltation due to erosion and deposition on the drains resulting in blockage, incoherent decision-making of the administration and lack of strategic planning, unawareness of the city dwellers and their unpredictable behavior, etc. However, according to the city dwellers the main causes of flooding have been identified as excessive rainfall, improper drainage system, and their poor maintenance, water rushed suddenly from the nearby hills and surroundings (Table 27.1). The residents of the Low Flood Intensity locations (42.2%) perceive excessive rainfall (high intensity) as the major cause responsible for flooding in their localities. Whereas, people inhabiting in the High Intensity areas absolutely discard

Table 27.1 Most Important Causes of Urban Flooding in Guwahati

Flood intensity zone	Percentage of Respondents to total Respondents under each Flood Intensity Zone				
	Excessive rain	Improper drainage system	Poor Maintenance of the drains	Water rushed from surrounding hills	Water rushed from nearby areas
High Intensity	0.00	88.89	100.0	51.11	66.67
Medium Intensity	11.11	60.0	97.78	42.22	8.89
Low Intensity	42.22	35.56	100.0	4.44	6.67

the idea and believe that even a medium or low intensity rainfall, if it continues for a long period of time; may result in flooding. Almost 90% of the people living in the High Intensity Zone hold improper drainage infrastructure and their unsystematic planning and maintenance responsible for the adverse flooding situations demand implementation of proper drainage network in the city (Table 27.1). However, some residents of the Low and Medium Intensity locations believe that the existing drainage network is more or less satisfactory, which only requires proper maintenance to deal with the flood problem in the city. During the field observation it was discovered that many areas of the city gets water-logged because the storm water lacks a proper evacuation route and migrates to nearby locations, expanding the spatial extent of flooding. In some localities, such as Anil Nagar, Nabin Nagar, Rajgarh, etc., the problem of migration of water from one place to another results in serious fluctuation in the flood depth making the problem more unpredictable. In recent years it has been observed that the level of water increases unusually during flood peaks as it is diverted from some other area inevitably, or is pumped out by the authorities and disposed elsewhere. It is a matter of serious concern especially in the High Intensity locations, where about 66.67 percent of the respondents also perceive it as one of the major causes of flooding (Table 27.1).

27.4.4   Consequences of Floods

The immediate consequences of urban floods in Guwahati city are of diverse nature including instant steps taken by the affected people during the floods to the nature and magnitude of damage of their belongings. According to the responses of the surveyed population it could be comprehended that the nature of after-effect of the floods may differ in different flood intensity locations (Fig. 27.9). Relocation of people from their houses during the flood period is found to be a commonly adopted measure in case of High Intensity zones (62.86%), which is otherwise uncommon for Medium and Low Intensity areas (Fig. 27.9). Structural damage of the houses such as the building walls and floors; and loss and damage of belongings such as

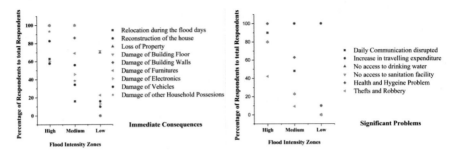

Fig. 27.9 After effects of flooding in Guwahati

electronic gadgets and various household items are most common consequences of flooding throughout the surveyed localities. Above 50 percent of the respondents in the High Intensity areas have also reported irreparable damage to their vehicles, in order to avoid which many people in recent times have constructed elevated garage facilities. Reconstruction of houses to avoid flooding is practically common in case of these areas, where some people have totally abandoned the ground floor of the building and constructed higher floors which could sustain flood water depth above 6 feet during worst cases. Raising the plinth of the building was observed as one of the most common practices to avoid flooding of the structure observed in almost all locations, which on the other hand results in rising of flood water level in the surrounding areas and the streets (Fig. 27.9).

Some of the other significant problems generally experienced by the people are disruption in communication, no access to basic amenities such as sanitation and drinking water, increase in expenditure level, health and hygiene problems, etc. (Fig. 27.9). Communication to some parts of the city lying within the High and Medium Intensity zones gets totally disrupted making the people entirely dependent on the external help and rescue operations provided to them, to fulfill their basic needs. Transportation becomes a big distress for the city-dwellers during floods with rare to no means of communication available, further worsened by the rate of travel expenditure. It is worth mentioning that all the respondents interviewed, agreed to the fact that cost of transportation increases manifold during the flood period (Fig. 27.9). Health and Hygiene is another major concern for the people, since the flood water which fills the streets and dwelling units is a blend of sewage water, seepage from the septic tanks and dustbins, mud, and various other deposits. All the people in high intensity zones and about 60 percent in medium intensity zones have reported to suffer from various types of skin irritations during prolonged flood periods (Fig. 27.9). Moreover, the stagnant water has also reportedly caused various other diseases, such as diarrhea, fever, malaria, cough, etc., in the affected localities. No access to safe drinking water and sanitation facility is yet another difficulty, leading many people in the high intensity zones to relocate from the areas of their residence; only due to non-accessibility of these facilities. Further, theft and robbery also dominate some parts of the flood prone areas during those days when the people relocate to other places and leave their houses vacant.

Therefore, consequences of flooding in Guwahati are multifaceted, associated with their own set of causes and implications. All the flood affected areas have an array of problems and issues specific to them, as they encounter the situation multiple times annually. Some problems take the center stage in particular localities which may not be that significant for some other areas. Some problems are highly exaggerated in certain localities due to the behavior of the local inhabitants. The most evident effect of flash floods in the city as a whole is the hindrance in its normal functioning, due to disruptions in mobility and daily affairs of the people. The whole city suffers from a great chaos and disturbance as the communication system stand paralyzed; some areas get totally disconnected, and accidents related to landslides, electrical short circuits, falling of people in the uncovered drains and manholes, drowning in the flood water, submergence of houses and streets, etc., almost become a common phenomenon. People face great difficulty in carrying out their regular activities and mere survival becomes a matter of concern in the worst hit areas. Among the groups of people who are most hit by this problem are the daily commuters (students, office goers, business people, daily wage earners, etc.); people living in the most vulnerable locations (high intensity flood zone) as most of their houses get flooded; poor people who do not have the capacity to shift or afford some cost-effective option (such as relocation, expensive transportation, etc.) to avoid the situation; and traders whose business establishments are located in the flood prone areas. Most of the flood affected locations remain under the spell of the problem for long even after the water is drained. It takes a considerable span of time for those areas, according to the seriousness of the problem to resume back to normal; and the people to adjust with the damages encountered after any flood instance.

27.4.5 Human Response to Urban Floods

The flood problem in Guwahati city has become a part of life for its people, who seem to have accepted to adjust the troubles associated with it. The city dwellers have found strategies and ways suiting them to deal with the problem since it is a combined effect of the constant and irreversible development process and lack of responsible urban planning. Most of the people irrespective of the flooding intensity in their localities, have responded that they prefer to stay indoors during the flood period. Besides reconstructing their dwelling units, replacing wooden furnitures by steel and iron ones to increase their durability even under prolonged submerged conditions. People travel in rubber boats provided by the government or even in the makeshift rafts constructed of local materials, in worst cases when water level goes up by several feet. Off late, the dwellers have also begun to take precautionary measures such as storing sufficient amount of food and other essentials; arranging their belongings in safe places; and managing the power supplies to avoid short circuits; etc, before the flood season starts. The residents of some localities even take collective or individual steps to minimize or mitigate the problem; such as spreading awareness, cleaning of their surroundings and drains, appealing to local administration to provide solutions, etc.

27.4.6 Possible Mitigation Measures

As has been discussed, some of the most important and direct causes are undulating physiography, unplanned drainage network, large-scale encroachment of the low-lying areas and water bodies, rapid concretization, high magnitude of siltation due to erosion resulting in blockage of drains, lack of strategic and responsible city planning and management, unawareness of the city dwellers and their irresponsible behavior, etc. Therefore, some of the most relevant approaches to tackle the problem are very obvious but difficult to be implemented in case of Guwahati city, due to its current structure and functioning. It is essential to develop and install a comprehensive and well planned drainage network with high capacity underground storm water drainage channels. Besides systematic and regular maintenance of the existing drainage channels by de-silting the drains and checking of soil erosion are important measures and would be easier to implement. Checking illegal encroachment and prevention of large-scale concretization of the low-lying areas and wetlands could be other significant mitigation measures. However, the city dwellers should also recognize their responsibilities and try to keep their surroundings clean, not dispose solid waste into the drains and in open, and refrain from illegitimate constructions.

27.5 Conclusion

The foregoing discussion reveals that although the city of Guwahati has been experiencing a decline in overall magnitude of rainfall, the instances of heavy rain events have significantly risen. A moderate to high intensity rainfall within a short span of time has the potential to cause severe flooding in some parts of the city, and its spatial dimension and complexity has been increasing over time. Along with it, inadequate and unplanned drainage network, hyper concretization, high rate of erosion and siltation and negligence of both administration and the city dwellers further aggravates the problem of flooding in Guwahati. This results in a chaotic situation which severely disrupts the normal functioning of the city even leading to loss of human lives in the worst cases. This entire phenomenon has become an annual event for almost two decades during the monsoon and pre-monsoon showers. Some areas consistently encounter severe urban floods even twice or thrice during peak of each season and conditions are deteriorating with every passing year. This situation calls for urgent and proper attention of the authorities concerned to address the whole issue in a more scientific and comprehensive manner.

Acknowledgements The authors would like to acknowledge the support extended by the different organizations who provided important secondary data and all the different authorities and officials who have helped in gathering information about the technical details and history of flooding, pertinent to this study. We are highly grateful to IMD, Guwahati, who provided all the necessary rainfall data; and extend our heartfelt gratitude to all the people of the affected localities who actively participated during the questionnaire survey and provided valuable information.

References

Barman P, Goswami DC (2009) Flood zone mapping of guwahati municipal corporation area using GIS technology. In: Geography in Action, 10th ESRI India User Conference

Batty M (2005) Cities and complexity: understanding cities through cellular automata, agent-based models, and fractals. MIT Press, Cambridge, MA

Batty M (2008) The size, scale, and shape of cities. Science 319:769–771

Bhattacharjee S (2015) Rainfall pattern and its influence on urban environment: a case study of Guwahati City. M.Phil Thesis, Dept. of Geography, Gauhati University, India

Bhattacharjee S, Kar BK (2015) Rainfall behaviour and urban environment of Guwahati City. North Eastern Geograph 38(1&2):18–34. ISSN–0973–0915

Bhattacharjee S, Kumar P, Thakur PK, Gupta K (2021) Hydrodynamic modelling and vulnerability analysis to assess flood risk in a dense Indian city using geospatial techniques. Nat Hazard 105(2):2117–2145. https://doi.org/10.1007/s11069-020-04392-z

Bhattacharyya NN (2001) Growth and changing face of Guwahati since, 1972. IN: Alam K, Das NC, Borah AK (eds) Guwahati: The Gateway to the East. Concept Publishing Company, New Delhi, pp 27–34

Chatterjee M (2010) Resilient Flood Loss Response Systems for Vulnerable Populations in Mumbai: A Neglected Alternative. PhD Thesis, Graduate School-New Brunswick Rutgers, The State University of New Jersey, USA

Dewi A (2007) Community based analysis of coping with urban flooding: a case study in Semarang, Indonesia. M.Sc Thesis, International Institute for Geo-Information Science and Earth Observation, Enschede, The Netherlands

Grimm NB, Faeth SH, Golubiewski NE, Redman CL, Wu JG, Bai XM, Briggs JM (2008) Global change and the ecology of cities. Science 319:756–760

Hidalgo J, Masson V, Baklanov A, Pigeon G, Gimenoa L (2008) Advances in urban climate

Imhoff ML, Bounoua L, DeFries R, Lawrence WT, Stutzer D, Tucker CJ (2004) The consequences of urban land transformation on net primary productivity in the United States. Remote Sens Environ 89:434–443

Kim K, Yoo C (2009) Hydrological modeling and evaluation of rainwater harvesting facilities: case study on several rainwater harvesting facilities in Korea. J Hydrol Eng 14(6):545–561

M¨uller A, Reiter J, Weiland U (2011) Assessment of urban vulnerability towards floods using an indicator-based approach—a case study for Santiago de Chile. Copernicus Publications on behalf of the European Geosciences Union Natural Hazards and Earth System Sciences, pp 2107–2123

McGrane SJ (2016) Impacts of urbanisation on hydrological and water quality dynamics, and urban water management: a review. Hydrol Sci J 61(13):2295–2311. https://doi.org/10.1080/02626667.2015.1128084

Mirzaei PA, Haghighat F (2010) Approaches to study urban heat Island—abilities and limitations. Build Environ 45(10):2192–2201. https://doi.org/10.1016/j.buildenv.2010.04.001

Nkwunonwo UC, Whitworth M, Baily B (2016) Review article: a review and critical analysis of the efforts towards urban flood risk management in the Lagos region of Nigeria. Nat Hazard 16:349–369. https://doi.org/10.5194/nhess-16-349-2016

Oke TR (1987) Boundary layer climates. Methuen, London and New York

Pappas EA, Smith DR, Huang C, Shuster WC, Bonta JV (2007) Impervious surface impacts to runoff and sediment discharge under laboratory rainfall simulation. Journal of CATENA 012–12:7

Sharma S (2001) Waterlogging and inundation problems of Guwahati: a geographical study Guwahati: The Gateway to the East, Alam K., Das N.C., Borah A.K. (eds), Concept Publishing Company, New Delhi, India, pp 97–115

Sharma P (2014) Genesis of a city: urban development in Guwahati. EBH Publishers, Guwahati, India, pp 152–155

Shi PJ, Yuan Y, Zheng J, Wang JA, Ge Y, Qiu GY (2007) The effect of land use/cover change on surface runoff in Shenzhen region China. Catena 69(1):31–35. https://doi.org/10.1016/j.catena.2006.04.015

Smith K, Ward R (1998) Floods, physical processes and human impacts. John Willey and Sons, Chichester

Tingsanchali T (2012) Urban flood disaster management. Proc Eng 32:25–37

Voogt JA, Oke TR (1997) Complete urban surface temperatures. J Appl Meteorol 36:1117–1132. https://doi.org/10.1175/1520-0450